Lectures in Applied Mathematics

Proceedings of the Summer Seminar, Boulder, Colorado, 1960

Proceedings of the Summer Seminar, Ithaca, New York, 1963

Proceedings of the Summer Seminar, Ithaca, New York, 1965

Proceedings of the Summer Seminar, Stanford, California, 1967

Applications of Group Theory
in Physics and Mathematical Physics

Volume 21
Lectures in Applied Mathematics

Applications of Group Theory
in Physics and Mathematical Physics

Edited by
Moshe Flato
Paul Sally
Gregg Zuckerman

1985

American Mathematical Society, Providence, Rhode Island

The proceedings of the Summer Seminar were prepared by the American Mathematical Society with partial support from National Science Foundation Grant MCS 8208734.

1980 Mathematics Subject Classification. Primary 16A58, 17B65, 20G45, 22-06, 22EXX, 35-02, 53B50, 53C15, 57S20, 81-XX, 83C45, 83EXX; Secondary 05A19, 15A66, 20F05, 35Q20, 81G20, 83C15.

Library of Congress Cataloging in Publication Data
Main entry under title:
  Applications of group theory in physics and mathematical physics.
  (Lectures in applied mathematics; v. 21)
  Papers from the Summer Seminar on Applications of Group Theory in Physics and Mathematical Physics, held in July 1982 in Chicago and sponsored by the American Mathematical Society and the Society for Industrial and Applied Mathematics.
  Includes bibliographies.
  1. Groups, Theory of—Congresses. 2. Mathematical physics—Congresses. I. Flato, M. (Moshé), 1937–   . II. Sally, Paul. III. Zuckerman, Gregg. IV. Summer Seminar on Applications of Group Theory in Physics and Mathematical Physics (1982; Chicago, Ill.) V. American Mathematical Society. VI. Society for Industrial and Applied Mathematics. VII. Series: Lectures in applied mathematics (American Mathematical Society); v. 21.
  QC20.7.G76A66   1984   530.1'5222   84-24191
  ISBN 0-8218-1121-5
  ISSN 0075-8485

Contents

vii

IV. Kac-Moody algebras and nonlinear theories

Appendix

Preface

The past decade has seen a renewal in the close ties between mathematics and physics. Completely integrable systems, classical and quantum field theories, in particular, gauge theories, supersymmetry and supergravity, and grand unified theories were developed with the aid of, and, in turn, brought new developments into, several diverse areas of mathematics. The Chicago Summer Seminar on Applications of Group Theory in Physics and Mathematical Physics, held in July, 1982, was organized to bring together a broad spectrum of scientists from theoretical physics, mathematical physics, and various branches of pure and applied mathematics in order to promote interaction and an exchange of ideas and results in areas of common interest. The spirit of the Seminar was typified by friendly cooperation among the participants.

This volume contains the papers submitted by speakers at the Seminar. The reader will find several groups of articles varying from the most abstract aspects of mathematics to a concrete phenomenological description of some models applicable to particle physics. We have divided the papers into four categories corresponding to the principal topics covered at the Seminar. This is only a rough division, and it should be clear from the brief descriptions we give below that some papers overlap two or more of these categories.

I. *General*

Y. Nambu. This paper should be of general interest to all readers. Nambu deals with topological excitation in physics, a topic which has been of great interest in theoretical physics in recent years. Nambu illustrates his ideas with well-chosen examples of current physical interest.

II. *Supersymmetry and supergravity*

1. **I. Bars.** Bars' contribution develops the structure theory and linear representation theory of supergroups and superalgebras. Along with a list of possible applications to physics, the reader will find definitions of particular families of superalgebras (e.g., $SU(N|M)$), the notion of corresponding supergroups, convenient instruments like supertraces and superdeterminants, invariant supercharacters, formulas for the dimensions of representations, etc. While the text does not contain a complete

mathematical theory, this paper will certainly be of interest to those who wish to become more familiar with the topics treated by the author.

2. **P. Freund.** In Freund's paper, space-time is written as a d-dimensional manifold ($d \geq 4$) which is the product of a 4-dimensional Lorentz manifold and a ($d - 4$)-dimensional compact Riemannian manifold. Starting from d-dimensional supergravity, which imposes the constraint $d \leq 11$, Freund analyzes the advantages of this type of generalized Kaluza–Klein approach and addresses some of the questions which it raises.

3. **M. K. Gaillard.** Here the perspective is much closer to phenomenology. Gaillard is interested in infinite-dimensional representations of noncompact groups and, especially, supergroups. She constructs explicit representations for $N = 4$ and $N = 8$ supergravities with possible applications to bound-state spectra in some (extended) supergravity theories.

4. **J. Schwarz.** Schwarz describes the old and new formulations of the supersymmetrical string action and discusses the rather complicated transformation that relates them. He then raises some related mathematical issues.

5. **P. van Nieuwenhuizen.** This paper is devoted to the problem of gauging groups and supergroups. Here, gauging is a procedure for building a Lagrangian field theory for a given group (or supergroup). The two main approaches to the gauging of groups are discussed by the author.

III. *Representations of noncompact Lie groups*

1. **C. Fronsdal.** This contribution deals with the connection between the representation theory of noncompact, semisimple Lie groups and physical theories. In particular, great emphasis is put on semisimple gauge theories: how to construct indecomposable representations relevant to gauge theories (Gupta–Bleuler triplets), how to construct conformal gravity, etc. This subject has gained new interest in recent years, both in the physical literature (especially in relation to q.e.d. and gravity), as well as in the mathematical literature (e.g., see J. Wolf's paper in this volume).

2. **R. Howe.** Motivated by the metaplectic representation of Shale and Weil, Howe introduces the notion of a dual pair of subgroups of a group. Dual pairs connect to physics in many respects: classical equations, the conformal groups, and, in general, massless particles and singletons, supersymmetry, and others.

3. **A. W. Knapp.** Knapp's paper (based on joint work with B. Speh) deals with the determination of the unitary dual of the conformal group of space-time, that is, SU(2, 2). Knapp approaches the problem by first outlining the Langlands classification of admissible representations and then finding the unitary representations within this class.

4. **G. W. Mackey.** In this paper, Mackey discusses the theory of group representations as it contributes to the conceptual foundations of modern elementary particle physics by providing an elegant, and more or less complete, mathematical model for the underlying physical theory. Mackey's intention is to describe the nature of quantum mechanics to mathematicians in the language of familiar mathematical concepts.

5. **D. Sternheimer.** The main problems encountered in Kostant's program on geometric quantization and unitary representations have to do with the choice of polarizations. Sternheimer describes an alternative program to quantization and unitary representations based on deformed brackets (products) of Lie (associative) algebras of C^∞ functions on convenient symplectic manifolds (coadjoint orbits). This program has developed quite recently thanks to contributions by several authors.

6. **D. Vogan.** For a connected semisimple Lie group G with maximal compact subgroup K, Vogan outlines his theory of the classification of irreducible representations of G in terms of their lowest K-types. He draws an analogy with the theory of highest weights for compact Lie groups, and he provides concrete examples to illustrate the theory.

7. **J. Wolf.** In close connection with Fronsdal's paper, Wolf describes his recent results obtained in collaboration with Rawnsley and Schmid. This work deals with singular representations of semisimple and reductive Lie groups and with the associated indefinite harmonic analysis (a situation typified by the indecomposables which are extensions of unitary irreducible representations).

8. **G. Zuckerman.** Zuckerman studies classical fields on space-times which are represented by coset spaces. The usefulness of the theory of induced modules and representations becomes obvious in this context. Examples of Dirac's conformal space and of free quantum fields (in Fock space) are then formulated group-theoretically.

IV. *Kac-Moody algebras and nonlinear theories*

1. **L. Dolan.** Dolan discusses the expectations for the use of Kac-Moody algebras in physics. The hope is that Kac-Moody algebras can serve as an important tool in constructing nonperturbative solutions to various nonlinear field theories.

2. **I. Frenkel.** Here, connections are made between Kac-Moody algebras and dual resonance models (in particular, the Virasoro algebra and vertex operators). The author also provides a starting point for the representation theory of Kac-Moody algebras and its connection with other parts of mathematics.

3. **B. Julia.** This paper deals with Kac-Moody (affine type 1) symmetries of sets of classical solutions of gravity and supergravity that depend on only two coordinates (out of four). Connections with completely

integrable systems and their nonlocal charges, dimensional reduction, and other theories are also mentioned. The facts presented indicate that Kac-Moody algebras have a connection with various physical models which is very deep and not yet fully understood.

4. **J. Lepowsky.** Lepowsky gives an expository account of certain constructions of the affine Lie algebra $A_1^{(1)}$. A particular operator which arises in the construction of $A_1^{(1)}$ as a Lie algebra of "concrete" operators turns out to be a vertex operator of the kind utilized in dual resonance models.

5. **J. Simon.** Simon presents a theory of nonlinear representations of Lie groups. In particular, the question of the linearization of such representations (exemplified here for the case of the affine group of the plane) is treated. This topic is connected to the problem of linearization of the nonlinear partial differential equations of physical theories, infinite conservation laws, and Kac-Moody algebras.

In addition to the authors whose papers are presented in this volume, the following individuals also lectured at the Seminar. For completeness, we give the titles of their talk(s) and the number of lectures by each speaker.

D. Freedman: *Positive energy in gauged, extended supergravity* (1)

D. Friedan: *Polyakov string theory* (1)

I. Kaplansky: *Algebras and superalgebras* (1)

B. Kostant: *Whittaker theory and quantization* (1)

W. Schmid: *Geometric constructions of representations* (1)

G. Segal: *Loop groups and the Sine–Gordon equation* (2)

E. Witten: *Supersymmetry and Morse theory* (2)

B. Zumino: *Supersymmetry: applications of superalgebras in physics* (3)

During the course of the Seminar, several informal talks were given by the participants. Some of these talks were reproduced in the Lecture Notes distributed at the Seminar. The Table of Contents of the Lecture Notes is reproduced at the end of this volume in the Appendix.

Along with the editors of this volume, the other members of the Organizing Committee were C. Fronsdal, I. Kaplansky, Y. Nambu, I. M. Singer, and J. Wolf. We express our thanks to these colleagues for their invaluable assistance. Thanks are also due to the American Mathematical Society and, in particular, Jan Ferreira, for their help in organizing the Seminar. Finally, we acknowledge the generous support of the National Science Foundation and the Departments of Mathematics and Physics of the University of Chicago.

MOSHE FLATO

PAUL SALLY

GREGG ZUCKERMAN

I. GENERAL

Lectures in Applied Mathematics
Volume **21**, 1985

Topological Excitations in Physics[1]

Y. Nambu

I. Topological field patterns. Physics is often described by fields, and fields often show distinct topological patterns. Here we consider two examples (in two dimensions, for simplicity).

1. *Electric field emanating from source P* (Figure 1). The field E has a fixed total flux (charge) Q, and no vorticity:

$$\oint E_n \, ds = Q \quad (\text{if } P \in D), \qquad \oint E_t \, ds = 0,$$

$$E = \frac{Q}{2\pi}\left(\frac{x}{r^2}, \frac{y}{r^2}\right) \qquad (r \neq 0). \tag{1}$$

E is uniquely determined (up to a constant) by Q. But usually we do not regard this as a topological configuration.

2. *Hydrodynamic flow* (Figure 2). The velocity field v has a fixed circulation Ω around P, and no source:

$$\oint v_t \, ds = \Omega \quad (\text{if } P \in D), \qquad \oint v_n \, ds = 0,$$

$$v = \frac{\Omega}{2\pi}\left(\frac{-y}{r^2}, \frac{x}{r^2}\right) \qquad (r \neq 0). \tag{2}$$

Again, v is uniquely determined up to a constant.

In either case, the total energy is similar and divergent unless we cut it off at small radius R_1 and large radius R_2:

$$\text{Energy} = \frac{1}{2}\int E^2 \, dx \, dy \quad \text{or} \quad \frac{1}{2}\int v^2 \, dx \, dy$$

$$\sim \ln \frac{R_2}{R_1}. \tag{3}$$

1980 *Mathematics Subject Classification.* Primary 81E99.
[1] Work supported in part by the NSF: PHY-79-23669.

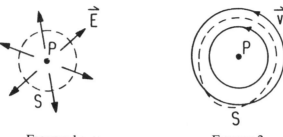

FIGURE 1   FIGURE 2

Sometimes the field distribution itself has no physical meaning. In other words, the energy does not depend on the field directly. Only the global topology, characterized by the nature of singularity P, matters. A typical example is electromagnetism. Replace v of the previous example by A, an electromagnetic potential (vector potential). A is unphysical, but the vorticity, or magnetic flux, located at P, is physical. Actually, in this case the second equation of (2) is not mandatory, so there is more arbitrariness in the choice of A (the gauge transformation $A \rightarrow A + \nabla\lambda$) for the same physical situation.

Nevertheless, the potential A is probed in quantum mechanics by a test particle (electron with electric charge e) obeying the Schrödinger equation

$$i\frac{\partial}{\partial t}\psi = \frac{1}{2m}(-i\nabla - eA)^2\psi \equiv \frac{1}{2m}(p - eA)^2\psi. \qquad (4)$$

Under the gauge transformation, the wave function changes as $\psi \rightarrow \exp[ie\lambda]\psi$ to keep the equation form-invariant. The wave function must be one valued, meaning that the angular momentum around P is quantized in the usual manner. The presence of a nontrivial topology will be felt unless the flux satisfies the quantization condition

$$e\oint A_t\,ds = e\Phi = 2\pi n. \qquad (5)$$

The physical effect that arises for nonquantized Φ is called the Aharonov-Bohm effect, which has recently been demonstrated in a beautiful electron holography experiment [1], in which the magnetic flux was confined within a torus made of ferromagnetic material.

The quantized flux situation can be realized when superconducting material is used to confine the flux. For example, a hollow superconducting tube can trap a flux inside (Figure 3). Outside of the inner space, the magnetic field $B = \nabla \times A$ is zero, but A is not zero. The energy of an electron in the tube is given by the eigenvalue ε:

$$\frac{1}{2m}(p - eA)^2\psi = \varepsilon\psi. \qquad (6)$$

To keep the energy as low as possible, we should set

$$(p - eA)^2 = 0,$$

or

$$0 = \oint (p - eA) \cdot dx = 2\pi nh - e\Phi \tag{7}$$

which is the quantization condition. Because of the Pauli principle, however, this is not allowed for all electrons in a normal conductor. In a superconductor, on the other hand, electrons pair up (Cooper pairs), and thus behave like Bose particles, all of which can condense into the quantized state. [So the charge e must be replaced by $2e$.]

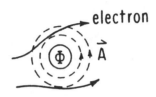

FIGURE 3

Examples of topological configurations (excitations) are given in the table below. ¡They are called excitations because the energy of such states are higher than in trivial situations.]

	Dimensionality of singular manifold in space-time R^4
Hydrodynamic vortices	2
Superfluid He^4II: vortex quanta	2
Superconductor: magnetic flux tubes	2
Superfluid He^3: various structures	
Hadronic string (gluon flux tube)	2
Domain wall (membrane)	3
Kink	1 in R^2
Magnetic monopole	1
Instanton	0

II. Description of extended objects. I will discuss a general description of the topological singularities as a manifold (or a family of manifolds) of dimension n embedded in space-time, i.e., an extended physical object of dimension $n - 1$. As this manifold is purely geometric (no internal structure), it is appropriate to introduce a skew symmetric tensor of rank

n representing its (oriented) volume element. Such a volume element also serves as a natural source for a "gauge field" of similar kind: the Maxwell gauge field A_μ ($\mu = 1,\ldots,4$) for a world line of a mass point, the Kalb-Ramond field $A_{\mu\nu}$ [2] for a world sheet of a string or a vortex, etc. [The K-R field is interpretable as velocity potential in relativistic hydrodynamics. The K-R and higher rank tensors also make their appearance in Kaluza-Klein type supergravity and other theories of similar nature.] The relevant mathematics seems to have been developed by mathematicians since the last century (Clebsch, Volterra, Carathéodory, Bateman, Rund, etc.) [3].

Let us define a hierarchy of n-forms ($n = 1, 2,\ldots$) according to

$$F^{(n)} = F_{\mu\nu\cdots\rho}\, dx^\mu \wedge dx^\nu \wedge \cdots \wedge dx^\rho$$
$$\equiv dA^{(n-1)} + \phi^0 d\phi^1 \wedge d\phi^2 \wedge \cdots \wedge d\phi^n,$$
$$dF^{(n)} = d\phi^0 \wedge d\phi^1 \wedge \cdots \wedge d\phi^n. \tag{8}$$

Here the ϕ's are generally independent fields, often called "Higgs fields". If a constraint

$$\sum_{a=0}^{n} \phi^a \phi^a = 1 \tag{9}$$

is imposed (for fixed n), we use the term nonlinear Higgs or O_n σ-model, for historical reasons. We will assume this from now on. Then

$$dF^{(n)} = 0 \tag{10}$$

except at singular points of mapping $R^{n+1} \to S^n$. These singularities are the "magnetic" source of the field F. Consider the case of $n = 2$, the usual Maxwell (gauge) field with possible monopoles added

$$F^{(2)} = dA^{(1)} + \phi^0 d\phi^1 \wedge d\phi^2,$$
$$d * F = d * dA = j \quad \text{(charge)},$$
$$dF = d\phi^0 \wedge d\phi^1 \wedge d\phi^2 \quad \text{(monopole)}. \tag{11}$$

The second and the third equations consitute general Maxwell equations. They can be generated by a Lagrangian

$$\mathcal{L} = -\frac{1}{4} F_{\mu\nu} F^{\mu\nu} - j_\mu A^\mu \tag{12}$$

where A_μ and the ϕ's are regarded as independent, but the ϕ's are constrained and the location of monopole singularities are given.

There are two ways to specify a manifold of singularities: (1) by parametrizing all the points of the manifold, or (2) by parametrizing a family of nonoverlapping manifolds. For a classical mass point, the former is the conventional description of its world line, the proper time being the parameter. It can be converted into the second type, which is the Hamilton-Jacobi description. In the latter, a family of world lines is represented as the family of normals to a scalar field (Hamilton's function) $S(x)$, with x^μ and $\partial_\mu S$ forming canonical pairs. For example, the world line $x^\mu(\tau)$ of a charge particle satisfies the geodesic (Lorentz) equation (metric $= - - - +$)

$$\frac{d^2 x^\mu}{d\tau^2} = eF^{\mu\nu}\frac{dx_\nu}{d\tau}, \qquad F_{\mu\nu} = \partial_\mu A_\nu - \partial_\nu A_\mu,$$

$$\left(\frac{dx}{d\tau}\right)^2 = \text{const} = m^2, \tag{13}$$

where m is interpreted as mass. This follows from the Lagrangian (with τ as evolution parameter)

$$\mathcal{L} = \frac{1}{2}\left(\frac{dx}{d\tau}\right)^2 + e\frac{dx}{d\tau} \cdot A. \tag{14}$$

In the Hamilton-Jacobi form, we set

$$p^\mu \equiv \frac{dx^\mu}{d\tau} + eA^\mu = \partial^\mu S,$$

$$\frac{dx^\mu}{d\tau} = \frac{\partial H}{\partial p_\mu}, \qquad \frac{dp^\mu}{d\tau} = -\frac{\partial H}{\partial x_\mu},$$

$$H \equiv \frac{1}{2}(p - eA)^2 = \frac{m^2}{2}. \tag{15}$$

The H-J form serves, at least historically, as a natural passage to quantum theory ($p^\mu \to -i\partial/\partial x_\mu$, $\exp[iS/h] \to \psi$). It is after this that the Dirac relation between charge and pole strengths comes in.

A similar construction in the case of a string goes as follows [4]. The world sheet of the string is parametrized as $x^\mu(\tau^1, \tau^2)$, and

$$\frac{\partial(x^\mu, x^\nu)}{\partial(\tau^1, \tau^2)} \equiv \{x^\mu, x^\nu\} \tag{16}$$

defines a field of tangential planes. The analogues of (13) and (14) are

$$\{\{x_\mu, x_\nu\}, x^\nu\} = eF_{\mu\nu\lambda}\{x^\nu, x^\lambda\}, \tag{17}$$

with $(\{x^\mu, x^\nu\})^2 = \text{const} = -\sigma^2$ and

$$\mathcal{L} = \frac{1}{4}(\{x^\mu, x^\nu\})^2 - e\{x^\mu, x^\nu\}A_{\mu\nu} \tag{18}$$

with two independent evolution parameters τ^1, τ^2. Here $A_{\mu\nu}$ and $F_{\mu\nu\lambda}$ are the Kalb-Ramond potential and field, respectively. The constant σ is interpreted as string tension. If $F_{\mu\nu\lambda}$ has a "magnetic" piece its source will be points in R^4 ("instantons"), according to (8).

The corresponding Hamilton form is given by

$$p^{\mu\nu} \equiv \{x^{\mu}, x^{\nu}\} + eA^{\mu\nu}, \qquad \{x^{\mu}, x^{\nu}\} = \frac{\partial H}{\partial p_{\mu\nu}},$$

$$\{p^{\mu\nu}, x_{\nu}\} = -\frac{\partial H}{\partial x_{\mu}}, \qquad H = \frac{1}{4}(p_{\mu\nu} - eA_{\mu\nu})^2 = -\frac{1}{4}\sigma^2. \quad (19)$$

The passage to the Jacobi form (family of geodesic surfaces) is rather complicated. Basically we need a pair of gauge fields $E^{(2)} \equiv dA^{(1)}$ and $D^{(2)} \equiv dB^{(1)}$ ($dE = dD = 0$), such that

$$p_{\mu\nu} = E_{\mu\nu}, \qquad p_{\mu\nu} - eA_{\mu\nu} \equiv \pi_{\mu\nu} = (* D_{\mu\nu})\chi \quad (20)$$

where χ is a scalar field, and

$$H = \tfrac{1}{4}\pi^2 = -\sigma^2, \qquad \pi * \pi = 0. \quad (21)$$

The quantum theory of strings looks much simpler, at least on the surface. We equate $p_{\mu\nu}$ with a loop derivative:

$$p_{\mu\nu}(x) \to -i\delta/\delta\sigma^{\mu\nu}(x) \quad (22)$$

parametrized by x^{λ} and acting on a wave function in the space of (oriented) loops (in the case of closed strings). Eigenfunctions of $p^{\mu\nu}$ are

$$\psi(C, A^{\mu}(x)) = \exp\left[i\oint_{C} A_{\mu} \, dx^{\mu}\right],$$

$$p^{\mu\nu}(x) \propto \left(n^{\mu}F^{\nu\lambda}n_{\lambda} - n^{\nu}F^{\mu\lambda}n_{\lambda}\right)\psi, \qquad x \in C,$$

$$= 0, \qquad x \notin C, \quad (23)$$

where n^{μ} is the unit tangent vector at $x \in C$. But the unsolved problem is how to define a proper measure in the loop space and whether the operators are well defined or not. At any rate, we can see here a possible connection between strings and gauge fields.

A different approach [5] is to quantize the string by integrating over all geometries of the internal parameter space, with a Lagrangian

$$\mathcal{L} = g^{ab} \frac{\partial x^{\mu}}{\partial \tau^{a}} \frac{\partial x_{\mu}}{\partial \tau^{b}} (\det g)^{1/2} \qquad (a, b = 1, 2). \quad (24)$$

As such, this should probably be regarded as a "first quantization". Although the integration over the parameter space can be carried out, it is not known whether we have a physically reasonable theory or not.

III. Monopoles [6, 7]. In (11) the monopoles are represented as an extra piece in the Maxwell field $F'_{\mu\nu}$, depending on the $O(3)$ Higgs fields ϕ^{a}, $a = 0, 1, 2$. For a fixed time, $\phi^1(x) = c^1$, $\phi^2(x) = c^2$ define a curve, or a string labeled by (c^1, c^2). When different strings converge on one point, we have a monopole there, and that singularity is all that matters physically. So in a way the φ's define a field of nonphysical strings. Two strings are equivalent if they share the same end point(s). Because of this equivalence, we might as well grab all the strings emanating from a common end and squeeze them into a single bunch. This is the Dirac description of a monopole. The usual Maxwell form of $F_{\mu\nu}$ holds everywhere except on a semi-infinite string. So the situation is like the magnetic flux discussed earlier except that the flux abruptly terminates and then leaks out (Figure 4). When a test charge is looped around the string, its wave function receives a phase factor

$$\psi \to U\psi = \exp\left[ie\oint A \cdot dx\right]\psi \tag{25}$$

in correspondence with an element of the first homotopy group of $U(1)$: $\pi_1(U(1)) = Z$. This phase factor must be $= 1$, because otherwise the shape of the string would be visible through the Aharonov-Bohm effect. Hence the Dirac quantization condition

$$U = 1 \quad \text{or} \quad e\Phi = 4\pi eg = 2\pi n. \tag{26}$$

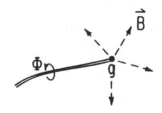

Figure 4

The same arguments also apply to non-Abelian gauge fields. In this case the phase factor U is an element of a Lie group G acting on a particular representation ψ. U must correspond to a subgroup H of G which is mapped to 1 when acting on ψ, and also be an element of $\pi_1(H)$. Then $U = 1$ can be reduced by a gauge transformation to a condition on the sum of diagonal phases of U,

$$e\sum \Lambda_i \Phi_i = 2\pi n. \tag{27}$$

In the case of 't Hooft-Polyakov monopole, H is a subgroup of G to which it is broken down by a Higgs mechanism, like in the Weinberg-Salam theory. If H contains a $U(1)$, monopoles are realized, but it depends on the dynamics. The Higgs field φ is nonzero and covariantly constant (at least sufficiently away from singularities) to minimize both potential and kinetic energy

$$\langle \phi \rangle = \phi_0, \qquad V(\phi_0) \sim 0,$$
$$D_\mu \phi_0 \sim 0, \qquad |\phi_0|^2 \sim \text{const}. \tag{28}$$

Here D_μ is the covariant derivative $\partial_\mu - ieA_\mu$, and A_μ is represented as a matrix. But then also

$$[D_\mu, D_\nu]\phi_0 \sim -ieF_{\mu\nu}\phi_0 \sim 0. \tag{29}$$

A nonzero magnetic field (εH) is possible only if ϕ_0 is zero eigenfunction of the corresponding $F_{\mu\nu}$. For example, if $G = \text{SU}(2)$, and ϕ_0 is its adjoint representation (2×2 Hermitian traceless matrices), A_μ is partially determined by ϕ_0 from (28).

$$eA_\mu \sim [\hat{\phi}_0, \partial_\mu \hat{\phi}_0] + a_\mu \hat{\phi}_0 \qquad (a_\mu \text{ arbitrary}),$$
$$eF_{\mu\nu} \sim [\partial_\mu \hat{\phi}_0, \partial_\nu \hat{\phi}_0] + f_{\mu\nu} \hat{\phi}_0 \qquad (f_{\mu\nu} = \partial_\mu a_\nu - \partial_\nu a_\mu),$$
$$eH_{\mu\nu} \equiv \text{tr } e(\hat{\phi}_0 F_{\mu\nu})$$
$$= \text{tr } \hat{\phi}_0 [\partial_\mu \hat{\phi}_0, \partial_\nu \hat{\phi}_0] + f_{\mu\nu} \qquad (\hat{\phi}_0 = \phi_0 / [\text{tr } \phi_0^2]^{1/2}). \tag{30}$$

The last equation defines the admissible monopole field $H_{\mu\nu}(x)$ living in a $U(1)$ subgroup H, which leaves the given ϕ_0 invariant at point x. Under the inversion $\phi_0 \to -\phi_0$, $H_{\mu\nu}$ reverses sign, leading to an antimonopole. A typical choice of ϕ_0 is the hedgehog configuration (Figure 5).

$$\hat{\phi}_0 \equiv \sum_{i=1}^{3} \sigma_i \phi_i, \qquad \phi_i \sim x_i/r,$$
$$eB_i \sim x_i/r^3 \quad \text{or} \quad eg = 4\pi \tag{31}$$

(the σ_i are Pauli spin matrices, B is the magnetic field). So ϕ_0 is in the orbit $G/H = \text{SU}(2)/U(1)$, and represents a mapping $\pi_2(G/H) = \pi_1(H) = Z$.

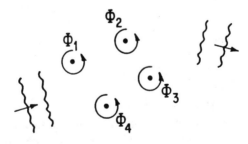

Electron wave

FIGURE 5

The difference between Dirac and 't Hooft-Polyakov monopoles is physically important. The latter is a natural consequence of a non-Abelian Higgs model, and its mass and other physical attributes are calculable from the parameters of the electric charge sector. The magnetic charge is twice the Dirac unit, due to the fact that ϕ_0 is in the adjoint representation (G is SO(3) rather than SU(2)). Use of the fundamental doublet representation runs into trouble with (29) because the eigenvalues of Pauli matrices are nonzero. The 't Hooft-Polyakov monopoles are realizable in the framework of conventional local field theory, and perhaps also in quantum field theory, but such is not the case yet with Dirac monopoles.

IV. GUT monopoles. The standard scenario of the grand unified theory (GUT) assumes a quintet (5) and a decimet (antisymmetric 5×5) of fermions (quarks q and leptons l). The quintet part is

$$f_5 \sim (q_1, q_2, q_3; l_4, l_5),$$
$$G = \mathrm{SU}(5) \to H$$
$$= \mathrm{SU}(3)_{\text{color}} \times \mathrm{SU}(2)_{\text{flavor}} \times U(1)_{\text{flavor}}. \tag{32}$$

There are also superheavy Higgs bosons Φ in the adjoint 24 which causes the above breaking of SU(5) according to the diagonal pattern

$$\Phi_0 \sim \begin{pmatrix} 2 & & & \vdots & \\ & 2 & & \vdots & \\ & & 2 & \vdots & \\ -&-&-&-&-&-&-&- \\ & & & \vdots & -3 & \\ & & & \vdots & & -3 \end{pmatrix} \phi_0, \quad \phi_0 \sim 10^{15} \text{ proton masses.} \tag{33}$$

The gauge bosons, again in the adjoint representation, are

$$G \sim \begin{pmatrix} \text{gluons} & \vdots & X, Y \\ -&-&-&-&- \\ X^+, Y^+ & \vdots & W, Z^0 \end{pmatrix},$$
$$M_{X,Y} \sim e\phi_0 \quad (\text{superheavy}), \quad e^2/4\pi \sim 1/50. \tag{34}$$

Recently Cabrera reported registering one event which has the characteristics of a monopole satisfying the Dirac condition $eg = 2\pi$. If it is a GUT monopole, its mass is expected to be $\sim g\phi_0 \sim 10^{16}$ protons $\sim 10^{-8}$ grams! But the inferred abundance seems too high, and causes all sorts of theoretical problems. It is by no means clear that a monopole has been found. Here, however, we will discuss theoretical aspects of SU(5) monopoles.

A monopole under consideration has a set of magnetic charges according to (27). In fact, to be compatible with one Dirac unit and the fact that the smallest electric charge is $e/3$ rather than e, it must have both $U(1)$ (electromagnetism) and SU(3) (color) magnetic charges as indicated by

$$F_{\mu\nu} \sim [\partial_\mu \Phi_0, \partial_\nu \Phi_0] \sim \begin{pmatrix} 1/3 & & & & \\ & 1/3 & & & \\ & & 1/3 & & \\ & & & -0 & \\ & & & & 0 \end{pmatrix}$$

$$+ \begin{pmatrix} 2/3 & & & & \\ & -1/3 & & & \\ & & -1/3 & & \\ & & & 0 & \\ & & & & 0 \end{pmatrix}$$

$$= \begin{pmatrix} 0 & & & & \\ & 0 & & & \\ & & 1 & & \\ & & & -1 & \\ & & & & 0 \end{pmatrix}. \tag{35}$$

The -1 corresponds to electron state, and is fixed by the Cabrera assumption. The position of the 1 can be at any of the first three quark diagonals, corresponding to Weyl transforms of each other. So there are three monopoles and three antimonopoles of this type.

To be a 't Hooft-Polyakov monopole, furthermore, the Higgs field Φ_0 must be on a nontrivial orbit $SU(2)/U(1)$, where $SU(2) \subset G$, $U(1) \subset H$. This is indeed all right since Φ_0 can take a hedgehog configuration regarding its 2×2 submatrix consisting of the 4th and 5th components in the above example.

There are some physical questions that have been raised recently. For example, what is the meaning and effect of the SU(3) color magnetic charge? Electric and magnetic charges must behave rather differently in quantum field theory, where charges become variable due to renormalization. To maintain the quantization condition, e and g will have to vary in opposite directions: $e \to \infty$, $g \to 0$ as scale length $\to \infty$. Color electric forces are confining, but color magnetic forces are not. Single monopoles may exist, with no color visible at large distances. This means that in some sense monopoles are like the hadrons, which have no net electric color.

Among other problems there is the effect of a monopole on other particles and fields [7]. If a particle moves in the field of a 't Hooft-Polyakov monopole, it sees the orientation of the internal SU(2) spin vary from place to place in a hedgehog pattern. Thus this internal spin plus the ordinary angular momentum becomes the quantum number that labels the states. Somehow the internal spin has got promoted to ordinary spin, and the statistics of the combined monopole-particle system changes accordingly. Furthermore, such a system of a 't Hooft-Polyakov monopole and a fermion admits zero energy bound states (zero modes), i.e., states having the same total energy as the monopole itself. These zero modes are a peculiarly quantum phenomenon, as bound states of a charge and a monopole (dyons) do not occur in classical theory.

Actually, zero modes are generally associated with restoration of spontaneously broken symmetries. There are zero modes to restore the Poincaré invariance of a fixed monopole configuration. The dyon zero modes mentioned above may have to do with some broken symmetry. In fact, fermion zero modes suggest the existence of a hidden supersymmetry involving monopoles.

A related issue is the recent claim [8] that a monopole can catalyze proton decay. A simple explanation of the effect will go as follows. It was mentioned above that the internal spin orientation varies around a monopole. This orientation is the reference frame with respect to which the physical identity of a test particle is to be made, e.g., whether it is a quark or a lepton. But the reference frame becomes ambiguous at the center of the hedgehog pattern. So a fermion will lose its identity if it reaches the center, hence the possibility of conversion of a proton, which is made up of three quarks, into an electron and a meson (quark and antiquark).

V. Concluding remark. I will close by posing two general questions. The first question is reminiscent of the Einstein-Infeld-Hoffmann program, in which mass points were to be regarded as singularities of the metric tensor itself, and therefore belong to the left-hand side of Einstein's equation. In the present case, the analogous process forms a closed cycle involving a dual pair of entities, as illustrated below:

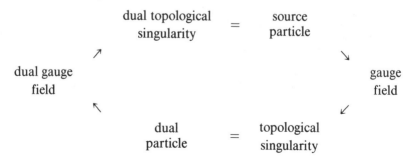

Only the right-hand half of the cycle is fairly well understood. The other half is more or less a matter of speculation [9]. In order to complete the cycle, it would be necessary to learn how to quantize a distribution of topological singularities (quantum mechanics of turbulent systems). Can they be represented by some dual fields, as ordinary particles are represented by quantized fields?

The second question concerns what I call Aharonov-Bohm medium. Imagine a 2-dimensional plane pierced by an arbitrary distribution of nonquantized vortex (flux) lines (Figure 5). How will the electrons behave in this plane? One vortex case has been solved by Aharonov and Bohm, but the general case has not. [Once Eguchi and I tried, and failed.] It boils down to that of a free Schrödinger equation (or diffusion equation with imaginary time) on a Riemann sheet:

$$i\frac{\partial}{\partial t}\psi = -\nabla^2\psi \qquad |\psi| \to 0 \quad \text{as } x \to x_i,$$

$$\oint_{c_i} \nabla \ln \psi \cdot dx = \Phi_i \quad \text{around each vortex at } x_i. \tag{36}$$

The apparent simplicity of the problem is very deceptive, but it may still be simple enough to admit rigorous mathematical statements. Some implied physical questions: with either regular lattice or random distribution of vortices, what will be the energy spectrum? Can the electrons be confined (localized)?

REFERENCES

1. A Tonomura et al., Phys. Rev. Lett. **48** (1982), 1443–1446.
2. See P. G. O. Freund and R. I. Nepomechie, Nucl. Phys. B **199** (1982), 482–494.
3. H. Rund, *The Hamilton-Jacobi theory in the calculus of variations*, Van Nostrand, London and New York, 1966; H. Rund, in *Topics in differential geometry* (H. Rund and W. F. Fields, eds.), Academic Press, New York, 1976, pp. 111–133; V. Volterra, Rend. Accad. Lincei (4) **7** (1890), 127–138; H. A. Kastrup and M. Rinke, Phys. Lett. B **105** (1981), 191–196.
4. Y. Nambu, Phys. Lett. B **92** (1980), 327–330; Phys. Lett B **102** (1981), 149–153; Y. Hosotani, Phys. Rev. Lett. **47** (1981), 399–401.
5. A. M. Polyakov, Phys. Lett. B **103** (1981), 207–210.
6. See the review article by A. Actor, Rev. Modern Phys. **51** (1979), 461–525.
7. See the review article by R. Jackiw, Rev. Modern Phys. **49** (1977), 681–706.
8. C. G. Callan, Phys. Rev. D **26** (1982), 2058–2068 and references therein.
9. See D. Olive, Phys. Rep. **49** (1979), 165–172.

THE ENRICO FERMI INSTITUTE AND THE DEPARTMENT OF PHYSICS, UNIVERSITY OF CHICAGO, CHICAGO, ILLINOIS 60637

III. REPRESENTATIONS OF NONCOMPACT LIE GROUPS

Lectures in Applied Mathematics
Volume **21**, 1985

Supergroups and Their Representations[1]

I. Bars[2]

CONTENTS

1980 *Mathematics Subject Classification.* Primary 20G45, 22E47, 22E70, 22E99.

[1]Reprinted by permission from *Introduction to Supersymmetry in Particle and Nuclear Physics* (O. Castaños, H. Frank, and L. Urrutia, eds.), Plenum, New York, 1984, pp. 107–184.

[2] Based on lectures delivered at

    (1) University of Louvain, Belgium, Nov. 1981;

    (2) School on Supersymmetry, Mexico, Dec. 1981; and

    (3) AMS-SIAM Summer Seminar, July 1982. Research supported in part by the U. S. Department of Energy under Contract No. DE-AC-02-76 ERO 3075.

1. Applications of supergroups and superalgebras in physics. About a decade ago a new kind of symmetry principle appeared in physics, namely supersymmetry. The novel feature of this symmetry is that it operates between bosons and fermions which have different space-time (or spin and statistics) properties. The generators of supersymmetry transformations form a Lie superalgebra whose even subalgebra is an ordinary Lie algebra and whose odd generators, which mix bosons and fermions, close under anticommutation to the even part.

The first applications of relativistic supersymmetry appeared in string theory [1] and field theory [2, 3]. Recent applications of static supersymmetry occurred in nuclear physics [4]. Recently a general theory that uses induced representations of noncompact supergroups in supersymmetric field theory has also been developed [5]. A different kind of application of superalgebras with a reinterpretation of (odd-even) $\rightarrow$ (left-right of Lorentz group) has been used in composite models of quarks and leptons to obtain theories of dynamically unbroken $SU(N)_L \times SU(N)_R$ chiral symmetries [6]. Finally, supergroup invariant integration has been developed in the context of path integrals for the quantized super "chiral" $SU(N/M)$ model analogous to the symmetric top problem [7]. There is one other early attempt which tries to use superalgebras for internal symmetries [8], but this approach is plagued by unphysical ghosts, for which so far only partial cures have been suggested. The cures are related to another application of superalgebras [9] in connection to the BRS symmetry and the Faddeev-Popov ghosts that appear in quantum gauge theories.

Most of these applications are to rather different problems and are unrelated to each other. We may, however, organize them according to the kinds of representations that they use: finite- and infinite-dimensional representations, as shown below.

I. *Compact supergroups. Finite-dimensional representations.* (1) Nuclear physics (static supersymmetry) [4] $SU(6/M)$.

(2) Composite quarks and leptons [6] $SU(N + 4/N)$, $SU(N/N) \times SU(M/M)$.

(3) Superchiral model ($\approx$ symmetric top) [7] $SU(N/M)$.

(4) Internal supergroups [8] $SU(N, 1)$.

(5) BRS symmetry in quantum gauge theories [9].

II. *Noncompact supergroups. Infinite-dimensional representations.*

(1)(a) Globally supersymmetric field theories [2, 3].

(b) Supersymmetric grand unification [10].

(c) Supersymmetric preon models [11] (super-Poincare group).

(2) Locally supersymmetric field theory—supergravity [12] (super-Poincare, $OSp(N/4)$, $SU(N/4)$, etc.).

(3) Superstring theory [1], super-Kac-Moody.

(4) General superfield theory with induced representations [5], super-Poincare in d-dimensions, OSp($N/2M$), SU(N/M).

Among these applications only nuclear physics has provided experimental support for the usefulness of superalgebras in nature. The theory is based on the superalgebra SU($6/M$) in which the 3-dimensional rotation group which assigns spins to nuclear states is embedded. This provides a classification scheme for many low lying nuclear states of several even-even (bosonic) and even-odd (fermionic) nuclei in the Platinum-Gold region. The classification in a single irreducible representation of SU($6/M$) predicts: (1) an energy formula for patterns of many nuclear levels, (2) relations among decay rates of excited states, and (3) relations between nucleon-transfer reactions among such nuclei. The predictions work not only qualitatively but also quantitatively within 10–20%. The 10–20% discrepancies notwithstanding, the fact that a large amount of data on different kinds of experiments is successfully described by this simple scheme is in itself remarkable. It has been shown that the supersymmetry scheme can be extended to situations which are far more complex than the original application [4]. Superalgebras have thus provided a rather general approach for correlating and organizing nuclear data as well as making many testable predictions. The whole approach is still under experimental investigation and theoretical development.

Supersymmetric field theory is generally more convergent than non-supersymmetric field theory, and there are hopes that some theories such as $N = 4$ super-Yang-Mills theory may even be finite to all orders of perturbation theory. The improvement of the convergence of field theory is a remarkable effect of supersymmetry. Unfortunately, there is no supersymmetric field theory in particle or gravitational physics which makes testable predictions of this kind of supersymmetry either directly or indirectly. There remains to be seen if more imaginative applications in field theory and particle physics will lead to measurable effects of supersymmetry. The formalism in [5] is being developed to look for other possibilities.

Superstring theory is one of the very early areas where superalgebras appear. However, strings are idealizations of what might happen in the theory of strong interactions, that is Quantum Chromodynamics (QCD). Strings have their own unphysical problems which are quite independent from the underlying field theory (QCD).

The use of superrepresentations in theories of composite quarks and leptons made from preons is very different [6]. The Lagrangians of these models are invariant under global internal symmetries U(N)$_L$ which transform left-handed fermions, and U(M)$_R$ which transform right-handed fermions. Furthermore, there is a precolor (or hypercolor) local gauge symmetry which binds the preons. There is no supersymmetry of the usual kind (although the model could be enlarged to include them). A supergroup or superalgebra $SU(N/M)$ is introduced which includes the symmetry of the Lagrangian $SU(N)_L \times SU(M)_R \times U(1)$ as the even subgroup. However, the even-odd grading of the generators is not according to bose-fermi, but rather, according to left-right. So, the odd super-generators (which are not symmetries) mix left with right. Even though $SU(N/M)$ is not a symmetry, it plays the role of a classification group. This classification of preons predicts sets of composite quarks and leptons which solves [6] certain constraints [13], called "decoupling", which are necessary for the dynamical survival of the chiral symmetries $SU(N)_L \times SU(M)_R \times U(1)$. The survival of chiral symmetries is needed to explain the small masses of composite quarks and leptons.

In these supergroup models there is a larger supergroup (again not a symmetry) which has an even subgroup which includes precolor together with the chiral symmetries:

$$\text{precolor} \times SU(N)_L \times SU(M)_R \times U(1).$$

The classification of antipreons together with the composite quarks and leptons in the same supermultiplet of this supergroup (not a symmetry) solves another constraint, called "anomaly matching", also needed for the dynamical survival of the chiral symmetry. A few more constraints needed for the same purpose are uniquely and simultaneously solved provided we restrict ourselves to just the supergroups $SU(N + 4/N)$ in the two-box superrepresentation ⊟ and $SU(N/N) \times SU(M/M)$ in the two-box superrepresentation $(⊠, ⊠)$. This notation will be explained below.

The BRS symmetry of quantum gauge theories can be viewed in the context of superalgebras [9]. This symmetry is used to prove the unitarity of the theory by generating Ward identities that show the cancellation of all ghost effects.

The toy model of the $SU(N/M)$ chiral model in 1-time dimension (like the symmetric top) is developed [7] for the purpose of understanding the meaning of supergroup invariant integration. This model can easily be solved by ordinary canonical quantization of its bosons and fermions and any Green function can be obtained. On the other hand, by considering also the path integral quantization (bosonic and fermionic integration), one must consider supergroup invariant integration via a super-Haar measure. Since we know all the answers already from the point of view of canonical quantization, the meaning and rules of supergroup integration can be developed, as in [7].

In the rest of the written version of these lectures I will not return to the applications. The interested reader can read the literature. I will now concentrate on the mathematical aspects of the representations of super-groups and superalgebras as developed in [14, 15]. Some new results not found in [14, 15] will also appear in this paper. Further developments [16] on Kac-Dynkin diagrams and on noncompact supergroups and unitary representations [17] will appear elsewhere. Other approaches to represen-tations of superalgebras can be found in the literature [18, 19].

2. The superalgebra $SU(N/M)$. Just as a complete classification of Lie algebras was obtained by Cartan, a complete classification of Lie super-algebras has been given by Kac [18]. It is useful to compare these two classifications and describe some analogies between the fundamental representations of Lie algebras and superalgebras.

Lie algebras fall into four infinite classes, A_n, B_n, C_n, D_n with $n = 1, 2, \ldots$, and five exceptional cases, G_2, F_4, E_6, E_7, E_8. The four infinite classes can be put in one-to-one correspondence with the classical groups $A_{n-1} \leftrightarrow SL(n)$ or $SU(n)$; $B_n \leftrightarrow SO(2n + 1)$; $C_n \leftrightarrow Sp(2n)$; $D_n \leftrightarrow SO(2n)$. Similarly, Lie superalgebras fall into five infinite classes: $A(n, m) \leftrightarrow SL(n + 1/m + 1)$; $B(n, m) \leftrightarrow OSp(2n + 1/2m)$; $D(n, m) \leftrightarrow OSp(2n/2m)$; $P(n)$; $Q(n)$, and the exceptional cases $D_\alpha(2, 1)$, $G(3)$, $F(4)$, plus the Cartan superalgebras. By comparing the structure of the matrices in the fundamental representation for Lie algebras and superalgebras, we can establish some useful analogies. Correspondences obtained in this way were used to develop the representation theory in terms of super-tableaux [14–17].

In these lectures we concentrate only on $SU(N/M)$, which is the compact form of $SL(N/M)$. We will describe here the compact version and its unitary finite-dimensional representations. A similar treatment was given for the finite-dimensional representations of $OSp(N/2M)$ and $P(N)$ [14].

Unitary representations of noncompact $SL(N/M)$ have also been constructed [17] but will not be discussed here.

In the $(N + M)$-dimensional fundamental representation the algebra of $SU(N/M)$ is given by hermitian $(N + M) \times (N + M)$ matrices of the form

$$\mathcal{H} = \left(\begin{array}{c|c} H_N & \theta \\ \hline \theta^\dagger & H_M \end{array} \right), \qquad \mathrm{Tr}\, H_N = \mathrm{Tr}\, H_M. \tag{1}$$

Here H_N (H_M) is an $N \times N$ ($M \times M$) hermitian matrix $H_N = H_N^\dagger$ ($H_M = H_M$) constructed from bosonic complex parameters. That is, H_{11}, H_{12}, etc., are complex bosonic numbers. On the other hand, θ is an $N \times M$ matrix filled with complex fermions or complex anticommuting Grassmann numbers. That is, θ together with $\theta^\dagger$ describe $2NM$ real anticommuting Grassmann numbers. Even powers of Grassmann num-bers will be included in the set of bosons, while odd powers will be included in the set of fermions. The traceless part of H_N (H_M) corresponds to an SU(N) (SU(M)) subgroup, while the trace part generates a $U(1)$. Thus the bosonic part of $\mathcal{K}$ forms a Lie subalgebra corresponding to SU(N) $\times$ SU(M) $\times$ $U(1)$.

As an example, consider SU(2/1) in the form

$$\mathcal{K} = \frac{1}{2}\left(\begin{array}{cc|c} \vec{\omega}\cdot\vec{\tau}+\omega^4\tau_4 & & \theta^1 - i\theta^2 \\ & & \theta^3 - i\theta^4 \\ \hline \theta^1 + i\theta^2 & \theta^3 + i\theta^4 & 2\omega^4 \end{array}\right), \tag{2}$$

$$\vec{\tau} = \text{Pauli matrices, } \tau_4 = \text{unity.}$$

This matrix contains four real bosonic parameters, ω^1, ω^2, ω^3, ω^4, and four real fermionic parameters θ^1, θ^2, θ^3, θ^4. The generators in the fundamental representation are given as the matrix coefficients of these real variables:

$$\vec{T} = \left(\begin{array}{c|c} \vec{\tau}/2 & 0 \\ \hline 0 & 0 \end{array}\right), \qquad T_4 = \left(\begin{array}{cc|c} \frac{1}{2} & & \\ & \frac{1}{2} & \\ \hline & & 1 \end{array}\right),$$

$$S_1 = \frac{1}{2}\left(\begin{array}{cc|c} 0 & & 1 \\ & & 0 \\ \hline 1 & 0 & 0 \end{array}\right), \qquad S_2 = \frac{1}{2}\left(\begin{array}{cc|c} 0 & & -i \\ & & 0 \\ \hline i & 0 & 0 \end{array}\right), \tag{3}$$

$$S_3 = \frac{1}{2}\left(\begin{array}{cc|c} 0 & & 0 \\ & & 1 \\ \hline 0 & 1 & 0 \end{array}\right), \qquad S_4 = \frac{1}{2}\left(\begin{array}{cc|c} 0 & & 0 \\ & & -i \\ \hline 0 & i & 0 \end{array}\right),$$

where T_m, $m = 1, 2, 3, 4$, are ordinary Lie generators forming an SU(2) $\times$ $U(1)$ subgroup, while S_μ, $\mu = 1, 2, 3, 4$, are the supergenerators.

Note the similarities between the SU(2/1) and SU(3) generators. However, despite appearances they are totally different. First, the S_μ close into the $\vec{T}$, T_4 under anticommutation rather than commutation as can be verified by explicit matrix multiplication,

$$\{S_\mu, S_\nu\} = C^m_{\mu\nu}T_m. \tag{4}$$

Second, T_4 instead of being a traceless 3×3 matrix is a supertraceless 3×3 matrix. The supertrace of a matrix such as $\mathcal{K}$ is defined as

$$\text{Str } \mathcal{K} = \text{Tr } H_N - \text{Tr } H_M, \tag{5}$$

and we have Str $\mathcal{K} = 0$; Str $T_4 = 0$.

The analogy between SU(2/1) and SU(3), or more generally SU(N/M) and SU($N + M$), is improved by combining the generators together with the parameters into the form $\mathcal{H}$. That is,

$$\mathcal{H} = \sum_{m=1}^{4} \omega^m T_m + \sum_{\mu=1}^{4} \theta^\mu S_\mu = \left(\begin{array}{c|c} H_N & \theta \\ \hline \theta^+ & H_M \end{array} \right). \tag{6}$$

In this form we need to consider only commutators and never anticommutators. This is because the product $\theta^\mu S_\mu$ essentially acts like a bosonic generator. Namely, consider two sets of fermionic anticommuting parameters θ^μ and $\tilde{\theta}^\mu$ and multiply equation (4) by the bosonic product $\theta^\mu \tilde{\theta}^\nu$:

$$\theta^\mu \tilde{\theta}^\nu \{S_\mu, S_\nu\} = i\theta^\mu \tilde{\theta}^\nu C_{\mu\nu}^a T_a. \tag{7}$$

Noting the anticommutation property of the fermions we can write, by changing their orders,

$$\theta^\mu \tilde{\theta}^\nu \{S_\mu, S_\nu\} = \theta^\mu \tilde{\theta}^\nu S_\mu S_\nu - \tilde{\theta}^\nu \theta^\mu S_\nu S_\mu = \left[\theta^\mu S_\mu, \tilde{\theta}^\nu S_\nu \right]. \tag{8}$$

Therefore, we see that closure is obtained with commutators

$$\left[\theta^\mu S_\mu, \tilde{\theta}^\nu S_\nu \right] = i\theta^\mu \tilde{\theta}^\nu f_{\mu\nu}^m T_m \sim i \left(\begin{array}{c|c} H_N & 0 \\ \hline 0 & H_M \end{array} \right). \tag{9}$$

From this one learns that, as long as the parameters are combined with the generators, *we only need to consider commutators in order to close the algebra. However, one must not forget to respect the order in which Grassmann numbers appear in a product.* In general, we find that for $SU(N/M)$ we have the Lie commutation rule

$$[\mathcal{H}, \mathcal{H}'] = i\mathcal{H}'', \tag{10a}$$

or, more explicitly,

$$\left[\left(\begin{array}{c|c} H_N & \theta \\ \hline \theta^\dagger & H_M \end{array} \right), \left(\begin{array}{c|c} H_N' & \theta' \\ \hline \theta'^\dagger & H_M' \end{array} \right) \right] = i \left(\begin{array}{c|c} H_N'' & \theta'' \\ \hline \theta^{\dagger''} & H_M'' \end{array} \right), \tag{10b}$$

where H_N'', H_M'', θ'', $\theta^{\dagger''}$ can be calculated by explicit matrix multiplication:

$$\begin{aligned} i\theta'' &= H_N \theta' + \theta H_M'' - H_N' \theta - \theta' H_M, \\ i\theta''^\dagger &= -\theta'^\dagger H_N - H_M' \theta^\dagger + \theta^\dagger H_N' + H_M \theta'^\dagger, \\ iH_N'' &= [H_N, H_N'] + \theta\theta'^\dagger - \theta'\theta^\dagger, \\ iH_M'' &= [H_M, H_M'] + \theta^\dagger \theta' - \theta'^\dagger \theta. \end{aligned} \tag{11}$$

It is easy to verify that $\operatorname{tr} H_N'' = \operatorname{tr} H_M''$. In the limit $\theta = \theta' = 0$ we obtain the $SU(N) \times SU(M) \times U(1)$ subgroup. In this form the analogy between $SU(N/M)$ and $SU(N + M)$ is complete. For $SU(N + M)$ the U(1) generator in $\mathcal{H}$ is traceless rather than supertraceless, and theta is bosonic rather than fermionic; otherwise the commutation rules look formally similar to $SU(N/M)$. Note that for iH_N'' to be an antihermitian

$N \times N$ matrix we must define the operation of hermitian conjugation to *change the orders of the Grassmann numbers* in a product. For example,

$$\left(\theta_1 \theta_2'^{\dagger} \right)^{\dagger} = \theta_2' \theta_1^{\dagger}. \tag{12}$$

This is consistent with a physicist's interpretation of a Grassmann number as being simply a fermion. This definition of hermitian conjugation is, in fact, necessary in order to apply supergroups in quantum theory.

3. Fermions and Grassmann numbers. At this point it may be useful to describe anticommuting Grassmann numbers in a physicist's language. It is nothing but an annihilation or creation operator of a fermion, but not both together. In this language it is also possible to obtain a matrix description of a Grassmann number. Consider the creation-annihilation operators of a single fermion, which satisfy

$$\{a, a\} = 0, \qquad \{a^{\dagger}, a^{\dagger}\} = 0, \qquad \{a, a^{\dagger}\} = 1. \tag{13}$$

From these one can construct only two states in a Fock space which is a direct product with the space on which the generators of the superalgebra act:

$$|1\rangle = |0\rangle = \text{vacuum}, \qquad |2\rangle = a^{\dagger}|0\rangle = \text{one-particle state}.$$

In this 2-dimensional vector space, one can calculate the matrix elements of the operator a:

$$\langle i|a|j\rangle \sim \begin{pmatrix} 0 & 1 \\ 0 & 0 \end{pmatrix}_{ij}. \tag{14}$$

The operator a, or its matrix representation taken with matrix multiplication, has all of the properties of a Grassmann number. To construct two Grassmann numbers consider two sets of creation-annihilation operators a_1, $a_1^{\dagger}$, a_2, $a_2^{\dagger}$. One can now construct four states, $|0\rangle$, $a_1^{\dagger}|0\rangle$, $a_2^{\dagger}|0\rangle$, $a_1^{\dagger}a_2^{\dagger}|0\rangle$, and obtain 4×4 matrix representations of the operators a_1, a_2. These two operators or their matrix representation have all the properties of two Grassmann numbers. Similarly with n sets of operators a_i, $a_i^{\dagger}$, $i = 1, 2, \ldots$, one obtains $2^n \times 2^n$ representations of the n operators a_i which can be considered as matrix representations of Grassmann numbers. From this physicist's description you learn that you should manipulate Grassmann numbers just like fermions, mainly respecting their anticommutation properties. In general, as will become evident below, we will need an *infinite "pool"* of Grassmann numbers in order to describe all possible supergroup elements, just as we need an infinite "pool" of bosonic commuting numbers (or an infinite number of values for the bosonic parameters). You can do differentiation and integration with Grassmann numbers, with the same rules as fermions. The differential operator is just the other half of the fermion, that is, the canonical conjugator to a_i, which is $a_i^\dagger \to \partial/\partial a_i$, and the integration rule is as in path integrals.

4. The supergroup $SU(N/M)$. We have seen that the algebra of $SU(N/M)$ closes with commutators as in (10) provided the generators are multiplied by the parameters. It is also true that in this form the ordinary Jacobi identities are satisfied trivially:

$$\left[[\mathcal{H}_1, \mathcal{H}_2], \mathcal{H}_3 \right] + \text{cyclic } 1, 2, 3 = 0. \tag{15}$$

This means that we obtain a group element by exponentiating $\mathcal{H}$ or by considering an infinite product of infinitesimal transformations

$$\mathcal{U}^B_{\ A} = \left(e^{\ i\mathcal{H}} \right)^B_{\ A} = \lim_{n \to \infty} \left[\left(1 + i\frac{\mathcal{H}}{n} \right)^n \right]^B_{\ A}. \tag{16}$$

The set of group elements $\mathcal{U}$ satisfy the ordinary group properties under matrix multiplication:

$$\mathcal{U}_1 \mathcal{U}_2 = \mathcal{U}_3 \quad \text{closure,}$$
$$(\mathcal{U}_1 \mathcal{U}_2)\mathcal{U}_3 = \mathcal{U}_1(\mathcal{U}_2 \mathcal{U}_3) \quad \text{associativity,} \tag{17}$$
$$\mathcal{U}^\dagger = \mathcal{U}^{-1} = e^{-i\mathcal{H}} \quad \text{inverse.}$$

This is the supergroup $SU(N/M)$. Because of the group property it is possible to consider it is as an invariance group of transformations on the variables of a physical theory, just as ordinary Lie groups. In the $SU(N/M)$ superalgebra we need $2NM$ real or NM complex fermionic parameters θ^α_a, $a = 1, 2, \ldots, N$; $\alpha = 1, 2, \ldots, M$, to describe a given general group element. In order to consider a different group element we need a different set of NM complex fermionic parameters θ', just as we would need a different set of bosonic parameters. Therefore, we should not limit the pool of available anticommuting parameters to just $2NM$ real fermions. In general, just as we need ordinary bosonic variables which can take an infinite number of values, we also need an infinite pool of anticommuting variables or fermions (which could be described by infinite-dimensional matrices). The $2NM$ fermionic variables in any given $SU(N/M)$ supergroup element can be picked from this infinite set. We can then use differentiation and integration rules for fermions, whenever needed, according to the same rules as in quantum theory, and path integrals.

5. The fundamental basis or module. Taking a set of states, "wavefunctions", or operators that transform like the basis states in the fundamental representation (i.e., a module) we can write the transformation rule

$$\Phi_A \to \Phi'_A = \mathcal{U}_A^{\ B} \Phi_B. \tag{18}$$

Such a basis will be denoted by the supertableau $\boxed{}$ in analogy to a

Young tableau. These states can be arranged in a column matrix in the form

$$\Phi_A = \begin{pmatrix} \phi_a \\ \psi_\alpha \end{pmatrix}, \qquad a = 1, 2, \ldots, N; \alpha = 1, 2, \ldots, M, \qquad (19)$$

such that ϕ_a are bosons and ψ_α are fermions. Such a wavefunction will be said to belong to a class I fundamental representation. A second wavefunction $\tilde{\Phi}_A$ can be introduced, such that

$$\tilde{\Phi}_A = \begin{pmatrix} \psi_a \\ \phi_\alpha \end{pmatrix}, \qquad a = 1, 2, \ldots, N; \alpha = 1, 2, \ldots, M, \qquad (20)$$

where $\psi_a =$ fermion and $\phi_\alpha =$ boson. $\tilde{\Phi}$ will be said to belong to class II. Mostly we will deal with class I because class II is obtained from class I by a "tilde" operation which commutes with the group element, as we will see.

Taking into account the fermionic property of θ, the transformed variables $\Phi'_A = (\phi'_a, \psi'_\alpha)$ preserve their bosonic or fermionic properties. With an infinitesimal transformation

$$\mathcal{U}_A{}^B \approx \delta_A{}^B + i\mathcal{H}_A{}^B \approx \left(e^{i\mathcal{H}} \right)_A{}^B, \qquad (21)$$

we can write

$$\begin{aligned} \delta\phi_a &= (H_N)_a{}^b \phi_b + \theta_a{}^\beta \psi_\beta = \text{boson}, \\ \delta\psi_\alpha &= (\theta^\dagger)_\alpha{}^b \phi_b + (H_M)_\alpha{}^\beta \psi_\beta = \text{fermion}, \end{aligned} \qquad (22)$$

where we note that the product of two fermions $\theta\psi$ is considered a boson, while the product of a boson and a fermion $\theta^\dagger\phi$ or $H_M\psi$ is considered a fermion. Hence the concept of boson includes even powers of fermions while the concept of fermion includes odd powers of fermions.

Under the $SU(N) \times SU(M) \times U(1)$ subgroup (i.e., $\theta = \theta^\dagger = 0$) it is clear that

$$\begin{aligned} \phi_a &\sim (\square, 1)_{1/N} \to \text{fundamental representation of } SU(N), \\ \psi_\alpha &\sim (1, \square)_{1/M} \to \text{fundamental representation of } SU(M). \end{aligned} \qquad (23)$$

The $U(1)$ charges are associated with the supertraceless $U(1)$ generator

$$T_0 \sim \begin{pmatrix} 1/N & 0 \\ \hline 0 & 1/M \end{pmatrix}. \qquad (24)$$

It is useful to define the grade of the index $A = (a, \alpha)$. This is given by

$$g(a) = 0 \quad \text{if } \phi_a = \text{boson},$$

$$g(A) \qquad\qquad\qquad\qquad\qquad (25)$$

$$g(\alpha) = 1 \quad \text{if } \psi_\alpha = \text{fermion}.$$

(The value of the grade is opposite for class II when ψ_a = fermion, ϕ_α = boson.) We can use the grade to interchange the orders of indices or wavefunctions taking into account the bosonic and fermionic property of various terms:

$$\Phi_A \Phi_B = (-1)^{g(A)/g(B)} \Phi_B \Phi_A,$$

$$\Phi_A \mathcal{U}_C{}^D = (-1)^{g(A)[g(C)-g(D)]} \mathcal{U}_C{}^D \Phi_A,$$

$$\mathcal{U}_A{}^B \mathcal{U}_C{}^D = (-1)^{[g(A)-g(B)][g(C)-g(D)]} \mathcal{U}_C{}^D \mathcal{U}_A{}^B, \qquad (26)$$

$$\mathcal{H}_{1A}{}^B \mathcal{H}_{2C}{}^D = (-1)^{[g(A)-g(B)][g(C)-g(D)]} \mathcal{H}_{2C}{}^D \mathcal{H}_{1A}{}^B,$$

etc.

6. Supertrace and superdeterminant. A useful concept is the supertrace, as defined above, which can be written as

$$\text{Str } \mathcal{U} = \sum_A (-1)^{g(A)} \mathcal{U}_A{}^A,$$

$$\text{Str } \mathcal{H} = \sum_A (-1)^{g(A)} \mathcal{H}_A{}^A = \text{Tr } H_N - \text{Tr } H_M. \qquad (27)$$

The supertrace has the cyclic property, as can be verified by using (26),

$$\text{Srt } \mathcal{H}_1 \mathcal{H}_2 = \text{Str } \mathcal{H}_2 \mathcal{H}_1,$$

$$\text{Str } \mathcal{H}_1 \mathcal{H}_2 \mathcal{H}_3 = \text{Str } \mathcal{H}_3 \mathcal{H}_1 \mathcal{H}_2 = \text{Str } \mathcal{H}_2 \mathcal{H}_3 \mathcal{H}_1, \qquad (28)$$

etc.

This is not true for the trace. Furthermore, the *supertrace is invariant* under similarity transformations

$$\mathcal{H}_A{}^B \rightarrow \mathcal{U}_A{}^{A'} \mathcal{H}_{A'}{}^{B'} \mathcal{U}^{-1}{}_{B'}{}^B. \qquad (29)$$

Using the cyclic property we can show easily that

$$\text{Str } \mathcal{U}_1 \mathcal{H} \mathcal{U}_1^{-1} = \text{Str } \mathcal{U}_1^{-1} \mathcal{U}_1 \mathcal{H} = \text{Str } \mathcal{H},$$

$$\text{Str } \mathcal{U}_1 \mathcal{U} \mathcal{U}_1^{-1} = \text{Str } \mathcal{U}_1^{-1} \mathcal{U}_1 \mathcal{U} = \text{Str } \mathcal{U}. \qquad (30)$$

The *superdeterminant* can be defined via the supertrace in analogy to the definition of determinant

$$\text{Sdet } \mathcal{U} = \exp(\text{Str ln } \mathcal{U}). \qquad (31)$$

If $\mathcal{U} = e^{i\mathcal{H}}$, then Sdet $\mathcal{U} = 1$ implies

$$\text{Str } \mathcal{H} = 0. \qquad (32)$$

The superdeterminant satisfies the factorization property

$$\text{Sdet } \mathcal{U}_1 \mathcal{U}_2 = \text{Sdet } \mathcal{U}_1 \text{ Sdet } \mathcal{U}_2, \qquad (33)$$

and is invariant under similarity transformations $\mathcal{U} \rightarrow \mathcal{U}_1 \mathcal{U} \mathcal{U}_1^{-1}$. For graded matrices of the form

$$\left(\begin{array}{c|c} A & 0 \\ \hline 0 & B \end{array}\right)$$

the superdeterminant reduces to a ratio of two ordinary determinants

$$\text{Sdet}\left(\begin{array}{c|c} A & 0 \\ \hline 0 & B \end{array}\right) = \frac{\det A}{\det B}. \tag{34}$$

Therefore Sdet $\mathcal{U}$ is not a polynomial in the matrix elements of $\mathcal{U}$, and could not be written by using an analogue of the completely antisymmetric Levi-Civita symbol.

7. The λ-matrices and Killing metric. Just as we identified the matrix representation of the generators for SU(2/1), as in (3), it is useful to identify a set of hermitian λ-matrices corresponding to the generators of SU(N/M) as the coefficients of bosonic and fermionic parameters:

$$(\mathcal{K})_A{}^B = (\lambda_I/2)_A{}^B \omega^I, \tag{35}$$

where ω^I describes bosons as well as fermions, in the adjoint representation, depending on the index I. It is clear that the bosonic generators are block diagonal while the fermionic ones are block off-diagonal. They can be chosen as in (3):

$$\text{SU}(N): \frac{1}{2}\left(\begin{array}{c|c} \lambda_n & 0 \\ \hline 0 & 0 \end{array}\right),$$

$$\text{SU}(M): \frac{1}{2}\left(\begin{array}{c|c} 0 & 0 \\ \hline 0 & \lambda_m \end{array}\right), \quad \text{U}(1): \frac{1}{2}\sqrt{\frac{2NM}{|N-M|}}\left(\begin{array}{c|c} 1/N & 0 \\ \hline 0 & 1/M \end{array}\right),$$

Supergenerators:

$$S_1 = \frac{1}{2}\left[\begin{array}{c|c} 0 & \begin{matrix} 1\,0\,\cdots \\ 0\,0\,\cdots \\ \vdots\;\vdots \end{matrix} \\ \hline \begin{matrix} 1\,0\,0\,\cdots \\ 0\,0\,0\,\cdots \\ \vdots\;\vdots\;\vdots \end{matrix} & 0 \end{array}\right], \quad S_2 = \frac{1}{2}\left[\begin{array}{c|c} 0 & \begin{matrix} -i\,0\,\cdots \\ 0\,0\,\cdots \\ \vdots\;\vdots \end{matrix} \\ \hline \begin{matrix} -i\,0\,\cdots \\ 0\,0\,\cdots \\ \vdots\;\vdots \end{matrix} & 0 \end{array}\right],$$

etc. $\tag{36}$

where λ_n (λ_m) are the SU(N) (SU(M)) traceless λ-matrices of Gell-Mann chosen to be orthogonal

$$\text{tr } \lambda_m \lambda_{m'} = 2\delta_{mm'}, \quad \text{etc.} \tag{37}$$

Note that for SU(N/N) we exclude the U(1) generator since it becomes proportional to unity and commutes with all other matrices. Hence we do not worry about the normalization involving $N - M$ in the denominator which is used only for $N \neq M$.

Using the fact that $\mathrm{Str}(\mathcal{K}_1\mathcal{K}_2)$ is invariant under similarity transformations (which induce the supergroup transformations on the adjoint representation), we can define the super Killing metric of the group as

$$2g_{IJ} = \mathrm{Str}\,\lambda_I\lambda_J, \tag{38}$$

which gives the invariant in the adjoint representation

$$\mathrm{Str}\,\mathcal{K}_1\mathcal{K}_2 = \tfrac{1}{2}\omega_1^I\omega_2^J g_{IJ}, \tag{39}$$

where the sum over I, J runs over bosonic and fermionic indices. We note that when both I and J are bosonic indices g_{IJ} is symmetric, but when both I and J are fermionic indices g_{IJ} is antisymmetric, that is,

$$g_{IJ} = g_{JI}(-1)^{g(I)\cdot g(J)}. \tag{40}$$

Explicitly, we find, from (36)–(38),

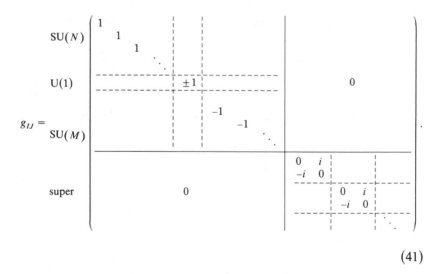

$$\tag{41}$$

The U(1) entry is $(+1)$ for $N - M > 0$, (-1) for $N - M < 0$, and omitted for $N - M = 0$. In a similar way we can define covariant tensors with the indices of the adjoint representation by using the invariance of $\mathrm{Str}(\mathcal{K}_1\mathcal{K}_2 \cdots \mathcal{K}_n)$. In particular, considering $\mathrm{Str}(\mathcal{K}^n)$ and factoring out $\omega^{I_1}\omega^{I_2}/\omega^{I_n}$, we define the covariant tensor

$$d^{(n)}_{I_1 I_2 \cdots I_n} = \frac{1}{n!}\mathrm{Str}\big[\lambda_{I_1}\lambda_{I_2}\cdots\lambda_{I_n} \pm \text{permutations}\big]. \tag{42}$$

The $\pm$ signs are given as $(-1)^{g(I_1)g(I_2)}$, etc., depending on the permutation. Then, under the interchange of any set of indices, $d^{(n)}_{I_1 I_2 \cdots I_n}$ is either symmetric or antisymmetric depending on the bosonic-fermionic property of the indices. For $n = 3$ this is the generalization of the d_{abc} coefficients of SU(N).

The metric g_{IJ} and its inverse g^{JK}, with property

$$g_{IJ}g^{JK} = \delta_I^{\ K} = g^{KJ}g_{JI}, \tag{43}$$

can be used in order to write invariants of the group. For example, if the generators of the group are denoted by X_I (e.g., represented by $\lambda_{I/2}$), then the quadratic Casimir invariant is obviously

$$C_2 = X_I g^{IJ} X_J \equiv X_I X^I = X^I X_I (-1)^{g(I)}, \tag{44}$$

where we have defined

$$X^I \equiv g^{IJ} X_J = X_J g^{IJ} = X_J g^{JI} (-1)^{g(I)g(J)}, \tag{45}$$

which follows from (40) and the form of g_{IJ} in (41).

The λ_I matrices listed in (36), excluding the identity matrix, form a complete set in terms of which any hermitian supertraceless matrix $\mathcal{H}$ can be expanded. The completeness relation (or super "Fierz" transformation) is written in terms of $\lambda_I g^{IK} \lambda_J \equiv \lambda_I \lambda^I$ and can be verified to be

$$N \neq M: \quad (\lambda_I)_A^{\ B} (\lambda^I)_C^{\ D} = 2\delta_A^{\ D}\delta_C^{\ B}(-1)^{g(C)g(B)} - \frac{2}{N-M}\delta_A^{\ B}\delta_C^{\ D}. \tag{46}$$

In the notation of (36), the sum on the left-hand side of (46) has the form

$$\left(\begin{array}{c|c} \lambda_n & 0 \\ \hline 0 & 0 \end{array}\right)_A^{\ B} \left(\begin{array}{c|c} \lambda_n & 0 \\ \hline 0 & 0 \end{array}\right)_C^{\ D} - \left(\begin{array}{c|c} 0 & 0 \\ \hline 0 & \lambda_m \end{array}\right)_A^{\ B} \left(\begin{array}{c|c} 0 & 0 \\ \hline 0 & \lambda_m \end{array}\right)_C^{\ D}$$

$$+ \frac{2NM}{N-M}\left(\begin{array}{c|c} 1/N & \\ \hline & 1/M \end{array}\right)_A^{\ B} \left(\begin{array}{c|c} 1/M & \\ \hline & 1/N \end{array}\right)_C^{\ D} + i(S_1)_A^{\ B}(S_2)_C^{\ D}$$

$$- i(S_2)_A^{\ B}(S_1)_C^{\ D} + \text{etc.}, \tag{47}$$

showing the origin of the $N - M$ in the denominator in (46). When $N = M$ the U(1) generator is omitted from the sum and then the completeness relation becomes

$$N = M: \quad (\lambda_I)_A^{\ B}(\lambda^I)_C^{\ D} = 2\delta_A^{\ D}\delta_C^{\ B}(-1)^{g(C)\,g(B)}$$

$$- \frac{1}{N}\delta_A^{\ B}\delta_C^{\ D}\left[(-1)^{g(A)\,g(B)} + (-1)^{g(C)\,g(D)}\right]. \tag{48}$$

The second term in this expression can be arrived at by direct computation or by moving the $U(1)$ generator in (47) to the right-hand side of (46) and then taking the limit as $M \to N$.

From (44) and (46) we compute the matrix elements of the quadratic Casimir operator when the generators X_I are in the fundamental representation $X_I = \lambda_{I/2}$ $(N \neq M)$,

$$(C_2)_A{}^D = \left(\frac{\lambda_I}{2}\right)_A^B \left(\frac{\lambda^I}{2}\right)_B^D = \frac{\frac{1}{2}\left[(N-M)^2 - 1\right]}{N-M} \delta_A{}^D, \qquad (49)$$

and is found proportional to the identity, as expected.

When $N = M$, $SU(N/N)$ does not exist in the fundamental representation, and its smallest representation is the adjoint [19]. This is because the matrix **1**, which is not one of the generators, is reproduced in the (anti-) commutation relations of the $SU(N/N)$ λ-matrices. That is, $[\mathfrak{K}, \mathfrak{K}']$ contains the matrix **1**, and therefore the correct commutation rules for $SU(N/N)$ are not represented by the fundamental λ-matrices. In fact, if we attempt to calculate $(\lambda_I \lambda^I)_A{}^D$ from (48), we find that it is not proportional to the identity, indicating that this is not a Casimir invariant. We may however define an algebra for the $SU(N/N)$ λ-matrices, by subtracting the part proportional to the identity. We introduce the star-bracket

$$[\mathfrak{K}, \mathfrak{K}']^* \equiv [\mathfrak{K}, \mathfrak{K}'] - \frac{1}{2N} \operatorname{tr}[\mathfrak{K}, \mathfrak{K}'] \times \mathbf{1} \equiv i\mathfrak{K}'', \qquad (50)$$

where we have used the ordinary trace, not the supertrace, since **1** is supertraceless for $SU(N/N)$. The star-bracket $[\mathfrak{K}, \mathfrak{K}']^* = i\mathfrak{K}''$ gives the correct commutation rules for the generators, and produces the correct structure constants, but it is no longer just a Lie commutator of the fundamental matrices. In the adjoint and higher-dimensional irreducible representations it will, of course, reduce to ordinary matrix commutation. Because of these complications we will ignore $SU(N/N)$ in these lectures.

8. Complex conjugate representation. We may define the complex or hermitian conjugate of the wavefunctions ϕ_A with an upper (contravariant) index as

$$\phi_A^\dagger \equiv \phi^A.$$

The transformation property in this basis is obtained by taking the hermitian conjugate of (18) and remembering to change the orders of the fermions in the product, as part of the definition of hermitian conjugation (see argument before (12)). Then we obtain

$$\phi^A \to \phi'^A = \phi^B (\mathfrak{U}^\dagger)_B{}^A, \qquad (51)$$

with $\mathcal{U}^\dagger$ appearing on the right. This contravariant basis will be denoted by a supertableau with a dot:

$$\phi^A \sim \boxed{\cdot}. \tag{52}$$

Unlike the group $SU(N)$, in which the contravariant basis is equivalent to an $(N-1)$-dimensional antisymmetric tensor, in $SU(N/M)$ the contravariant basis cannot be written in terms of the covariant one. This is because there is no invariant tensor analogous to the Levi-Civita completely antisymmetric tensor. This difference between $SU(N)$ and $SU(N/M)$ is related to the fact that the superdeterminant is not a polynomial (see (34)).

9. Harmonic oscillator representation. We introduce a set of bosonic and fermionic harmonic oscillators with the indices of the fundamental representation

$$\boxed{\cdot} \sim \xi_A = \begin{pmatrix} b_a \\ f_\alpha \end{pmatrix}, \qquad \boxed{\cdot} \sim \xi^{\dagger A} = (b^{\dagger a} f^{\dagger \alpha}), \tag{53}$$

with commutation-anticommutation rules

$$\left[b_a, b^{\dagger c} \right] = \delta_a^{\ c}, \qquad \{ f_\alpha, f^{\dagger \beta} \} = \delta_\alpha^{\ \beta}, \tag{54}$$

all other obvious (anti-) commutators being zero. Equivalently, we may write

$$\left[\xi_A, \xi^{\dagger B} \right\} = \delta_A^{\ B}, \tag{55}$$

where the graded commutator $[\cdot, \cdot \}$ is defined as

$$\xi_A \xi^{\dagger B} - (-1)^{g(A)g(B)} \xi^{\dagger B} \xi_A = \left[\xi_A, \xi^{\dagger B} \right\} = \delta_A^{\ B}. \tag{56}$$

We may combine these oscillators with $\mathcal{K}_A^{\ B}$ in order to form quantum generators X_I as

$$\xi^{\dagger A} \mathcal{K}_A^{\ B} \xi_B \equiv X_I \omega^I, \tag{57}$$

where (up to some minus signs coming from anticommutaing fermions f_α with the fermions ω^I) we obtain

$$X_I = \left(\frac{\lambda_I}{2} \right)_A^{\ B} \xi_B \xi^{\dagger A} (-1)^{g(A)} = \mathrm{Str} \left(\xi \xi^\dagger \frac{\lambda_I}{2} \right). \tag{58}$$

However, since we always obtain simplifications by combining parameters with generators we will always use the forms

$$\xi^\dagger \mathcal{K} \xi \equiv X, \qquad \xi^\dagger \mathcal{K}' \xi = X', \quad \text{etc.} \tag{59}$$

Note that the anticommuting fermions appearing in $\mathcal{K}_A^{\ B}$ must be taken to anticommute with fermionic oscillators as well, because an infinitesimal transformation of the oscillators under the group, as in (22), must maintain the commutation rules of the oscillators. In fact, a supergroup transformation with $\mathcal{U}_A{}^B$ on the oscillators, as in (18), is a canonical transformation because it leaves the commutation rules of (55) invariant. Now, consider commuting two generators by using the quantum commutators of (54)–(56). We find a cancellation of minus signs leading to the simple result

$$[X, X'] = [\xi^\dagger \mathcal{H}\xi, \xi^\dagger \mathcal{H}'\xi] = \xi^\dagger [\mathcal{H}, \mathcal{H}']\xi = i\xi^\dagger \mathcal{H}''\xi = iX'', \quad (60)$$

where $\mathcal{H}''$ has the same form as $\mathcal{H}$ or $\mathcal{H}'$ as demonstrated in (11) by using only commutators. We see that the quantum generators X (including parameters) close again under *commutation and not anticommutation*.

We may introduce many sets of creation-annihilation operators by adjoining an extra index on ξ_{Ai}, $i = 1, 2, \ldots, k$. To save indices we will use a vector notation $\vec{\xi}_A$ instead of the index i. The generators formed from each ξ_{Ai} close among themselves for each i. Therefore, their sums defined as

$$X = \vec{\xi}^\dagger \mathcal{H}\vec{\xi}, \qquad X' = \vec{\xi}^\dagger \mathcal{H}'\vec{\xi} \quad (61)$$

(where the vector index i is summed over) will also close just as above:

$$[X, X'] = iX''. \quad (62)$$

It is useful to consider the quantum operator group element $\hat{\mathcal{U}}$ obtained by exponentiating X:

$$\hat{\mathcal{U}} = \exp(iX) = \exp(i\vec{\xi}^\dagger \mathcal{H}\vec{\xi}). \quad (63)$$

From (60) we learn that these operators form a group

$$\hat{\mathcal{U}}\hat{\mathcal{U}}' = \hat{\mathcal{U}}''.$$

Using the quantum commutators of (55), (56) we derive by explicit quantum commutation

$$[X, \vec{\xi}_A] = -\mathcal{H}_A{}^B \vec{\xi}_B, \qquad [X, \vec{\xi}^{\dagger A}] = \vec{\xi}^{\dagger B}\mathcal{H}_B{}^A, \quad (64)$$

where the order of $\mathcal{H}$ and ξ are important to avoid extra (-1) signs. Furthermore, it follows immediately that

$$\hat{\mathcal{U}}^\dagger \vec{\xi}_A \hat{\mathcal{U}} = (e^{i\mathcal{H}})_A{}^B \vec{\xi}_B \sim \boxtimes,$$
$$\hat{\mathcal{U}}^\dagger \vec{\xi}^{\dagger A} \hat{\mathcal{U}} = \vec{\xi}^{\dagger B}(e^{-i\mathcal{H}})_B{}^A \sim \boxtimes, \quad (65)$$
$$\hat{\mathcal{U}}^\dagger (\xi^\dagger \mathcal{H}'\xi)\hat{\mathcal{U}} = \xi^\dagger [e^{-i\mathcal{H}}\mathcal{H}'e^{i\mathcal{H}}]\xi.$$

We define the covariant generators

$$\overline{X}_A{}^B \equiv X_I(\lambda^I)_A{}^B, \qquad \text{Str}(\overline{X}\mathcal{H}) = X. \quad (66)$$

By using the completeness of the λ's we can write

$$\overline{X}_A{}^B = \vec{\xi}_A \cdot \vec{\xi}^{\dagger B} - \delta_A{}^B(\vec{\xi}^\dagger \cdot \vec{\xi} + 1). \tag{67}$$

Then we note that

$$\hat{\mathfrak{U}}^\dagger \overline{X}_A{}^B \hat{\mathfrak{U}} = (e^{i\mathfrak{K}})_A{}^{A'} \overline{X}_{A'}{}^{B'} (e^{-i\mathfrak{K}})_{B'}{}^B, \tag{68}$$

and therefore

$$\sum_A \overline{X}_A{}^B \overline{X}_B{}^A (-1)^{g(A)} = \mathrm{Str}(\overline{X}^2) \tag{69}$$

is an invariant. The quantum quadratic Casimir operator is now given by

$$\hat{C}_2 = X_I X^I = \tfrac{1}{2}\,\mathrm{Str}(\overline{X}^2). \tag{70}$$

This can be rewritten in the form

$$\hat{C}_2 = \frac{1}{2}\Big[\big(\xi_i^\dagger \cdot \xi_j\big)\big(\xi_j^\dagger \cdot \xi_i\big) \\ + (N - M - K)\big(\xi_i^\dagger \cdot \xi_i\big) - \frac{1}{N-M}\big(\xi_i^\dagger \cdot \xi_i\big)^2\Big], \tag{71}$$

where $i, j = 1, 2, \ldots, K$ run through the copies of oscillators. The dot products

$$\xi_i^\dagger \cdot \xi_j \equiv \xi_i^{\dagger A} \xi_{jA} \tag{72}$$

are invariant under a group transformation according to

$$\hat{\mathfrak{U}}^\dagger\big(\xi_i^\dagger \cdot \xi_j\big)\hat{\mathfrak{U}} = \xi_i^\dagger e^{-i\mathfrak{K}} e^{i\mathfrak{K}} \xi_j = \xi_i^\dagger \cdot \xi_j. \tag{73}$$

This is another way of seeing that the quadratic Casimir operator indeed commutes with $\hat{\mathfrak{U}}$ or with all the generators X for any set of parameters $\mathfrak{K}$.

If we specialize to only one set of oscillators $i = j = k = 1$ then $\hat{C}_2$ takes the simple form

$$\hat{C}_2 = \frac{N - M - 1}{2(N - M)}\hat{n}(N - M + \hat{n}), \tag{74}$$

where $\hat{n}$ is the number operator which counts bosons plus fermions:

$$\hat{n} = \xi^\dagger \cdot \xi = \hat{n}_b + \hat{n}_f, \qquad \hat{n}_b = b^\dagger \cdot b, \hat{n}_f = f^\dagger \cdot f. \tag{75}$$

Thus, $\hat{n}$ rather than $\hat{C}_2$ can be taken as the Casimir invariant. Note that even though $\hat{C}_2$ could not be defined for $N = M$, the invariant operator $\hat{n}$ certainly exists and is meaningful for $SU(N/N)$.

The nth higher order quantum Casimir operators can be constructed by using the covariant tensors $\overline{X}_A{}^B$ of (66). Thus $\mathrm{Str}(\overline{X}^P)$ is an invariant, and we can write the pth order Casimir invariant as

$$\hat{C}_p = \mathrm{Str}(\overline{X}^P) = X_{I_1} X_{I_2} \cdots X_{I_p} \mathrm{Str}(\lambda^{I_1} \lambda^{I_2} \cdots \lambda^{I_p}), \tag{76}$$

where the index in λ^I is raised by using the killing metric as in (45). In certain applications it is more convenient to define another pth order Casimir operator by using the covariant tensors of (42)

$$\hat{C}'_p = X_{I_1} X_{I_2} \cdots X_{I_p} d^{(p)I_1 I_2 \cdots I_p}, \tag{77}$$

where the indices are raised by using g^{IJ}. The $\hat{C}'_p$ are linear combinations of $\hat{C}_p$ and lower order Casimirs of type $\hat{C}_{p-1}, \hat{C}_{p-2}, \ldots$, as can be seen by using $[X_{I_1}, X_{I_2}] \sim X_{I_3}$ to change the orders in the various terms of $d^{(p)I_1 I_2 \cdots I_p}$. In the special case of a single set of creation-annihilation operators ξ_i, $\xi_j^\dagger$, $i = j = k = 1$, all of these Casimir operators become a function of only the number operator,

$$\hat{n} = \xi^{\dagger A} \xi_A = \hat{n}_b + \hat{n}_f,$$

as can easily be seen from the form $\text{Str}(\overline{X}^p)$. Thus in the case of a single set of harmonic oscillators there is only one independent Casimir invariant, namely the number operator $\hat{n}$. It is clear that in order to construct the most general representation one must take many copies of oscillators.

The harmonic oscillators could also be assigned to the complex conjugate representation

$$\boxtimes \sim \eta^A = (c^a g^\alpha); \qquad \boxtimes \sim \eta_A^\dagger = \begin{pmatrix} c_a^\dagger \\ g_\alpha^\dagger \end{pmatrix}, \tag{78}$$

where c^a is a boson and g^α is a fermion. These differ from (53) in the assignment of lower and upper indices. They satisfy the quantum rules

$$[\eta_A^\dagger, \eta^B] = -\delta_A{}^B (-1)^{g(A)} \tag{79}$$

as opposed to (55), in which the order is important. We can now define generators as in (57),

$$\eta^A (-\mathcal{H})_A{}^B \eta_B^\dagger \equiv Y_I \omega^I \equiv Y, \tag{80}$$

and find that the Y_I satisfy the same commutation rules as the X_I, that is,

$$[\eta(-\mathcal{H})\eta^\dagger, \eta(-\mathcal{H}')\eta^\dagger] = \eta(-[\mathcal{H}, \mathcal{H}'])\eta^\dagger = i\eta(-\mathcal{H}'')\eta^\dagger. \tag{81}$$

Therefore, by taking many copies of ξ's and η's we may introduce a total generator G_I constructed as

$$G_I = X_I + Y_I = \text{Str}\left(\vec{\xi}\,\vec{\xi}^\dagger \frac{\lambda_I}{2}\right) - \text{Str}\left(\vec{\eta}^\dagger \eta \frac{\lambda_I}{2}\right),$$

$$G = \vec{\xi}^\dagger \mathcal{H} \vec{\xi} + \vec{\eta}(-\mathcal{H})\vec{\eta}^\dagger, \tag{82}$$

$$G_A^B = G_I (\lambda^I)_A{}^B = \vec{\xi}_A \cdot \vec{\xi}^{\dagger B} - \vec{\eta}_A^\dagger \cdot \vec{\eta}^B - \frac{\delta_A{}^B}{N - M}(\vec{\xi}^\dagger \cdot \vec{\xi} - \vec{\eta} \cdot \vec{\eta}^\dagger).$$

Similarly, defining a total group operator that induces group transformations of all quantum oscillators,

$$\hat{\mathcal{U}} = e^{iG} = \exp^i\{\vec{\xi}^\dagger \mathcal{K} \vec{\xi} + \vec{\eta}(-\mathcal{K})\vec{\eta}^\dagger\}, \tag{83}$$

we compute, by using the quantum commutators as in (65) for any oscillator $\vec{\xi}_A$,

$$\hat{\mathcal{U}}^\dagger \vec{\xi}_A \hat{\mathcal{U}} = (e^{i\mathcal{K}})_A{}^B \vec{\xi}_B \sim \boxtimes,$$
$$\hat{\mathcal{U}}^\dagger \eta^\dagger_A \hat{\mathcal{U}} = (e^{i\mathcal{K}})_A{}^B \eta^\dagger_B \sim \boxtimes,$$
$$\hat{\mathcal{U}}^\dagger \xi^{\dagger A} \hat{\mathcal{U}} = \xi^{\dagger B}(e^{-i\mathcal{K}})_B{}^A \sim \boxtimes, \tag{84}$$
$$\hat{\mathcal{U}}^\dagger \eta^A \hat{\mathcal{U}} = \eta^B(e^{-i\mathcal{K}})_B{}^A \sim \boxtimes,$$
$$\hat{\mathcal{U}}^\dagger G_A{}^B \hat{\mathcal{U}} = (e^{i\mathcal{K}})_A{}^{A'} G_{A'}{}^{B'}(e^{-i\mathcal{K}})_{B'}{}^B \sim \boxtimes\boxtimes \quad \text{(adjoint)},$$

as expected. The orders of the factors are important, otherwise a multitude of minus signs will appear.

In the presence of both $\vec{\xi}$ and $\vec{\eta}$ the Casimir operators are constructed as

$$\hat{C}_p = \mathrm{Str}(G^p). \tag{85}$$

They clearly commute with all G_I, since $\hat{\mathcal{U}}^\dagger \hat{C}_p \hat{\mathcal{U}} = C_p$ as a result of (84).

10. The tilde operator. There is a further generalization concerning the introduction of oscillators in a class II basis. We will distinguish them from class I with a tilde sign:

$$\boxtimes \sim \tilde{\rho}_A = \begin{pmatrix} v_a \\ r_\alpha \end{pmatrix}; \qquad \boxtimes \sim \tilde{\rho}^{\dagger A} = (v^{\dagger a}, r^{\dagger \alpha}), \tag{86}$$

with $v =$ fermion, $r =$ boson, as distinguished from (58):

$$\{v_a, v^{\dagger b}\} = \delta_a{}^b, \qquad [r_\alpha, r^{\dagger \beta}] = \delta_\alpha{}^\beta. \tag{87}$$

These can be combined to the same form as (55),

$$[\tilde{\rho}_A, \tilde{\rho}^{\dagger B}\} = \delta_A{}^B, \tag{88}$$

provided the grades of the indices are interpreted properly. The class II oscillators transform under the group just in the same way as class I oscillators.

Out of this set we may again construct generators by combining with $\mathcal{K}_A{}^B$,

$$\tilde{X} = \tilde{\rho}^\dagger \mathcal{K} \tilde{\rho}, \qquad \tilde{X}' = \tilde{\rho}^\dagger \mathcal{K}' \rho, \quad \text{etc.,} \qquad [\tilde{X}, \tilde{X}'] = i\tilde{X}'', \tag{89}$$

in analogy to (60). Then, taking $\hat{\mathcal{U}} = \exp iX$, we have

$$\hat{\mathcal{U}}^\dagger \tilde{\rho}_A \hat{\mathcal{U}} = (e^{i\mathcal{K}})_A{}^B \tilde{\rho}_B \sim \boxtimes, \tag{90}$$

etc., as in (65), (84), as expected.

Similarly, we also have class II oscillators in a contravariant basis

$$\boxtimes \sim \tilde{\sigma}^A = (w^a s^\alpha); \qquad \boxtimes \sim \tilde{\sigma}_A^\dagger = \begin{pmatrix} w_a^\dagger \\ s_\alpha^\dagger \end{pmatrix}, \tag{91}$$

where w is a fermion and s is a boson. The generators

$$\tilde{Y} = \tilde{\sigma}(-\mathcal{K})\tilde{\sigma}^\dagger \tag{92}$$

will have the usual properties.

The total quantum generator constructed from all oscillators in (82), (89) and (92) is

$$Q = \vec{\xi}^\dagger \mathcal{K} \vec{\xi} - \vec{\eta}\, \mathcal{K} \vec{\eta}^\dagger + \vec{\rho}^\dagger \mathcal{K} \vec{\rho} - \vec{\sigma} \mathcal{K} \vec{\sigma}^\dagger, \tag{93}$$

where we have included multiple copies of the oscillators though the vector notation $\vec{\xi}$, etc. Then the full quantum operator group element which acts on all oscillators is

$$\mathcal{U} = e^{iQ}. \tag{94}$$

It satisfies equations similar to (84), including the class II oscillators. Casimir invariants in the full space are constructed from Q.

Now we are ready for the *tilde operator* $\hat{t}$. Consider the $(N + M)$-dimensional matrix

$$t = \begin{pmatrix} \theta_0 & & & & & & & \\ & \theta_0 & & & & & & \\ & & \ddots & & & & & \\ & & & \theta_0 & & & & \\ \hline & & & & -\theta_0 & & & \\ & & & & & -\theta_0 & & \\ & & & & & & \ddots & \\ & & & & & & & -\theta_0 \end{pmatrix}, \tag{95}$$

$$(\theta_0)^2 = 1 \leftrightarrow t^2 = 1,$$

where θ_0, considered as a fermionic parameter, anticommutes with all the Grassmann numbers θ_a^b in $\mathcal{K}_A^B$ as well as with all the quantum fermionic oscillators. We will however take $(\theta_0)^2 = 1$, implying that θ_0 is the sum of a fermionic creation and annihilation operator in the language of Grassmann numbers described before. This makes θ_0 a Clifford number, constructed from a Grassmann number plus its derivative $\theta_0 = \theta + \partial/\partial\theta$; $(\theta_0)^2 = 1$. We emphasize that θ_0 with these properties is just a convenience but not necessary for the final point to be made below. If θ_0 is omitted from the argument a combination of commutators and anticommutators will have to be used as usual. The matrix t is fermionic; its supertrace is *not* zero, and therefore is not included in $\mathcal{K}$.

By direct matrix multiplication and the use of the anticommutation property of Grassmann numbers, it can be seen that t *and* $\mathcal{K}$ *commute!*

$$[t, \mathcal{K}] = 0. \tag{96}$$

We may now construct the quantum tilde operator $\hat{t}$ as

$$\hat{t} = \vec{\xi}^\dagger t \vec{\rho} + \vec{\rho}^\dagger t \vec{\xi} - \vec{\eta} t \vec{\sigma}^\dagger - \vec{\sigma} t \vec{\eta}^\dagger. \tag{97}$$

Using (96) and the quantum oscillator commutation rules introduced before, we find

$$[\hat{t}, \vec{\xi}_A] = -(t\vec{\rho})_A = (-1)^{g(A)} \vec{\rho}_A \theta_0,$$
$$[\hat{t}, \vec{\rho}_A] = -(t\vec{\xi})_A = -(-1)^{g(A)} \vec{\xi}_A \theta_0, \tag{98}$$
etc.,

such that

$$[\hat{t}, Q] = 0, \tag{99}$$

implying that the operator $\hat{t}$ is group invariant. We now construct the unitary transformation

$$\hat{T} = e^{i\hat{t}\pi/2}, \qquad \hat{T}^\dagger \begin{pmatrix} \xi \\ \tilde{\rho} \end{pmatrix} \hat{T} = \begin{pmatrix} 0 & it \\ it & 0 \end{pmatrix} \begin{pmatrix} \xi \\ \tilde{\rho} \end{pmatrix}, \tag{100}$$

where the indices are suppressed. A similar equation holds for η and $\tilde{\sigma}$. The transformation interchanges class I with class II oscillators whenever they appear and as a result of (99) commutes with the group transformations $\hat{\mathcal{U}} = \exp iQ$ which acts on the indices of oscillators. Therefore, class I and class II representations of any dimension will transform in identical ways under the group transformation. When the tilde operation $\hat{t}$ is applied on all the states in a representation, it will simply switch all bosons into fermions and vice versa:

$$\text{boson} \overset{\hat{t}}{\leftrightarrow} \text{fermion},$$

and will insert appropriate minus signs as required by a representation of the matrix of (95). The new set of states thus obtained (fermions + bosons) have identical transportation properties as the old set (bosons + fermions) under the action of the group, because of (99).

Because of the existence of the operator $\hat{t}$, it will be sufficient to construct group elements on pure class I bases, since the group element on class II or mixed [14] class I + class II bases will always be the same as pure class I. (The mixed tensor products of class I + class II bases often yield states whose bose-fermi content is switched relative to pure class I or pure class II. For example, you can obtain a basis for an adjoint representation whose bose-fermi content is opposite to the generators. Some examples are found in [14].)

11. Higher representations from tensor products and supertableaux. Let us consider the transformation properties of the direct product of two wavefunctions following the rule of (18):

$$\Phi_A^{(1)}\Phi_B^{(2)} \to \Phi_A'^{(1)}\Phi_B'^{(2)} = \mathcal{U}_A{}^{A'}\Phi_{A'}^{(1)}\mathcal{U}_B{}^{B'}\Phi_{B'}^{(2)}. \tag{101a}$$

Note that the order of factors could not be changed without introducing some minus signs, as in (26).

$$\Phi_A'^{(1)}\Phi_B'^{(2)} = (-1)^{g(A')[g(B)-g(B')]}\mathcal{U}_A{}^{A'}\mathcal{U}_B{}^{B'} \cdot \Phi_{A'}^{(1)}\Phi_{B'}^{(2)}. \tag{101b}$$

In terms of supertableaux this direct product can be written as

$$\boxed{\diagdown} \times \boxed{\diagdown}.$$

The direct product is reducible. To obtain irreducible tensors we supersymmetrize, and introduce the supertableau notation

$$\boxed{\diagdown\diagdown} \sim \Phi_{AB}^{(+)} = \Phi_A^{(1)}\Phi_B^{(2)} + \Phi_A^{(2)}\Phi_B^{(1)},$$

$$\boxed{}\!\!\boxed{} \sim \Phi_{AB}^{(-)} = \Phi_A^{(1)}\Phi_B^{(2)} - \Phi_A^{(2)}\Phi_B^{(1)}. \tag{102}$$

Note the supertableau rule that the indices A, B are kept in the same order, but the wavefunctions $\Phi_A^{(1)}$, $\Phi_B^{(2)}$ are interchanged. This is necessary so that both terms in the sum transform in the same way with the overall factors

$$(-1)^{g(A')[g(B)-g(B')]}\mathcal{U}_A{}^{A'}\mathcal{U}_B{}^{B'}, \tag{103}$$

as in (101). Then the symmetric (antisymmetic) sum stays symmetric (antisymmetric) after the supergroup transformation. If we wish to keep the order of wavefunctions the same, we may use (26) to write also

$$\Phi_{AB}^{(+)} = \Phi_A^{(1)}\Phi_B^{(2)} + (-1)^{g(A)g(B)}\Phi_B^{(1)}\Phi_A^{(2)},$$

$$\Phi_{AB}^{(-)} = \Phi_A^{(1)}\Phi_B^{(2)} - (-1)^{g(A)g(B)}\Phi_B^{(1)}\Phi_A^{(2)}. \tag{104}$$

These arrangements of indices transform irreducibly by the above argument. Thus, under the interchange of indices we have the superantisymmetric tensors

$$\Phi_{AB}^{(+)} = (-1)^{g(A)g(B)}\Phi_{BA}^{(+)},$$

$$\Phi_{AB}^{(-)} = -(-1)^{g(A)g(B)}\Phi_{BA}^{(-)}. \tag{105}$$

By specializing the indices $A = (a, \alpha)$, i.e. $\phi_A = \binom{\phi_a}{\psi_\alpha}$, to bosons and fermions we identify the various components

$$
\boxed{\diagdown} \quad \Phi^{(+)}_{AB} = \begin{cases} \phi_{ab} = \phi_{ba} \sim (\square\square, 1)_{2/N}; & \tfrac{1}{2}N(N+1) \text{ bosons,} \\ \phi_{a\beta} = \phi_{\beta a} \sim (\square, \square)_{1/N+1/M}; & NM \text{ fermions,} \\ \phi_{\alpha\beta} = -\phi_{\beta\alpha} \sim (1, \square\!\square)_{2/M}; & \tfrac{1}{2}M(M-1) \text{ bosons,} \end{cases}
$$

(106)

$$
\boxed{\square} \quad \Phi^{(-)}_{AB} = \begin{cases} \phi_{ab} = -\phi_{ba} \sim (\square\!\square, 1)_{2/N} \sim \tfrac{1}{2}N(N-1) \text{ bosons,} \\ \phi_{a\beta} = -\phi_{\beta a} - (\square, \square)_{1/N+1/M} \sim NM \text{ fermions,} \\ \phi_{\alpha\beta} = +\phi_{\beta\alpha} \sim (1, \square\square)_{2/M} \sim \tfrac{1}{2}M(M+1) \text{ bosons.} \end{cases}
$$

(107)

The tensors with an odd number of α indices are fermions by construction. However, we may construct class II tensors $\tilde{\Phi}^{(+)}_{AB}$, $\tilde{\Phi}^{(-)}_{AB}$ by substituting in (104) the class II first-rank tensor $\tilde{\Phi}^{(1)}_B = \binom{\psi_b}{\phi_\beta}$ instead of the class I $\Phi^{(2)}_B = \binom{\phi_b}{\psi_\beta}$ (keeping $\Phi^{(1)}_A$ the same) where these two differ in the nature of anticommuting variables (all ϕ's bosons and all ψ's fermions). Then the tilde transformed tensors $\tilde{\Phi}_{AB}$ will have boson or fermion components switched relative to the old ones:

$$
\tilde{\Phi}^{(+)}_{AB} = \begin{cases} (\square\square, 1) = \text{fermions,} \\ (\square, \square) = \text{bosons,} \\ (1, \square) = \text{fermions,} \end{cases}
$$

(108)

and similarly for $\tilde{\Phi}^{(-)}_{AB}$. Even though the bose $\leftrightarrow$ fermi properties are interchanged, Φ_{AB} and $\tilde{\Phi}_{AB}$ transform identically under the group. This is a manifestation analogous to

$$
[\hat{t}, Q] = 0
$$

(109)

that was described in the last section in terms of the quantum oscillators. Because we can always construct class II bases from those of class I via the tilde operation, it will be sufficient to concentrate only on class I.

To obtain the matrix repsentation of the group element we write the transformation of the new basis as

$$
\Phi'^{(+)}_{AB} = \mathcal{U}^{A'B'}_{AB} \Phi^{(+)}_{A'B'},
$$

(110)

and extract a properly symmetrized group element in this higher-dimensional representation ($\boxed{\diagdown}$)

$$
\mathcal{U}^{A'B'}_{AB} = \tfrac{1}{2}\big\{ (-1)^{g(A')[g(B)-g(B')]} \mathcal{U}^{A'}_A \mathcal{U}^{B'}_B \\ + (-1)^{g(A)g(B)}(-1)^{g(A')[g(A)-g(B')]} \mathcal{U}^{A'}_B \mathcal{U}^{B'}_A \big\},
$$

(111)

which satisfies the superpermutation properties indicated by the super-tableau

$$\boxed{\diagdown\diagdown}: \quad \mathcal{U}_{AB}^{A'B'} = (-1)^{g(A)g(B)}\mathcal{U}_{BA}^{A'B'} = (-1)^{g(A')g(B')}\mathcal{U}_{AB}^{B'A'}. \tag{112}$$

Before we go on, let us also construct states in a Fock superspace which have the same transformation properties as the tensors above. Let us start with the class I oscillators of (53). Define the vacuum state

$$\xi_A|0\rangle = 0, \qquad \langle 0|\xi^{\dagger A} = 0. \tag{113}$$

Then the one-particle bra state,

$$\langle 0|\xi_A \equiv \langle A|, \tag{114}$$

transforms as the fundamental basis $\boxed{\diagdown}$:

$$\langle A| \rightarrow \langle A|\hat{\mathcal{U}} = \langle 0|\hat{\mathcal{U}}\hat{\mathcal{U}}^{\dagger}\xi_A\hat{\mathcal{U}} = \langle 0|(e^{i\mathcal{K}})_A{}^B\xi_B = (e^{i\mathcal{K}})_A{}^B\langle B|. \tag{115}$$

Similarly the two-particle state

$$\langle AB| \equiv \langle 0|\xi_A\xi_B \tag{116}$$

is automatically supersymmetric, since the oscillators satisfy

$$\xi_A\xi_B = (-1)^{g(A)g(B)}\xi_B\xi_A, \tag{117}$$

and therefore it transforms as $\boxed{\diagdown\diagdown}$, with the matrix of (111):

$$\langle AB|\hat{\mathcal{U}} = \mathcal{U}_{AB}{}^{A'B'}\langle A'B'|. \tag{118}$$

We may in fact relate the two approaches of wavefunctions and Fock superspace states by defining

$$\Phi^{(1)} = \Phi_A^{(1)}\xi^{\dagger A}, \quad \Phi^{(2)} = \Phi_A^{(2)}\xi^{\dagger A}, \quad \text{etc.}, \tag{119}$$

and applying them on the vacuum

$$|\Phi^{(1)}\rangle \equiv \Phi^{(1)}|0\rangle, \quad \text{etc.},$$

$$|\Phi^{(1)}\Phi^{(2)}\rangle \equiv \Phi^{(1)}\Phi^{(2)}|0\rangle, \quad \text{etc.} \tag{120}$$

Then, it is clear that the wavefunctions can be defined as

$$\Phi_A^{(1)} = \langle A|\Phi^{(1)}\rangle \sim \boxed{\diagdown}, \qquad \Phi_{AB}^{(+)} = \langle AB|\Phi^{(1)}\Phi^{(2)}\rangle \sim \boxed{\diagdown\diagdown}, \tag{121}$$

and they will have the same transformation properties as the states in the Fock superspace.

This method of obtaining higher-dimensional supertensors is obviously generalized:

$$\overset{n}{\boxed{\diagdown\diagdown\diagdown\diagdown\diagdown}} \sim \langle 0|\xi_{A_1}\xi_{A_2}\cdots\xi_{A_n} = \langle A_1A_2\cdots A_n|,$$

$$\Phi_{A_1A_2\cdots A_n}^{(+)} = \langle A_1A_2\cdots A_n|\Phi^{(1)}\Phi^{(2)}\cdots\Phi^{(n)}\rangle. \tag{122}$$

To make supertableaux with more than one row, we must use more than one set of creation-ahhihilation operators. In fact with K sets

$$\vec{\xi}_A = \left(\xi_A^{(1)}, \xi_A^{(2)}, \ldots, \xi_A^{(K)}\right),$$

we can construct up to K rows. Actually, with more than one set the method becomes cumbersome and we may instead simply use products of wavefunctions as in (102). For example, the tensor corresponding to the supertableau ▨ is

$$\Phi_{AB,C} = \Phi_A^{(1)}\Phi_B^{(2)}\Phi_C^{(3)} + \Phi_A^{(2)}\Phi_B^{(1)}\Phi_C^{(3)}$$
$$- \Phi_A^{(3)}\Phi_B^{(2)}\Phi_C^{(1)} - \Phi_A^{(2)}\Phi_B^{(3)}\Phi_C^{(1)}. \tag{123}$$

Note again the supertableau rule that the indices keep the same order and the wavefunctions are symmetrized-antisymmetrized according to the tableau. Note further that, as in usual Young tableaux, *first* the symmetrization is done for every *row* and next the antisymmetrization is done for every column. In the above example the net tensor has simple supersymmetry properties in the interchange of indices $A \leftrightarrow B$ but not in $A \leftrightarrow C$ or $B \leftrightarrow C$, as in usual Young tableaux.

Matrix elements of the supergroup in this higher representation are now obtained by applying a transformation on every $\Phi_A^{(1)}$, $\Phi_B^{(2)}$, $\Phi_C^{(3)}$ and writing the result as

$$\Phi'_{AB,C} = \mathcal{U}_{AB,C}{}^{A'B',C'}\Phi_{A'B',C'}, \tag{124}$$

analogously to (110). It is left as an exercise to obtain these matrix elements in terms of the fundamental $\mathcal{U}_A{}^B$ (see [14]).

There are more representations that involve the direct products of covariant ▨ with contravariant ▨ bases. For example, the adjoint representation

▨

which we have already encountered in (84) is such an example. As already emphasized (see §8), unlike ordinary $SU/(N)$ the contravariant indices ▨ could not be reproduced by taking direct products of covariant ones. The only thing to notice is that for a tensor

$$\Phi_{A_1 A_2 \cdots A_n}{}^{B_1 B_2 \cdots B_m} \tag{125}$$

to be irreducible it must satisfy the following conditions.

(1) The lower indices must be supersymmetrized according to a supertableau.

(2) The upper indices must independently be supersymmetrized according to some other supertableau.

(3) The tensor must be supertraceless in all pairs of upper-lower indices

$$\sum_{A_1}(-1)^{g(A_1)}\Phi_{A_1A_2\cdots A_n}{}^{A_1B_2\cdots B_m} = 0, \quad \text{etc.} \tag{126}$$

These rules are summarized by the general supertableau in Figure 1.

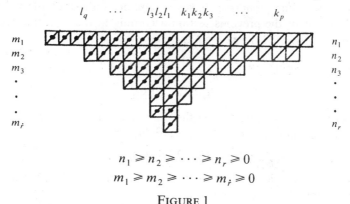

$$n_1 \geqslant n_2 \geqslant \cdots \geqslant n_r \geqslant 0$$
$$m_1 \geqslant m_2 \geqslant \cdots \geqslant m_{\hat{r}} \geqslant 0$$

FIGURE 1

The dotted boxes correspond to the upper indices and are drawn as mirror images of the usual Young tableaux. The same rules could be applied to $SU(N)$ and it will give an irreducible tensor. If one wishes, one can transform the $SU(N)$ dotted tableau to an undotted one by the simple replacement of every column with k-dotted boxes, by a column with $N - k$ undotted boxes:

$$k \;\longrightarrow\; N - k \qquad \begin{aligned}&\text{true for } SU(N),\\[1em]&\text{false for } SU(N/M).\end{aligned}$$

This property, which is based on the Levi-Civita tensor, does not apply for supertableaux, since no such invariant tensor exists.

After obtaining an irreducible tensor, we arrive at the matrix representation of the group elements

$$\mathfrak{A}^{(A_1'A_2'\cdots A_n')(B_1B_2\cdots B_m)}_{(A_1A_2\cdots A_n)(B_1'B_2'\cdots B_m')} \tag{127}$$

by making a transformation on every ϕ_A and rearranging the order of the factors as in (101b), (111). Further examples of these can be found in [14].

12. The invariant characters. We have already seen in (30) that the supertrace of the fundamental group element $\mathfrak{A}_A{}^B$ is invariant under similarity transformations with other group elements.

$$\text{Str } \mathcal{U}_1 \mathcal{U} \mathcal{U}_1^\dagger = \text{Str } \mathcal{U}. \tag{128}$$

This is the invariant character in the fundamental representation

$$\chi_\square(\mathcal{U}) = \text{Str } \mathcal{U}. \tag{129}$$

To construct the invariant characters in the higher-dimensional representations, we simply contract lower and upper indices of the matrix representations of the previous section according to the supertrace rule. For example, using (111) we obtain the character

$$\chi_{\square\square}(\mathcal{U}) = \sum_{A,B} (-1)^{g(A)}(-1)^{g(B)} \mathcal{U}_{AB}{}^{AB}$$
$$= \tfrac{1}{2}(\text{Str } \mathcal{U})^2 + \tfrac{1}{2}\text{Str}(\mathcal{U}^2), \tag{130}$$

which is obviously invariant under the similarity transformations. Applying the same method we obtain for the adjoint representation

$$\chi_{\square\square}(\mathcal{U}) = (\text{Str } \mathcal{U}^\dagger)(\text{Str } \mathcal{U}) - 1. \tag{131}$$

As an exercise, I leave it to the reader to calculate with the methods described above the following quantities:

(a) the matrix representations and

(b) the characters

corresponding to the following supertableaux, and obtain the given answers. The procedure that yields these results can be extended, of course, to any supertableau.

$$\chi_\square = \text{Str } \mathcal{U},$$

$$\chi_{\square\square} = \tfrac{1}{2}(\text{Str } \mathcal{U})^2 + \tfrac{1}{2}\text{Str}(\mathcal{U}^2),$$

$$\chi_{\overset{\square}{\square}} = \tfrac{1}{2}(\text{Str } \mathcal{U})^2 - \tfrac{1}{2}\text{Str}(\mathcal{U}^2),$$

$$\chi_{\square\square\square} = \tfrac{1}{6}(\text{Str } \mathcal{U})^3 + \tfrac{1}{2}\text{Str } \mathcal{U}\,\text{Str}(\mathcal{U}^2) + \tfrac{1}{3}\text{Str}(\mathcal{U}^3),$$

$$\chi_{\overset{\square\square}{\square}} = \tfrac{1}{3}(\text{Str } \mathcal{U})^3 - \tfrac{1}{3}\text{Str}(\mathcal{U}^3),$$

$$\chi_{\overset{\square}{\overset{\square}{\square}}} = \tfrac{1}{3}(\text{Str } \mathcal{U})^3 - \tfrac{1}{2}\text{Str } \mathcal{U}\,\text{Str}(\mathcal{U}^2) + \tfrac{1}{3}\text{Str}(\mathcal{U}^3), \tag{132}$$

$$\chi_{\square\square\square\square} = \tfrac{1}{24}(\text{Str } \mathcal{U})^4 + \tfrac{1}{4}(\text{Str } \mathcal{U})^2\,\text{Str}(\mathcal{U}^2)$$
$$+ \tfrac{1}{3}\text{Str } \mathcal{U}\,\text{Str}(\mathcal{U}^3) + \tfrac{1}{8}\big(\text{Str}(\mathcal{U}^2)\big)^2 + \tfrac{1}{4}\text{Str}(\mathcal{U}^4),$$

$$\chi_{\overset{\square}{\overset{\square}{\overset{\square}{\square}}}} = \tfrac{1}{24}(\text{Str } \mathcal{U})^4 - \tfrac{1}{4}(\text{Str } \mathcal{U})^2\,\text{Str}(\mathcal{U}^2)$$
$$+ \tfrac{1}{3}\text{Str } \mathcal{U}\,\text{Str}(\mathcal{U}^3) + \tfrac{1}{8}(\text{Str } \mathcal{U}^2)^2 - \tfrac{1}{4}\text{Str}(\mathcal{U}^4),$$

$$\chi_{\overset{\square\square}{\square\square}} = \tfrac{1}{8}(\text{Str } \mathcal{U})^4 + \tfrac{1}{4}(\text{Str } \mathcal{U})^2\,\text{Str}(\mathcal{U}^2) - \tfrac{1}{8}(\text{Str } \mathcal{U}^2)^2 - \tfrac{1}{4}\text{Str}(\mathcal{U}^4),$$

$$\chi_{\tiny\yng(2,1,1)} = \tfrac{1}{8}(\text{Str }\mathcal{U})^4 - \tfrac{1}{4}(\text{Str }\mathcal{U})^2\,\text{Str}(\mathcal{U}^2) - \tfrac{1}{8}(\text{Str }\mathcal{U}^2)^2 + \tfrac{1}{4}\text{Str}(\mathcal{U}^4),$$

$$\chi_{\tiny\yng(2,2)} = \tfrac{1}{12}(\text{Str }\mathcal{U})^4 - \tfrac{1}{3}\text{Str }\mathcal{U}\,\text{Str}(\mathcal{U}^3) + \tfrac{1}{4}(\text{Str }\mathcal{U}^2)^2.$$

$$\chi_{\tiny\yng(1)} = \text{Str }\mathcal{U}^{\dagger},$$

$$\chi_{\tiny\yng(1,1)} = \text{Str }\mathcal{U}^{\dagger}\,\text{Str }\mathcal{U} - 1,$$

$$\chi_{\tiny\yng(2,1)} = \tfrac{1}{2}\text{Str }\mathcal{U}^{\dagger}\big[(\text{Str }\mathcal{U})^2 + \text{Str}(\mathcal{U}^2)\big] - \text{Str }\mathcal{U},$$

$$\chi_{\tiny\yng(1,1,1)} = \tfrac{1}{2}\text{Str }\mathcal{U}^{\dagger}\big[(\text{Str }\mathcal{U})^2 - \text{Str}(\mathcal{U}^2)\big] - \text{Str }\mathcal{U}, \tag{133}$$

$$\chi_{\tiny\yng(3,1)} = \tfrac{1}{4}\big[(\text{Str }\mathcal{U}^{\dagger})^2 + \text{Str}(\mathcal{U}^{\dagger 2})\big]\big[(\text{Str }\mathcal{U})^2 + \text{Str}(\mathcal{U}^2)\big]$$
$$\qquad - \text{Str }\mathcal{U}^{\dagger}\,\text{Str }\mathcal{U},$$

$$\chi_{\tiny\yng(2,2)} = \tfrac{1}{4}\big[(\text{Str }\mathcal{U}^{\dagger})^2 + \text{Str}(\mathcal{U}^{\dagger 2})\big]\big[(\text{Str }\mathcal{U})^2 - \text{Str}(\mathcal{U}^2)\big]$$
$$\qquad - \text{Str }\mathcal{U}^{\dagger}\,\text{Str }\mathcal{U} + 1,$$

$$\chi_{\tiny\yng(2,1,1)} = \tfrac{1}{4}\big[(\text{Str }\mathcal{U}^{\dagger})^2 - \text{Str}(\mathcal{U}^{\dagger 2})\big]\big[(\text{Str }\mathcal{U})^2 - \text{Str}(\mathcal{U}^2)\big] - \text{Str }\mathcal{U}^{\dagger}\,\text{Str }\mathcal{U}.$$

Note that if the same calculations are repeated for SU(N) *the answers will formally look the same except for the replacement of supertrace by trace.* From this observation, you learn that if you can write the SU(N) answer in terms of tr($\mathcal{U}^n$) for some Young tableaux you only need to change formally tr to Str to obtain the character for the analogous supertableau!

$$\text{SU}(N) \to \text{SU}(N/M), \qquad \text{tr }\mathcal{U}^n \to \text{Str }\mathcal{U}^n. \tag{134}$$

There is a systematic way of writing down the answer for any given supertableau. First, from the above examples in (132), note that for a supertableau with a single row with n boxes you can write

$$\chi_{\tiny\underbrace{\yng(6)}_{n}} = \frac{1}{n!}(\text{Str }\mathcal{U})^n + \frac{(\text{Str }\mathcal{U})^{n-2}}{(n-2)!}\frac{\text{Str}(\mathcal{U}^2)}{2} + \cdots + \text{Str}\left(\frac{\mathcal{U}^n}{n}\right)$$

$$= \sum_{k_1, k_2 \cdots k_n} \frac{\delta(n - k_1 - 2k_2 - \cdots - nk_n)}{k_1! k_2! \cdots k_n!}$$

$$\cdot \left[\left(\frac{\text{Str }\mathcal{U}}{1}\right)^{k_1}\left(\frac{\text{Str }\mathcal{U}^2}{2}\right)^{k_2}\cdots\left(\frac{\text{Str }\mathcal{U}^n}{n}\right)^{k_n}\right]. \tag{135}$$

This sum can be rewritten as an integral which depends only on $\mathcal{U}$ and the number of superboxes n. We will denote

$$\chi_n \equiv \chi_{\boxed{}_n},$$

$$\boxed{}^n \quad \chi_n(\mathcal{U}) = \int_0^{2\pi} \frac{d\phi}{2\pi} \frac{e^{-in\phi}}{\text{Sdet}(1 - e^{i\phi}\mathcal{U})}; \qquad n \geq 0, \quad (136)$$

where for $n = 0$, $\chi_0 = 1$ and for negative integers $\chi_n = 0$ ($n < 0$) automatically.

If all the boxes in a single row are replaced by dotted boxes (upper indices), then every $\mathcal{U}$ is replaced by $\mathcal{U}^\dagger$ in the transformation property of the tensor. Thus, the character of the corresponding group element will have the same form as $\chi_n(\mathcal{U})$, except for $\mathcal{U}$ replaced by $\mathcal{U}^\dagger$.

We will denote this expression by $\dot{\chi}_{-n}(\mathcal{U})$ as a convenient notation:

$$\dot{\chi}_{-n}(\mathcal{U}) \equiv \chi_n(\mathcal{U}^\dagger), \quad \dot{\chi}_{-n}(\mathcal{U}) = \int_0^{2\pi} d\phi \cdot \frac{e^{-in\phi}}{\text{Sdet}(1 - e^{i\phi}\mathcal{U}^\dagger)}.$$

$$(137)$$

Returning to the examples above in (132), we observe that for two and three rows of undotted boxes the answer is reproduced correctly by the following determinants:

$$\chi_{n_1,n_2}(\mathcal{U}) = \begin{vmatrix} \chi_{n_1} & \chi_{n_2-1} \\ \chi_{n_1+1} & \chi_{n_2} \end{vmatrix},$$

$$\chi_{n_1,n_2,n_3}(\mathcal{U}) = \begin{vmatrix} \chi_{n_1} & \chi_{n_2-1} & \chi_{n_3-2} \\ \chi_{n_1+1} & \chi_{n_2} & \chi_{n_3-1} \\ \chi_{n_1+2} & \chi_{n_2+1} & \chi_{n_3} \end{vmatrix}. \quad (138)$$

It is clear from the systematics of these expressions that for r rows we need to evaluate an $r \times r$ determinant of a matrix whose (i, j) elements are

$$\chi_{(n_1,n_2,\ldots,n_r)} = \det(\chi_{n_j+i-j}). \quad (139)$$

This form, which was derived for supergroups [14] on the basis of the above observations, is also a familiar form [20] for ordinary $SU(N)$ and is appropriate for the analogy drawn in (134): For $SU(N)$, indeed, the correct character is reproduced by (139) and hence for $SU(N/M)$ as well.

Now we also consider the mixed dotted-undotted supertableaux. We observe from the examples above that the correct answers are reproduced by the following determinants:

m_1 n_1

$$\chi_{(-m_i; n_1)} = \begin{vmatrix} \dot{\chi}_{-m_1} & \chi_{n_1-1} \\ \dot{\chi}_{-m_1+1} & \chi_{n_1} \end{vmatrix}, \tag{140}$$

m_1 n_1
n_2

$$\chi_{(-m_1; n_1, n_2)} = \begin{vmatrix} \dot{\chi}_{-m_1} & \chi_{n_1-1} & \chi_{n_2-2} \\ \dot{\chi}_{-m_1+1} & \chi_{n_1} & \chi_{n_2-1} \\ \dot{\chi}_{-m_1+2} & \chi_{n_1+1} & \chi_{n_2} \end{vmatrix},$$

m_1 n_1
m_2 n_2

$$\chi_{(-m_2, -m_1; n_1, n_2)} = \begin{vmatrix} \dot{\chi}_{-m_2} & \dot{\chi}_{-m_1-1} & \chi_{n_1-2} & \chi_{n_2-3} \\ \dot{\chi}_{-m_2+1} & \dot{\chi}_{-m_1} & \chi_{n_1-1} & \chi_{n_2-2} \\ \dot{\chi}_{-m_2+2} & \dot{\chi}_{-m_1+1} & \chi_{n_1} & \chi_{n_2-1} \\ \dot{\chi}_{-m_2+3} & \dot{\chi}_{-m_1+2} & \chi_{n_1+1} & \chi_{n_2} \end{vmatrix}. \tag{141}$$

The systematics of the mixed tableaux are seen to be the same as the undotted tableaux: On the *diagonal* we write the character of a row starting with the shortest dotted row and ending with the shortest undotted row. The rest of the determinant is filled by increasing the index downward or decreasing it upward. Note again the definition of $\dot{\chi}_{-m}(\mathcal{U}) = \chi_m(\mathcal{U}^\dagger)$ as in (137). Using this prescription we can now compute the character for *any supertableau* with r undotted and $\dot{r}$ dotted rows:

$$\chi_{(-m_{\dot{r}}, \ldots, -m_1; n_1, \ldots, n_r)} = \begin{vmatrix} \dot{\chi}_{-m_{\dot{r}}} & & \text{decrease} & & \chi_{n_r-r-\dot{r}+1} \\ & \ddots & & \Uparrow & \\ & & \dot{\chi}_{-m_1} & & \\ & & & \chi_{n_1} & \\ & & \Downarrow & & \ddots \\ \dot{\chi}_{-m_{\dot{r}}+\dot{r}+r-1} & & \text{increase} & & \chi_{n_r} \end{vmatrix}. \tag{142}$$

This formula gives a very explicit answer, such as the examples of (132), (133), which is very useful in practical calculations. This new expression for characters for mixed tableaux can also be applied in

SU(N) because of the analogy of (134). *It appears that this is a new form* [14] *not only for* SU(N/M) *but also for* SU(N). Its validity is checked by comparing to alternative forms of the characters and is seen to work, as expected.

In the above expressions the supertableaux were specified by the number of boxes in each row and the characters were calculated accordingly. A similar procedure works if the labeling is done by the number of boxes in each column. Returning to the examples in (132) we observe that for a single column of undotted boxes we can write

$$\chi_{\substack{\boxed{}\\n}} = (-1)^n \sum_{k_1,\ldots,k_n} \frac{\delta(n - k_1 - 2k_2 - \cdots - nk_n)}{k_1! k_2! \cdots k_n!} \cdot \left[\left(\frac{-\operatorname{Str} \mathcal{U}}{1} \right)^{k_1} \left(\frac{-\operatorname{Str} \mathcal{U}^2}{2} \right)^{k_2} \cdots \left(\frac{-\operatorname{Str} \mathcal{U}^n}{n} \right)^{k_n} \right]. \tag{143}$$

This result can also be obtained [14] by evaluating an $n \times n$ determinant according to the formula in (139). We will denote this expression by $A_n(\mathcal{U})$, and note that it can be written as an integral, in analogy to (136),

$$A_n(\mathcal{U}) \equiv \chi_{\substack{\boxed{}\\n}}(\mathcal{U}),$$

$$A_n(\mathcal{U}) = (-1)^n \int_0^{2\pi} \frac{d\phi}{2\pi} e^{-in\phi} \operatorname{Sdet}(1 - e^{i\phi}\mathcal{U}). \tag{144}$$

As in the case of rows, $A_0 = 1$ and $A_n = 0$ if $n < 0$, follows automatically from this equation. It is clear that it would be very cumbersome to use an $n \times n$ determinant instead of this simple form. Similar simplifications would occur if we develop a formula appropriate to tableaux with long columns. Then following the steps above and substituting columns instead of rows we verify that the explicit examples in (132) are correctly reproduced by

$$A_{k_1,k_2}(\mathcal{U}) = \begin{vmatrix} A_{k_1} & A_{k_2-1} \\ A_{k_1+1} & A_{k_2} \end{vmatrix}, \quad \text{etc.} \tag{145}$$

The systematics with columns are the same as with the rows, and the character for the general mixed supertableau of Figure 1 can be evaluated according to the number of boxes in each *column* in Figure 1:

$$A_{(-l_q,\ldots,-l_1;k_1,\ldots,k_p)} = \chi_{(-m_r,\ldots,-m_1;n_1,\ldots,n_r)},$$

$$A_{(-l_q,\ldots,l_1;k_1,\ldots k_p)} = \begin{vmatrix} \dot{A}_{-l_q} \Uparrow & \text{decrease} & \\ & \ddots & \\ \Downarrow & \dot{A}_{-l_1} \Uparrow & \\ & A_{k_1} & \Uparrow \\ & \Downarrow & \ddots \\ & \text{increase} & \Downarrow A_{k_p} \end{vmatrix}, \qquad (146)$$

where $\dot{A}_{-m}(\mathcal{U}) = A_m(\mathcal{U}^\dagger)$ is a definition. The answer is, of course, the same when the character is evaluated via the rows or columns of the supertableau. As exercises the reader can return to (132), (133).

13. Dimensions. In a given representation we are often interested in the number of states in the vector space. For supergroups, we will need the numbers of bosons and fermions separately. Let us define

$$\begin{aligned} \text{number of bosons} &\equiv B, \\ \text{number of fermions} &\equiv F, \\ \text{difference} &\equiv d = B - F, \\ \text{sum} &\equiv S = B + F. \end{aligned} \qquad (147)$$

We will be able to compute d and S rather simply. In some simple cases above, we have already counted these states. For example, in the fundamental representation

$$d_\square = N - M, \qquad S_\square = N + M. \qquad (148)$$

Also, from (106) and (107) we have

$$\begin{aligned} d_{\square\square} &= \frac{1}{2}(N - M)(N - M + 1), \\ S_{\square\square} &= \frac{1}{2}\left[(N + M)^2 + N - M\right], \end{aligned} \qquad (149)$$

and

$$\begin{aligned} d_{\square\atop\square} &= \frac{1}{2}(N - M)(N - M - 1), \\ S_{\square\atop\square} &= \frac{1}{2}\left[(N + M)^2 - (N - M)\right]. \end{aligned} \qquad (150)$$

Rather than counting the states as in (106), (107), we can do our computation much more efficiently by realizing that d and S follow from the evaluation of the character for the identity matrix $\mathcal{U} = 1$ and for the matrix

$$\mathcal{U} = J = \left(\begin{array}{c|c} 1 & \\ \hline & -1 \end{array}\right)$$

respectively. For example, for the fundamental representation

$$d_{\square} = \text{Str } 1 = N - M, \qquad S_{\square} = \text{Str } J = N + M. \qquad (151)$$

Similarly, using the character formulas in (130), we have

$$d_{\square\square} = \frac{1}{2}\left[(\text{Str } 1)^2 + \text{Str}(1^2)\right] = \frac{1}{2}\left[(N - M)^2 + (N - M)\right],$$
$$S_{\square\square} = \frac{1}{2}\left[(\text{Str } J)^2 + \text{Str}(J^2)\right] = \frac{1}{2}\left[(N + M)^2 + (N - M)\right], \qquad (152)$$

in agreement with (149). Similarly, for the adjoint representation

$$d_{\square\square} = (N - M)^2 - 1 = (\text{Str } 1)^2 - 1,$$
$$S_{\square\square} = (N + M)^2 - 1 = (\text{Str } J)^2 - 1, \qquad (153)$$

which coincide with the counting of bose + fermi parameters in the superalgebra given by $\mathcal{H}$ in (1), as expected. The same can be checked explicitly for all the representation of (132), (133). For an arbitrary representation R we can now make the statement

$$d_R = \chi_R(1), \qquad S_R = \chi_R(J). \qquad (154)$$

These can be computed from the character formulas above provided we first compute $\chi_n(1)$ and $\chi_n(J)$ for one row (and/or for one column).

Thus, in general, we need to evaluate d and S for the single column and single row supertableaux and simply substitute the result in the general character formulas of (142) or (146). We define the dimensions for one row and one column as

$$\overset{n}{\boxed{\diagdown\diagdown\diagdown\diagdown\diagdown}} \rightarrow d_n, S_n,$$

$$(155)$$

$$n\ \boxed{\begin{array}{c}\diagdown\\\diagdown\\\diagdown\\\diagdown\end{array}} \rightarrow \tilde{d}_n, \tilde{S}_n.$$

In an obvious notation that follows (137) we obtain for a dotted row or column

$$
\boxed{}^{\,n} \;\to\; \dot{d}_{-n},\,\dot{S}_{-n}, \quad \text{with} \quad \begin{cases} \dot{d}_{-n} = d_n, \\ \dot{S}_{-n} = S_n, \end{cases}
$$

$$
\left.\begin{matrix}\end{matrix}\right\}^{\,n} \;\to\; \dot{\tilde{d}}_{-n},\,\dot{\tilde{S}}_{-n}, \quad \text{with} \quad \begin{cases} \dot{\tilde{d}}_{-n} = \tilde{d}_n, \\ \dot{\tilde{S}}_{-n} = \tilde{S}_n. \end{cases}
\tag{156}
$$

We evaluate them directly from the integral representations in (136) for one row:

$$
\begin{aligned}
d_n &= \int_0^{2\pi} \frac{d\phi}{2\pi} e^{-in\phi}(1 - e^{i\phi})^{M-N} \\
&= \frac{1}{n!}(N - M)(N - M + 1) \cdots (N - M + n - 1),
\end{aligned}
\tag{157}
$$

$$
S_n = \int_0^{2\pi} \frac{d\phi}{2\pi} e^{-in\phi} \frac{(1 + e^{i\phi})^M}{(1 - e^{i\phi})^N} = \sum_k \binom{N + n - k - 1}{N - 1}\binom{M}{k},
$$

where

$$
\binom{b}{a} = \frac{b!}{a!\,(b - a)!}
$$

is the binomial coefficient. Similarly $\tilde{d}_n$, $\tilde{S}_n$ for one column are given via (144):

$$
\begin{aligned}
\tilde{d}_n &= (-1)^n \int_0^{2\pi} \frac{d\phi}{2\pi} e^{-in\phi}(1 - e^{i\phi})^{N-M} \\
&= \frac{1}{n!}(N - M)(N - M - 1) \cdots (N - M - n + 1),
\end{aligned}
\tag{158}
$$

$$
\begin{aligned}
\tilde{S}_n &= (-1)^n \int_0^{2\pi} \frac{d\phi}{2\pi} e^{-in\phi} \frac{(1 - e^{i\phi})^N}{(1 + e^{i\phi})^M} \\
&= \sum_k \binom{N}{k}\binom{M + n - k - 1}{M - 1}.
\end{aligned}
$$

It is striking that the formulas for d_n and $\tilde{d}_n$ coincide with those of the dimension formulas for the group SU($N - M$) for the one row and column *ordinary* Young tableaux!

$$d_n(\mathrm{SU}(N/M)) = \text{dimension for } \overset{n}{\boxed{\boxed{\boxed{\boxed{\boxed{\boxed{}}}}}}} \text{ of } \mathrm{SU}(N-M),$$

$$\tilde{d}_n(\mathrm{SU}(N/M)) = \text{dimension for } \boxed{} n \text{ of } \mathrm{SU}(N-M). \tag{159}$$

While the formulas for S_n, $\tilde{S}_n$ coincide with the sum of dimensions of the $\mathrm{SU}(N) \times \mathrm{SU}(M)$ tableaux,

$$S_n(\mathrm{SU}(N/M))$$
$$= \text{dimension for } \sum_k \left(\overset{n-k}{\boxed{\boxed{\boxed{\boxed{\boxed{}}}}}} , \boxed{} k \right) \text{ for } \mathrm{SU}(N) \times \mathrm{SU}(M),$$

$$\tilde{S}_n(\mathrm{SU}(N/M))$$
$$= \text{dimension for } \sum_k \left(\boxed{} k , \overset{n-k}{\boxed{\boxed{\boxed{\boxed{\boxed{}}}}}} \right) \text{ for } \mathrm{SU}(N) \times \mathrm{SU}(M). \tag{160}$$

The general dimension formulas for the supertableau of Figure 1 can now be given via (142) or (146) and (155)–(160), by using columns or rows:

$$d = \left| \begin{array}{ccccccc} \dot{d}_{-m_f} & & & \text{decrease} & & & \\ & \ddots & \Uparrow & & & & \\ & \Downarrow & d_{-m_1} & & & & \\ & & & d_{m_1} & \Uparrow & & \\ & & & & \Downarrow & \ddots & \\ & & & \text{increase} & & & d_{n_r} \end{array} \right|$$

$$= \left| \begin{array}{ccccccc} \dot{\tilde{d}}_{-l_q} & & \text{decrease} & & & \\ & \ddots & \Uparrow & & & \\ & \Downarrow & \dot{\tilde{d}}_{-l_1} & & & \\ & & & d_{k_1} & \Uparrow & \\ & & & & \Downarrow & \ddots \\ & & \text{increase} & & & d_{k_p} \end{array} \right|, \tag{161}$$

$$
S = \begin{vmatrix}
\dot{S}_{-m_f} & & \text{decrease} & & \\
& \ddots & \Uparrow & & \\
& \Downarrow & \dot{S}_{-m_1} & & \\
& & & S_{n_1} & \Uparrow \\
& & & \Downarrow & \ddots \\
& & \text{increase} & & S_{n_r}
\end{vmatrix}
$$

$$
= \begin{vmatrix}
\tilde{\dot{S}}_{-l_q} & & \text{decrease} & & \\
& \ddots & \Uparrow & & \\
& \Downarrow & \tilde{\dot{S}}_{-l_1} & & \\
& & & \tilde{S}_{k_1} & \Uparrow \\
& & & \Downarrow & \ddots \\
& & \text{increase} & & \tilde{S}_{k_p}
\end{vmatrix} .
$$

$$(162)$$

It is very useful to note that, because of (159), the number of bosons *minus* fermions d_R in a supertableau R is numerically identical to the dimension of a representation of the ordinary group $SU(N - M)$, provided the same shape ordinary Young tableaux is used. The following few striking examples will show the significance of this observation.

(1) For $U(1)$ the only possible nonvanishing tableaux have n boxes:

$$\boxed{}\boxed{}\boxed{}\boxed{}\boxed{}\boxed{}\boxed{}$$

The dimension of the representation is 1 for any n. Two or more rows will have vanishing dimension. Therefore for $SU(N + 1/N)$ we have

n
bosons $-$ fermions $= 1$, $\qquad n \geqslant 0$,

n_1 bosons $-$ fermions $= 0$, $\qquad n_1 \geqslant n_2 \geqslant 1$,
n_2

etc., for more rows.

Thus the number of bosons is equal to the number of fermions for two or more rows. Note that the two or more row supertableaux do not vanish for $SU(N + 1/N)$, unlike $U(1)$. The total number of (bosons + fermions) is nonzero as can be checked from (162).

(2) For SU(2) the nonvanishing tableaux are

SU(2)   ⊞⊞⊞⊞⊞⊞ n_1 dimension $= n_1 - n_2 + 1$.
               n_2

Therefore, for SU($N + 2/N$) the same shape supertableau has

bosons $-$ fermions $= n_1 - n_2 + 1$ ▨▨▨▨ n_1
                                        n_2

For three or more rows in the supertableau we have

$$\text{bosons} - \text{fermions} = 0,$$

while the total S is nonvanishing.

(3) For SU(5)

        ⊟   is 10-dimensional;

therefore for SU($N + 5/N$)

$$\text{bosons} - \text{fermions} = 10 \quad ⬓ \ .$$

The examples can go on indefinitely. An important result is that for SU(N/M), if the supertableau has more than $N - M$ rows, then the number of bosons is equal to the number of fermions in that representation. These include the "typical" representations that Kac has described in a very different language. The dimension formula for

$$\text{bosons} + \text{fermions} = S$$

for ($d = 0$) as well as nontypical ($d \neq 0$) representations is given via (162). "Typical" representations (19) have a SU(N/M) dimension formula $d = 0$ and $S = 2^{NM} \times D$, where D is the dimension of the irreducible SU(N) $\times$ SU(M) representation that contains the highest weight of the superrepresentation. Not all of our $d = 0$ representations satisfy this formula for S so that although some such representations are typical, others are "nontypical" even though they contain equal numbers of bosons and fermions ($d = 0$).

For example, for SU($N + 1/N$) or SU($N/N + 1$) the adjoint representation

$$▨▨ \text{ satisfies } d_{▨▨} = 0$$

while $S = (2N + 1)^2 - 1$. Furthermore, the adjoint representations of SU($N + 1/N$) satisfy the following branching to SU($N + 1$) $\times$ SU(N) $\times$ U(1):

$$▨▨ = (\square, \boxdot) + (\boxed{\boxdot}, 1) + (1, 1) + (1, \boxed{\boxdot})_0 + (\boxdot, \square)$$

with the first piece $(\square, \boxdot)$ containing the highest weight. The dimension of this piece is $D = (N + 1)N$. Therefore, we may ask for which $SU(N + 1/N)$ or $SU(N/N + 1)$ is the adjoint representation a typical representation? We must satisfy the requirement for the form of the dimension formula $S = 2^{(N+1)N}D$, giving the condition

$$S = (2N + 1)^2 - 1 = 2^{(N+1)N}(N + 1)N,$$

which is satisfied for $N = 1$. Thus only for $SU(2/1)$ or $SU(1/2)$ the adjoint representation is a "typical" representation. For all other $SU(N + 1/N)$ or $SU(N/N + 1)$ the adjoint representation is not "typical", but still contains an equal number of bosons and fermions.

14. Eigenvalues of Casimir operators. The quantum Casimir operators of order n, $\hat{C}_n$ were constructed in (70), (71), (74), (75), (76), (85). By building the states of a representation via the quantum oscillators as in (113)–(122) and further generalizations involving all the oscillators ξ, η, $\tilde{\rho}$, $\tilde{\sigma}$, we may compute the eigenvalues. This works very efficiently for representations of the type

$$\boxed{\diagup\diagup\diagup\diagup\diagup\diagup}$$

as in (122), since, as we argued following (77), in this case there really is a single independent invariant, which is $\hat{n} = \xi^{\dagger} \cdot \xi$. For example, applying the quadratic Casimir operator $\hat{C}_2$ in (74) to the state in (122) we obtain immediately the eigenvalue for the supergroup $SU(N/M)$:

$$C_2(\overset{n}{\boxed{\diagup\diagup\diagup\diagup}}) = \frac{N - M - 1}{2(N - M)}n(N - M + n); \quad N \neq M. \quad (163)$$

Note that for $N = M$, C_2 could not be defined, but the invariant operator $\hat{n}$ and its eigenvalue exists. The computation of $C_2(R)$ is not so easy if a few copies of oscillators are involved as can be seen from (71).

A different general approach was described in [**14**], which uses the characters (142), (146) as the starting point: Recall that the group parameters are described by the matix $\mathcal{H}$ in the fundamental representation, as in (1), (2), (35),

$$\mathcal{H}_A{}^B = \left(\frac{\lambda_I}{2}\right)_A{}^B \omega^I, \qquad \mathcal{U} = e^{i\mathcal{H}}, \quad (164)$$

where $(\lambda_I/2)_A{}^B$ represent the generators χ_I. In an arbitrary representation R the generator will have a matrix representation $R(X_I)$.

Combining generators with parameters, we can then write the representation of the matrix $\mathcal{H}$ as

$$R_p{}^q(\mathcal{H}) = R_p{}^q(X_I)\omega^I, \quad (165)$$

where the indices p, q run over the bosonic and fermionic states of the representation space. The matrix representation of the generators $\overline{X}_A{}^B$ or $\overline{G}_A{}^B$ defined in (66) or (82) can now be obtained by differentiating with respect to the parameters $\mathcal{K}_A{}^B$, as in (1), (2),

$$R_p{}^q(\overline{X}_A{}^B) = \frac{\partial}{\partial \mathcal{K}_B{}^A} R_p{}^q(\mathcal{K}). \tag{166}$$

Writing the nth order Casimir operator as (76)

$$\hat{C}_n = \mathrm{Str}(\overline{X}^n) = \overline{X}_{A_1}{}^{A_2}\overline{X}_{A_2}{}^{A_3} \cdots \overline{X}_{A_n}{}^{A_1}(-1)^{g(A_1)}, \tag{167}$$

we see that its matrix representation will be given by

$$R_p{}^q(\hat{C}_n) = (-1)^{g(A_1)} \frac{\partial}{\partial \mathcal{K}_{A_1}{}^{A_n}} \cdots \frac{\partial}{\partial \mathcal{K}_{A_3}{}^{A_2}} \frac{\partial}{\partial \mathcal{K}_{A_2}{}^{A_1}} R_p{}^q(\mathcal{K}^n). \tag{168}$$

The matrix $R_p{}^q(\hat{C}_n)$ is found proportional to the identity for $d(R) \neq 0$ representations. While it probably has nondiagonal parts for $d_R = 0$ representations, it can be argued that Shur's lemma works only in representations in which the number of bosons is different from the number of fermions ($d_R \neq 0$). Then we may write

$$R_p{}^q(\hat{C}_n) = C_n(R)\delta_p{}^q, \qquad d_R \neq 0. \tag{169}$$

Our aim is to obtain the number $C_n(R)$. For $d(R) = 0$ representations since there could be a nondiagonal piece our method will not work.

Now we describe the method of computation and give the answer. By taking the supertrace of (169), we obtain

$$\mathrm{Str}\, R(\hat{C}_n) = C_n(R)d_R,$$

or

$$C_n(R) = \mathrm{Str}\, R(\hat{C}_n)/d_R; \qquad d_R \neq 0, \tag{170}$$

so that we only need to compute $\mathrm{Str}\, R(\hat{C}_n)$ which is given, from (168), by

$$\mathrm{Str}\, R(\hat{C}_n) = \left[\mathrm{Str}\left(\frac{\partial}{\partial \mathcal{K}}\right)^n\right][\mathrm{Str}\, R(\mathcal{K}^n)], \tag{171}$$

where we wrote the differential operators symbolically. To calculate this quantity we return to the character which we have already computed. The aim is to bypass the need of constructing the matrix element $R_p{}^q(\mathcal{K})$ explicitly, and use only the character which is much simpler.

The characters of (142), (146), which depend explicitly on the parameters $\mathcal{K}_A{}^B$, can also be expressed in a different form by computing the supertrace of the representation of supergroup elements in the notation of (165):

$$\text{group:} \quad R_p{}^q(\mathfrak{U}) = \left[e^{iR(\mathfrak{H})}\right]_p{}^q,$$
$$\text{character:} \quad \chi_R(\mathfrak{U}) = \text{Str}\left[e^{iR(\mathfrak{H})}\right]. \tag{172}$$

Then, expanding the exponential in powers of $R(\mathfrak{H})$ it is clear that (171) follows from

$$\text{Str}\, R(\hat{C}_n) = \frac{n!}{(i)^n}\left\{\left[\text{Str}\left(\frac{\partial}{\partial \mathfrak{H}}\right)^n\right]\chi_R(\mathfrak{U})\right\}_{\mathfrak{H}=0}, \tag{173}$$

where $\mathfrak{H}_A{}^B = 0$ after differentiating n times. But now the desired quantity has been expressed in terms of $\chi_R(\mathfrak{U})$ and we no longer need $R(\mathfrak{H})$.

Simply, we can substitute (142) or (146) for $\chi_R(\mathfrak{U})$ which is quite explicit in terms of $\mathfrak{H}$. Thus, we use our explicit character formulas, substitute $\mathfrak{U} = e^{i\mathfrak{H}}$ and expand in powers of $\mathfrak{H}$. (As an exercise you can do this with a few examples provided by (132), (133).) This expansion may be written in two forms, the first following from (172) and the second from (142), (146):

$$\chi_R(e^{i\mathfrak{H}}) \Bigg\langle
\begin{aligned}
& \text{Str}\, R(1) + \frac{i^2}{2!}\text{Str}\left[R(\mathfrak{H})\right]^2 + \frac{i^3}{3!}\text{Str}\left[R(\mathfrak{H})\right]^3 \\
& \qquad + \frac{i^4}{4!}\text{Str}\left[R(\mathfrak{H})\right]^4 + \cdots \\[6pt]
& d_R + q_R\frac{\text{Str}(i\mathfrak{H})^2}{2!} + k_R\frac{\text{Str}(i\mathfrak{H})^3}{3!} \\
& \qquad + \left\{v_R\frac{\text{Str}(i\mathfrak{H})^4}{4!} + w_R\frac{\left(\text{Str}\left((i\mathfrak{H})^2/2\right)\right)^2}{2!}\right\} + \cdots,
\end{aligned}
\tag{174}$$

where we hve taken into account that $\text{Str}\,\mathfrak{H} = 0$. Thus, in order to compute (173) we will need to first find the coefficients $q(R)$, $k(R)$, $v(R)$, $w(R)$, etc., via (142), (146), which will be given below.

From this form we show how to compute, say, the quadratic Casimir. From (170), (173), (174) we have for a representation with $d_R \neq 0$

$$C_2(R) = \frac{1}{d_R}\cdot q_R \cdot \left[\text{Str}\left(\frac{\partial}{\partial \mathfrak{H}}\right)^2\right]\left[\text{Str}\,\mathfrak{H}^2\right]. \tag{175}$$

The last factor, as defined through (170), (171), is simply the Casimir in the fundamental representation $C_2(\boxdot)$ times $d(\boxdot)$, which we can compute easily, as in (49). Therefore

$$C_2(R) = \frac{q_R}{d_R} \cdot C_2(\boxdot)d(\boxdot). \tag{176}$$

Similarly, we obtain the cubic Casimir as

$$C_3(R) = \frac{k_R}{d_R} \cdot C_3(\boxdot)d(\boxdot), \tag{177}$$

and so on for the higher Casimirs.

Now, we turn to the computation of $q(R)$, $k(R)$, etc. First we need to compute the expansion of χ_n for one row and/or A_n for one column, and then substitute them in the general character formulas (142), (146). For one row or column in (136), (144) we substitute $\mathfrak{A} = e^{i\mathcal{H}}$ and expand in powers of $\mathcal{H}$. By doing some simple integrals we get [14]:

$$q \; \underbrace{\boxed{\diagdown\diagdown\diagdown\diagdown\diagdown\diagdown}}_{n} \equiv q_n = \frac{n(N - M + n)}{(N - M)(N - M + 1)} \cdot d_n,$$

$$k \; \underbrace{\boxed{\diagdown\diagdown\diagdown\diagdown\diagdown\diagdown}}_{n} \equiv k_n = \frac{n(N - M + n)(N - M + 2n)}{(N - M)(N - M + 1)(N - M + 2)} \cdot d_n, \tag{178}$$

$$q \; \underbrace{\boxed{\begin{array}{c}\diagdown\\\diagdown\\\diagdown\\\diagdown\end{array}}}_{n} \equiv \tilde{q}_n = \frac{n(N - M - n)}{(N - M)(N - M - 1)} \cdot \tilde{d}_n,$$

$$k \; \underbrace{\boxed{\begin{array}{c}\diagdown\\\diagdown\\\diagdown\\\diagdown\end{array}}}_{n} \equiv \tilde{k}_n = \frac{n(N - M - n)(N - M - 2n)}{(N - M)(N - M - 1)(N - M - 2)} \cdot \tilde{d}_n, \tag{179}$$

and so on for v_n, w_n, etc., where d_n, $\tilde{d}_n$ are already given in (157), (158). Thus, for one row or column, (174) takes the form

$$\chi_n(e^{i\mathcal{H}}) = d_n + q_n \frac{\mathrm{Str}(i\mathcal{H})^2}{2!} + k_n \frac{\mathrm{Str}(i\mathcal{H})^3}{3!} + \cdots,$$

$$A_n(e^{i\mathcal{H}}) = \tilde{d}_n + \tilde{q}_n \frac{\mathrm{Str}(i\mathcal{H})^2}{2!} + \tilde{k}_n \frac{\mathrm{Str}(i\mathcal{H})^2}{3!} + \cdots. \tag{180}$$

Substituting these in the general forms (142), (146), re-expanding in powers of $\mathcal{H}$ and comparing to (174) we obtain the general $q(R)$, $k(R)$, etc., in the form of a sum of determinants. Symbolically this looks like

$$
X(R) = \begin{vmatrix} X & d & d & \cdots \\ X & d & d & \cdots \\ X & d & d & \cdots \\ \vdots & \vdots & \vdots & \end{vmatrix} + \begin{vmatrix} d & X & d & \cdots \\ d & X & d & \cdots \\ d & X & d & \cdots \\ \vdots & \vdots & \vdots & \end{vmatrix}
$$

$$
+ \begin{vmatrix} d & d & X & \cdots \\ d & d & X & \cdots \\ d & d & X & \cdots \\ \vdots & \vdots & \vdots & \end{vmatrix}, \tag{181}
$$

where $X(R)$ stands for $q(R)$ or $k(R)$ if (142) is used, or for $\tilde{q}(R)$ or $\tilde{k}(R)$ if (146) is used. In the case of $v(R)$, $w(R)$ and higher coefficients, the pattern of determinants is more complicated, because the Taylor expansion of the determinant in powers of $\mathcal{H}$ is more involved for $\mathcal{H}^4$ and higher powers.

Here we can make an important observation [14, 15] that all of the coefficients $d(R)$, $q(R)$, $k(R)$, $v(R)$, etc., for all representations R are coming out as functions of only $N - M$ as seen from (178)–(181). Indeed had we calculated all the analogous coefficients $d(R)$, $q(R)$, etc., and Casimir eigenvalues for the ordinary group $SU(N - M)$ for the *same shape Young tableaux*, we would get identical numerical expressions as above by using exactly the same formalism step by step. This supplements the previous observation in (159), (161) about $d(R) = $ (bosons $-$ fermions). Thus, we have the correspondence between $SU(N - M)$ and $SU(N/M)$:

$$
SU(N - M) \rightarrow \left(\begin{array}{l} d_R, q_R, q_R, v_R, w_R, \text{ etc., and} \\ \text{Casimir eigenvalues for } R \ (d_R \neq 0) \end{array} \right) \tag{182}
$$
$$
\rightarrow SU(N/M).
$$

Note that the correspondence for the expansion coefficients $d(R)$, $q(R)$, $k(R)$, etc., hold for $d_R = 0$ as well as $d_R \neq 0$ representations, since they occur in the expansion of the character and are independent from the validity of (169) and (170) for the Casimir operators in which we need to divide by d_R.

From the above statements we now observe [14,15] *that if we have an expression for the Casimir eigenvalues for the group* $SU(N)$, *written as a function of* N, *we could simply substitute* $N \rightarrow N - M$ *in order to obtain the Casimir eigenvalues for* $SU(N/M)$. Indeed, making use of this trick, all Casimir eigenvalues for up to the $(N - M)$th order Casimir have been computed [15] by simply translating $N \to N - M$ in the expressions available in the literature on SU(N). The answer is expressed in the form of a generating functional that yields the eigenvalues of the pth order SU(N/M) Casimir operator for $p \leqslant |N - M|$,

$$C_p = \mathrm{Str}(\overline{X}^p), \qquad p \leqslant |N - M|,$$

supertableau

$$R = (n_1, n_2, n_3, \ldots);$$

then the generating functional is [21]

$$\sum_{p=0}^{\infty} C_p(R)z^p = (N - M)e^{-f(z)} + \frac{1 - e^{-f(z)}}{z},$$

where

$$f(z) = \sum_{k=z}^{\infty} \sum_{j=1}^{k-1} \frac{z^k}{k} \binom{k}{j} S_j,$$

$$S_j = \sum_{i=1}^{N-M} \sum_{l=0}^{j-1} \binom{j}{l} \left(n_i - \frac{\sum_r n_r}{N - M} \right)^{j-l} (N - M - i)^l.$$

In comparison to the determinant forms (181), the quadratic Casimir eigenvalues computed in this way are simpler, and given by [15]

$$C_2(R) = \frac{1}{2} \sum_{i=1}^{\infty} \left[n_i^2 - 2in_i + (N - M + 1)n_i \right] - \frac{1}{2(N - M)} \left(\sum_{r=1}^{\infty} n_r \right)^2.$$

Note that, unlike SU(N), the sum extends to infinity. We have generalized this formula to the general supertableau in Figure 1 containing dotted boxes as well. Thus, for the quadratic Casimir we obtain (for $d_R \neq 0$) the new result

$$C_2(R) = \frac{1}{2} \sum_{i=1}^{\infty} \left\{ n_i^2 + m_i^2 + (n_i + m_i)(N - M + 1 - 2i) \right\}$$

$$- \frac{1}{2(N - M)} \left[\sum_{r=1}^{\infty} (n_r - m_r) \right]^2.$$

15. The branching rules SU(N/M) $\to$ SU(N) $\times$ SU(M) $\times$ U(1). This branching rule of the superrepresentations is of primary importance in physical applications [4, 6]. This decomposition is also the key for establishing the relation between supertableaux and Kac-Dynkin diagrams. The irreducibility (or reducibility but indecomposability in some cases) properties of our representations, which were discussed to a limited extent in our previous work, become evident after this connection.

As in the previous sections, we continue to make progress by further exploiting the relationship between $SU(N/M)$ and $SU(N + M)$. Here we will compare the branching rules for

$$SU(N + M) \supset SU(N) \otimes SU(M) \otimes U(1),$$
$$SU(N/M) \supset SU(N) \otimes SU(M) \otimes U(1). \tag{183}$$

In the $(N + M)$-dimensional fundamental representation, the $U(1)$ generator is a traceless matrix for $SU(N + M)$ and a supertraceless one for $SU(N/M)$. Up to an overall constant, it is given as

$$U(1): \begin{pmatrix} 1/N & 0 \\ \hline 0 & -1/M \end{pmatrix} \quad \text{for } SU(N + M),$$

$$U(1): \begin{pmatrix} 1/N & 0 \\ \hline 0 & 1/M \end{pmatrix} \quad \text{for } SU(N/M). \tag{184}$$

In the following we will work our way up by starting with a few simple examples and eventually arrive at some general observations that hold for any representation.

The fundamental representation $\phi_A \sim \square$ (or $\boxslash$) can be split into the direct sum $\phi_A = \phi_a \oplus \psi_\alpha$. Here the N-dimensional piece ϕ_a, $a = 1, 2, \ldots, N$, transforms like the fundamental representation of $SU(N)$, is singlet under $SU(M)$, and carries the $U(1)$ charge $1/N$. We denote this part by $\phi_a \sim (\square, 1)_{1/N}$. Similarly the M-dimensional piece ψ_α, $\alpha = 1, 2, \ldots, M$, belongs to the fundamental representation of $SU(M)$, is a singlet under $SU(N)$, and carries the $U(1)$ charge $(-1/M)$ for $SU(N + M)$ and $1/M$ for $SU(N/M)$. We denote it by $\psi_\alpha \sim (1, \square)_{-1/M}$ or $(1, \square)_{1/M}$ respectively. For $SU(N + M)$, ϕ_a and ψ_α are both bosons (or both fermions). For $SU(N/M)$ one of them is a boson and the other is a fermion. In representations of class I we chose $\phi = $ boson and $\psi = $ fermion. It is sufficient to restrict ourselves to only class I representations since all other representations (class II and mixed cases) can be obtained from those of pure class I representations by simply switching bosons and fermions in the final *basis* without changing the matrix representation of the *group element*.

From the above explanation, the branching equation (183) for the fundamental representation $\phi_A = \phi_a \oplus \psi_\alpha$ can be expressed in terms of tableaux as

$$\square = (\square, 1)_{1/N} \oplus (1, \square)_{-1/M} \quad \text{for } SU(N + M),$$
$$\boxslash = (\square, 1)_{1/N} \oplus (1, \square)_{1/M} \quad \text{for } SU(N/M). \tag{185}$$

Next consider the *completely symmetric* (completely supersymmetric) tensor with n indices $\phi_{(A_1 A_2 \cdots A_n)}$. By specializing each index $A_i = a_i \oplus \alpha_i$, where $a_i = 1, 2, \ldots, N$ and $\alpha_i = 1, 2, \ldots, M$, we can find the various $SU(N) \otimes SU(M) \otimes U(1)$ components of this tensor:

$$\phi_{(A_1 A_2 \cdots A_n)} = \phi_{(a_1 a_2 \cdots a_n)} \oplus \phi_{(a_1 a_2 \cdots a_{n-1})}(\alpha_n) \oplus \cdots$$

$$\oplus \phi_{(a_1 a_2)(\alpha_3 \alpha_4 \cdots \alpha_n)} \oplus \phi_{(a_1)(\alpha_2 \alpha_3 \cdots \alpha_n)} \oplus \phi_{(\alpha_1 \alpha_2 \cdots \alpha_n)}. \quad (186)$$

The $U(1)$ charges can be computed by assigning $1/N$ to each a_i and $-1/M(+1/M)$ to each α_i. For $SU(N + M) \phi_{(a_1 a_2 \cdots a_{n-k})(\alpha_{n-k+1} \cdots \alpha_N)}$ must be completely symmetric in both sets of indices a_i or α_i, since the original indices $(A_1, A_2 \cdots A_n)$ were completely symmetrized. Thus, it transforms as the direct product representation

$$(\; \overset{n-k}{\boxed{\boxed{\boxed{\boxed{\boxed{}}}}}} \; , \; \overset{k}{\boxed{\boxed{\boxed{}}}} \;)$$

of $SU(N) \times SU(M)$. But for $SU(N/M)$, since supersymmetrization of $(A_1 A_2 \cdots A_n)$ implies symmetrization of the bosons ϕ_{a_i} and antisymmetrization of the fermions ψ_{α_i}, the bosonic indices $(a_1 a_2 \cdots a_{n-k})$ are symmetric but the fermionic indices $(\alpha_{n-k+1} \cdots \alpha_n)$ are antisymmetric. Thus, from (186) we can write the branching rule

$$\overset{n}{\boxed{\boxed{\boxed{\boxed{\boxed{\boxed{\boxed{}}}}}}}} = \sum_{k=0}^{n} (\; \overset{n-k}{\boxed{\boxed{\boxed{\boxed{}}}}} \; , \; \overset{k}{\boxed{\boxed{\boxed{}}}} \;)_{(n-k)/N-k/M}$$

$$\text{for } SU(N + M),$$

$$(187)$$

$$\overset{n}{\boxed{\boxed{\boxed{\boxed{\boxed{\boxed{\boxed{\diagdown}}}}}}}} = \sum_{k=0}^{n} \left(\; \overset{n-k}{\boxed{\boxed{\boxed{\boxed{}}}}} \; , \; \boxed{\boxed{}}\,k \; \right)_{(n-k)/N+k/M}$$

$$\text{for } SU(N/M).$$

The number of terms appearing in the first of these equations is $n + 1$. However, in the second equation some of the Young tableaux

$$\boxed{\boxed{}}\,k$$

for $SU(M)$ vanish if $k > M$. Thus, if $n \leqslant M$ there will be $n + 1$ terms, but if $n > M$ there will be fewer terms.

Note that the *pictures* in (187) are independent of the value of N and M. They are completely determined by the permutation symmetry of the original one-row tableau. Therefore, the comparison of the branching rule for Lie groups and Lie supergroups need not be restricted to groups that have identical subgroups. For example, instead of comparing the branching rules of SU(6/4) to those of SU(10), we may just as well compare them to those of SU(75), since the *pictures* of SU(10) and SU(75) are identical except that Young tableaux for SU(10) with more than ten rows vanish. Similarly, for a given shape of the supertableau the pictures of SU(6/4) are identical with those of any other SU(N/M) except for the illegal SU(N) or SU(M) tableaux that vanish. Thus, in branching rule calculations, we will always consider N and M to be as large as necessary so that, for both SU($N + M$) and SU(N/M), none of the SU(N) or SU(M) tableaux vanish . Thus, the branching rule *pictures* that are obtained will be generic to the tableau and independent of the values of N and M. They will depend only on the numbers m_i and n_i of the dotted and undotted boxes in the original tableau as for examples in Figure 1. After obtaining the branching rule for a *given tableau* specified by m_i and n_i (not for given M and N), we can specialize to any desired values of N and M and eliminate, if necessary, and illegal SU(N) or SU(M) tableau.

A legal SU(N) tableau which contains dotted boxes is specified in Figure 2. Note that the number of dotted *plus* undotted boxes cannot exceed N:

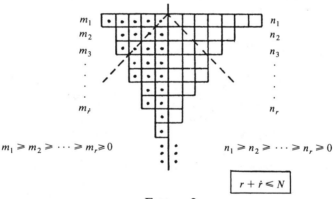

In this way we obtain branching rules for the whole series of SU(N/M) supergroups rather than specific N, M. (187), which is the first such example, was obtained in [**14**].

Next we consider the SU($N + M$) tableau

and SU(N/M) supertableau

which correspond to a tensor $\phi_{(A_1A_2A_3);B_1}$. By specializing the indices $A_i = a_i \oplus \alpha_i$, $B_1 = b_1 \oplus \beta_1$, we obtain the various components. The *independent* components are specified by considering the indices a_i to be *lower* than the indices b_i within SU(N) and both a_i and b_i to be lower than the α_i or β_i within SU($N + M$), when they are allowed to take values $A = 1, 2, \ldots, N + M$. Then we obtain

$$\phi_{(A_1A_2A_3);B_1} = \phi_{(a_1a_2a_3);b_1} \oplus \phi_{(a_1a_2a_3);\beta_1} \oplus \phi_{(a_1a_2)(\alpha_3);b_1} \oplus \phi_{(a_1a_2)(\alpha_3);\beta_1}$$
$$\oplus \phi_{(a_1)(\alpha_2\alpha_3);b_1} \oplus \phi_{(a_1)(\alpha_2\alpha_3);\beta_1} \oplus \phi_{(\alpha_1\alpha_2\alpha_3);\beta_1}. \tag{188}$$

Note that we did not include $\psi_{(\alpha_1\alpha_2\alpha_3);b_1}$, and some others, even though they could appear as possible components of the tensor. This is because of the ordering rule $a_i < b_i < \alpha_i < \beta_i$ which allows us to select the independent components of the tensor *only once*. According to this rule, we cannot allow b_1 to appear in the second row when the first row contains only α's. The component with the symmetries of $\phi_{(\alpha_1\alpha_2\alpha_3);b_1}$ is already counted as seen below.

For SU($N + M$), (188) can be written in terms of SU(N) $\otimes$ SU(M) Young tableaux as

$$\phi_{(a_1a_2a_3);b_1} = \left(\young[⊞⊔]\,,\, 1 \right),$$

$$\phi_{(a_1a_2a_3);\beta_1} = \left(\square\square\square\,,\, \square \right),$$

$$\phi_{(a_1a_2)(\alpha_3);b_1} = \left(\young[⊞⊔]\,,\, \square \right),$$

$$\phi_{(a_1a_2)(\alpha_3);\beta_1} = \left(\square\square\,,\, \square\square \right) + \left(\square\square\,,\, \young[⊟] \right), \tag{189}$$

$$\phi_{(a_1)(\alpha_2\alpha_3);b_1} = \left(\young[⊟]\,,\, \square\square \right)$$

$$\phi_{(a_1)(\alpha_2\alpha_3);\beta_1} = \left(\square\,,\, \young[⊞⊔] \right) + \left(\square\,,\, \square\square \right),$$

$$\phi_{(\alpha_1\alpha_2\alpha_3);\beta_1} = \left(1\,,\, \young[⊞⊔] \right).$$

Note that $\phi_{(a_1a_2)(\alpha_3);\beta_1}$ has two irreducible SU(N) components since the α_3 index in the first row and the β_1 index in the second row are not forced to be in any symmetry relation relative to each other. Thus, we obtain the two irreducible pieces, because for SU(M),

$$\square \otimes \square = \square\square \oplus \young[⊟]\,. \tag{190}$$

Similarly, $\phi_{(a_1)(\alpha_2\alpha_3);\beta_1}$ has two irreducible components since

$$\square\square \otimes \square = \young[⊞⊔] \oplus \square\square\square\,. \tag{191}$$

The second piece in (191) corresponds to the component $\phi_{(\alpha_1\alpha_2\alpha_3);b_1}$ mentioned above.

Thus, for $SU(N + M) \supset SU(N) \otimes SU(M) \otimes U(1)$ the branching rule for this tableau is

$$
\begin{aligned}
\text{[tableau]} = &\left(\text{[tableau]}, 1 \right)_{4/N} \oplus \left(1, \text{[tableau]} \right)_{-4/M} \\
&\oplus \left(\text{[tableau]}, \text{[tableau]} \right)_{3/N-1/M} \oplus \left(\text{[tableau]}, \text{[tableau]} \right)_{1/N-3/M} \\
&\oplus \left(\text{[tableau]}, \text{[tableau]} \right)_{3/N-1/M} \oplus \left(\text{[tableau]}, \text{[tableau]} \right)_{1/N-3/M} \\
&\oplus \left(\text{[tableau]}, \text{[tableau]} \right)_{2/N-2/M} \oplus \left(\text{[tableau]}, \text{[tableau]} \right)_{2/N-2/M} \\
&\oplus \left(\text{[tableau]}, \text{[tableau]} \right)_{2/N-2/M}.
\end{aligned} \tag{192}
$$

Note that in the final result the *pictures* are symmetric under the interchange of $SU(M)$ and $SU(N)$, as they should be, since the permutation symmetry of the original tableau does not distinguish between $SU(N)$ and $SU(M)$ indices. Identical results would be obtained by considering the a_i, b_i indices to be higher compared to the α_i indices. The $SU(N) \leftrightarrow SU(M)$ interchangeability of the branching rule reflects this fact. Thus, for every irreducible component (X, Y) there exists another irreducible component (Y, X) where X and Y represent the *pictures* of Young tableaux.

The same reasoning can be applied step by step to the $SU(N/M)$ group. The only difference is that whenever the α_i's were symmetrized within $SU(N + M)$ they should be antisymmetrized within $SU(N/M)$ and vice-versa. This is required by the supersymmetrization indicated by the supertableau. The a_i's have the same permutation properties in both $SU(N + M)$ and $SU(N/M)$ as in the previous example. Thus, for $SU(N/M)$, (189) will be modified by changing every $SU(M)$ row into a column and vice-versa. This means that every irreducible component (X, Y) that appeared for $SU(N + M)$ will have a counterpart $(X, \tilde{Y})$ for $SU(N/M)$ where $\tilde{Y}$ is an $SU(M)$ Young tableaux reflected along the diagonal relative to Y. Thus, the analogue of (192) for $SU(N/M) \supset SU(N) \otimes SU(M) \otimes U(1)$ becomes

$$
\begin{aligned}
\text{[tableau]} = &\left(\text{[tableau]}, 1 \right)_{4/N} \oplus \left(1, \text{[tableau]} \right)_{4/M} \oplus \left(\text{[tableau]}, \text{[tableau]} \right)_{3/NM+1/M} \\
&\oplus \left(\text{[tableau]}, \text{[tableau]} \right)_{1/N+3/M} \oplus \left(\text{[tableau]}, \text{[tableau]} \right)_{3/N+1/M} \oplus \left(\text{[tableau]}, \text{[tableau]} \right)_{1/N+3/M} \\
&\oplus \left(\text{[tableau]}, \text{[tableau]} \right)_{2/N+2/M} \oplus \left(\text{[tableau]}, \text{[tableau]} \right)_{2/N+2/M} \\
&\oplus \left(\text{[tableau]}, \text{[tableau]} \right)_{2/N+2/M}.
\end{aligned} \tag{193}
$$

Note that the $1/M$ pieces in the U(1)'s have switched signs. Furthermore the pictures are now symmetric under the SU(N) $\leftrightarrow$ SU(M) interchange only after being reflected also the diagonal. That is, for every irreducible component ($X, \tilde{Y}$) there exists also ($Y, \tilde{X}$).

We use the same methods as above for tableaux containing n_1 boxes in the first row and n_2 boxes in the second row, where $n_1 \geqslant n_2 \geqslant 0$. The result for SU($N + M$) is

$$
\begin{matrix} n_1 \\ n_2 \end{matrix}
$$

$$
(194)
$$

$$
= \sum_{k_2=0}^{n_2} \sum_{j=0}^{k_2} \sum_{k_1=k_2-j}^{n_1-n_2+k_2-j} \left(\begin{matrix} n_1 - k_1 \\ n_2 - k_2 \end{matrix} \quad , \quad \begin{matrix} k_1 + j \\ k_2 - j \end{matrix} \right)
$$

while for SU(N/M) we only need to reflect the SU(M) tableau and obtain

$$
\begin{matrix} n_1 \\ n_2 \end{matrix}
$$

$$
= \sum_{k_2=0}^{n_2} \sum_{j=0}^{k_2} \sum_{k_1=k_2-j}^{n_1-n_2+k_2-j} \left[\begin{matrix} n_1 - k_1 \\ n_2 - k_2 \end{matrix} \quad , \quad \begin{matrix} k_1+j & k_2-j \end{matrix} \right].
$$

$$
(195)
$$

The U(1) charges were not indicated for lack of space, but they are computed by counting the number of boxes for SU(N) and SU(M) tableaux and given as $(n_1 + n_2 - k_1 - k_2)/N \mp (k_1 + k_2)/M$, with the upper sign for SU($N + M$) and the lower one for SU(N/M). The number of independent irreducible terms appearing in the sum of (194), (195) is easily computed to be

$$
\tfrac{1}{2}(n_1 - n_2 + 1)(n_2 + 1)(n_2 + 2). \tag{196}
$$

Some of these terms may vanish if N or M are too small and the tableau becomes illegal. Equations (194), (195) reduce to (18) for the special case $n_2 = 0$.

After obtaining the result for one and two rows, it is immediate to arrive at the branching rule for one and two columns, simply by reflecting each tableau along the diagonal. These and a few other cases not containing dotted boxes are summarized in Table 1.

Returning to tableaux containing dotted boxes, we begin with the fundamental contravariant representation, $\phi^A = \phi^a \oplus \psi^a$, which may be written as

$$\boxed{\cdot} = (\boxed{\cdot}, 1)_{-1/N} \oplus (1, \boxed{\cdot})_{1/M} \quad \text{for } SU(N+M),$$
$$\boxed{\cdot} = (\boxed{\cdot}, 1)_{-1/N} \oplus (1, \boxed{\cdot})_{-1/M} \quad \text{for } SU(N/M). \tag{197}$$

TABLE 1

SU(N+M) IRREP / SU(N/M)	SU(N) ⊗ SU(M) IRREP	U(1) CHARGE	LIMITS ON NUMBERS OF BOXES	TOTAL NUMBER OF TERMS IF ALL TABLEAUX ARE LEGAL
[tableau, length n]	[tableaux: $n-k$, k]	$\dfrac{n-k}{N} + \dfrac{k}{M}$	$0 \le k \le n$	$n+1$
[tableau, column length n]	[tableaux: column $n-k$, column k] / [column $n-k$, k]	$\dfrac{n-k}{N} + \dfrac{k}{M}$	$0 \le k \le n$	$n+1$
[tableaux n_1, n_2]	$\left[\begin{smallmatrix} n_1-k_1 \\ n_2-k_2 \end{smallmatrix}\ \Box\ , \ \Box\begin{smallmatrix}k_1+j\\k_2-j\end{smallmatrix}\right]$	$\dfrac{n_1+n_2-k_1-k_2}{N} + \dfrac{k_1+k_2}{M}$	$0 \le k_2 \le n_2$, $\;0 \le j \le k_2$, $\;k_2-j \le k_1 \le n_1-n_2+k_2-j$	$\tfrac{1}{2}(n_1-n_2+1)(n_2+1)\times(n_2+2)$
[tableaux n_1, n_2]	$\left[n_1-k_1\begin{smallmatrix}n_2-k_2\\ \\k_1+j\end{smallmatrix}, \ \Box\begin{smallmatrix}k_2-j\end{smallmatrix}\right]$	$\dfrac{n_1+n_2-k_1-k_2}{N} + \dfrac{k_1+k_2}{M}$	$0 \le k_2 \le n_2$, $\;0 \le j \le k_2$, $\;k_2-j \le k_1 \le n_1-n_2+k_2-j$	$\tfrac{1}{2}(n_1-n_2+1)(n_2+1)(n_2+2)$
[tableau, n, m]	[tableaux: $n-k$, $m-\ell$, $k+j$, $\ell-j$]	$\dfrac{n+m-k-\ell}{N} + \dfrac{k+\ell}{M}$	$j = 0,1$; $\;j \le \ell \le m$; $\;1-j-\delta_{\ell,0} \le k \le n-1+(1-j)\delta_{\ell,m}$; $\;\delta_{\ell_i} = \text{Kronecker delta}$	$2nm+n-m+1$
[tableau, n, m]	[tableaux: $n-k_1$, $n-k_2$, ..., $n-k_m$, k_m, ..., k_1]	$\dfrac{nm}{N} - \left[\dfrac{1}{N} + \dfrac{1}{M}\right]\displaystyle\sum_{i=1}^{m} k_i$	$0 \le k_1 \le k_2 \ldots \le k_m \le n$	$\dfrac{(n+m)!}{n!\,m!}$

The next simplest case is the adjoint representation $\phi_A{}^B$ which is a traceless matrix for $SU(N + M)$ and supertraceless for $SU(N/M)$. Specializing the indices $A = 2 \oplus \alpha$, $B = b \oplus \beta$, we obtain the components

$$\phi_A{}^B = \phi_a{}^b \oplus \phi_a{}^\beta \oplus \phi_\alpha{}^b \oplus \phi_\alpha{}^\beta. \tag{198}$$

The pieces $\phi_a{}^b$ and $\phi_\alpha{}^\beta$ are not irreducible with respect to $SU(N)$ and $SU(M)$ respectively, since we have not yet insured that they are traceless. Thus

$$\phi_a{}^b = \tilde{\phi}_a{}^b + \frac{1}{N}\delta_a{}^b\phi = (\boxdot, 1) \oplus (1,1),$$
$$\phi_\alpha{}^\beta = \tilde{\phi}_\alpha{}^\beta \mp \frac{1}{M}\delta_\alpha{}^\beta\phi = (1,\boxdot) \oplus (1,1), \tag{199}$$

where the singlet part $\phi = (1,1)$ is identical in both pieces so that the tracelessness (supertracelessness) condition on $\phi_A{}^B$ is satisfied. The result for $SU(N + M)$ is

$$\boxdot = (\boxdot, 1)_0 \oplus (1, \boxdot)_0 \oplus (1,1)_0$$
$$\oplus (\square, \boxdot)_{1/N+1/M} \oplus (\boxdot, \square)_{-1/N-1/M}, \tag{200}$$

while for $SU(N/M)$ we have

$$\boxed{\diagdown} = (\boxdot, 1)_0 \oplus (1, \boxdot)_0 \oplus (1, 1_0)$$
$$\oplus (\square, \boxdot)_{1/N-1/M} \oplus (\boxdot, \square)_{-1/N+1/M}. \tag{201}$$

Next we consider the tensor $\phi_{A_1 A_2 \cdots A_n}{}^{B_1 B_2 \cdots B_m}$ with both lower and upper indices symmetrized (or supersymmetrized) and satisfying the trace (or supertrace) condition. Specializing the indices, we have

$$\phi_{A_1 A_2 \cdots A_n}{}^{B_1 B_2 \cdots B_m} = \sum_{k=0}^{n} \sum_{l=0}^{m} \phi_{(a_1 a_2 \cdots a_{n-k})(\alpha_{n-k+1} \cdots \alpha_n)}{}^{(b_1 b_2 \cdots b_{n-l})(\beta_{m-l+1} \cdots \beta_m)}. \tag{202}$$

The $U(1)$ quantum numbers can be calculated by assigning $1/N$ for each a_i, $-1/N$ for each b_i, $\mp 1/M$ for each α_i, and $\pm 1/M$ for each β_i, where the upper sign is for $SU(N + M)$ and the lower signs for $SU(N/M)$. The various terms in the sum are, in general, reducible with respect to $SU(N) \otimes SU(M)$. For example, for $n = 2$ and $m = 2$,

$$\phi_{(a_1 a_2)}{}^{(b_1 b_2)} = (\boxed{\cdot \cdot}, 1) \oplus (\boxed{\cdot}, 1) \oplus (1,1),$$
$$\phi_{(a_1 \alpha_2)}{}^{(b_1 b_2)} = (\boxed{\cdot \cdot}, \square) \oplus (\boxdot, \square), \quad \text{etc.} \tag{203}$$

Note that, for each reduction which is achieved by using a Kronecker delta $\delta_a{}^b$ (or $\delta_\alpha{}^\beta$), we get to eliminate one dotted and one undotted box from the picture of an $SU(N)$ (or an $SU(M)$) tableau. Just as the ϕ in (199), we must be aware that the tracelessness of the original tensor $\phi_{A_1 \cdots A_n}^{B_1 \cdots B_m}$ imposes that some of the pieces in the various traces are identical and should not be counted more than once. To insure this feature we count only the traces calculated by contracting with the $SU(M)$ $\delta_\alpha^{\ \beta}$, and ignore those obtained with $\delta_a^{\ b}$ since they are the same

TABLE 2

SU(N+M) IRREP / SU(N/M)	SU(N) ⊗ SU(M) IRREP	U(1) CHARGE	LIMITS ON NUMBERS OF BOXES	TOTAL NUMBER OF TERMS IF ALL TABLEAUX ARE LEGAL
(tableaux, labels m, n)	$\left[\begin{array}{cc} m-\ell & n-k \\ \end{array} , \begin{array}{cc} \ell-i & k-i \\ \end{array} \right]$ $\left[\begin{array}{cc} m-\ell & n-k \\ \end{array} , \begin{array}{cc} \ell-i & k-i \\ \end{array} \right]$	$\dfrac{n-k-m+\ell}{N} + \dfrac{k-\ell}{M}$	$0 \le k \le n$ $0 \le \ell \le m$ $0 \le i \le \min(k,\ell)$	$\frac{1}{6}(m+1)(m+2)(3n-m+3)$ if $n \ge m$ $\frac{1}{6}(n+1)(n+2)(3m-n+3)$ if $n \le m$
(tableaux, labels m, n)	$\left[\begin{array}{cc} m-\ell & n-k & \ell-i & k-i \\ \end{array} \right]$ $\left[\begin{array}{cc} m-\ell & n-k & \ell-i & k-i \\ \end{array} \right]$	$\dfrac{n-k-m+\ell}{N} + \dfrac{k-\ell}{M}$	$0 \le k \le n$ $0 \le \ell \le m$ $0 \le i \le \min(k,\ell)$	$\frac{1}{6}(m+1)(m+2)(3n-m+3)$ if $n \ge m$ $\frac{1}{6}(n+1)(n+2)(3m-n+3)$ if $n \le m$
(tableaux, labels n, m)	$\left[m-\ell \begin{array}{cc} n-k & k-i \\ \ell-i & \end{array} \right]$ $\left[m-\ell \begin{array}{cc} n-k & \ell-i \\ & k-i \end{array} \right]$	$\dfrac{n-k-m+\ell}{N} + \dfrac{k-\ell}{M}$	$i = 0,1$ $i \le k \le n$ $i \le \ell \le m$	$2mn+m+n+1$
(tableaux, labels m, n)	$\left[\begin{array}{cc} m-\ell & \ell-i \\ n-k & k-i \end{array} \right]$ $\left[\begin{array}{cc} m-\ell & k-i \\ n-k & \ell-i \end{array} \right]$	$\dfrac{n-k-m+\ell}{N} + \dfrac{k-\ell}{M}$	$i = 0,1$ $i \le k \le n$ $i \le \ell \le m$	$2mn+m+n+1$
(tableaux, labels m_1, n_1, n_2)	$\left[\begin{array}{ccc} m_1-\ell_1 & n_1-k_1 & \ell_1-i_1-i_2 \ k_1+j-i_1 \\ n_2-k_2 & & k_2-j-i_2 \end{array} \right]$ $\left[\begin{array}{ccc} & \ell_1-i_1-i_2 \ k_1+j-i_1 \\ m_1-\ell_1 & n_1-k_1 & k_2-j-i_2 \\ n_2-k_2 & \end{array} \right]$	$\dfrac{n_1+n_2-m_1}{N}$ $- \left[\dfrac{1}{N} - \dfrac{1}{M}\right][k_1+k_2-\ell_1]$	$0 \le \ell_1 \le m_1$ $0 \le k_2 \le n_2$ $0 \le j \le k_2$ $k_2-j \le k_1 \le n_1-n_2+k_2-j$ $0 \le i_2 \le \min(k_2-j, \ell_1)$ $0 \le i_1 \le \min(k_1-k_2+2j, \ell_1-i_2)$	
(tableaux, labels m_1, n_1, n_2)	$\left[m_1-\ell_1 \begin{array}{ccc} n_1-k_1 & \ell_1-i_1-i_2 \ k_1+j-i_1 \\ n_2-k_2 & k_2-j-i_2 \end{array} \right]$ $\left[m_1-\ell_1 \begin{array}{ccc} n_1-k_1 & \ell_1-i_1-i_2 \ k_1+j-i_1 \\ n_2-k_2 & k_2-j-i_2 \end{array} \right]$	$\dfrac{n_1+n_2-m_1}{N}$ $- \left[\dfrac{1}{N} + \dfrac{1}{M}\right][k_1+k_2-\ell_1]$	$i_1 = 0,1 ; \ i_2 = 0,1$ $i_1+i_2 \le \ell_1 \le m_1$ $i_2 \le k_2 \le n_2$ $0 \le j \le k_2-i_2$ $i_1-i_2+k_2-j \le k_1 \le n_1-n_2+k_2-j$	$\frac{1}{2}(n_2+1) \times$ $\left[(3m_1n_2+2)(n_1-n_2) + 2m_1(2n_1+1)+2 \right]$ $m_1 \ge 1$

From I. Bars and A. Baha Balentekin, *op. cit.* Reprinted by permission.

ones. With these conditions, we arrive at the SU($N + M$) branching rule

$$= \sum_{k=0}^{n} \sum_{l=0}^{m} \sum_{i=0}^{\min(k,l)} \left(\boxed{m-l \quad n-k} , \boxed{l-i \quad k-i} \right). \tag{204}$$

The sum over i takes care of the traces and produces the pieces similar to the $\phi \sim (1, 1)$ of (199). The U(1) charge, which depends only on the number of boxes, is given as $(n - k - m + l)/N - (k - l)/M$.

For SU(N/M) the reasoning is identical, however, the α_i's and β_i's should now be antisymmetrized as opposed to being symmetrized in the previous case. Therefore, the SU(M) tableaux should be changed relative to the previous case by substituting columns instead of rows. Otherwise, every step can be repeated to obtain

$$= \sum_{k=0}^{n} \sum_{l=0}^{m} \sum_{i=0}^{\min(k,l)} \left(\boxed{m-l \quad n-k} , \; l-i \; \boxed{} \; k-i \right). \tag{205}$$

We see that in going from SU($N + M$) to SU(N/M), the SU(M) tableaux get reflected independently for the dotted and undotted boxes along their respective diagonals. The diagonals are shown in Figure 2.

Our method should be quite clear to the reader by now. Without giving any more details, we list our results for a few tableaux containing dotted boxes in Table 2. We emphasize that these correspond to arbitrarily large representations of arbitrarily large groups. Together with Table 1 we expect that these concrete results should be quite sufficient for a variety of physical applications that we can now foresee. More complicated cases can be worked out, if necessary, with the same methods.

16. The branching rules for SU($N_1 + N_2/M_1 + M_2$) $\supset$ SU(N_1/M_1) $\otimes$ SU(N_2/M_2) $\otimes$ U(1). This branching is again obtained from that of SU($N + M$) $\supset$ SU(N) $\otimes$ SU(M) $\otimes$ U(1) by a reinterpretation of the boxes in the tableaux. Let us first identify the fundamental representation $\phi_A = \phi_a \oplus \phi_\alpha$ as follows: ϕ_a contains N_1 bosons and M_1 fermions and belongs to the fundamental representation $\phi_a \sim \square$ of SU(N_1/M_1); similarly, ϕ_α contains N_2 bosons and M_2 fermions and belongs to the fundamental representation $\phi_\alpha \sim \square$ of SU(N_1/M_2).

The U(1) generator, which is a $(N_1 + M_1 + N_2 + M)$-dimensional diagonal matrix in the fundamental representation, is identified as

$$
\left|
\begin{array}{c|c}
\dfrac{1}{N_1 - M_1} & 0 \\
\hline
0 & \dfrac{-1}{N_2 - M_2}
\end{array}
\right|
\begin{array}{l}
\Big\} \;\; N_1 + M_1 \\[1em]
\Big\} \;\; N_2 + M_2
\end{array}
\cdot
\tag{206}
$$

Thus, for each index a or α we obtain the U(1) charges $1/(N_1 - M_1)$ or $-1/(N_2 - M_2)$ respectively. Therefore, $\phi_A = \phi_a \oplus \phi_\alpha$ may be written in terms of tableaux as

$$
\boxed{/} = (\boxed{/}, 1)_{1/(N_1 - M_1)} \oplus = (1, \boxed{/})_{-1/(N_2 - M_2)}.
\tag{207}
$$

Note the formal similarity to (185) for $SU(N + M)$ except that *every box* is replaced by a slashed box, and the values of the U(1) charges are computed by different assignments to a and α, as explained above.

The completely supersymmetric tensor $\phi_{(A_1 A_2 \cdots A_n)}$ can be decomposed by specializing each index $A_i = a_i \oplus \alpha_i$, just as in (186). However, the meaning of $\phi_{(a_1 a_2 \cdots a_n)}$, etc., now differs from the $SU(N + M)$ case in that the indices $(a_1 a_2 \cdots)$ or $(\alpha_1 \alpha_2 \cdots)$ are supersymmetrized rather than simply symmetrized. Therefore, (187) now gets replaced by

$$
\overset{n}{\boxed{//////}} = \sum_{k=0}^{n} (\overset{n-k}{\boxed{///}} , \overset{k}{\boxed{//}}),
\tag{208}
$$

with the U(1) charges given by $[(n - k)/(N_1 - M_1) - k/(N_2 - M_2)]$.

The analysis is the same for any other supertableau and the result for the new branching rule is obtained from the known cases of $SU(N + M)$ by simply replacing every $SU(N)$ or $SU(M)$ box $\square$ by slashed boxes $\boxed{/}$ belonging to $SU(N_1/M_1)$ or $SU(N_2/M_2)$ respectively. Similarly the U(1) is computed by replacing every $1/N$ or $-1/M$ in the old expressions by $1/(N_1 - M_1)$ or $-1(N_2 - M_2)$ respectively. Therefore, all the results listed in Tables 1 and 2 are directly generalized to the new branching.

17. Comments on branching rules. We have established a one-to-one correspondence between the branching rules $SU(N + M) \supset SU(N) \otimes SU(M) \otimes U(1)$ and $SU(N/M) \supset SU(N) \otimes SU(M) \otimes U(1)$ as well as

$$
SU((N_1 + N_2)/(M_1 + M_2)) \supset SU(N_1/M_1) \otimes SU(N_2/M_2) \otimes U(1).
$$

Some concrete examples which we expect to be sufficient for most physical applications have been explicitly worked out and listed in Tables 1 and 2. More complicated cases can be analyzed with the methods given here.

For SU($N + M$) branching rules, there are useful lists available in the literature [22]. Our SU($N + M$) results are in complete agreement with these known cases. One virtue of our approach is that we are not limited by large dimensions of representations or groups. Thus, in our Tables 1 and 2 one finds large dimensions not covered in the extensive lists of [23].

In making comparisons with these lists one must be aware that some Young tableau for SU(N) or SU(M) become illegal (see Figure 2) and vanish if either N or M are too small. One may use the available SU($N + M$) lists to derive additional practical SU(N/M) results not covered in this paper explicitly, provided N and M are large enough to insure no tableau vanishes. Useful branching rules can be obtained from [23] or similar lists by noting the following general observations which follow from our analysis above.

Consider an arbitrary Young tableau T for SU($N + M$) and the SU(N) $\otimes$ SU(M) $\otimes$ U(1) branching rule

$$T = \sum \bigoplus (X, Y), \qquad (209)$$

where X and Y denote Young tableaux for SU(N) and SU(M) respectively. In general, T, X, Y contain both dotted and undotted boxes. In tables such as [13], Dynkin indices are used to label a representation. They must be converted to Young tableau notation in order to apply our method below.

For every SU($N + M$) Young tableau T, as in Figure 1, we can define an SU(N/M) supertableau $\mathbf{T}$, with identical numbers n_i, m_i for its rows, except that every box $\square$ or $\boxdot$ is replaced by a slashed box $\boxslash$ or $\boxbslash$. We may then consider the branching rule for SU(N/M) $\rightarrow$ SU(N) $\otimes$ SU(M) $\otimes$ U(1) as

$$\mathbf{T} = \sum \bigoplus (X, \tilde{Y}), \qquad (210)$$

where $\tilde{Y}$ is the reflection of Y along its diagonals. The diagonals are shown in Figure 2. Similarly, the branching rule for

$$\text{SU}((N_1 + N_2)/(M_1 + M_2)) \supset \text{SU}(N_1/M_1) \otimes \text{SU}(N_2/M_2) \otimes \text{U}(1)$$

can be written as

$$\mathbf{T} = \sum \bigoplus (\mathbf{X}, \mathbf{Y}), \qquad (211)$$

where $\mathbf{X}$ and $\mathbf{Y}$ are the supertableaux analogous to X and Y. (210) and (211) follow from (209) if one uses the tensor language $\phi_{A_1 A_2 \cdots}^{\ B_1 B \cdots}$ and specializes each index $A_i = a_i \oplus \alpha_i$, etc. It is necessary to consider the meaning of supersymmetrization in relation to ordinary symmetrization, and then extracting irreducible SU(N) and SU(M) components. These statements can be understood by following the examples in Tables 1, 2.

Equation (209) contains the terms $(1, T)$ and $(T, 1)$ where T is the SU(N) and SU(M) representation with the most boxes that could appear in the branching rule. The *pictures* that represent the decomposition $T = \Sigma \oplus (X, Y)$ are independent of N or M. Therefore, we will assume that N and M are large enough so that the Young tableau T or the reflection from its diagonals $\tilde{T}$ does not vanish for SU(N) or SU(M). This insures that every (X, Y) or $(X, \tilde{Y})$ that could appear in the sum does not vanish for SU($N + M$) or SU(N/M). After obtaining the branching rules for such large N, M, we can apply the result to smaller N, M, as it may be necessary in some practical application. Then, we only need to eliminate the illegal SU(N) or SU(M) tableaux according to Figure 2.

The following statements hold for any decomposition noted in (209)–(211), and can be used as a check in any calculation.

(1) The total number of undotted *minus* dotted boxes is identical in every term of the sum and equal to the same quantity for T or **T**.

(2) For a term (X, Y) that appears in (209) there should be another term (Y, X) provided it does not vanish according to Figure 2. For (210) this implies that for every term $(X, \tilde{Y})$ there should be a $(Y, \tilde{X})$, while in (211) for every $(\mathbf{X}, \mathbf{Y})$ there should exist a $(\mathbf{Y}, \mathbf{X})$.

(3) The maximum number of rows and columns that can appear in any (X, Y) or $(X, \tilde{Y})$ or $(\mathbf{X}, \mathbf{Y})$ as well as the general shape of these Young tableaux are predetermined by the number of rows and columns and general shape of T or $\mathbf{T}$. See the examples in Tables 1, 2.

(4) The dimension of **T** or T on the left-hand side of these equations should match with the sum of the dimensions on the right-hand side. For this purpose one can use the practical dimension formulas for the numbers of bosons and fermions developed in [14] and in previous sections. On the right-hand side of (210) a fermion is obtained when the SU(M) representation $\tilde{Y}$ contains an odd total number of dotted *plus* undotted boxes (independent of X). (This is a fermion in a class I representation. For class II we demand an odd number of boxes in X rather than in Y. While in mixed (class I)-(class II) representations, the roles of bosons and fermions may be interchanged. The representation of the group element is independent from the class.)

18. Branching rules SU($(N_1 N_2 + M_1 M_2)/(N_1 M_2 + M_1 N_2)$) → SU($N_1/M_1$) ⊗ SU($N_2/M_2$). This branching is analogous to SU(NM) → SU(N) × SU(M), and as in the previous section we will use such analogies. Consider the fundamental bases of SU(N_1/M_1), $\Phi_{A_1} = (\phi_{a_1}, \psi_{\alpha_1})$ ($a_1 = 1, 2, \ldots, N_1$; $\alpha_1 = 1, 2, \ldots, M_1$) and of SU($N_2/M_2$), $\Phi'_{A_2} = (\phi'_{a_2}, \psi'_{\alpha_2})$ with ($a_2 = 1, 2, \ldots, N$; $\alpha_2 = 1, 2, \ldots, M_2$). Take their product and count the number of bosons and fermions.

bosons: $\phi_{a_1}\phi'_{a_2} \oplus \psi_{\alpha_1}\psi'_{\alpha_2} \to N_1 N_2 + M_1 M_2,$

fermions: $\phi_{a_1}\psi'_{\alpha_2} \oplus \psi_{\alpha_1}\phi'_{a_2} \to N_1 M_2 + M_1 N_2.$ (212)

We assign these bosonic and fermionic states to the fundamental basis of $SU((N_1 N_2 + M_1 M_2)/(N_1 M_2 + N_2 M_1))$. Then we can write the branching rule

$$\square = (\square, \square), \quad (213)$$

which simply expresses what we have just described. Here the first box describes $SU(N_1/M_1)$ and the second one describes $SU(N_2/M_2)$.

To obtain the branching rules for higher-dimensional representations we take direct products and supersymmetrize. For example,

$$\square\square = [(\square, \square) \otimes (\square, \square)]_S, \quad (214)$$

where S stands for symmetric product under superpermutations. A superpermutation is one in which wavefunctions are permuted but indices kept fixed in the same order, as in (102). Another way of describing it is as a permutation of the indices in a tensor, provided $\pm$ signs are inserted, depending on whether a pair of bosonic or fermionic indices are permuted, as in (104). In any case, superpermutations are built-in in our notation of supertableaux. The pictures that describe a symmetric product under superpermutations, such as $\square\square$, look the same as the ordinary symmetric Young tableau except for the slashes which summarize the supersymmetrization rule. This notation, of course, extends to the direct product. The result will look the same as in the case of $SU(NM) \to SU(N) \times SU(M)$ except for the slashes through the boxes which automatically take into account the superpermutation rule. Thus, we obtain

$$\square\square = (\square\square, \square\square) + \left(\begin{array}{c}\square\\\square\end{array}, \begin{array}{c}\square\\\square\end{array}\right), \quad (215)$$

where each term in the sum is symmetric according to superpermutations. Similarly, for the antisymmetric product the result is

$$\begin{array}{c}\square\\\square\end{array} = \left(\begin{array}{c}\square\\\square\end{array}, \square\square\right) + \left(\square\square, \begin{array}{c}\square\\\square\end{array}\right), \quad (216)$$

where each term is antisymmetric under superpermutations.

It is now straightforward to obtain the branching for higher representations by simply copying the pictures that would apply to $SU(NM) \to SU(N) \times SU(M)$, except for the slashes. Thus, we write a few cases:

$$\square\square\square = (\square\square\square, \square\square\square) + \left(\boxplus, \boxplus\right) + \left(\begin{array}{c}\square\\\square\\\square\end{array}, \begin{array}{c}\square\\\square\\\square\end{array}\right), \quad (217)$$

$$\begin{array}{c}\square\\\square\end{array} = \left(\square\square\square, \begin{array}{c}\square\\\square\\\square\end{array}\right) + \left(\boxplus, \boxplus\right) + \left(\begin{array}{c}\square\\\square\\\square\end{array}, \square\square\square\right), \quad (218)$$

$$\boxed{}\,= (\boxed{}\,,\boxed{}\,) + \left(\boxed{}\,,\boxed{}\,\right) + \left(\boxed{}\,,\boxed{}\,\right)$$

$$+ \left(\boxed{}\,,\boxed{}\,\right) + \left(\boxed{}\,,\boxed{}\,\right), \tag{219}$$

$$\boxed{}\,= (\boxed{}\,,1) + (1,\boxed{}\,) + (\boxed{}\,,\boxed{}\,). \tag{220}$$

In general, for a completely symmetric row we have

$$\overset{n}{\underbrace{\boxed{}}} = \sum_{\gamma} (\mathbf{Y}_n, \mathbf{Y}_n), \tag{221}$$

where on the right-hand side both entries have identical-looking super-tableaux $\mathbf{Y}_n$ with n boxes in each, and the sum extends over *all possible n-box tableaux*. Similarly for a column we have

$$n\,\boxed{} = \sum_{\gamma} (\mathbf{Y}_n, \tilde{\mathbf{Y}}_n), \tag{222}$$

where the second entry $\tilde{\mathbf{Y}}_n$ is reflected along the diagonal relative to the first entry $\mathbf{Y}_n$, and the sum extends over all supertableaux containing n-boxes.

Of course, as in the case of $\mathrm{SU}(N/M) \to \mathrm{SU}(N) \times \mathrm{SU}(M) \times \mathrm{U}(1)$, available tables [23] for ordinary groups can be used to arrive at the branching of more complicated supertableaux, provided that the result is first expressed in terms of Young tableaux, and *one insures that M, N were large enough so that none of the pictures that would have entered have been eliminated*. After this step, by simply replacing each box by a superbox we arrive at the correct branching for the supergroup. The pictures that describe such branchings, provided they are complete, are inherent to the tableau and do not depend on the size of the group (i.e., N_1, M_1, N_2, M_2) so that the branchings for arbitrarily large supergroups and representations are readily obtained in this way.

19. Kac-Dynkin diagrams and supertableaux. We begin by recalling the relation between ordinary Young tableaux and Dynkin diagrams. Any of the states described by a Young tableau is assigned a weight or a collection of "charges". The weight is given by the eigenvalues of the commuting Cartan generators (commuting charge operators) on that state. In the fundamental representation of $\mathrm{SU}(N)$ the $N - 1$ commuting generators are chosen as

$$
H_1 = \begin{pmatrix} 1 & & & & & \\ & -1 & & & & \\ & & 0 & & & \\ & & & 0 & & \\ & & & & 0 & \\ & & & & & \ddots \end{pmatrix},
$$

$$
H_2 = \begin{pmatrix} 0 & & & & & \\ & 1 & & & & \\ & & -1 & & & \\ & & & 0 & & \\ & & & & 0 & \\ & & & & & \ddots \end{pmatrix},
$$

$$
H_3 = \begin{pmatrix} 0 & & & & & \\ & 0 & & & & \\ & & 1 & & & \\ & & & -1 & & \\ & & & & 0 & \\ & & & & & \ddots \end{pmatrix},
$$

$$
\vdots
$$

$$
H_{N-1} = \begin{pmatrix} 0 & & & & & \\ & 0 & & & & \\ & & \ddots & & & \\ & & & 0 & & \\ & & & & 1 & \\ & & & & & -1 \end{pmatrix}.
$$

(223)

Then, their eigenvalues on the states $\Box \sim \phi_i$, $i = 1, 2, \ldots, N$, are computed via matrix multiplication. For example, if we take the basis

$$
\boxed{i} \sim |i\rangle \sim \begin{pmatrix} 0 \\ 0 \\ \vdots \\ 1 \\ 0 \\ \vdots \\ 0 \end{pmatrix} \leftarrow i\text{th } entry,
$$

(224)

then

$$H_l |i\rangle \sim \sum_j (H_l)_i^j |j\rangle, \tag{225}$$

immediately yields the eigenvalues. Some examples are given:

$$\begin{array}{llll}
H_1 |1\rangle = 1|1\rangle, & H_2 |1\rangle = 0|1\rangle, & H_3 |1\rangle = 0|1\rangle, & \text{etc.,} \\
H_1 |2\rangle = -1|2\rangle, & H_2 |2\rangle = 1|2\rangle, & H_3 |2\rangle = 0|2\rangle, & \text{etc.,} \\
H_1 |3\rangle = 0|3\rangle, & H_2 |3\rangle = -1|3\rangle, & H_3 |3\rangle = 1|a\rangle, & \text{etc.}
\end{array} \tag{226}$$

So, for any state, a collection of eigenvalues of H_e, such as $(h_1, h_2, \ldots, h_{N-1})$, can be given. Thus, we write the weights corresponding to the states of the fundamental representation

$$\begin{aligned}
|1\rangle &\to (1, 0, 0, 0, \ldots, 0), \\
|2\rangle &\to (-1, 1, 0, 0, \ldots, 0), \\
|3\rangle &\to (0, -1, 1, 0, \ldots, 0), \\
&\vdots \\
|N\rangle &\to (0, 0, 0, \ldots, 0, -1).
\end{aligned} \tag{227}$$

The weights are given an order such that the eigenvalues of H_1 are specified first, next the eigenvalues of H_2, etc., all the way to H_{N-1}.

The highest weight is the one that has the largest *positive* eigenvalues first relative to H_1, then H_2, then H_3, etc. Thus, in the fundamental representation the highest weight is

$$\square \to (1, 0, 0, \ldots, 0), \tag{228}$$

and the highest state is then $|1\rangle$.

It is easy to compute the highest weight of a higher representation in the tensor notation or in terms of Young tableaux. For example, for the symmetric tensor

$$\phi_{ij} \sim \boxed{i}\boxed{j}$$

the highest state corresponds to

$$i = 1, j = 1 \to \phi_{11} \sim \boxed{1}\boxed{1}$$

and the highest weight is then simply obtained by applying H_l to each index $i = 1, j = 1$, giving

$$\boxed{}\boxed{} \to (2, 0, 0, \ldots, 0). \tag{229}$$

This is computed by applying the matrices $(H_l)_i^j$ on ϕ_{ij} in the form of an infinitesimal transformation (adding the charges)

$$(H_l \phi)_{ij} \equiv (H_l)_i^{i'} \phi_{i'j} + (H_l)_j^{j'} \phi_{ij'}, \tag{230}$$

which immediately yields $(2,0,0,\ldots,0)$ on $\phi_{11} \sim$ ▢▢ . Similarly for the n'th order symmetric tensor

$$\phi_{i_1 i_2 \cdots i_n} \sim \boxed{}$$

the highest state is specified by $i_1 = i_2 = \cdots = i_n = 1$, and the highest weight is

$$(n,0,0,\ldots,0) \sim \boxed{} \tag{231}$$

The eigenvalues in the highest weight coincide with the Dynkin indices $(a_1, a_2, \ldots, a_{N-1})$ for $SU(N)$. Thus, we can establish a one-to-one correspondence between Young tableaux and Dynkin diagrams via the highest weight, as in (228), (229), (231),

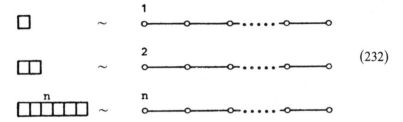

(232)

Generalizing to more complicated Young tableaux it can be seen that for undotted boxes the highest state is found by assigning $i = 1$ to every box on the first row, $i = 2$ to every box on the second row, $i = 3$ to every box in the third row, etc:

$$
\begin{array}{l}
\boxed{1\,1\,1\,1\,1\,1\,1\,1\,1\,1\,1} \quad n_1 \\
\boxed{2\,2\,2\,2\,2\,2\,2} \quad\quad n_2 \\
\boxed{3\,3\,3\,3} \quad\quad\quad\; n_3 \\
\boxed{4\,4\,4} \quad\quad\quad\quad n_4 \\
\quad\cdot \quad\quad\quad\quad\quad\;\; \cdot \\
\quad\cdot \quad\quad\quad\quad\quad\;\; \cdot \\
\quad\cdot \quad\quad\quad\quad\quad\;\; \cdot
\end{array}
\tag{233}
$$

The eigenvalues of the "charges" $H_1, H_2, H_3, \ldots$, etc., on this state are

$$(n_1 - n_2, n_2 - n_3, n_3 - n_4, \ldots). \tag{234}$$

This is so since H_1 assigns $+1$ to $|1\rangle$ and -1 to $|2\rangle$, zero to all other states; similarly H_2 picks nonzero charges from $|2\rangle$ and $|3\rangle$, etc. This provides the translation from Young tableaux to Dynkin diagrams. Thus for the tableau of (233) the Dynkin diagram is

$$
\overset{\displaystyle n_1-n_2 \quad n_2-n_3 \quad n_3-n_4}{\circ\!-\!-\!-\!\circ\!-\!-\!-\!\circ\!-\!\cdots\!-\!\circ\!-\!-\!\circ}
\tag{235}
$$

If the tableau has only dotted boxes, then the "charges" (H_l) are applied with the opposite sign on the states of the tensor with upper indices (antiparticles). Then a little thought will show that the highest state is obtained by assigning the values of indices $i = N$ to every box in the first row, $i = N - 1$ to every box in the second row, etc:

$$
\begin{array}{l}
m_1 \\
m_2 \\
m_3 \\
\cdot \\
\cdot \\
\cdot
\end{array}
\quad
\begin{array}{|c|c|c|c|c|c|c|}
\hline
\cdot N & \cdot N & \cdot N & \cdot N & \cdot N & \cdot N & \cdot N \\
\hline
\multicolumn{2}{c|}{} & \cdot N\!-\!1 & \cdot N\!-\!1 & \cdot N\!-\!1 & \cdot N\!-\!1 & \cdot N\!-\!1 \\
\cline{3-7}
\multicolumn{4}{c|}{} & \cdot N\!-\!2 & \cdot N\!-\!2 & \cdot N\!-\!2 \\
\cline{5-7}
\multicolumn{5}{c|}{} & \cdot N\!-\!3 & \cdot N\!-\!3 \\
\cline{6-7}
\end{array}
\tag{236}
$$

Then, the highest weight, computed by applying H_l on this state (with the opposite sign!), is

$$(0, 0, \ldots, m_3 - m_4, m_2 - m_3, m_1 - m_2). \tag{237}$$

Finally, a general tableau of the type of Figure 2 has a highest state specified by the assignments of (233) and (236) on the undotted and dotted boxes. So its highest weight is

$$\left(n_1 - n_2, n_2 - n_3, \ldots, m_2 - m_3, m_1 - m_2\right), \tag{238}$$

and the Dynkin diagram corresponding to Figure 2 is then

$$
\begin{array}{ccccccc}
n_1\!-\!n_2 & n_2\!-\!n_3 & n_3\!-\!n_4 & & m_3\!-\!m_4 & m_2\!-\!m_3 & m_1\!-\!m_2 \\
\circ\!\!-\!\!\!-\!\!\!-\!\!\circ\!\!-\!\!\!-\!\!\!-\!\!\circ\!\!-\!\!\!-\!\!\!- & & \cdots & \!\!-\!\!\circ\!\!-\!\!\!-\!\!\!-\!\!\circ\!\!-\!\!\!-\!\!\!-\!\!\circ
\end{array}
\tag{239}
$$

Now, we return to supergroups. In the fundamental representations the commuting Cartan matrices are chosen as

$$
H_1 = \left(
\begin{array}{c|c}
\begin{matrix} 1 & & & \\ & -1 & & \\ & & 0 & \\ & & 0 & \ddots \\ & & & & 0 \end{matrix} & 0 \\
\hline
0 & 0
\end{array}
\right),
\tag{240}
$$

$$
H_2 = \left(
\begin{array}{c|c}
\begin{matrix} 0 & & & \\ & 1 & & \\ & & -1 & \\ & & & 0 & \ddots \\ & & & & 0 \end{matrix} & 0 \\
\hline
0 & 0
\end{array}
\right), \ldots, \quad
H_{N-1} = \left(
\begin{array}{c|c}
\begin{matrix} 0 & & & \\ & \ddots & & \\ & & 0 & \\ & & & 1 & \\ & & & & -1 \end{matrix} & 0 \\
\hline
0 & 0
\end{array}
\right),
$$

(240) continued

$$L_0 = \begin{pmatrix} \begin{matrix} 0 & & & \\ & \ddots & & \\ & & 0 & \\ & & & 1 \end{matrix} & \begin{matrix} & 0 & \\ & & \\ & & \end{matrix} \\ \hline \begin{matrix} & & \\ & 0 & \\ & & \end{matrix} & \begin{matrix} 1 & & \\ & 0 & \\ & & \ddots \\ & & & 0 \end{matrix} \end{pmatrix} ,$$

$$K_1 = \begin{pmatrix} \begin{matrix} 0 & \\ & \\ & 0 \end{matrix} & \begin{matrix} & 0 & \\ & & \\ & & \end{matrix} \\ \hline \begin{matrix} & & \\ & 0 & \end{matrix} & \begin{matrix} 1 & & \\ & -1 & \\ & & 0 \\ & & & \ddots \\ & & & & 0 \end{matrix} \end{pmatrix} ,$$

$$K_2 = \begin{pmatrix} \begin{matrix} 0 & \\ & 0 \end{matrix} & \begin{matrix} 0 \\ \\ \end{matrix} \\ \hline \begin{matrix} & & \\ & 0 \end{matrix} & \begin{matrix} 0 & & \\ & 1 & \\ & & -1 \\ & & & 0 \\ & & & & \ddots \\ & & & & & 0 \end{matrix} \end{pmatrix} , \dots, K_{M-1} = \begin{pmatrix} \begin{matrix} 0 & \\ & 0 \end{matrix} & \begin{matrix} 0 \\ \end{matrix} \\ \hline \begin{matrix} & \\ 0 \end{matrix} & \begin{matrix} 0 & & \\ & \ddots & \\ & & 0 \\ & & & 1 \\ & & & & -1 \end{matrix} \end{pmatrix} .$$

Note that they must be supertraceless, and this fixes L_0 with its two positive entries in contrast to all the others.

Using the same approach as above, we compute the eigenvalues of these operators on the states of any representation. If we identify the highest state and highest weight by the same conventions as above, then the eigenvalues that specify the highest weight correspond to the indices in the Kac-Dynkin diagram

$$\circ\!\!-\!\!\circ\!\cdots\!-\!\circ\!\!-\!\!\otimes\!\!-\!\!\circ\!\!-\!\!\circ\!\!-\!\!\circ \tag{241}$$

The circle with the cross corresponds to the eigenvalue of L_0.

To find the highest weight we analyze the $SU(N) \times SU(M) \times U(1)$ content of any $SU(N/M)$ representation as is done in §§15–17, Tables 1, 2 and (210),

$$\bar{\mathbf{X}} = \sum (X, \tilde{Y})_q, \tag{242}$$

where X, Y represent the pictures of $SU(N)$ and $SU(M)$ Young tableaux respectively, and q specifies the $U(1)$ eigenvalues of all the states described by X, Y. Generally, we find that the highest state is contained in the $(X, \tilde{Y})$ component that carries the smallest value of q (in a convention $N > M$). q is given as

$$q = \alpha/N + \beta/M, \qquad (243)$$

where α is the number of undotted minus dotted boxes in X, and β is the corresponding quantity in $\tilde{Y}$. Thus, to obtain the smallest value of q we must choose the $(X, \tilde{Y})$ component that has the most $SU(N)$ undotted boxes in X and the most $SU(M)$ dotted boxes in $\tilde{Y}$. This is easy to find pictorially directly from the original supertableaux as follows:

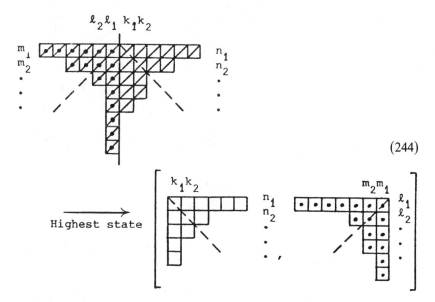

$$(244)$$

We simply have assigned all undotted boxes to $SU(N)$ and all dotted boxes to $SU(M)$ (recall that the $SU(M)$ tableau is reflected along the diagonal relative to the original tableau). If N and M are sufficiently large so that this highest component $(X, \tilde{Y})$ is nonvanishing (i.e. they are legal $SU(N) \times SU(M)$ tableaux), then the highest state is contained here. The weight is now computed by applying the Cartan generators $H_1, H_2, \ldots, H_{N-1}$; L_0; $K_1, K_2, \ldots, K_{M-1}$ of (240). It is clear that the highest weight will then be

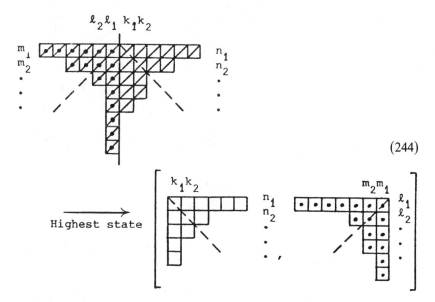

$$(245)$$

The eigenvalue l clearly must be related to q (equation (243)). We can find the relation between q and l through the fundamental representation, where the operator Q corresponding to the eigenvalue q is identified as the supertraceless matrix

$$
Q = \begin{pmatrix}
1/N & & & & & & \\
 & 1/N & & & & & \\
 & & \ddots & & & & \\
 & & & 1/N & & & \\
\hline
 & & & & 1/M & & \\
 & & & & & 1/M & \\
 & & & & & & \ddots & \\
 & & & & & & & 1/M
\end{pmatrix}. \tag{246}
$$

Then we see that Q can be written in terms of the commuting Cartan generators as

$$
Q = \sum_{i=1}^{N-1} \frac{i}{N} H_i - \sum_{j=1}^{M-1} \frac{j}{M} K_{M-j} + L_0. \tag{247}
$$

Since we already know the eigenvalues of H_i, K_j, L_0 as specified in (245), (243), we immediately deduce the value of l from

$$
\frac{\alpha}{N} + \frac{\beta}{M} = \sum_{i=1}^{N-1} \frac{i}{N}(n_i - n_{i+1}) - \sum_{j=1}^{M-1} \frac{j}{M}(m_j - m_{j+1}) + l, \tag{248}
$$

where α and β were given in (243). If the tableaux of (244) are nonvanishing, then α = number of boxes in SU(N) tableaux and β = number of boxes in SU(M) tableaux. *Note that α, β must be computed before eliminating any completed columns which may have exactly N boxes for* SU(N) *and M boxes for* SU(M).

Let us look at some examples, where we assume that N, M are very large so that the highest state is easily identified as in (244):

$$
\boxtimes \;\longrightarrow\; (\,\square\,,\,1) \;\longrightarrow\; \overset{1}{\underset{}{\circ\!\!-\!\!-\!\!\circ\!\!-\!\!-\!\!\circ\!\!-\cdots-\!\!\otimes\!\!-\cdots-\!\!\circ\!\!-\!\!-\!\!\circ}} \tag{249}
$$

$$
\boxtimes \;\longrightarrow\; (1,\,\boxtimes) \;\longrightarrow\; \circ\!\!-\!\!-\!\!\circ\!\!-\!\!-\!\!\circ\!\!-\cdots-\!\!\overset{1}{\otimes}\!\!-\cdots-\!\!\circ\!\!-\!\!-\!\!\circ \tag{250}
$$

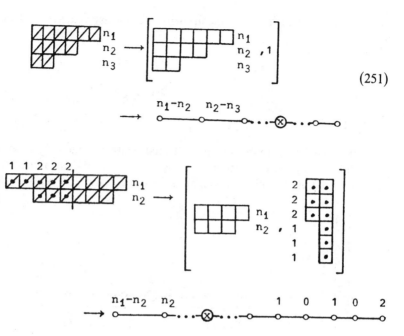

$$(251)$$

$$(252)$$

When there are a large number of undotted rows or a large number of dotted columns then the pictures of (244) become illegal. We must then return to the branching rules of the previous section, and Tables 1, 2, in order to identify the first nonvanishing component with the lowest value of the U(1) charge q. This is best explained with an example. Consider the supertableau

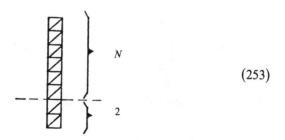

$$(253)$$

whose undotted boxes in a column exceed N. The simple prescription of (244) now fails. However, all we need to do, after the naive step of (244), is simply transfer the extra two boxes to SU(M) in order to find the first nonvanishing set of Young tableaux with the highest value of q. This is just

$$\left[\begin{array}{c} \Box_N \end{array} , \quad \overset{2}{\Box\Box} \right]_{N/N + 2/M} \Rightarrow (1, \; \Box\Box)_{1+2/M} \quad \cdot \quad (254)$$

Note that the SU(M) tableaux are reflected relative to the original ones. Now the SU(N) representation is a singlet and the SU(M) representation is $\Box\Box$. Thus, the highest weight is identified and the Kac-Dynkin diagram is

$$(255)$$

A similar situation would occur with a row of dotted superboxes whose length exceeds M:

$$(256)$$

With arguments similar to the above, we identify the highest state as the first nonvanishing one with the smallest U(1) charge q. This is given by first considering the naive step of (244) and then transferring the extra three boxes to SU(N):

$$\left[\begin{array}{c} \overset{3}{\boxed{\cdot\,\cdot\,\cdot}} \end{array} , \quad \boxed{\begin{array}{c}\cdot\\\cdot\\\cdot\\\cdot\\\cdot\end{array}}_M \right]_{-3/N - M/M} \Rightarrow (\boxed{\cdot\,\cdot\,\cdot}, \; 1)_{-3/N-1} \cdot \quad (257)$$

The highest weight and the Kac-Dynkin diagram is now written as

$$(258)$$

    I list a few more examples without explanation, to be figured out by the reader.

Exercise 1.

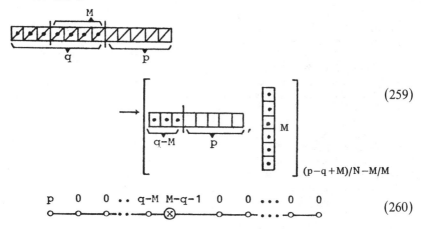

$$(259)$$

$$
\begin{array}{ccccccccccc}
\text{p} & 0 & 0 & .. & \text{q-M} & \text{M-q-1} & 0 & 0 & ... & 0 & 0
\end{array}
$$

$$(260)$$

Exercise 2.

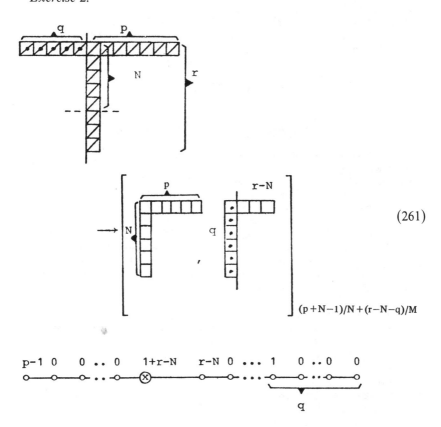

$$(261)$$

$$
\begin{array}{cccccccccccc}
\text{p-1} & 0 & 0 & .. & 0 & \text{1+r-N} & \text{r-N} & 0 & ... & 1 & 0 & ..0 & 0
\end{array}
$$

Further discussion can be found in a future publication [16].

REFERENCES

1. A. Neveu and J. H. Schwarz, Nucl. Phys. B **31** (1971), 86; P. Ramond, Phys. Rev. D **3** (1971), 2415.

2. J. Wess and B. Zumino, Nucl. Phys. B **70** (1974), 39.

3. For a review see P. Fayet and S. Ferara, Phys. Rep. **32** (1977), 69; P. Van Nieuwenhuzien, Phys. Rep. C **68** (1981), 189.

4. A. B. Balantekin, I. Bars and F. Iachello, Phys. Rev. Lett. **47** (1981), 19; Nucl. Phys. A **370** (1981), 284; A. B. Balantekin, I. Bars, I Bijker and F. Iachello, Phys. Rev. C **27** (1983), 2430.

5. A. Salam and J. Strathdee, Nucl. Phys. B **79** (1974), 477.

6. T. Banks, A. Schwimmer, S. Yankielowicz, Phys. Lett. B **96** (1980), 67; I. Bars and S. Yankielowicz B **101** (1981), 159; I. Bars, Phys. Lett. B **106** (1981), 105; Phys. Lett. B **114** (1982), 118; Nucl. Phys. B **208** (1982), 77; Yale preprint YTP-82-24, Proc. Rencontre de Moriond 1982 (J. Tran Thahn Van, ed.), p. 541; A. Schwimmer, Rutgers preprint RU-81-49.

7. I. Bars, to be published.

8. Y. Ne'eman, Phys. Lett. B **81** (1979), 190 and Tel Aviv preprint TAUP-134-81.

9. C. Becchi, A. Rouet and A. Stora, Commun. Math. Phys. **42** (1975), 127; J. Thierry-Mieg and Y. Ne'eman, Ann. Physics **123** (1979), 247; R. Delbourgo and P. Jarvis, J. Phys. A **15** (1982), 611.

10. P. Fayet, Phys. Lett. B **69** (1977), 489; **70** (1977), 461; S. Dimopoulos and S. Raby, Nucl. Phys. B **192** (1981), 353; M. Dine, W. Fishler and M. Srednicki, Nucl. Phys. B **189** (1981), 575; S. Dimopoulos and H. Georgi, Nucl. Phys. B **193** (1981), 150; S. Weinberg, Phys. Rev. D **26** (1982), 287.

11. I. Bars, G. Veneziano and S. Yankielowicz, to be published.

12. S. Deser and B. Zumino, Phys. Lett. B **62** (1976), 335; D. Z. Freedman, P. van Nieuwenhuzien and S. Ferara, Phys. Rev. D **13** (1976), 3214.

13. G. 't Hooft, Recent Developments in Gauge Theories (G. 't Hooft et al., eds.), Plenum, New York, 1980.

14. A. B. Balantekin and I. Bars, J. Math. Phys. **22** (1981), 1149; **22** (1981), 1810; **23** (1982), 1239.

15. A. B. Balantekin, J. Math. Phys. **23** (1982), 486.

16. I. Bars, B. Morel and H. Ruegg, J. Math. Phys. **24** (1983), 2253.

17. I. Bars and M. Günarydin, Comm. Math. Phys. **91** (1983), 31.

18. V. G. Kac, Differential Geometrical Methods in Math. Phys. (K. Bleuler, H. R. Petry and A. Reetz, eds.), Springer-Verlag, Berlin, 1978.

19. Y. Ne'eman and S. Sternberg, Proc. Nat. Acad. Sci. U.S.A. **77** (1980), 3127; P. H. Dondi and P. Jarvis, Z. Phys. C **4** (1980), 201; J. Phys. A **14** (1981), 547; M. Scheunert, W. Nahm and V. Rittenberg, J. Math. Phys. **18** (1977), 155; M. Marcu, J. Math. Phys. **21** (1980), J. Thierry-Mieg and B. Morel, Harvard preprint HUTMP 80/8100.

20. H. Weyl, *Classical groups*, Princeton Univ. Press, Princeton, N. J., 1946.

21. A. M. Perelomov and V. S. Popov, Soviet J. Nuclear Phys. **3** (1966), 676 and **5** (1967), 489.

22. See e.g. W. G. McKay and J. Patera, *Tables of dimensions, indices and branching rules for representations of simple Lie algebras*, Marcel Dekker, New York, 1981; B. G. Wybourne, *Symmetry principles and atomic spectroscopy*; E. B. Dynkin, Amer. Math. Soc. Transl. **6** (1957), 353; H. Freudental, Nederl. Akad. Wetensch. Indag. Math. **16** (1954), 490.

DEPARTMENT OF PHYSICS, UNIVERSITY OF SOUTHERN CALIFORNIA, LOS ANGELES, CALIFORNIA 90089-0484

Lectures in Applied Mathematics
Volume **21**, 1985

Topics in Dimensional Reduction[1]

Peter G. O. Freund

According to Einstein, space-time is a dynamically determined Lorentz manifold M. The relevant dynamics is provided by the Einstein equations, and the dimension d of M is set at $d = 4$ for "phenomenological" reasons. Could it be that the value of d can also be theoretically understood? Values of d other than $d = 4$, then have to be considered. To conform with experimental evidence, at present $d \geqslant 4$ and M must have the product structure

$$M = M_4 \times M_{d-4}, \tag{1}$$

with M_4 a 4-dimensional Lorentz manifold, and M_{d-4} a $(d-4)$-dimensional compact Riemann manifold of very small "size" ($\ll 10^{-3} \, \mathrm{GeV}^{-1}$).

How would the $N = d - 4$ extra dimensions manifest themselves in 4-dimensional physics? In d-dimensions we certainly have to include gravity with its $\frac{1}{2}d(d-3) = \frac{1}{2}(N+4)(N+1)$ degrees of freedom. In 4-dimensions these get rearranged into gravity (2 degrees of freedom), N massless vector fields ($2N$ degrees of freedom) and $\frac{1}{2}N(N+1)$ scalar fields ($\frac{1}{2}N(N+1)$ degrees of freedom). This is unrealistic in that fermions are absent from the spectrum, but that is easily corrected by starting from d-dimensional supergravity [1, 2] rather than d-dimensional gravity. The interesting and powerful constraint

$$d \leqslant 11 \tag{2}$$

1980 *Mathematics Subject Classification.* Primary 53B50; Secondary 83C15.
[1] Work supported in part by NSF Grant No. PHY-79-23669.

is then automatically introduced [3] ($d = 11$, $d = 10$ being the most important cases) [2, 3, 4].

Before going into details, a few virtues of this generalized Kaluza-Klein approach [5], as well as some questions it raises, deserve mention. First the virtues:

(VI) Gravity is included in this unification from the outset.

(V2) For the first time, a theoretical determination of the space-time dimensionality is in sight.

(V3) The higher-dimensional theory determines the gauge symmetry and the "matter" fields of its 4-dimensional reduction.

Now to the questions:

(Q1) Are the extra dimensions timelike or spacelike?

(Q2) Why do the extra dimensions compactify, and how does this bear on cosmology?

(Q3) What is the characteristic size of these extra dimensions?

(Q4) Why is the Lorentzian part of the product manifold (1) 4-dimensional?

We shall now attempt to answer these questions and to further explore the implications of this approach.

Corresponding to the product manifold M of (1) we have the metric [5]

$$\gamma_{MN} = \begin{pmatrix} g_{\mu\nu} + e^2\kappa^2 g_{pq} A_\mu^p A_\nu^q & e\kappa g_{np} A_\mu^p \\ e\kappa g_{mp} A_\nu^p & g_{mn} \end{pmatrix}. \tag{3}$$

Here $M, N, \ldots = 1, 2, \ldots, d$; $\mu, \nu, \ldots = 1, 2, \ldots, d - N$; $m, n = d - N + 1, d - N + 2, \ldots, d$ (the interesting case being $d - N = 4$). We have assumed for simplicity that the manifold of the "extra" dimensions M_N is that of an N-dimensional Lie group G. A_μ^p are then the $(d - N)$-dimensional gauge fields corresponding to this Lie group, e the gauge coupling constant, $g_{\mu\nu}$ is $(d - N)$-dimensional gravity and g_{mn} are scalar fields in $d - N$ dimensions. All these fields depend only on the first $d - N$ coordinates x^p (any dependence on the x^r's involves the scale of the extra dimensions). Finally, κ is a constant required to adjust dimensions which, as we shall presently see, is related to the size of the extra dimensions.

If we start from pure gravity in d-dimensions, then the action is, up to a constant factor, the integral over M of the d-dimensional scalar curvature density $R_d\sqrt{-g_d}$

$$S_d = -\frac{1}{16\pi G_d} \int d^d x \sqrt{-g_d}\, R_d \tag{4}$$

(here G_d is a d-dimensional gravitational constant). On a product manifold (1) with the metric (3), after carrying out N integrations, one finds [6]

$$S_d = \int d^{d-N}x (\mathcal{L}_E + \mathcal{L}_{YM} + \mathcal{L}_S)\sqrt{|g_{d-N}|},$$

$$\mathcal{L}_E = -\frac{1}{16\pi G} R_{d-N}\sqrt{|g_N|},$$

$$\mathcal{L}_{YM} = -\frac{1}{4}\frac{e^2\kappa^2}{16\pi G}(-g_{ab})g^{\mu\rho}g^{\nu\sigma}F^a_{\mu\nu}F^b_{\rho\sigma}\sqrt{|g_N|},$$

$$\mathcal{L}_S = -\frac{\sqrt{|g_N|}}{16\pi G}\left(\frac{1}{4}g^{ab}g^{cd}g^{\mu\nu}(\nabla_\mu g_{ac}\nabla_\nu g_{bd} - \nabla_\mu g_{ab}\nabla_\nu g_{cd})\right.$$

$$-\frac{1}{2}g^{\mu\nu}(\nabla_\mu g^{ab}\nabla_\nu g_{ab} + 2g^{ab}\nabla_\mu\nabla_\nu g_{ab})$$

$$\left.-\frac{1}{2\kappa^2}\left(f^c_{ad}f^d_{bc}g^{ab} + \frac{1}{2}f^i_{ac}f^j_{bd}g^{ab}g^{cd}g_{ij}\right)\right). \tag{5}$$

Here R_{d-N} is the $(d-N)$-dimensional scalar curvature, $g_{d-N} = \det(g_{\mu\nu})$, $g_N = \det(g_{mn})$, f^c_{ab} are the structure constants of the gauge group g^{ab} and $g^{\mu\nu}$ are inverse matrices to g_{bc} and $g_{\nu\rho}$. G is the $(d-N)$-dimensional gravitational constant, and $\nabla_\mu g_{ab}$ is the Yang-Mills and gravitational covariant derivative. We recognize $\mathcal{L}_E$ as the 4-dimensional Einstein lagrangian with variable gravitational constant (à la Jordan, Brans and Dicke) [7] and $\mathcal{L}_{YM}$ as the 4-dimensional Yang-Mills lagrangian (somewhat modified on account of the presence of the scalar fields $g_{ab}\sqrt{|g_N|}$ contracting the field strengths). S_S is the action of a 4-dimensional nonlinear $GL(N, R)/O(N, R)$ σ-model, that provides the $O(N)$ part of the $O(N) \times O(d-N-1, 1)$ gauge invariance still available from the original d-dimensional gauged Lorentz group $O(d-1, 1)$. The $O(d-N-1, 1)$ gauge invariance is enforced by the $(d-N)$-dimensional gravity $g_{\mu\nu}$.

The correct sign (positive energy) for $\mathcal{L}_{YM}$ emerges if the N extra dimensions are all spacelike, in line with the causality requirement that there be at most one time dimension so as to avoid closed timelike curves. This answers (Q1).

Expanding the scalar fields g_{ab} around the unit matrix, we can enforce not only the correct sign, but also the correct normalization of the Yang-Mills lagrangian. This yields

$$\frac{e^2\kappa^2}{16\pi G} = 1. \tag{6}$$

Now consider the case of interest, $d - N = 4$. A scalar field with harmonic dependence on the coordinate characterizing a $U(1)$ subgroup of the symmetry group of M_{d-4} appears in 4-dimensions as a charged scalar field. The proper value of its charge is obtained for

$$\kappa = l/2\pi, \tag{7}$$

where l is the periodicity distance in the chosen $U(1)$ coordinate. Substituting this into equation (6), we find

$$l = 4\pi(G/\alpha)^{1/2} \tag{8}$$

where $\alpha = e^2/4\pi$. For $\alpha \approx \alpha_{GUT} \approx 0.01$ this means [8] $l \approx 10^{-17}\,\mathrm{GeV}^{-1}$ (this answers (Q3) very close to the supersymmetric grand unification scale $l_{SUSY} \sim 10^{-16}\,\mathrm{GeV}^{-1}$ (for two Higgs supermultiplets). Thus in generalized Kaluza-Klein theories, by the time the grand unification scale is reached, 4-dimensional physics becomes inapplicable and the full higher-dimensional physics comes into play (in 11-dimensional supergravity this argument depends on the detailed embedding of

$$SU(3)_{color} \times (SU(2) \times U(1))_{electroweak}$$

in the gauge group $G_{Kaluza-Klein} \times SU(8)_{Cremmer-Julia}$).

The proximity of l to l_{SUSY} is irrelevant for estimates of proton decay, but it may seriously affect the cosmology of the very early universe. One is thus led to cosmologies in which the "effective" dimension of space changes with time. (Q2) and (Q4) now both become relevant. Let us first address ourselves to (Q4) as to why the "large" space dimensions number three. Were one to start from pure gravity in d-dimensions, one could find classical solutions in which any number N of the $d - 1$ space dimensions compactifies, thus leaving $d - N = 4$ a mystery. To achieve $d - N = 4$ one has to add matter fields [9] and à priori this leaves a lot of arbitrariness. In supergravity theories though, one cannot add matter fields at will, and it may happen that the matter fields dictated by supersymmetry are such that a preferential compactification of all but three space dimensions occurs [10]. In 11-dimensional supergravity [2] there is precisely one Bose matter field, an antisymmetric tensor field A_{MNP} of rank three. The correspoinding field strengths F_{MNPQ} span an antisymmetric tensor of rank four. If both M_{11-N} and M_N are maximally symmetric spaces (more general cases can also be considered) [10], then F_{MNPQ} can live as a form-invariant tensor on one of them only if $N = 4$ or $N = 7$. This form-invariant antisymmetric tensor field induces a cosmological term with one value of the cosmological constant on M_N and a different value on M_{11-N}. The signs of these induced cosmological constants are such as to compactify the purely spacelike manifold M_N but not M_{11-N}. A preferential compactification of $N = 4$ or $N = 7$ spacelike dimensions is thus generated. The case $N = 7$ is, of course, precisely what we want. The case $N = 4$ could possibly be ruled out by remarking that physical space-time would then be 7-dimensional and as such could not support any renormalizable quantum field theories to describe "low-energy" physics.

We concentrate on the case $N = 7$. For this case interesting cosmological solutions have been found [11, 12], which we now describe. We assume the 11-dimensional space-time manifold to have a generalized Friedmann-Robertson-Walker form in that the *space-part* of the manifold is not one 10-dimensional maximally symmetric manifold but the product of an N- and of a $(10 - N)$-dimensional manifold $M_N \times M_{10-N}$. We use the antisymmetric tensor field to fix $N = 7$ and the field equations to determine the cosmological scale factors $R_3(t)$ and $R_7(t)$ of M_3 and M_7, as well as the time dependence of the antisymmetric tensor field. Thus we find a solution in which both M_3 and M_7 expand but M_3 expands much faster than M_7:

$$R_3(4) = r_3(t + t_0), \qquad R_7 = r_7(t + t_0)^{1/7} \qquad (9)$$

(where r_3, r_7, t_0 are constants). M_7 is a torus and M_3 a pseudosphere. The effective 4-dimensional gravitational constant is proportional to $(R_7)^{-7}$ and as such decreases like $1/t$ as proposed by Dirac [13].

An alternative solution has M_7 a sphere and M_3 an anti-de Sitter universe with cosmological factor $R_3(t) = R_3(0)\cos \alpha t$, with α determined by the 11-dimensional gravitational constant and by the here time-independent scale of the antisymmetric tensor field.

These solutions do not involve the Fermi-matter fields and as such should not be valid in the matter dominated era. They also break down at too early times where quantum gravity matters, but they can reasonably be expected to be relevant in some time interval around the dimensional transition where the 11-dimensional manifold splits into $M_4 \times M_7$. Other cosmological solutions along similar lines have been discussed in references [11, 12 and 14]. New scales are introduced in these solutions: the time t_s at which all d-dimensions have comparable sizes, the actual size $l(t_s)$ of the dimensions at time t_s and the strength of gravity at that time. Depending on the details of the evolution, these scales can considerably exceed the present day Planck ($10^{-19}\,\mathrm{GeV}^{-1}$) and Kaluza-Klein ($10^{-17}\,\mathrm{GeV}^{-1}$) scales. The cosmology of the very early universe is thus seriously affected. We remark that in these solutions already, the space manifolds M_N and M_{d-N} were not necessarily group manifolds (e.g., the 3-sphere is a group but not the 7-sphere). Yet other compactified manifolds have been considered. In particular, the interesting observation has been made [15] that for $d = 11$, $N = 7$ and $M^7 = P_2(C) \times S^2 \times S^1$ the gauge group is precisely SU(3) $\times$ SU(2) $\times$ U(1). It should be noted that the preferential compactification of 7 space dimensions has been achieved at the classical level. The alternative could be entertained that the choice of four large dimensions occurs at the quantum level, by somehow summing over all possible dimensionalities, and for dynamical reasons ending up with four large physical dimensions in some approximation. Such a possibility was considered by others as well [16].

Finally, I would like to briefly address another interesting feature of generalized Kaluza-Klein theories. In 4-dimensional field theories three discrete symmetry operations are important: space inversion (P), time-reversal (T) and charge conjugation (C). Whereas only CPT invariance is guaranteed, all three operations are very interesting. P and T have obvious geometrical meaning, but C, which switches matter and anti-matter, does not. The same would be true in any dimension. Yet in gravity and $N = 1$ supergravity theories charge conjugation becomes a trivial operation. For example, in 11-dimensions gravity and the antisymmetric tensor field both have definite charge-conjugation parity $C = +1$ and the gravitino is a Majorana (i.e., self-charge-conjugate fermion). Yet the reduced 4-dimensional theory has SU(8) symmetry and contains fields that do not transform trivially under 4-dimensional charge conjugation. In theories of this type the questions can be asked as to the origin of the 4-dimensional charge conjugation operation C_4. The higher-dimensional charge conjugation being trivial, C_4 must originate in some other symmetry of the higher-dimensional theory. It is easy to show [17] that this symmetry is a reflection of some of the extra space coordinates. What we call antimatter in 4-dimensions is really matter mirrored in some of the extra dimensions.

References

1. D. Z. Freedman, P. van Nieuwenhuizen and S. Ferrara, Phys. Rev. D (3) **13** (1976), 1314; S. Deser and B. Zumino, Phys. Lett. B **62** (1976), 335.

2. E. Cremmer, B. Julia and J. Scherk, Phys. Lett. B **76** (1978), 409.

3. W. Nahm, Nucl. Phys. B **135** (1978), 149.

4. F. Gliozzi, J. Scherk and D. Olive, Nucl. Phys. B **122** (1977), 253; L. Brink, J. Schwarz and J. Scherk, Nucl. Phys. B **121** (1977), 77; A. Chamseddine, Nucl. Phys. B **185** (1981), 403.

5. Th. Kaluza, Sitzungsber. Preuss. Akad. Wiss. Phys. Math. Kl. (1921), 966; O. Klein, Z. Phys. **37** (1926), 895; B. de Witt, *Dynamical theories of groups and fields*, Gordon & Breach, New York, 1965, p. 139; R. Kerner, Ann. Inst. H. Poincaré **9** (1968), 143; A. Trautman, Rep. Math. Phys. **1** (1970), 29; Y. M. Cho and P. G. O. Freund, Phys. Rev D **12** (1975), 1711. Interesting recent papers include: A. Salam and J. Strathdee, ICTP Trieste preprint IC/81/211, 1981; M. J. Duff and D. J. Toms, CERN preprint TH 3259, 1982.

6. Y. M. Cho and P. G. O. Freund [5].

7. P. Jordan, *Schwerkraft und Weltall*, Vieweg u. Sohn, Braunschweig, 1955; Z. Phys. **157** (1957), 112; C. Brans and R. H. Dicke, Phys. Rev. D (3) **124** (1961), 925.

8. P. G. O. Freund, Phys. Lett. B **120** (1983), 335.

9. E. Cremmer and J. Scherk, Nucl. Phys. B **108** (1976), 409; Nucl. Phys. B **118** (1977), 61; J. F. Luciani, Nucl. Phys. B **135** (1978), 111.

10. P. G. O. Freund and M. A. Rubin, Phys. Lett. B **97** (1980), 233.

11. P. G. O. Freund, Nucl. Phys. B **199** (1982), 482.

12. P. Ramond, University of Florida preprint UFTP-82-21.

13. P. A. M. Dirac, Proc. Roy. Soc. London Ser. A **165** (1935), 199.

14. A. Chodos and S. Detweiler, Phys. Rev. D **21** (1980), 2167.

15. E. Witten, Nucl. Phys. B **186** (1981), 412.

16. R. Geroch, private communication; S. Shenker, private communication.

17. P. G. O. Freund, preprint EFI 81/53, Festschrift dedicated to F. Gürsey.

THE ENRICO FERMI INSTITUTE AND THE DEPARTMENT OF PHYSICS, UNIVERSITY OF CHICAGO, CHICAGO, ILLINOIS 60637

Lectures in Applied Mathematics
Volume 21, 1985

Bound State Spectra in
Extended Supergravity Theories

Mary K. Gaillard

ABSTRACT. After reviewing the motivations for considering infinite representations of the noncompact group and supersymmetry algebra of extended supergravity theories, we display explicit representations of the algebra for $N = 4$ and $N = 8$ supergravity.

1. Introduction. There is by now an impressive body of experimental evidence supporting the theoretical description of the strong, weak and electromagnetic interactions in terms of renormalizable gauge field theories. On the other hand, there is no evidence whatsoever that supersymmetry [1] is relevant to the description of elementary particles. Nevertheless the idea that supersymmetry has some bearing on physical phenomena has recently become very popular among elementary particle theorists. The reasons for this are twofold. On the one hand, it is hoped that supersymmetry, however badly broken it may be, may eliminate some of the arbitrariness inherent to our present formulation of gauge theories. Secondly, supersymmetry, or more specifically supergravity [2], offers a possibility for the ultimate unification of the strong and electroweak forces with gravity. The prescription for writing down a renormalizable gauge field theory has by now become text book material. The rules are:

(1) Pick a gauge group. Examples are: $U(1)_{em}$ for quantum electrodynamics (QED); $SU(3)_{color}$ for quantum chromodynamics (QCD), the currently accepted theory of the strong interactions; $SU(2)_{Left} \times U(1)$

1980 *Mathematics Subject Classification.* Primary 22-06, 22E70, 81G20, 83E50.

for the "electroweak" theory [3] which in some sense unifies the electromagnetic and weak interactions. The unification of all of these interactions can be achieved in the context of a simple group encompassing $SU(3)_c \times SU(2)_L \times U(1)$, such as $SU(5)$, $SO(10)$, $E_6, \ldots$.

(2) Assign vector fields to the adjoint representation of the group.

(3) Assign spin-$\frac{1}{2}$ fermion fields to arbitrary representations of the group.

At this point the main degree of arbitrariness[1] is in the choice of the gauge group and of the fermion representations. Given these, interactions are completely specified in terms of a single coupling constant if the gauge group is simple, or in terms of a number of coupling constants equal to the number of factor groups for a direct product.

However, nature does not respect the symmetries which we find necessary for the description of elementary particle interactions. Thus we must introduce symmetry breaking, and the only way we know how to do this while retaining renormalizability, i.e. the cancellation of infinities up to terms which can be reabsorbed into the definition of the basic parameters (coupling constants) of the theory, is through the introduction of scalar fields. It is this feature of the theories which introduces a high degree of arbitrariness: scalar couplings to one another and to fermions, as well as scalar masses, are highly arbitrary. Since supersymmetry is an invariance under transformations among fields of different spin, one would expect that scalar couplings to one another and to fermions should be related to gauge couplings in a supersymmetric version of the theory.

Unfortunately, supersymmetry as applied to the ordinary gauge theories listed above does not really remove any arbitrariness because none of the known particle spectrum can be grouped into supermultiplets; instead we must introduce a new superpartner for each particle. In addition, only simple supersymmetry, which does not relate any scalar self-coupling to gauge couplings, can be used for chiral gauge theories like $SU(2)_c \times U(1)$ and $SU(5)$ which are required by experiment. Simple supersymmetry does provide mechanisms for preventing some scalars from acquiring large masses, which accounts for its current popularity among gauge theorists.

In addition to displaying a high degree of arbitrariness, conventional gauge theories cannot accommodate gravity because quantum gravity requires the introduction of a field of spin 2. With increasing spin, quantum corrections to amplitudes for physical processes become increasingly divergent. The incorporation of spin-1 fields was achieved [4]

---

[1] In addition there may be fermion masses, which, however necessarily vanish if fermion representations are "chiral", or helicity dependent, as in the standard $SU(2)_L \times U(1)$ and $SU(5)$ models.

by demanding a local symmetry [5], assured by coupling spin-1 fields to scalar and fermionic matter fields only through the minimal gauge coupling. This amounts to modifying the kinetic energy by replacing the ordinary derivative by the "covariant" derivative

$$\partial_\mu = \frac{\partial}{\partial x_\mu} \to D_\mu = \partial_\mu + \vec{V}_\mu(x) \cdot \vec{T}, \tag{1.1}$$

where $V_\mu(x)$ is a gauge field and $\vec{T}$ generates an infinitesimal transformation under the group. Then $\psi^\dagger D_\mu \psi$ is invariant under

$$\psi \to g(x)\psi = e^{i\vec{\lambda}(x) \cdot \vec{T}}\psi, \tag{1.2}$$

$$\vec{V}_\mu \vec{T} \to g(x)\left(\vec{V}_\mu \vec{T}\right)g^{-1}(x) - \left[\partial_\mu g(x)\right]g^{-1}(x). \tag{1.3}$$

This local symmetry assures cancellations among divergent contributions to quantum corrections, rendering amplitudes for physical processes finite and calculable.

The introduction of spin-2 fields would entail a still higher degree of divergence and so a higher degree of cancellation—which might be provided by a still higher degree of symmetry—is needed. Supersymmetry appears to provide at least some of the required cancellations. Supersymmetric theories are characterized by the number N of independent supersymmetries. The largest supersymmetric theory of spin-1 fields, the $N = 4$ Yang-Mills theory, is apparently finite [6], and it has been conjectured that the largest supergravity theory, with $N = 8$, may also be finite, but this is not yet known [7].

$N = 8$ supergravity has, in fact, several attractive features as a candidate theory for the unification of all interactions. As the largest supergravity theory it possesses the highest degree of internal symmetry. Its particle spectrum and their couplings are uniquely specified. In fact, these constraints are so tight that it was shown some time ago [8] that a straightforward identification of the elementary fields of supergravity with the particle spectrum of conventional gauge theories is not possible. In particular, to obtain a gauge theory of the conventional type requires spin-1 fields in the adjoint representation of the gauge group. The multiplicity of spin-1 fields appearing in the N-supergravity multiplet corresponds to the adjoint of $SO(N)$, and the largest of these, $SO(8)$, does not contain the product group $SU(3)_c \times SU(2)_L \times U(1)$ required for the description of the strong and electroweak interactions, let alone $SU(5)$ or a larger group which could unify them. It has been noted however, that $SO(8)$ does embed the possibly exact invariance groups $SU(3)_c \times U(1)_{em}$ of the strong and electromagnetic interactions, and a possible point of view is that only the gauge couplings associated with these interactions are elementary, especially since the electroweak $SU(2)_L \times U(1)$ theory still awaits direct confirmation in the form of detection of the associated massive spin-1 particles, $W^{\pm}$ and Z, and a unifying theory awaits even indirect confirmation in the form of proton decay. Results of these experiments are immanent; I shall assume here that they will be positive.

One is then led to assume that at least a part of the spectrum of particles needed for the formulation of conventional gauge theories must be composite—a conclusion which follows also from the observation [8] that the fermion spectrum of $N = 8$ supergravity does not contain all of the quarks and leptons that have already been detected. It was therefore conjectured [9] that conventional gauge theories are *effective* theories describing the effective couplings of composite fields whose quanta are bound states of the elementary preons of $N = 8$ supergravity. The bound state radius is assumed to be of the order of the Planck length, so that these composite fields appear local on energy scales relevant to high energy experiments. The conjectured bound state spectrum was determined by exploiting the symmetries of $N = 8$ supergravity which were discovered by Cremmer and Julia [10], who studied the theory without introducing [11] an explicit SO(8) gauge symmetry associated with the elementary vector fields.

Specifically, Cremmer and Julia showed [10] that the ungauged form of extended supergravity theories with $4 \leqslant N \leqslant 8$ is invariant under a symmetry group of the form

$$\mathcal{G}_{\text{rigid}} \times \mathcal{K}_{\text{local}} \tag{1.4}$$

where $\mathcal{G}$ is a noncompact group of rigid, or global, transformations and $\mathcal{K}$ is its maximal compact subgroup. The local symmetry is implemented by the scalars which are valued on the coset space $\mathcal{G}/\mathcal{K}$. To make the full invariance under (1.4) manifest, one introduces scalar fields valued on the full parameter space of $\mathcal{G}$; the fields may be represented by a group element [10, 12]

$$g(x) = e^{Q(x) + P(x)}, \tag{1.5}$$

$$Q(x) = \vec{X} \cdot \vec{\phi}; \qquad P(x) = \vec{Y} \cdot \vec{\phi} \tag{1.6}$$

where $\vec{X}$ is a set of matrices which represent the generators of infinitesimal transformations under the compact part $\mathcal{K}$ of $\mathcal{G}$, and $\vec{Y}$ represents the generators of the remaining transformations under $\mathcal{G}$. We denote the vector field strengths by $F_{\mu\nu}$, their duals by $\tilde{F}_{\mu\nu} \equiv \varepsilon_{\mu\nu\rho\sigma} F^{\rho\sigma}$, the spin-$\frac{1}{2}$ and spin-$\frac{3}{2}$ fermion fields by ψ. Then if g_0 is a group element of $\mathcal{G}_{\text{rigid}}$ and $k(x)$ is a group element of $\mathcal{K}_{\text{local}}$ the supergravity Lagrangian is invariant under (1.4) with transformations defined by [10, 12]:

$$\begin{pmatrix} F \\ \tilde{F} \end{pmatrix} \to g_0 \begin{pmatrix} F \\ \tilde{F} \end{pmatrix}, \quad \psi \to k(x)\psi, \quad g(x) \to g_0 g(x) k^{-1}(x). \quad (1.7)$$

Invariance under the local gauge group $\mathcal{K}$ is assured by the replacement

$$\partial_\mu \to D_\mu = \partial_\mu + Q_\mu, \quad (1.8)$$

analogous to (1.1), where Q_μ is a composite field which is the part in $\mathcal{K}$ of the element

$$g(x)^{-1}\partial_\mu g(x) \equiv Q_\mu + P_\mu, \quad (1.9)$$

of the Lie algebra of $\mathcal{G}$. It follows immediately from (1.7) that Q_μ transforms under (1.4) according to

$$Q_\mu \to k(x)Q_\mu k(x)^{-1} - \left[\partial_\mu k(x)\right]k^{-1}(x) \quad (1.10)$$

in analogy with (1.3).

In writing (1.5), we introduce additional, unphysical, scalar degrees of freedom. These may be removed [12] by a transformation under $\mathcal{K}$ specified by a group element $\hat{k}(x) = e^{\hat{Q}(x)}$ chosen such that

$$g(x)\hat{k}^{-1}(x) = e^{P'(x)} \equiv g \cdot (x). \quad (1.11)$$

The form (1.11) is not preserved by the general transformations (1.7); in this "physical" gauge the group of invariance reduces to

$$\begin{pmatrix} F \\ \tilde{F} \end{pmatrix} \to g_0 \begin{pmatrix} F \\ \tilde{F} \end{pmatrix}, \quad \psi \to k[g_0, P'(x)]\psi, \quad (1.12a)$$

$$e^{P'(x)} \to g_0 e^{P'(x)} k^{-1}[g_0, P'(X)] \equiv e^{P''(x)}, \quad (1.12b)$$

where (1.12b) serves to define $k^{-1}[g_0, P'(x)]$. In particular, if $g_0 \equiv k_0$ is an element of the compact part of $\mathcal{G}$ then $k[g_0, P'(x)] = k_0$. In the physical gauge the only manifest symmetry is $\mathcal{G}$. Its compact part is the diagonal of $\mathcal{K}_{\text{rigid}} \in \mathcal{G}_{\text{rigid}}$ and $\mathcal{K}_{\text{local}}$ for constant parameters, and its noncompact part is realized nonlinearly in this gauge [12].

Cremmer and Julia [10] conjectured that, in analogy with effects discovered [13] in 2-dimensional CP^n models, the propagators of the composite operators (1.10) might develop zero-mass poles, which suggests that the spin-1 particles of conventional gauge theories may arise dynamically as bound states of the elementary fields of supergravity. It was subsequently conjectured [9] (also in analogy with supersymmetric CP^n models [14]) that the full fermionic and bosonic content of conventional gauge theories arises as supergravity bound states, and it was shown that the $N = 8$ supermultiplet which contains the zero-mass-shell projection of the operators (1.10) also contains spin-$\frac{1}{2}$ and spin-0 states which can be identified with all of the particles needed to formulate a viable unified gauge theory [9], and even a simply supersymmetric one [15].

If this picture is correct, most of the symmetries of $N = 8$ supergravity must be broken at an energy scale of the order of the Planck mass, but some subset of symmetries, namely an effective gauge group [which in this picture can be [9] as large as SU(5), but not, say, SO(10) or E_6] and perhaps a simple supersymmetry [15], may remain manifest at lower energies. In addition, the full symmetry of the underlying supergravity theory may determine the particle spectrum. In the first attempts [9, 16, 17] to extract a bound state spectrum it was assumed only that $N = 8$ supersymmetry—which implies automatically an SU(8) multiplet structure—need be respected [18]. The spectra extracted in this way contain a large number of high spin states which must disappear—through whatever mechanism breaks $N = 8$ supersymmetry down to $N = 1$ at most—from the low energy sector of the theory which is to be described by an effective renormalizable theory. The most conventional assumption would be that the unwanted states acquire masses of order of the Planck mass, but it is impossible [9] to achieve this with a single $N = 8$ supermultiplet and appeared [16] nearly so with any finite number [19]. The reason is that a massive particle of spin J has $2J + 1$ helicity states which must transform identically under the surviving gauge group, and these cannot be found among the states of a single massless multiplet for the gauge groups required to fit observation.

It was therefore suggested [20, 15] that the bound state spectrum might be infinite—as, in fact, should be anticipated on the basis of the symmetry (1.4) of the underlying supergravity theory. It has been shown [21] that in CP^n models with a symmetry group of the type (1.4), the bound state spectrum realizes the larger group $\mathcal{G}$. In the case where $\mathcal{G}$ is noncompact, as in supergravity, its unitary realizations are either nonlinear, as for the preons, or infinite. If the scalar preons do not appear explicitly in the effective interactions of composite fields, then the bound state spectrum must necessarily form a linear, infinite representation of the noncompact group $\mathcal{G}$. The problem then is to find irreducible representations, not necessarily of $\mathcal{G}$, but of the full algebra of the symmetries of the theory including supersymmetry.

In the following section we illustrate the program by discussing the algebraically simpler case of $N = 4$ supergravity, and in §3 we discuss the physically interesting case of $N = 8$. For both cases we shall exhibit explicit representations of the full supersymmetry and noncompact group algebra. The solution found for $N = 8$ permits [15], at least in a technical sense, a solution to the problem of assigning masses to high spin particles which are invariant under the effective surviving gauge group. However, there is no immediately apparent criterion for determining which states of this infinite set will actually remain massless, so as to form the particle content of the conventional gauge theory of observed physics. This last problem is of course the ultimate objective of the present program.

2. $N = 4$ supergravity and linear representations of SU(1, 1). For $N = 4$ supergravity [22], we have the identification

$$\mathcal{G} = SU(4) \times SU(1, 1), \qquad \mathcal{K} = U(4), \qquad (2.1)$$

so that the coset space is simply

$$\mathcal{G}/\mathcal{K} = SU(1, 1)/U(1).$$

The nonabelian part of the compact subgroup factors out, leaving a trivial $U(1)$ factor as the maximally compact subgroup of the noncompact group SU(1, 1), whose representations are well known [23].

The full algebra of the $N = 4$ supersymmetry and of (2.1) has been given previously [15, 24]. It was derived by applying successive infinitesimal transformations of SU(1, 1) and supersymmetry on the preon fields as defined [22] in the special gauge in which no unphysical degrees of freedom appear. This gauge has the advantages that the supersymmetry transformations are simple, and that the spectrum of elementary supergravity states represents the full algebra in a transparent manner. In this gauge there are two scalar degrees of freedom which can be represented as a group element of SU(1, 1) by

$$g^{\cdot}(x) = \exp\left(\frac{1}{\sqrt{2}} \begin{pmatrix} 0 & \phi^* \\ \phi & 0 \end{pmatrix} \right), \qquad (2.2)$$

and SU(1, 1) transformations are as defined in (1.12). We denote by L_3 the infinitesimal generator of compact $U(1)$ transformation and by $L_{\pm}$ the remaining generators of SU(1, 1);

$$L_+ = L_-^{\dagger}, \qquad [L_+, L_-] = -2L_3, \qquad [L_3, L_{\pm}] = \pm L_{\pm}. \quad (2.3)$$

Then taking for the group element of an infinitesimal SU(1, 1) transformation

$$g_0 = 1 + \varepsilon_- L_+ + \varepsilon_+ L_-,$$

we may use (1.12b) to determine the transformation properties

$$\delta_{\pm} \phi = i[L_{\pm}, \phi], \qquad (2.4)$$

of the scalar field and the group elements $k(g_0, \phi)$ which determine the transformation properties of the fermions. The results are most simply expressed by introducing a scalar field

$$Z = \frac{\phi}{|\phi|} \tanh \frac{|\phi|}{\sqrt{2}}, \qquad (2.5)$$

which satisfies

$$[L_3, Z] = -Z, \quad [L_3, \bar{Z}] = \bar{Z}, \quad 0 \leqslant |Z|^2 \leqslant 1. \tag{2.6}$$

Then one obtains [15, 24]

$$[L_+, Z] = 1, \qquad [L_-, Z] = Z^2,$$
$$[L_+, \bar{Z}] = -\bar{Z}^2, \qquad [L_-, \bar{Z}] = -1 \tag{2.7}$$

and

$$k(g_0, \phi) = 1 - \varepsilon_- \bar{Z} L_3 - \varepsilon_+ Z L_3, \tag{2.8}$$

so that, for example, the spin-$\frac{3}{2}$ fields ψ_μ which satisfy

$$[L_3, \psi_\mu] = \tfrac{1}{4}\gamma_5 \psi_\mu, \tag{2.9}$$

further satisfy

$$[L_+, \psi_\mu] = -\frac{\bar{Z}}{4}\gamma_5 \psi_\mu, \qquad [L_-, \psi_\mu] = -\frac{Z}{4}\gamma_5 \psi_\mu. \tag{2.10}$$

Since the supersymmetric charge operators, when applied to the $U(1,1)$ invariant states of helicity ± 2 give states of helicity $\pm \frac{3}{2}$, they must transform under $U(1,1)$ like the fields ψ_μ. Defining Q_A and $\bar{Q}^A$, where $A = 1,\ldots,4$ is an SU(4) index, as the helicity lowering and raising operators, respectively, we obtain

$$[L_3, Q] = \tfrac{1}{4}Q, \qquad\quad [L_3, \bar{Q}] = -\tfrac{1}{4}\bar{Q},$$
$$[L_+, Q] = -\tfrac{1}{4}\bar{Z}Q, \qquad [L_+, \bar{Q}] = \tfrac{1}{4}\bar{Z}\bar{Q},$$
$$[L_-, Q] = -\tfrac{1}{4}ZQ, \qquad [L_-, \bar{Q}] = \tfrac{1}{4}Z\bar{Q}, \tag{2.11}$$

where here Z and $\bar{Z}$ are to be interpreted [15, 24] as constant operators which are the asymptotic values of the scalar fields as $x \to \infty$. Using the commutation relations (2.6) and (2.7) and

$$\{Q, Q\} = \{\bar{Q}, \bar{Q}\} = [Q, P] = [\bar{Q}, P] = 0, \tag{2.12a}$$

$$\{\bar{Q}^A, Q_B\} = \gamma^\mu P_\mu \delta_{AB}, \tag{2.12b}$$

$$[Z, \bar{Z}] = 0, \tag{2.12c}$$

and the fact that the commutators of Q and P with Z and $\bar{Z}$ vanish asymptotically, we generate by successive commutation the full algebra which contains an infinite number of fermionic and bosonic generators:

$$Q_{A,n,m} = \bar{Z}^n Z^m Q_A, \qquad \bar{Q}^A_{n,m} = \bar{Z}^n Z^m \bar{Q}^A = (Q_{A,m,n})^\dagger, \tag{2.13a}$$

$$P_{\mu,n,m} = \bar{Z}^n Z^m P_\mu, \tag{2.13b}$$

which satisfy the commutation relations (2.12a) and

$$\{\overline{Q}_{nm}^A, Q_{Blk}\} = \delta_B^A \gamma^\mu P_{\mu, n+l, m+k}, \tag{2.14}$$

$$[L_3, A_{n,m}] = (\eta_A + n - m)A_{n,m},$$

$$[L_+, A_{n,m}] = -(\eta_A + n)A_{n+1,m} + mA_{n,m-1},$$

$$[L_-, A_{n,m}] = -nA_{n-1,m} + (m - \eta_A)A_{n,m+1} \tag{2.15}$$

with $\eta_A = \frac{1}{4}, -\frac{1}{4}, 0$ for $A = Q_B, \overline{Q}^B, P_\mu$, respectively.

The above construction is similar to that by which infinite Kac-Moody algebras [25] are obtained from ordinary Lie algebras. In the Kac-Moody case one finds c-number terms which are necessary for the construction of certain unitary representations. In the generalized supersymmetry algebra however, the Q's and P's have nontrivial Lorentz transformation properties, so that c-numbers could arise only in the commutators $\{Q, Q\}$ and $\{\overline{Q}, \overline{Q}\}$.

The algebra defined by (2.3), (2.12a), (2.14) and (2.15), obtained here using plausibility arguments, is identical to that obtained [15, 24] by explicit construction from the transformation properties of the $N = 4$ supergravity fields under supersymmetry and SU(1, 1). The problem is now to find its unitary, linear and therefore infinite representations. To represent the generalized momenta (2.13a) on a subspace of the Hilbert space with fixed momentum, it is necessary and sufficient to represent the operators Z and $\overline{Z}$. Once states $|\psi\rangle$ are found on which the operators $Z, \overline{Z}$ and $L_\pm, L_3$ can be represented, it is trivial to represent the full algebra by

$$P_{\mu,n,m}|\psi\rangle = P_\mu \overline{Z}^n Z^m |\psi\rangle, \qquad Q_{A,n,m}|\psi\rangle = Q_A \overline{Z}^n Z^m |\psi\rangle,$$

$$\overline{Q}_{n,m}^A |\psi\rangle = \overline{Q}^A \overline{Z}^n Z^m |\psi\rangle. \tag{2.16}$$

Such a representation will correspond to an infinite series of ordinary $N = 4$ supermultiplets with fixed maximum helicity λ_m. There may be other, possibly more interesting, representations with an infinite helicity spectrum as suggested by the presence of an infinite number of super-symmetry generators, but it is clear that states representing the P's and the SU(1, 1) generators will also represent the Q's as in (2.16).

The problem is now reduced to finding a representation of the operators $Z, \overline{Z}, L_+, L_-$ and L_3 subject to the commutation relations (2.3), (2.6), (2.7) and (2.12c). If we work in a basis in which L_3 is diagonal, then $\overline{Z}$ and L_+ can have nonvanishing matrix elements between the same states, as can Z and L_-. Clearly Z and $\overline{Z}$ can be represented in the same space as L_+ and L_- if one pair of operators can be expressed as

functions of the other. We therefore look for representations on which
the algebra reduces as follows[2] (note that $[L_3, Z\bar{Z}] = 0$):

$$L_+ = \bar{Z}\big[A(Z\bar{Z})L_3 + B(Z\bar{Z})\big],$$
$$L_- = L_+^\dagger = \big[A^*(Z\bar{Z})L_3 + B^*(Z\bar{Z})\big]Z. \qquad (2.17)$$

Then the commutation relations (2.7) require

$$A(Z\bar{Z}) = -1, \qquad Z\bar{Z}A(Z\bar{Z}) = -1, \qquad (2.18)$$

and (2.3) requires

$$A(Z\bar{Z}) = 1/Z\bar{Z}, \qquad 2\,\mathrm{Re}\,B(Z\bar{Z}) = A(Z\bar{Z}). \qquad (2.19)$$

Putting these results together we obtain the forms

$$L_+ = -\bar{Z}\big(L_3 + \tfrac{1}{2} + i\alpha\big),$$
$$L_- = -\big(L_3 + \tfrac{1}{2} - i\alpha\big)Z = -Z\big(L_3 - \tfrac{1}{2} - i\alpha\big), \qquad (2.20)$$

with the condition $Z\bar{Z} = 1$, which means that the variable Z is restricted
to the unit circle which is an $SU(1, 1)$ invariant subspace of the unit disc
on which Z is defined. Using the construction (2.20) we can evaluate the
$SU(1, 1)$ Casimir operator as

$$K \equiv \tfrac{1}{2}(L_+L_- + L_-L_+) - L_3^2 = +\tfrac{1}{4} + \alpha^2. \qquad (2.21)$$

This class of $SU(1, 1)$ representations corresponds to the principal series
[23]. The $N = 4$ supergravity algebra may be represented as follows. If we
consider an ordinary supermultiplet whose state of highest helicity λ_m
has L_3 eigenvalue η_m, then we generate an infinite series of supermultiplets characterized by λ_m and $\eta_m + a, a = 0, \pm 1, \pm 2,\ldots$:

$$P_{n,m}|\lambda, \eta\rangle = P|\lambda, \eta + n - m\rangle,$$
$$Q_{An,m}|\lambda, \eta\rangle = Q_A|\lambda, \eta + n - m\rangle,$$
$$Q_{n,m}^A|\lambda, \eta\rangle = Q^A|\lambda, \eta + n - m\rangle \qquad (2.22)$$

and

$$L_3|\lambda, \eta\rangle = \eta|\lambda, \eta\rangle,$$
$$L_\pm|\lambda, \eta\rangle = -\big(\eta + \tfrac{1}{2} + i\alpha\big)|\lambda, \eta \pm 1\rangle, \qquad (2.23)$$

with $\eta = \eta_m + (\lambda_m - \lambda)/2$. Notice that the invariant condition $Z\bar{Z} = 1$
reduces the doubly infinite algebra to a simply infinite one:

$$P_{n,m} \equiv P_{n-m}, \quad Q_{Anm} \equiv Q_{A,n-m}, \quad Q_{nm}^A \equiv Q_{n-m}^A,$$
$$[L_3, A_a] = (\eta_A + a)A_a, \qquad [L_\pm, A_a] = -(\eta_A + a)A_{a\pm 1}. \qquad (2.24)$$

---

[2] One might consider a more general function of L_3, but it seems unlikely that a higher
order polynomial can be made to satisfy (2.7).

The supermultiplets containing the zero-mass-shell projection of the SU(4) $\times$ U(1) currents are the multiplet with $\lambda_m = \frac{3}{2}$, $\eta_m = -\frac{1}{4}$, and its TCP conjugate multiplet $\lambda_m = \frac{1}{2}$, $\eta_m = -\frac{3}{4}$. Notice that the helicity zero states have $\eta = \pm\frac{1}{2}$. For the choice $\alpha = 0$, the SU(1, 1) multiplets to which they are assigned split into two irreducible multiplets of the discreet series with $\eta = \frac{1}{2} + |n|$ and $\eta = -\frac{1}{2} - |n|$, respectively. Otherwise the multiplet for each helicity is irreducible.

Are there other possible representations of the algebra given above? The ansatz (2.17) need not be generally valid, and one might have guessed, for example, that the gauge vectors should be SU(1, 1) singlets since the gauge transformation (1.10) is abelian in this case; the transverse part of Q_μ is invariant under SU(1, 1). However, it is easy to see that the algebra does not permit an SU(1, 1) singlet as part of a discreet spectrum: If

$$L_\pm |0\rangle = 0. \tag{2.25}$$

Then, for example,

$$0 = \langle 0 | [L_-, P_{\mu 1,0}] | 0 \rangle = \langle 0 | P_\mu | 0 \rangle = p_\mu \langle 0 | 0 \rangle, \tag{2.26}$$

which can be satisfied only for the vacuum state. On the other hand, if the state defined by (2.25) is part of a continuum of states with $L_3 = 0$, characterized by a continuous parameter ν, and normalized such that

$$\langle 0, \nu' | 0, \nu \rangle = \delta(\nu - \nu'), \tag{2.27}$$

then it may be possible to represent the algebra on these states. However, these would not correspond to bound states which should form a discrete spectrum.

More generally, irreducible representations of SU(1, 1) can be classified [23] according to the eigenvalue of the Casimir operator (2.21):

$$K | l_3, \nu \rangle = -\nu(\nu - 1) | l_3, \nu \rangle, \tag{2.28}$$

and the eigenvalues of L_3:

$$l_3 = \eta + a, \quad a = 0, \pm 1, \pm 2, \ldots, 0 < \eta \leq 1. \tag{2.29}$$

Because the matrix element

$$\langle l_3, \nu | L_+ L_- | l_3, \nu \rangle = (l_3 - \nu)(l_3 + \nu - 1), \tag{2.30}$$

is positive definite, the values of ν are restricted to three classes:[3]

(a) The principal series: $\nu = \frac{1}{2} + i\alpha$, which is the one found above, (2.21). It admits the eigenvalues (2.29) for any value of η.

---

[3] There are further restrictions [23] on the eigenvalues of L_3 and K, if one wishes to represent the group and not only the algebra.

(b) The supplementary series: $0 < \nu < \frac{1}{2}$, which admits eigenvalues (2.29) for $\eta < \nu$ and $\eta > 1 - \nu$. If $L_3 = 0$, the limit $\nu \to 0$ corresponds to an SU(1, 1) singlet.

(c) The discreet series: $\nu = |l_3|_{\min}$, which has L_3 eigenvalues $l_3 = \nu + a$ or $l_3 = -\nu - a, a = 0, 1, \ldots, \infty$.

The eigenfunctions of K and L_3 can be represented as functions of Z and $\bar{Z}$ on the unit disc,[4] and $L_\pm$, L_3 are differential operators [24]. The solutions of the eigenvalue equation (2.28) which converge inside the unit disc are hypergeometric functions multiplied by an appropriate power of Z or $\bar{Z}$ and by $(1 - Z\bar{Z})^\nu$. The scalar products are defined by integration with the invariant measure $dZd\bar{Z}(1 - Z\bar{Z})^{-2}$. For the supplementary series and principal series the normalization is infinite [23]; for the principal series functions have been constructed [23] which are normalized as in (2.27). For the discreet series with $\nu > \frac{1}{2}$ the integral is well defined, and one might guess that such representations would be natural candidates for the bound state spectrum. However, in practice it appears difficult [24] to represent the full algebra on these states; if a unitary representation does exist it will involve states which are degenerate with respect to helicity and L_3.

What we have done in this section is to display one class of representations of the SU(1, 1) and $N = 4$ supersymmetry algebra which appears to be a minimal one; we will draw on this result to find an analogous class of representations of the $E_{7(7)}$ and $N = 8$ supersymmetry algebra of $N = 8$ supergravity.

3. $N = 8$ supergravity and linear representations of $E_{7(7)}$. For $N = 8$ supergravity, the group of invariance [10] is (1.4) with

$$\mathcal{K} = \mathrm{SU}(8), \qquad \mathcal{G} = E_{7(7)}. \tag{3.1}$$

Rather than deriving the algebra directly from the transformation properties [10] of the elementary supergravity fields, we shall follow the line of reasoning used in the previous section. Consider an infinitesimal $E_{7(7)}$ transformation as defined by (1.12) in the special gauge with

$$g_0 = 1 + \vec{\alpha} \cdot \vec{Y}, \tag{3.2}$$

where $Y_i = -Y_i^\dagger$ and α_i is real $(i = 1, \ldots, 70)$. The spin-$\frac{3}{2}$ fields which transform as $(8 + \bar{8})$ under SU(8), transform under (3.2) in the same way

---

[4] More generally, these are represented as functions of Z and $\bar{Z}$ multiplying a fixed eigenfunction of SU(1), as for induced representations. The form of the differential operators [24] which represent L depends on the inducing representations of SU(1). Note that for $Z\bar{Z} \to 1$ these operators can be expressed in the form (2.17), but with $A = 1$, $B = 0$; this discrepancy with (2.20) may be related to the fact that the solutions (2.21) of the eigenvalue equation (2.28), expressed as functions of $Z, \bar{Z}$ on the unit disc, are divergent on the unit circle.

as under an infinitesimal SU(8) transformation

$$k(\phi_i, \alpha_i) = 1 + \vec{\mathcal{Q}}(\phi_i, \alpha_i) \cdot \vec{X}, \qquad (3.3)$$

with $X_a = -X_a^\dagger$ and $\mathcal{Q}_a(\phi_i, \alpha_i)$ real $(a = 1, \ldots, 63)$. It follows that the helicity lowering and raising supersymmetry operators Q_A and $\overline{Q}^A$ $(A = 1, \ldots, 8)$, respectively, transform under (3.2) as

$$\begin{aligned}
\delta Q_A &= \alpha_i[Y_i, Q_A] = \mathcal{Q}^a(\alpha_i, \phi_i) X_A^{aA'} Q_{A'}, \\
\delta \overline{Q}^A &= \alpha_i[Y_i, \overline{Q}^A] = - \mathcal{Q}^a(\alpha_i, \phi_i) \overline{Q}^{A'} X_{A'}^{aA}.
\end{aligned} \qquad (3.4)$$

The problem of determining the algebra then consists in determining the parameters $\mathcal{Q}(\phi_i, \alpha_i)$ and the variation $\delta\phi_i$ of ϕ_i under (3.2). The results (see Appendix) can be expressed as

$$\delta Z_i = \alpha_j\left[\tfrac{1}{2}\delta_{ij}(1 + NZ^2) - NZ_i Z_j\right] = \alpha_j[Y_j, Z_i], \qquad (3.5)$$

$$\mathcal{Q}^a(\phi, \alpha) = x_{ij}^a Z_i \alpha_j. \qquad (3.6)$$

with

$$Z_i = \frac{\phi_i}{\sqrt{N \Sigma \phi_i^2}} \tanh\left(\frac{\sqrt{N \Sigma \phi_i^2}}{2}\right), \qquad 0 \leqslant Z^2 = \sum Z_i^2 \leqslant \frac{1}{N}, \quad (3.7)$$

where $x_{ij}^a = -x_{ji}^a$ is the matrix element of X^a in the 70-dimensional SU(8) representation and $N = \tfrac{3}{23}$ is related to the eigenvalue of the Casimir operator $\Sigma_a(X^a)^2$ for the same representation: $K_{70} = 69N = 9$. Thus the algebra can be formulated in terms of a variable $\vec{Z}$ valued in a 70-dimensional sphere whose surface

$$\sum_i Z_i^2 = \frac{1}{N}, \qquad (3.8)$$

is an $E_{7(7)}$ invariant subspace. By taking successively the commutators (3.4), (2.12), (3.5) and

$$[Y_i, Y_j] = -x_{ji}^a X_a, \quad [X^a, Y_i] = x_{ji}^a Y_j, \quad [X^a, X^b] = -x_{abc} X^c, \quad (3.9)$$

one generates the infinite algebra, including the operators

$$\begin{aligned}
Q_{A i_1, \ldots, i_n} &= Z_{i_1} \cdots Z_{i_n} Q_A, \\
\overline{Q}^A_{i_1, \ldots, i_n} &= Z_{i_1} \cdots Z_{i_n} \overline{Q}^A, \\
P_{\mu i_1, \ldots, i_n} &= Z_{i_1} \cdots Z_{i_n} P_\mu,
\end{aligned} \qquad (3.10)$$

which are totally symmetric in the indices $i_1, \ldots, i_n$ and which satisfy (2.12) and

$$
\left[Y_j, A_{A i_1, \ldots, i_n} \right] = \sum_{k=1}^{n} \left(\frac{1}{2} \delta_{j i_k} \left[Q_{A i_1, \ldots, i_{k-1}, i_{k+1}, \ldots, i_n} + N Q_{A l i_1, \ldots, i_{k-1}, i_{k+1}, \ldots, i_n} \right] \right.
$$
$$
\left. + N Q_{A j i_1, \ldots, i_{k-1}, i_{k+1}, \ldots, i_n} \right)
$$
$$
+ \sum_a x_{lj}^a X_A^{A'} Q_{A' l i_1, \ldots, i_n}, \tag{3.11}
$$

and similarly for $\bar{Q}$ and P. Again, representation of the P's is equivalent to representation of the Z_i which are again to be interpreted as constant operators corresponding to the asymptotic values of the fields. We therefore look for a unitary representation of the operators X, Y and Z subject to the commutation relations (3.5), (3.9) and

$$
\left[Z_i, Z_j \right] = 0, \qquad \left[X^a, Z_i \right] = x_{ji}^a Z_j. \tag{3.12}
$$

Since Y_i and Z_i have the same transformation properties under SU(8) we look for representations on which we can make the identification (note that $[X^a, Z^2] = 0$)

$$
Y_i = A(Z^2) Z_i + B(Z^2) x_{ij}^a Z_j X^a, \tag{3.13}
$$

which is the only form linear in $\vec{X}$ with the correct SU(8) transformation property, and is analogous to (2.17). The antihermiticity of Y and X

$$
Y_i^\dagger = A^*(Z^2) Z_i - B^*(Z^2) x_{ij}^a X^a Z_j
$$
$$
= (A^* - 69 N B^*) Z_i - B^* x_{ij}^a Z_j X^a = -Y_i, \tag{3.14}
$$

where we have used (3.12) and (A.16) of the Appendix, require

$$
B^*(Z^2) = B(Z^2), \qquad 2 \operatorname{Re} A(Z^2) = 69 N B(Z^2), \tag{3.15}
$$

and the commutator (3.5) requires

$$
B = -1, \qquad Z^2 = 1/N, \tag{3.16}
$$

so that $\vec{Z}$ is restricted to the surface of the sphere on which it is defined. Now setting

$$
Y_i = -x_{ij}^a Z_j X^a - Z_i N(69/2 + i\omega), \tag{3.17}
$$

one can verify that the commutators (3.9) are satisfied, provided the condition (3.16) is imposed. Thus if we can find unitary representations of the Z_i, subject to (3.16), they will represent the full algebra. They are not expected to be irreducible with respect to $E_{7(7)}$ and, in fact, the $E_{7(7)}$ Casimir operator $X^a X^a - Y_i Y_i$ does not reduce to a multiple of the identity operator as it did for the analogous construction in the SU(1, 1) case.

As an example of a unitary representation of $\vec{Z}$, consider the following construction. Let $|\alpha, n, r\rangle$ be a state which transforms according to an irreducible representation r of SU(8) which is contained in the totally symmetric product of n 70-plets with all traces, i.e. all configurations for which two 70-plets combine to form an SU(8) singlet, removed. The reducible representation containing all such r for fixed n can be represented by a totally symmetric, traceless tensor of rank n

$$T^n_{i_1 \cdots i_n} = \sum_{r,\alpha} S^{\alpha,r,n}_{i_1 \cdots i_n} T^\alpha_{r,n}. \tag{3.18}$$

We normalize the tensors $S^{r,\alpha}$ such that

$$\sum_{\{i\}_n} S^{\alpha,r,n}_{i_1 \cdots i_n} S^{\beta,p,n}_{i_1 \cdots i_n} = \delta_{rp} \delta_{\alpha\beta}, \tag{3.19}$$

and

$$\sum_\alpha S^{\alpha,r,n}_{i_1 \cdots i_n} S^{\alpha,r,n}_{j_1 \cdots j_n} = \delta_{\{i\}_n \{j\}_n} \frac{M_r}{M(n)}, \tag{3.20}$$

where M_r is the multiplicity of the representation r $(\alpha = 1, \ldots, M_r)$, and

$$M(n) = \frac{(67 + n)!}{68!} (68 + 2n) = \sum_{r \in n} M_r, \tag{3.21}$$

is the total multiplicity of the traceless, symmetric tensors (3.18); $\delta_{\{i\}_n \{j\}_n}$ is the projection operator in $(70)^n$ dimensions onto the space of these tensors:

$$\delta_{\{i\}_n \{j\}_n} = \frac{1}{n!} \Big(\delta_{i_1 j_1} \cdots \delta_{i_n j_n} + \text{permutations in } \{j\}_n \Big)$$

$$- \frac{2}{n(n-1)} \frac{1}{66 + 2n} \Big\{ \delta_{i_1 i_2} \delta_{j_1 j_2} \big(\delta_{i_3 j_3} \cdots \delta_{i_n j_n}$$

$$+ \text{perms}\{j\}_{n-1} \big) + \text{perms} \Big\}$$

$$+ \cdots,$$

$$\sum_{\{i\}_n} \delta_{\{i\}_n \{i\}_n} = M(n),$$

$$\sum_i \delta_{\{i, i_1 \cdots i_{n-1}\}, \{i, j_1 \cdots j_{n-1}\}} = \frac{M(n)}{M(n-1)} \delta_{\{i\}_{n-1} \{j\}_{n-1}}, \quad \text{etc.} \tag{3.22}$$

We define operation with Z_i on the state $|\alpha, n, r\rangle$

$$Z_i|\alpha, n, r\rangle$$

$$= \frac{1}{\sqrt{N}}\left\{\sqrt{\frac{n}{68+2n}}\sum_{\beta,p,i_k} S^{\alpha,r,n}_{i\,i_1\cdots i_{n-1}}S^{\beta,p,n-1}_{i_1\cdots i_{n-1}}|\beta, n-1, p\rangle\right.$$

$$\left.+\sqrt{\frac{n+1}{70+2n}}\sum_{\gamma,g,j_k} S^{\alpha,r,n}_{j_1\cdots j_n}S^{\gamma,q,n+1}_{i\,j_1\cdots j_n}|\gamma, n+1, q\rangle\right\}.$$

$$(3.23)$$

With the help of (3.18)–(3.22), one can verify explicitly that the condition (3.8) is satisfied, namely

$$\left\langle\beta, m, q\left|\sum_i Z_i^2\right|\alpha, n, r\right\rangle = \frac{1}{N}\delta_{\alpha\beta}\delta_{nm}\delta_{qr}, \qquad (3.24)$$

that the Z_i commute, and that the hermicity condition

$$\langle\beta, m, q|Z_i|\alpha, n, r\rangle = \langle\alpha, n, r|Z_i|\beta, m, q\rangle^*, \qquad (3.25)$$

is satisfied by the definition (3.23). The effect of operation on $|\alpha, n, r\rangle$ by the Y_i then follows immediately from (3.17). Clearly, other representations can be constructed on states $|M, R; n, r; R_0\rangle$ constructed from a fixed irreducible representation R_0 of SU(8): $R \in r \times R_0$, $M = 1, \ldots, M(R)$, where $M(R)$ is the multiplicity of the irreducible representation R of SU(8).

Then we can represent the full algebra on an infinite set of $N = 8$ multiplets of common highest helicity and with the states of highest helicity λ_m belonging to the representation of the algebra (3.5), (3.9) and (3.12) which is generated in the above way from some SU(8) representation R_m. The states of helicity $\lambda_m - \frac{1}{2}$ will be in the representations generated from $R_m \times 8$, and so on, with

$$Q_{i_1\cdots i_n}|\psi\rangle = Z_{i_1}\cdots Z_{i_n}Q|\psi\rangle, \quad \text{etc.} \qquad (3.26)$$

Notice that as in the SU(1, 1) case the infinite algebra reduces to a smaller infinite algebra because of the condition (3.16) which implies that the operators (3.10) are not all independent, since they are related by trace conditions.

The representations obtained in this way contain, for fixed helicity states generated from some (reducible for $\lambda \neq \lambda_{max}$ or λ_{min}) SU(8) representation R_λ, will contain a given SU(8) representation as many times as it occurs in

$$\sum_{n=0}^{\infty}\sum_{r\in n} r \times R_\lambda. \qquad (3.27)$$

Note that the degeneracy of each SU(8) representation here is smaller than that for the $E_{7(7)}$ representations constructed using the oscillator method [26] because the condition (3.16) removes a part of that degeneracy. However the representations found here are still not expected to be irreducible under $E_{7(7)}$, as can be seen by noting that the $E_{7(7)}$ Casimer constructed from (3.17)

$$-\sum_i Y_i Y_i + \sum_a X^a X^a = \left(\frac{69}{2}\right)^2 + \omega^2 + \sum_a (X^a)^2 + Z_i x_{ij}^a x_{jk}^b Z_k X^a X^b,$$

$$(3.28)$$

is not a multiple of the identity, as was the case for SU(1, 1), (2.21). Furthermore, it does not seem possible to reduce the representation constructed explicitly above, by, for example, assuming some of the representations $r \in n$ are absent, while maintaining the requirements (3.24), (3.25) and (3.12).

In contrast to the group SU(1, 1), little is known about $E_{7(7)}$ representations. One can construct[5] induced representations as functions of the Z_i multiplying some fixed representation R_0 of SU(8). These are apparently [27] the same as those constructed [26] using the oscillator method. Irreducible representations can be constructed [26, 28] on spaces which are smaller than the 70-dimensional space of the variable $\vec{Z}$, but it is difficult to see how the Z's, and consequently the generalized momenta of (3.10), can be represented on these smaller spaces. In addition, the general SU(8) structure of supermultiplets, which require that if there is a state of maximum helicity λ_m in a representation R_M, there must be a state of helicity $\lambda_m - \frac{1}{2}$ in the reducible representation $R_M \times 8$, and so on, suggests that states of fixed helicity which belong to a single irreducible representation of $E_{7(7)}$ representable on a smaller space cannot represent the full algebra.

In conclusion, while we have not shown that it is a unique solution to the problem, we have explicitly displayed a class of representations of the $E_{7(7)}$ and $N = 8$ supersymmetry algebra of $N = 8$ supergravity. They have the property that states of fixed helicity include all the SU(8) representations contained in the totally symmetrized products of 70-plets with some fixed inducing SU(8) representation. As discussed previously [15], each of these contains all representations of the SU(6) subgroup of SU(8) an infinite number of times, and they therefore allow, in principle

---

[5] An SU(8) singlet is an eigenfunction of the Casimir operator (3.28). Solutions of the corresponding differential equation for eigenvalues $K = (69/2)^2 + \omega^2$ diverge on the invariant surface of the sphere and are similar to those of the principal series for SU(1, 1); other solutions appear to be more strongly divergent when square integrated over the invariant measure $d^{70}Z(1 - Z^2)^{-70}$. We have not studied these functions for representations which are induced from nontrivial SU(8) representations.

at least, a mechanism by which unwanted high spin states may acquire masses. What is still missing in this picture is a criterion for determining which states remain massless. It would also be interesting to know what, if any, relation there is between the infinite spectra suggested here on the grounds of the symmetries of supergravity and analogy with CP^n models, and those suggested by dynamical arguments [7, 29].

Appendix. The algebra of $E_{7(7)}$. The scalars can be represented [10] by a 56 × 56 matrix

$$g^{\cdot}(x) = \exp\begin{pmatrix} 0 & \bar{\phi} \\ \phi & 0 \end{pmatrix}, \tag{A.1}$$

in the fundamental $E_{7(7)}$ representation whose basis vectors transform as $28 + \overline{28}$ under SU(8):

$$\psi = \begin{pmatrix} \psi_{[AB]} \\ \psi^{[AB]} \end{pmatrix}, \qquad A, B = 1, \ldots, 8, \tag{A.2}$$

and $\bar{\phi}_{[ABCD]}$ is interpreted as a 28 × 28 matrix with 70 independent components. In this representation the generator of an infinitesimal SU(8) transformation is of the form

$$\vec{\mathcal{Q}} \cdot \vec{X} = \begin{pmatrix} A_{[AB]}{}^{[CD]} & \\ & -A^{[AB]}{}_{[CD]} \end{pmatrix}, \tag{A.3}$$

where

$$A_{[AB]}^{[CD]} = \vec{\mathcal{Q}}_{[A}^{[C} \delta_{B]}^{D]}, \tag{A.4}$$

is an antihermitian 28 × 28 matrix with 63 independent components. An infinitesimal generator of noncompact $E_{7(7)}$ transformations takes the form

$$\vec{\alpha} \cdot \vec{Y} = \begin{pmatrix} 0 & \bar{\alpha}^{[ABCD]} \\ \alpha_{[ABCD]} & 0 \end{pmatrix}. \tag{A.5}$$

Under an infinitesimal $E_{7(7)}$ transformation we have, from (1.12b),

$$g^{\cdot}(x) \to g^{\cdot\prime}(x) = g^{\cdot}(x) + \delta g^{\cdot}(x),$$
$$\delta_{\alpha} g^{\cdot}(x) = (\vec{\alpha} \cdot \vec{Y})g^{\cdot}(x) - g^{\cdot}(x)(\vec{\mathcal{Q}} \cdot \vec{Y}). \tag{A.6}$$

Writing

$$g^{\cdot}(x) = \begin{pmatrix} \sqrt{1 + \eta\bar{\eta}} \,_{[AB]}{}^{[CD]} & \bar{\eta}^{[AB][CD]} \\ \eta_{[AB][CD]} & \sqrt{1 + \eta\bar{\eta}}^{[AB]}{}_{[CD]} \end{pmatrix}, \tag{A.7}$$

where

$$\eta = \phi \sum_n \frac{(\bar{\phi}\phi)^n}{(2n+1)!} = \phi(\bar{\phi}\phi)^{-1/2}\sinh(\bar{\phi}\phi)^{1/2},$$

$$\sqrt{1+\eta\bar{\eta}} = \sum_n \frac{(\bar{\phi}\phi)^n}{2n!} = \cosh(\bar{\phi}\phi)^{1/2}, \qquad (A.8)$$

we may use (A.6) to obtain two independent equations which serve to determine $\mathcal{C}$ and $\delta\phi$. They are most concisely expressed in the form

$$\delta_\alpha\chi = \alpha - \chi\bar{\alpha}\chi,$$

$$\{A, (1-\chi\bar{\chi})^{-1/2}\} = (1-\chi\bar{\chi})^{-1/2}\chi\bar{\alpha} - \alpha\bar{\chi}(1-\chi\bar{\chi})^{-1/2}. \quad (A.9)$$

where

$$\chi = (1+\eta\bar{\eta})^{-1/2}\eta = \phi(\bar{\phi}\phi)^{-1/2}\tanh(\bar{\phi}\phi)^{1/2},$$

$$(1+\eta\bar{\eta})^{1/2} = (1-\chi\bar{\chi})^{-1/2}. \qquad (A.10)$$

The matrix equations (A.9) are cumbersome, and rather than solve them directly, we shall make use of general arguments to obtain (3.5) and (3.6).

First note that $\delta_\alpha\phi$ transforms under SU(8) according to the reducible representation 70×70; it is linear in α and an arbitrary function of ϕ. In terms of real variables ϕ_i, α_i, $i = 1,\ldots,70$,

$$\delta_\alpha\phi_i = A\alpha_i + \alpha_i\phi^2 F_1(\phi^2) - \phi_i(\alpha\cdot\phi)F_2(\phi^2), \quad \phi^2 = \sum_i \phi_i^2. \quad (A.11)$$

This form may be further constrained by comparison with (A.9) and (A.10) which give immediately $A = 1$. Next we invert (A.10) to write

$$\phi = \chi(\bar{\chi}\chi)^{-1/2}\arctanh(\bar{\chi}\chi)^{1/2} = \chi \sum_{n=0}^{\infty} \frac{(\bar{\chi}\chi)^n}{2n+1}. \qquad (A.12)$$

We may then use (A.9) to show that

$$\mathrm{Tr}\,\bar{\phi}\delta_\alpha\phi = \sum_i \phi_i\delta_\alpha\phi_i = \mathrm{Tr}\,\bar{\phi}\alpha = \phi\cdot\alpha, \qquad (A.13)$$

which requires $F_1(\phi^2) = F_2(\phi^2)$:

$$\delta_\alpha\phi_i = \alpha_i + [\alpha_i\phi^2 - \phi_i(\alpha\cdot\phi)]F(\phi^2). \qquad (A.14)$$

The commutation relations (3.9) require

$$\delta_\beta\delta_\alpha\phi_i - \delta_\alpha\delta_\beta\phi_i = -\alpha_j x_{jk}^a \beta_k x_{mi}^a \phi_m, \qquad (A.15)$$

which may be rewritten using

$$\sum_a x_{ij}^a x_{kl}^a = N(\delta_{ik}\delta_{jl} - \delta_{il}\delta_{jk}), \qquad (A.16)$$

which follows from the fact that the quartic form,

$$(\phi_1 \vec{x} \phi_2)(\phi_3 \vec{x} \phi_4) \propto (\phi_1 \cdot \phi_3)(\phi_2 \cdot \phi_4) - (\phi_1 \cdot \phi_4)(\phi_2 \cdot \phi_3),$$

is an SU(8) invariant which is antisymmetric under $(1 \leftrightarrow 2)$ or $(3 \leftrightarrow 4)$, and

$$\sum_a x_{ij}^a x_{jl}^a = -K_{70}\delta_{il} = -69N\delta_{il}, \tag{A.17}$$

where $K_{70} = 9$ is the SU(8) Casimir for the representation 70. We thus get

$$\delta_\beta \delta_\alpha \phi_i - \delta_\alpha \delta_\beta \phi_i = N[\alpha_i(\phi \cdot \beta) - \beta_i(\phi \cdot \alpha)], \tag{A.18}$$

and comparison with the result obtained using (A.14) gives a constraint on $F(\phi^2)$. We next express the parameters $\vec{\mathcal{Q}}$ in terms of real variables $\mathcal{Q}^a(\alpha, \psi)$, $a = 1,\ldots,63$, and, since $\vec{\mathcal{Q}}$ is linear in α, SU(8) covariance requires

$$\vec{\mathcal{Q}}^a(\alpha, \phi) = \alpha_i x_{ij}^a \phi_j G(\phi^2). \tag{A.19}$$

Using

$$\delta_\alpha Q_A = \vec{\mathcal{Q}}^a(\alpha, \phi)\left(\frac{i\lambda^a}{2}\right)_A^{A'} Q_{A'}, \tag{A.20}$$

where λ^a is an SU(8) Gell-Mann matrix in 8 dimensions: $\mathrm{Tr}\,\lambda^a\lambda^b = 2\delta_{ab}$, and

$$[\delta_\beta, \delta_\alpha]Q_A = \left(\alpha_i x_{ij}^a \beta_j\right)\left(\frac{i\lambda^a}{2}\right)_A^{A'} Q_{A'}, \tag{A.21}$$

we obtain further constraints on the functions $F(\phi^2)$ and $G(\phi^2)$ by comparing (2.21) with the result obtained using (A.14), (A.19) and (A.20), namely,

$$1 + xF(x) = \frac{1}{2G(x)} + \frac{NxG(x)}{2}, \tag{A.22}$$

and

$$2G'(x) = \frac{1}{2x} - \frac{G(x)}{x} - \frac{NG^2(x)}{2}. \tag{A.23}$$

(A.23) has as solution

$$G(x) = \frac{1}{\sqrt{Nx}} \tanh\left(\frac{\sqrt{Nx}}{2}\right),$$

and (3.5)–(3.7) follow.

Note added in proof. Equations (2.20) and (3.17) are special cases of a procedure known in the literature as "expansion" of a group to a larger group. It can also be used as a decontraction formula to reconstruct a group from its contraction in the sense of Wigner-Inönü. It was first used by A. Sankaranarayanan and R. H. Good, Jr., Phys. Rev. B **140** (1965), 509. Subsequent work by Sankaranarayanan and others is reviewed in R. Gilmore, *Lie groups, Lie algebras and some of their applications*, Wiley, New York, 1974. The procedure was independently known to Y. Dothan, M. Gell-Mann and Y. Ne'eman (unpublished); their work is described by Y. Dothan and Y. Ne'eman in *Symmetry groups*, F. Dyson, ed., Benjamin, New York, 1966 and by R. Hermann, *Lie groups for physicists*, where a construction similar to (2.20) and (3.17) is referred to as "Gell-Mann's formula". However, the formula in Hermann's book contains misprints.

I wish to thank M. Gell-Mann and Y. Ne'eman for pointing out some of the earlier literature to me.

Acknowledgment. This work was begun in collaboration with John Ellis and Bruno Zumino; its subsequent development has benefited from numerous discussions with B. Zumino. I have also enjoyed conversations with Murray Gell-Mann and John Schwarz. I understand that M. Günaydin has independently worked out aspects of the $N = 8$ supergravity algebra.

Part of this work was completed at Fermi Lab and at the Aspen Center of Physics; their hospitality is gratefully acknowledged. This work was suppored in part by the National Science Foundation under Research Grant No. Phy-82-03424 and the U. S. Department of Energy under Contract DE-AC02-76CHO-3000.

REFERENCES

1. D. V. Volkov and V. P. Akulov, Phys. Lett. B **46** (1973), 109; J. Wess and B. Zumino, Phys. Lett. B **62** (1976), 335.

2. S. Ferrara, D. Z. Freedman and P. van Niewenheusen, Phys. Rev. D (3) **13** (1976), 3214; S. Deser and B. Zumino, Phys. Lett. B **62** (1976), 335.

3. S. L. Glashow, Nucl. Phys. **22** (1961), 579; S. Weinberg, Phys. Rev. Lett. **19** (1967), 1264; A. Salam, Proc. 8th Nobel Symposium (N. Svarthholm, ed.), Almqvist & Wiksell, Stockholm, 1968, p. 367.

4. G. 't Hooft, Nucl. Phys. B **35** (1971), 167.

5. C. N. Yang and R. L. Mills, Phys. Rev. (3) **96** (1954), 191.

6. S. Mandelstam, Nucl. Phys. B **213** (1983), 149.

7. M. Green and J. Schwarz, Nucl. Phys. B **181** (1981), 502, and private communication, argue that $N = 8$ supergravity is a singular limit in 4-dimensions of a finite string theory in higher dimensions.

8. M. Gell-Mann, Talk, 1977 Washington Meeting of the Amer. Phys. Soc. (unpublished).

9. J. Ellis, M. K. Gaillard, L. Maiani and B. Zumino, Unification of the Fundamental Particle Interactions (S. Ferrara, J. Ellis and P. van Nieuwenhuizen, eds.), Plenum Press, New York, 1980, p. 69; J. Ellis, M. K. Gaillard and B. Zumino, Phys. Lett. B **94** (1980), 343.

10. E. Cremmer and B. Julia, Nucl. Phys. B **159** (1979), 141.

11. B. De Wit and H. Nicolai, Nucl. Phys. B **188** (1981), 98 and B **208** (1982), 323.

12. M. K. Gaillard and B. Zumino, Nucl. Phys. B **193** (1981), 221.

13. A. D'Adda, P. Di Vecchia and M. Lüscher, Nucl. Phys. B **146** (1978), 63; E. Witten, Nucl. Phys. B **149** (1979), 285.

14. _____, Nucl. Phys. B **152** (1979), 125; E. R. Nussimov and S. J. Pacheva, C. R. Acad. Bulgare Sci. **32** (1979), 1475 and Lett. Math. Phys. **5** (1981), 67, 333; D. Amati, R. Barbieri, A. C. Davis and G. Veneziano, Phys. Lett. B **102** (1981), 408.

15. J. Ellis, M. K. Gaillard and B. Zumino, *Superunification*, Acta Phys. Pol. B **13** (1982), 343.

16. P. H. Frampton, Phys. Rev. Lett. **46** (1981), 881; J. P. Derendinger, S. Ferrara and C. A. Savoy, Nucl. Phys. B **188** (1981), 77.

17. J. E. Kim and H. S. Song, Phys. Rev. D **25** (1982), 2996.

18. The earliest work in this direction exploited only the SU(8) symmetry: T. Curtright and P. G. O. Freund, Supergravity (P. van Nieuwenheusen and D. Z. Freedman, eds.), North-Holland, Amsterdam, 1979.

19. An example of this mechanism using a finite number of supersymmetry multiplets has recently been found, but it does not yield a phenomenologically acceptable spectrum: G. Altarelli, N. Cabibbo and L. Maiani, Nucl. Phys. B **206** (1982), 397.

20. B. Zumino, High Energy Physics—1980 (Proc. Twentieth Internat. High Energy Physics Conf., Madison, Wisconsin, 1980, L. Durand and L. G. Pondrom, eds.), AIP, New York, 1981; W. Bardeen and E. Rabinovici, private communication.

21. H. E. Haber, I. Hinchliffe and E. Rabinovici, Nucl. Phys. B **172** (1980), 458.

22. E. Cremmer, J. Scherk and S. Ferrara, Phys. Lett. B **74** (1978), 61.

23. V. Bargmann, Ann. of Math. (2) **48** (1947), 568; A. O. Barut (Proc. 1966 Boulder Summer Institute), Lectures in Theoretical Physics, Vol. 9A, p. 125; W. J. Holman and L. C. Biedenharn, Ann. Physics **39** (1966), 1 and **47** (1968), 205.

24. A more detailed analysis is given in J. Ellis, M. K. Gaillard, M. Günaydin and B. Zumino, Nucl. Phys. B **224** (1983), 427.

25. V. Kac, Math. U.S.S.R.-Izv. **32** (1968), 1271; R. Moody, Bull. Amer. Math. Soc. **73** (1967), 217 and J. Algebra **10** (1968), 211.

26. M. Günaydin and C. Saçlioḡlu, Phys. Lett. B **108** (1982), 169; Comm. Math. Phys. **87** (1982), 267.

27. M. Günaydin, private communication.

28. M. Gell-Mann, private communication; R. Herman, *Lie groups for physicists*, Benjamin, New York, 1966.

29. M. T. Grisaru and H. J. Schnitzer, Phys. Lett. B **107** (1981), 196 and Nucl. Phys. B **204** (1982), 267.

LAWRENCE BERKELEY LABORATORY, BERKELEY, CALIFORNIA 94720

DEPARTMENT OF PHYSICS, UNIVERSITY OF CALIFORNIA, BERKELEY, CALIFORNIA 94720

Lectures in Applied Mathematics
Volume 21, 1985

Mathematical Issues in Superstring Theory[1]

John H. Schwarz

ABSTRACT. The old and new formulations of the supersymmetrical string action are described, and the rather complicated transformation that relates them is described. Three mathematical issues are raised: (1) What is the mathematical meaning of the transformation formula and how can one complete the proof that it has the alleged properties? (2) Why can the old formalism be obtained starting from a covariant gauge-invariant action principle, whereas the new one apparently requires using a physical gauge from the outset? (3) Is it possible to prove the uniqueness of the superstring theory, or else to construct new examples?

1. Introduction. The study of dual resonance models began in 1968 with the discovery [1] of an explicit four-particle amplitude that combines the narrow-resonance approximation with Regge behavior and crossing symmetry. Soon multiparticle generalizations that have consistent factorization properties on a well-defined spectrum of positive-definite states were developed. The resulting dual model, usually referred to as the "Veneziano model" (VM), contains bosons only. In 1971 a second dual model, originally called the "dual pion model" (DPM), containing a spectrum of fermions [2], was developed [3]. A third dual model was developed in 1976 [4], but it is a somewhat degenerate case requiring two-dimensional spacetime.

The motivation for the work on dual models was a desire to develop a realistic phenomenological model of hadrons. This program did, in fact, achieve a certain degree of success. By doctoring the models, so as to

1980 *Mathematics Subject Classification.* Primary 83C45, 83E15; Secondary 81G20, 83E50.

[1] Work supported in part by the U.S. Department of Energy under Contract No. DE-ACO3-81-ER40050.

make them more realistic, but less theoretically consistent, it was possible to qualitatively describe a good deal of data. More important, perhaps, was the recognition that these models could be understood as the quantum theory of one-dimensional extended objects, called "strings" [5, 6]. This fact has influenced the way one thinks about the modern theory of hadrons, quantum chromodynamics (QCD). The fact is, however, that no dual string model was found that is exactly right for describing hadrons. It is generally agreed that QCD should reduce to a string theory in the infinite-color limit. The problem is that this limit is very difficult to perform analytically, and as a result the corresponding string model has not been explicitly formulated. It appears that some progress is being made in this program by considering the dynamics of Wilson loop operators, but there is still some way to go.

One feature that convinced us that the known dual string models could not be modified in a consistent way to give a realistic description of hadrons (corresponding to the infinite-color limit of QCD) was the persistent appearance of massless states. Both the VM and the DPM contain massless vector and tensor states, and all attempts to move them to positive mass invariably led to the introduction of ghost states, or worse yet, a complete breakdown of factorization. In 1974 a radical change in philosophy [7] was proposed. We suggested that a dual string model be used to describe elementary particles such as quarks, gluons, leptons, $W^{\pm}$, photon, graviton, and so forth. The motivation for this suggestion was the realization that the massless vector particles behave precisely as Yang-Mills (YM) gauge fields [8, 9], while the massless tensor state interacts in exactly the right way to be identified as a graviton [7, 10]. This suggested to us that the natural length scale[2] for such strings would be the Planck length, 20 orders of magnitude smaller than hadronic strings.

Another important issue in the study of dual string models concerns the dimension of spacetime. Each model has a "critical dimension" in which the analysis is most straightforward and unambiguous—26 for the VM, 10 for the DPM, 2 for the model of [4]. In each case, one dimension is time and the others are spatial. It has been proved that the spectrum of single particle states forms a positive-definite Hilbert space in the critical dimension [11–14]. There has been a good deal of discussion and debate whether these models are consistent in the critical dimension only, or if they can also be consistently interpreted in lower dimensions. (Everyone agrees that higher dimensions are excluded.) There is one method that certainly does work. In this approach one incorporates the old idea of

---

[2] The length scale of a string is $\sqrt{\alpha'}$, where α' is the universal Regge-slope parameter. It is related to the string tension T by $\alpha' = (2\pi T)^{-1}$.

Kaluza and Klein [15, 16] of having extra spatial dimensions form a small compact space so that they are unobservable at energies low compared to the scale determined by the size R of the compact dimensions. For this method to work, it is essential that the compact dimensions be of finite size. The limit in which $R \to 0$ at fixed α' is singular (for loop amplitudes).

The DPM is an extension of the VM that contains additional spin degrees of freedom. As originally formulated, it has a tachyonic ground state (that was identified as the pion). This state is odd with respect to a multiplicative quantum number that we called "G-parity." It was always evident that one could define a model free from tachyons by restriction to the "even G sector." However, for a long time we were not seriously tempted to pursue this possibility because of our insistence on identifying the tachyon as a slightly misplaced "pion." However, as the move away from a hadronic interpretation developed, we became liberated from this inflexible position.

An important step was taken in [17], where it was pointed out that by restricting the DPM spectrum to the even G parity bosons and to fermions satisfying simultaneous Majorana and Weyl conditions, the number of boson and fermion states becomes the same at each mass level. (This result depends critically on the choice $D = 10$.) This observation was a compelling argument, if not a proof, that the model with this restricted spectrum is a supersymmetrical theory in 10-dimensional spacetime. In particular, the massless closed-string states form a supergravity multiplet. Implicit in this remark is the fact that the DPM with the odd -G states included is inconsistent since it contains a gravitino (supersymmetry gauge field) that interacts with states (such as the tachyon) that do not belong to supersymmetry multiplets. The restricted DPM—which I refer to as "superstring theory"—is, therefore, the *only* candidate for a consistent theory of interacting strings free from ghosts and tachyons. More precisely, it is a class of theories, because (as we will discuss) there are "type I" superstring theories (SST I) based on interacting open and closed strings for which one can choose a YM gauge group and a "type II" superstring theory (SST II) based entirely on closed strings.

In this paper we discuss string actions that describe the quantum mechanics of free strings. (These actions can also be used in a path-integral description of interactions.) In the case of the 10-dimensional models, two formalisms are described—in §2 the old one that gives the DPM including the odd -G sector and in §3 a new one [18] that has 10-dimensional supersymmetry and therefore gives the superstring theories directly. We show that SST I has one supersymmetry in 10 dimensions, whereas SST II has two supersymmetries in 10 dimensions. When six dimensions are compactified, these become four and eight supersymmetries in four dimensions, respectively. One can define limits α', $R \to 0$ in which SST I reduces to $N = 4$ YM theory, and SST II reduces to $N = 8$ supergravity (SG) [19]. The emergence of these particular field theories is very encouraging since they are maximally supersymmetrical and have a number of nice theoretical features. Neither $N = 8$ SG nor $N = 4$ YM coupled to $N = 4$ SG appears likely to give a satisfactory quantum theory of gravity, but the infinities that occur are apparently absent or under control (by renormalization) in the corresponding superstring theories, which may therefore provide candidates for a complete synthesis of fundamental interactions.

The correspondence between the operators of the old and new formulations is presented in §4. The transformation is quite nontrivial, and it is an interesting mathematical challenge to complete the proof that it is correct and to understand its significance. In §5 some additional unsettled mathematical issues are discussed.

This paper is extracted from §2 of a review article that I have written recently [20]. The reader is referred to it for additional details, including a discussion of how to base an interacting quantum theory on the free strings described here. Earlier reviews of dual string models can be found in [21–23].

2. Old superstring formalism. The identification of the VM spectrum as the excitations of a one-dimensional system (string) was given in [5, 6]. The formulation of a reparametrization-invariant action principle [24–26] proved useful for a deeper understanding, culminating in the analysis [27] of the quantum mechanics of free strings in a light-cone gauge. Later the various string theories were formulated in terms of covariant gauge-invariant action principles containing auxiliary two-dimensional "gravity" and "supergravity" fields [28–30].

It is pedagogically useful to describe the VM string first, since it does not involve all the complications of the superstring theory. The entire dynamics of a VM string is contained in $X^\mu(\sigma, \tau)$, the position in space and time of a string. The spatial coordinate σ labels points along the string and is conveniently chosen to run from 0 to π, while τ is a timelike evolution parameter. It is puzzling at first to have both X^0 and τ playing the role of time. This paradox is resolved by requiring that X^μ satisfy a gauge-invariant action principle that allows, in particular, the gauge choice $X^0 \propto \tau$. Identifying the appropriate gauge invariance as precisely reparametrization invariance of the world sheet of the string and the suitable action as the geometric area of the world sheet were important insights [24].

One method of achieving coordinate invariance is well known in general relativity and may be applied to the problem of formulating a reparametrization-invariant string action. Since the world sheet is two-dimensional, we introduce a two-dimensional auxiliary metric $g_{\alpha\beta}(\sigma, \tau)$. The indices α and β take the values 0 and 1, referring to the τ and σ directions. Correspondingly, derivatives ∂_α stand for $\partial/\partial\tau$ and $\partial/\partial\sigma$. Letting g and $g^{\alpha\beta}$ represent the determinant and inverse of $g_{\alpha\beta}$, as usual, we may write the manifestly reparametrization-invariant action [28–30]

$$S = -\frac{1}{4\pi\alpha'} \int d\sigma \, d\tau \, \eta_{\mu\nu} \sqrt{-g} \, g^{\alpha\beta} \partial_\alpha X^\mu \partial_\beta X^\nu. \tag{1}$$

The parameter α', identified as a Regge-slope, has dimensions of length squared, since X^μ is length, and σ and τ are chosen to be dimensionless. $\eta_{\mu\nu}$ represents a spacetime metric that is conveniently chosen to be the flat Minkowski metric

$$\eta_{\mu\nu} = \text{diag}(-1, 1, 1, \ldots, 1). \tag{2}$$

Varying (1) with respect to $g^{\alpha\beta}$ gives

$$h_{\alpha\beta} \equiv \partial_\alpha X \cdot \partial_\beta X = \tfrac{1}{2} g_{\alpha\beta} g^{\gamma\delta} \partial_\gamma X \cdot \partial_\delta X. \tag{3}$$

Taking the square root of the determinant of each side shows that $g_{\alpha\beta}$ may be eliminated algebraically via its equation of motion leaving the Nambu action

$$S = -\frac{1}{2\pi\alpha'} \int d\sigma \, d\tau \sqrt{-\det h_{\alpha\beta}} \,. \tag{4}$$

This elimination is not so straightforward in the quantum theory. Indeed, [31] shows that as a result of the combined effects of trace anomalies and Fadeev-Popov ghosts, a remnant of the g field must survive if one hopes to make sense of the quantum theory for $D \neq 26$. These subtleties may be ignored here since we only consider the theory in the critical dimension.

From the analysis of [27], which is also reviewed in [22, 23], one knows that the reparametrization invariance of (4) allows one to choose an orthonormal gauge in which

$$\partial_\sigma X \cdot \partial_\tau X = 0, \tag{5a}$$

$$\partial_\sigma X \cdot \partial_\sigma X + \partial_\tau X \cdot \partial_\tau X = 0. \tag{5b}$$

These equations correspond to putting $g_{\alpha\beta} \propto \eta_{\alpha\beta}$ in (3). Once this is done there is still a residual gauge invariance, which taken together with the equation of motion for the light-cone coordinate X^+,[3] allows one to choose a "light-cone gauge" in which

$$X^+ (\sigma, \tau) = x^+ (\tau) = x^+ + 2\alpha' p^+ \tau. \tag{6}$$

---

[3] In our conventions $\eta_{\mu\nu} f^\mu g^\nu = f^i g^i - f^+ g^- - f^- g^+$.

We have introduced the notation

$$x^\mu(\tau) = \frac{1}{\pi} \int_0^\pi X^\mu(\sigma, \tau) \, d\sigma \tag{7}$$

for the "center of mass" coordinates of the string. The equations of motion imply that these are linear functions of τ

$$x^\mu(\tau) = x^\mu + 2\alpha' p^\mu \tau, \tag{8}$$

where p^μ is the total D-momentum of the string. The justification of this gauge choice is rather delicate but has been adequately discussed in the references cited. It does lead to a consistent picture in the critical dimension.

Once these choices are made, (5a) and (5b) may be used to solve for $X^-(\sigma, \tau)$ in terms of the other coordinates (except for the integration constant x^-). As a consequence all the dynamics of string oscillations resides in the transverse coordinates X^i, which in this gauge satisfy free wave equations

$$\left(\frac{\partial^2}{\partial \sigma^2} - \frac{\partial^2}{\partial \tau^2} \right) X^i(\sigma, \tau) = 0. \tag{9}$$

This must be supplemented by boundary conditions, which are just the obvious periodicity in the case of closed strings

$$X^i_{\text{closed}}(0, \tau) = X^i_{\text{closed}}(\pi, \tau) \tag{10}$$

and

$$\left. \frac{\partial}{\partial \sigma} X^i_{\text{open}} \right|_{\sigma=0} = \left. \frac{\partial}{\partial \sigma} X^i_{\text{open}} \right|_{\sigma=\pi} = 0 \tag{11}$$

for open strings. The latter is necessary in order to drop surface terms in obtaining the equations of motion. These results are summarized by the "light-cone-gauge action" for the VM string

$$S^{\text{l.c.}} = -\frac{1}{4\pi\alpha'} \int d\sigma \, d\tau \, \partial_\alpha X^i \partial^\alpha X^i, \tag{12}$$

which serves as a starting point for the analysis of string amplitudes as path integrals [32].

Equations (9) and (11) are solved by the open-string normal-mode expansion

$$X^i(\sigma, \tau) = x^i + p^i\tau + i \sum_{n \neq 0} \frac{1}{n} \alpha_n^i \cos n\sigma \, e^{-in\tau}, \tag{13}$$

where we have set $\alpha' = \frac{1}{2}$ as a choice of length scale. Canonical quantization gives

$$[x^\mu, p^\nu] = i\eta^{\mu\nu}, \tag{14}$$

$$\left[\alpha_m^i, \alpha_n^j\right] = m\delta_{m+n,0}\delta^{ij}. \tag{15}$$

The string excitations are therefore described by an infinite number of harmonic oscillators with lowering operators

$$a_n^i = \frac{1}{\sqrt{n}}\alpha_n^i, \qquad n = 1, 2, \ldots, \tag{16a}$$

and raising operators

$$\left(a_n^i\right)^\dagger = \frac{1}{\sqrt{n}}\alpha_{-n}^i, \qquad n = 1, 2, \ldots. \tag{16b}$$

Since we are working in a physical gauge, there are no extra timelike oscillators and the Hilbert space of single-particle states is trivially positive definite. Therefore, in this formalism the absence of ghost states is manifest. The nontrivial issue is Lorentz invariance.

The mass-shell condition is obtained by integrating (5a) and (5b) over σ, substituting (6) and (13). An additive constant is not determined due to ordering ambiguities. Therefore, one has

$$\alpha'(\text{mass})^2 = N \tag{17}$$

$$N = \sum_{n=1}^{\infty} \alpha_{-n}^i \alpha_n^i - c. \tag{18}$$

The constant c is determined by a very simple consideration. Namely, the first excited level consists of states $\alpha_{-1}^i|0\rangle$ that describe the $D - 2$ modes of a massless vector particle. Lorentz invariance therefore requires the choice $c = 1$ in order that $N = 0$ for these states. It then follows that the ground state, $|0\rangle$, is a tachyon.

The light-cone-gauge action in (12) is invariant under rotations of the transverse coordinates. The corresponding Lorentz generators can be deduced by the Noether procedure. Expressed in terms of the oscillators, they are

$$J^{ij} = l^{ij} - i\sum_{n=1}^{\infty} \frac{1}{n}\left(\alpha_{-n}^i\alpha_n^j - \alpha_{-n}^j\alpha_n^i\right), \tag{19}$$

where

$$l^{\mu\nu} = x^\mu p^\nu - x^\nu p^\mu. \tag{20}$$

The J^{ij}'s are easily seen to form an SO($D - 2$) algebra. The remaining generators are not found so readily. However, following [27] we set

$$J^{+-} = l^{+-}, \tag{21}$$

$$J^{i+} = l^{i+}, \tag{22}$$

$$J^{i-} = l^{i-} - i(p^+)^{-1} \sum_1^\infty \frac{1}{n} \left(\alpha_{-n}^i \alpha_n^- - \alpha_{-n}^- \alpha_n^i \right), \tag{23}$$

where

$$\alpha_n^- = \frac{1}{2} \sum_{m=-\infty}^\infty \alpha_{n-m}^i \alpha_m^i, \qquad n \neq 0, \tag{24}$$

are modes occurring in the expansion of X^- as determined by (4)–(6). One must now check the rest of the Lorentz algebra. Except for the commutator

$$[J^{i-}, J^{j-}] = 0, \tag{25}$$

all of the algebra is easily seen to work out properly irrespective of D. Checking the condition in (25), on the other hand, requires a very delicate calculation that only gives the indicated result for on-mass-shell states and $D = 26$. The Lorentz generators tell us how to combine the SO(24) multiplets at a given massive level, corresponding to various oscillator excitations, into irreducible SO(25) multiplets.

Solving the wave equation (9) with the closed-string boundary condition (10) gives the normal-mode expansion

$$X^i(\sigma, \tau) = x^i + p^i\tau + \frac{i}{2} \sum_{n \neq 0} \frac{1}{n} \left(\alpha_n^i e^{-2in(\tau-\sigma)} + \tilde{\alpha}_n^i e^{-2in(\tau+\sigma)} \right). \tag{26}$$

In this case canonical quantization gives (14) and (15) as well as

$$\left[\tilde{\alpha}_m^i, \tilde{\alpha}_n^j \right] = m\delta_{m+n,0}\delta^{ij}, \tag{27}$$

$$\left[\alpha_m^i, \tilde{\alpha}_n^j \right] = 0. \tag{28}$$

The conditions for physical on-mass-shell states are again given (up to a constant) by σ integrals of (5):

$$\tfrac{1}{4}\alpha'(\text{mass})^2 = N = \tilde{N}, \tag{29}$$

where N is given in (18) and

$$\tilde{N} = \sum_{n=1}^\infty \tilde{\alpha}_{-n}^i \tilde{\alpha}_n^i - c. \tag{30}$$

One sees from (29) that the universal slope of closed-string Regge trajectories is one-half that of the open-string trajectories (since the maximum angular momentum is given by $N + \tilde{N}$). This still leaves two possibilities for closed-string theories. All states can be decomposed into ones that are symmetric under interchange of α and $\tilde{\alpha}$ oscillators and ones that are antisymmetric. Type I closed strings consist of symmetric states only, as these states can form a consistent interacting theory by themselves. The interacting theory of type I closed strings is sometimes referred to as the restricted Shapiro-Virasoro model (RSVM). Type II closed strings comprise the entire set of symmetric and antisymmetric states which can also form a consistent interacting theory. In this theory (sometimes referred to as the extended Shapiro-Virasoro model or ESVM) the antisymmetric states are odd with respect to a multiplicatively conserved quantum number, and therefore any nonzero amplitude must involve an even number of them. The Lorentz invariance of the closed-string theories works very similarly to that of the open-string theory. The first excited level involves states $\alpha^i_{-1}\tilde{\alpha}^j_{-1}|0\rangle$ that cannot form $SO(25)$ multiplets and therefore must be massless. It follows that the parameter c in (18) and (30) must again be unity. In the RSVM the massless states are a symmetric traceless second-rank tensor ("graviton") and a singlet scalar. The ESVM massless sector contains an antisymmetric second-rank tensor of $SO(24)$ as well.

The supersymmetric strings contain spinor degrees of freedom in addition to the coordinates $X^\mu(\sigma, \tau)$. In the old formalism of this section, they are described by D two-component spinors $\lambda^{\mu A}(\sigma, \tau)$ ($A = 1, 2$). Thus $\lambda^{\mu A}$ transforms as a spacetime vector and a world-sheet spinor. In order to understand the subsidiary conditions that allow negative norm states to be avoided, two local symmetries are required to compensate for the time components of X^μ and λ^μ, respectively. The appropriate trick is to extend (1) to an expression that not only has general-coordinate invariance in two dimensions but local supersymmetry as well. This was achieved in [28, 29], a relatively easy construction following upon the discovery of four-dimensional supergravity [33–35]. As in supergravity theory, it is necessary to introduce a "zweibein" field V^a_α, related to the metric $g_{\alpha\beta}$ in the usual way and a Majorana Rarita-Schwinger field ψ^A_α. Then, letting ρ^a denote two-dimensional Dirac matrices, the action

$$S = -\frac{1}{4\pi\alpha'}\int d\sigma\, d\tau\, \eta_{\mu\nu} V\left\{ g^{\alpha\beta}\partial_\alpha X^\mu \partial_\beta X^\nu + iV^\alpha_a \bar{\lambda}^\mu \rho^a \partial_\alpha \lambda^\nu \right.$$

$$\left. + 2V^\alpha_a V^\beta_b \bar{\psi}_\alpha \rho^b \rho^a \lambda^\mu \left(\partial_\beta X^\nu + \frac{1}{2}\bar{\lambda}^\nu \psi_\beta \right) \right\}$$ 31)

has the necessary local supersymmetry.

The gauge invariances of (31) allow the analysis of the previous section to be generalized in such a way that all the dynamics ends up in the transverse modes X^i and λ^{Ai}. The result is summarized by the light-cone-gauge action

$$S^{1.c} = -\frac{1}{4\pi\alpha'}\int d\sigma\, d\tau \{\partial_\alpha X^i \partial^\alpha X^i + i\bar{\lambda}^i \rho^\alpha \partial_\alpha \lambda^i\}.$$ (32)

This gives rise to the same mode expansion of X^i as in (13). Choosing the Majorana representation

$$\rho^0 = \begin{pmatrix} 0 & -i \\ i & 0 \end{pmatrix} \quad \text{and} \quad \rho^1 = \begin{pmatrix} 0 & i \\ i & 0 \end{pmatrix}, \tag{33}$$

the λ equations of motion are

$$\left(\frac{\partial}{\partial \sigma} + \frac{\partial}{\partial \tau} \right) \lambda^{1i} = 0, \tag{34a}$$

$$\left(\frac{\partial}{\partial \sigma} - \frac{\partial}{\partial \tau} \right) \lambda^{2i} = 0. \tag{34b}$$

The open-string boundary conditions are

$$\lambda^{1i}(0, \tau) = \lambda^{2i}(0, \tau) \tag{35}$$

$$\lambda^{1i}(\pi, \tau) = \begin{cases} \lambda^{2i}(\pi, \tau) & \text{for } \textit{fermions} \\ -\lambda^{2i}(\pi, \tau) & \text{for } \textit{bosons}. \end{cases} \tag{36}$$

Therefore fermion strings involve normal-mode expansions

$$\lambda^{1i} = \sum_{n=-\infty}^{\infty} d_n^i e^{-in(\tau-\sigma)}, \tag{37a}$$

$$\lambda^{2i} = \sum_{n=-\infty}^{\infty} d_n^i e^{-in(\tau+\sigma)}, \tag{37b}$$

where the summation index n runs over all integers. For boson strings the extra minus sign in (36) results in expansions

$$\lambda^{1i} = \sum_{r=-\infty}^{\infty} b_r^i e^{-ir(\tau-\sigma)}, \tag{38a}$$

$$\lambda^{2i} = \sum_{r=-\infty}^{\infty} b_r^i e^{-ir(\tau+\sigma)}, \tag{38b}$$

where the index r takes all half-integer values, i.e., $\pm 1/2, \pm 3/2, \ldots$. The quantization conditions in the two cases are

$$\{d_m^i, d_n^j\} = \delta_{m+n,0} \delta^{ij}, \tag{39}$$

$$\{b_r^i, b_s^j\} = \delta_{r+s,0} \delta^{ij}. \tag{40}$$

It should be emphasized that the d oscillators act only on fermion states to give new fermion states, and the b oscillators act only on boson states to give new boson states.

For the boson sector the analog of (17) and (18) is

$$\alpha'(\text{mass})^2 = N = \sum_{n=1}^{\infty} \alpha_{-n}^i \alpha_n^i + \sum_{r=1/2}^{\infty} r b_{-r}^i b_r^i - c. \tag{41}$$

The lowest excited states are given by $b^i_{-1/2}|0\rangle$, which once again must describe a massless vector. Therefore, Lorentz invariance in this case requires that $c = 1/2$. The boson-sector Lorentz generators analogous to (19–23) are

$$J^{ij} = l^{ij} - i \sum_{n=1}^{\infty} \frac{1}{n} \left(\alpha^i_{-n} \alpha^j_n - \alpha^j_{-n} \alpha^i_n \right) + K^{ij}_0, \tag{42a}$$

$$J^{+-} = l^{+-}, \tag{42b}$$

$$J^{i+} = l^{i+}, \tag{42c}$$

$$J^{i-} = l^{i-} - i(p^+)^{-1} \sum_{n=1}^{\infty} \frac{1}{n} \left(\alpha^i_{-n} \alpha^-_n - \alpha^-_{-n} \alpha^i_n \right) + (p^+)^{-1} \sum_{-\infty}^{\infty} K^{ij}_{-n} \alpha^j_n, \tag{42d}$$

where

$$K^{ij}_0 = -i \sum_{r=1/2}^{\infty} \left(b^i_{-r} b^j_r - b^j_{-r} b^i_r \right), \tag{43a}$$

$$K^{ij}_m = -i \sum_{r=-\infty}^{\infty} b^i_{m-r} b^j_r, \qquad m \neq 0, \tag{43b}$$

$$\alpha^-_n = \frac{1}{2} \sum_{m=-\infty}^{\infty} \alpha^i_{n-m} \alpha^i_m + \frac{1}{2} \sum_{r=-\infty}^{\infty} \left(r - \frac{n}{2} \right) b^i_{n-r} b^i_r, \qquad n \neq 0, \tag{44}$$

and

$$\alpha^i_0 \equiv p^i. \tag{45}$$

(25) is only satisfied in this case if $D = 10$.

In the fermion sector there are only a few minor changes. The b^i_r oscillators are replaced by d^i_n oscillators throughout, and the constant c in (41) must be chosen equal to zero, so that the fermion ground state is massless.

The number of boson states at mass levels $N = 0, 1, 2, \ldots$ is the same as the number of fermion states at the corresponding levels provided that one chooses the ground-state spinor to be simultaneously Majorana and Weyl so as to have eight physical helicity states. [17] also showed that these two conditions for a spinor in Minkowski space are compatible whenever the dimension of spacetime satisfies

$$D = 2 \ (\mathrm{mod}\,8). \tag{46}$$

Therefore, restricting the spinors in this way and throwing away the "odd G" bosons for which N is half-integer (including the tachyonic ground state!), leaves a system for which the number of bosons and fermions is identical at every mass level. This version of the theory was plausibly conjectured to possess supersymmetry in 10-dimensional spacetime. This symmetry is terribly obscure in the "old" formalism described here, which prompted us to develop a new one in which the supersymmetry is clearer [18, 19], presented in the next section.

3. New superstring formalism. A fundamental property of superstrings is their supersymmetry in 10-dimensional spacetime. In the old formalism discussed in the preceding subsection, this symmetry is obscure, and its relationship (if any) to the two-dimensional supersymmetry of (31) is, to say the least, puzzling. A new formalism exhibiting spacetime supersymmetry from the outset is clearly desirable. For a time we sought a covariant gauge-invariant action analogous to (1) and (31) but also possessing spacetime supersymmetry. Having failed to find one, we tend to believe that none exists. A covariant action principle is not essential. If in the case of the VM one only knew (6) and the light-cone-gauge action in (12), it would still be possible to establish Lorentz invariance by "guessing" the Lorentz generators in (19)–(23) and verifying that they satisfy the Lorentz algebra in the critical dimension. Furthermore, this is all the information required to develop the theory of interacting strings. One would simply be working in a physical light-cone gauge from the outset rather than deducing it from a manifestly covariant formalism. Some of the mystery has been removed from this approach by the observation [36] that there are classical relativistic field theories that cannot be formulated in a manifestly covariant form, but that can be described in a light-cone gauge.

The supersymmetric light-cone-gauge action [18] involves the transverse coordinates $X^i(\sigma, \tau)$ as well as additional fermion degrees of freedom $S^{Aa}(\sigma, \tau)$. The S's replace the λ^{Ai} variables of the preceding subsection. Both S and λ have the index A indicating that they are two-component spinors as regards their transformation properties on the two-dimensional world sheet. The important difference is that whereas λ transforms as a spacetime vector, S transforms as a spinor. Thus the index a is 32-valued, as appropriate for a Dirac index in 10-dimensional spacetime. Because S is a spacetime spinor, it connects bosons to fermions. This removes one of the mysterious features of the old formalism: even though the λ's satisfy anticommutation rules they are, in fact, boson operators. In the new formalism based on the S's, anything that anticommutes is genuinely a fermion operator so there are no "spin and statistics" paradoxes.

To obtain a supersymmetric description, the number of physical S modes should be equal to the number of X^i components, which is eight—the number of transverse directions in 10-dimensional spacetime. This is achieved by first requiring that as a 32-spinor (a index) S is

Majorana and Weyl. These restrictions reduce the 64 complex components of S^{Aa} to 32 real components. We use the notation $(\gamma^\mu)^{ab}$ for
spacetime Dirac matrices (choosing a Majorana representation for convenience). The Dirac matrices for the two-dimensional world sheet are
denoted $(\rho^\alpha)^{AB}$ and given explicitly in (33). The Weyl condition is

$$h_A^{ab} S^{Ab} = 0, \qquad A = 1, 2, \tag{47}$$

where h_A^{ab} denotes Weyl projections $\frac{1}{2}(1 \pm \gamma_{11})$. There are two physically
distinct possibilities: either S^1 and S^2 have the same handedness or
opposite handedness. Next, the light-cone-gauge condition in (6) is
supplemented with a corresponding one for the S variables:

$$(\gamma^+)^{ab} S^{Ab} = 0. \tag{48}$$

(48) further restricts S^{Aa} to 16 real components, which is the desired
number. (The last factor of two is provided by the Dirac equation.)

With S^{Aa} restricted in the ways described above, the supersymmetrical
light-cone-gauge string action takes the form

$$S^{\text{l.c.}} = \int d\sigma \, d\tau \left(-\frac{1}{4\pi\alpha'} \partial_\alpha X^i \partial^\alpha X^i + \frac{i}{4\pi} \bar{S}\gamma^- \rho^\alpha \partial_\alpha S \right). \tag{49}$$

All that has been done is to adjoin a free Dirac action for the S variables
to the expression in (12). The notation $\bar{S}$ means

$$\bar{S}^{Aa} = S^{\dagger Bb} (\gamma^0)^{ba} (\rho^0)^{BA}. \tag{50}$$

If the factor γ^- were not included in the Dirac term, the expression
would be identically zero due to the projection property in (48). One
should not be disturbed by this lack of covariance; the X term also only
has manifest invariance for transverse rotations.

The action (49) is invariant under the supersymmetry transformations

$$\delta X^i = (p^+)^{-1/2} \bar{\varepsilon} \gamma^i S, \tag{51}$$

$$\delta S = i(p^+)^{-1/2} \gamma_- \gamma_\alpha (\rho \cdot \partial X^\alpha) \varepsilon \tag{52}$$

up to surface terms whose cancellation depends on the choice of boundary
conditions (discussed later). The normalization factor $(p^+)^{-1/2}$ is included for later convenience. The parameters ε^{Aa} are Majorana-Weyl
spinors in 10 dimensions. Global transformations are of primary interest,
even though some σ, τ dependence in ε is possible. ε^{Aa} may be regarded
as parametrizing either two 10-dimensional supersymmetries or 16 two-
dimensional supersymmetries. In a sense both interpretations are correct
because commuting two global supersymmetry transformations with

infinitesimal parameters $\varepsilon^{(1)}$ and $\varepsilon^{(2)}$ (and using the equation of motion $\rho \cdot \partial S = 0$), one finds that

$$[\delta_1, \delta_2] X^i = \xi^\alpha \partial_\alpha X^i + a^i \tag{53}$$

and

$$[\delta_1, \delta_2] S = \xi^\alpha \partial_\alpha S, \tag{54}$$

where

$$\xi^\alpha = -2i(p^+)^{-1} \bar{\varepsilon}^{(1)} \gamma_- \rho^\alpha \varepsilon^{(2)}, \tag{55}$$

$$a^i = -2i \bar{\varepsilon}^{(1)} \rho^0 \gamma^i \varepsilon^{(2)}. \tag{56}$$

ξ^α describes an infinitesimal translation of the coordinates σ and τ, while a^i describes a translation of the X^i variables, both of which correspond to symmetries of (49). The supersymmetry charges may be deduced from (49), (51) and (52) by the Noether procedure

$$Q^{Aa} = \frac{i}{\pi\sqrt{p^+}} \int_0^\pi \left[\gamma_\mu (\rho \cdot \partial X^\mu) \rho^0 S \right]^{Aa} d\sigma. \tag{57}$$

Note that as a consequence of (48), the undefined component X^- does not occur in this expression.

Let us now analyze the consequences of this approach for open strings. The equation of motion and boundary conditions of the X^i variables are given in (9) and (11). The normal-mode expansion and canonical commutation rules remain the same as in (13)–(15). The S variables satisfy the two-dimensional Dirac equation, which written out in components gives

$$\left(\frac{\partial}{\partial \tau} + \frac{\partial}{\partial \sigma} \right) S^{1a} = 0, \tag{58a}$$

$$\left(\frac{\partial}{\partial \tau} - \frac{\partial}{\partial \sigma} \right) S^{2a} = 0 \tag{58b}$$

in analogy with (34). For boundary conditions one must choose

$$S^{1a}(0, \tau) = S^{2a}(0, \tau), \tag{59a}$$

$$S^{1a}(\pi, \tau) = S^{2a}(\pi, \tau). \tag{59b}$$

These are analogous to (35) and (36), except that the minus sign that resulted in half-integer moding of the boson sector in the old formalism is excluded in this case by the requirement of global supersymmetry. The normal-mode expansions implied by (58) and (59) are

$$S^{1a} = \sum_{n=-\infty}^{\infty} S_n^a e^{-in(\tau-\sigma)}, \tag{60a}$$

$$S^{2a} = \sum_{n=-\infty}^{\infty} S_n^a e^{-in(\tau+\sigma)}. \tag{60b}$$

Note that only oscillators with integer mode numbers (n) occur. It should also be remarked that (59) requires that S^1 and S^2 have the *same* handedness ($h_1 = h_2 \equiv h$), so this is the only possibility for open superstrings. Canonical quantization gives

$$\{S_m^a, \bar{S}_n^b\} = (\gamma^+ h)^{ab}\delta_{m+n,0}, \tag{61}$$

$$[\alpha_m^i, S_n^a] = 0. \tag{62}$$

Our convention is that the bar on S_n represents multiplication by γ^0 without hermitian conjugation.

Substituting the expansions (13) and (60) into the supercharge formula, (57), shows that the two supercharges Q^{1a} and Q^{2a} depend on the parameter τ and therefore are not conserved. The reason for this is that there are surface terms that must be considered in the proof that (49) is invariant under the transformations in (51) and (52). For the open-string boundary conditions, these surface terms do not cancel *unless* $\varepsilon^{1a} = -\varepsilon^{2a}$. Therefore only the single supercharge

$$Q^a = \tfrac{1}{2}(Q^{1a} + Q^{2a}) \tag{63}$$

is conserved. Explicitly,

$$Q^a = i\sqrt{p^+}\,(\gamma_+ S_0)^a + i(p^+)^{-1/2} \sum_{-\infty}^{\infty} (\gamma_i S_{-n})^a \alpha_n^i, \tag{64}$$

which evidently has no τ dependence. Note that Q^a and S_n^a have opposite handedness. The supersymmetry anticommutation relation is

$$\{Q^a, \bar{Q}^b\} = -2(h\gamma \cdot p)^{ab}. \tag{65}$$

Since we are using a physical gauge, this equation is only satisfied when evaluated between on-mass-shell states.

The Lorentz generators for open strings may again be written in the form given in (42). However, in this case (43) and (44) are replaced by

$$K_m^{ij} = \frac{i}{8} \sum_{n=-\infty}^{\infty} \bar{S}_{m-n}\gamma^{ij-}S_n, \tag{66}$$

$$\alpha_n^- = \frac{1}{2} \sum_{m=-\infty}^{\infty} \alpha_{n-m}^i \alpha_m^i + \frac{1}{4} \sum_{m=-\infty}^{\infty} \left(m - \frac{n}{2}\right)\bar{S}_{n-m}\gamma^- S_m, \qquad n \neq 0. \tag{67}$$

We use the convenient (and fairly standard) notation $\gamma^{\mu_1\mu_2\cdots\mu_N}$ for a totally antisymmetric product of Dirac matrices, normalized so that $\gamma^{12\cdots} = \gamma^1\gamma^2\cdots$. The proof that the Lorentz algebra is satisfied, including the fact that Q^a transforms as a spinor,

$$[J^{\mu\nu}, Q^a] = -\frac{i}{2}(\gamma^{\mu\nu})^{ab}Q^b, \tag{68}$$

were given in [19]. In verifying the super-Poincaré algebra, it is necessary
to use the mass-shell condition

$$\alpha'(\text{mass})^2 = N = \sum_{n=1}^{\infty} \left(\alpha_{-n}^i \alpha_n^i + \frac{n}{2} \bar{S}_{-n} \gamma^- S_n \right). \tag{69}$$

Therefore in this formalism the Fock-space ground state corresponds to
massless string modes. The tachyon of the old formalism cannot be
described in the new one. The reason, of course, is that its occurrence
would imply a breakdown of supersymmetry, which is built into the new
formalism from the outset.

The boundary conditions for closed strings are just the obvious periodicity requirements, (10) and

$$S^{Aa}(0, \tau) = S^{Aa}(\pi, \tau). \tag{70}$$

The normal-mode expansion and quantization conditions for the X^i
variables are again given by (26)–(28). The conditions in (58) and (70)
lead to the expansions

$$S^{1a} = \sum_{-\infty}^{\infty} S_n^a e^{-2in(\tau-\sigma)}, \tag{71a}$$

$$S^{2a} = \sum_{-\infty}^{\infty} \tilde{S}_n^a e^{-2in(\tau+\sigma)}. \tag{71b}$$

The canonical quantization rules are as given in (61) and (62), with
corresponding equations for the operators with tildes. Also

$$\left[\alpha_m^i, \tilde{\alpha}_n^j \right] = \left[\alpha_m^i, \tilde{S}_n^a \right] = \left[\tilde{\alpha}_m^i, S_n^a \right] = \left\{ S_m^a, \tilde{S}_n^b \right\} = 0. \tag{72}$$

The physical on-mass-shell states are given, as in (29), by

$$\tfrac{1}{4} \alpha'(\text{mass})^2 = N = \tilde{N}, \tag{73}$$

but now N is given by (69), and $\tilde{N}$ is the corresponding expression with
tildes.

Substituting the expansions of (26) and (71) into the supercharges in
(57) gives

$$Q^{1a} = i\sqrt{p^+} \left(\gamma_+ S_0 \right)^a + 2i(p^+)^{-1/2} \sum_{-\infty}^{\infty} \left(\gamma_i S_{-n} \right)^a \alpha_n^i, \tag{74a}$$

$$Q^{2a} = i\sqrt{p^+} \left(\gamma_+ \tilde{S}_0 \right)^a + 2i(p^+)^{-1/2} \sum_{-\infty}^{\infty} \left(\gamma_i \tilde{S}_{-n} \right)^a \tilde{\alpha}_n^i. \tag{74b}$$

Both Q^1 and Q^2 are independent of τ, and therefore in the closed-string
case there are two conserved supercharges ("$N = 2$" supersymmetry). By
the same sort of algebra as for the open-string case, one can then show
that

$$\left\{ Q^{Aa}, \bar{Q}^{Bb} \right\} = -2(h_A \gamma \cdot p)^{ab} (\rho^0)^{AB}, \tag{75}$$

which is the $N = 2$ supersymmetry algebra. In the case of type II superstrings, analogous to the ESVM, no further conditions are imposed. Therefore this is a theory with $N = 2$ supersymmetry in 10 dimensions (which gives $N = 8$ on reduction to $D = 4$). There are actually two inequivalent possibilities, since one may choose the two Weyl projections h_1 and h_2 to be the same or opposite. The type I closed superstrings that are the analogue of the RSVM, are obtained once again by restricting the spectrum to states that are symmetrical under interchange of the tilde and no tilde spaces. (This requires that $h_1 = h_2$.) The restricted model no longer has two supersymmetries since the action of Q^1 or Q^2 alone maps states out of the symmetrized sector. The operator Q^a as defined in (63) is still a symmetry generator since it preserves the symmetry of the states. Thus type I closed superstrings only have $N = 1$ supersymmetry.

4. Relationship between old and new oscillators. This section describes the relationship between the "old" oscillator description of superstrings in §2 and the "new" oscillator description in §3. Since the old formalism is capable of describing a larger space of states than the new one, it is impossible to express the b and d oscillators in terms of the S oscillators. However, if restricting the old scheme to the supersymmetrical sector is really equivalent to the new one, then it should be possible to express the S oscillators in terms of the b and d oscillators. The solution to this problem sketched here is actually the route by which the new formalism was discovered [19]. The presentation given here differs from that of [19] in two minor respects—some notational changes and the identification of the transverse oscillators as those of a light-cone-gauge formulation rather than as elements of a spectrum generating algebra.

Since the S oscillators connect boson states to fermion states and vice versa, the natural starting point is to look for such operators in the old formalism. The obvious candidates are the vertex operators that were developed to describe the emission of a fermion [37–40]. All these references use a covariant formalism rather than the light-cone gauge, but it is easy to see (at least in special frames) that the light-cone-gauge vertex has essentially the same structure as the covariant one restricted to transverse oscillators only. The fermion-emission vertex consists of a factor involving α oscillators times a factor involving b and d oscillators. The part involving the b and d oscillators is

$$X_{FB}^a(\vartheta)e^{i\vartheta/2}\langle 0_b|\exp\left(e^{i\vartheta}D_{-1}\right)\sin I(\vartheta)\exp E(\vartheta)|0_d\rangle \tag{76}$$

where

$$D_m = \frac{1}{2}\sum_{n=-\infty}^{\infty}\left(n - \frac{m}{2}\right):d_{m-n}^i d_n^i: \tag{77}$$

$$I(\vartheta) = \sum_{m,n=0}^{\infty} (-1)^{m+n} e^{i(n-m-1/2)\vartheta} b_{m+1/2}^i d_{-n}^i, \qquad (78)$$

$$E(\vartheta) = -\frac{1}{4} \sum_{m,n=0}^{\infty} C_{mn} e^{-i(m+n+1)\vartheta} b_{m+1/2}^i b_{n+1/2}^i \qquad (79)$$

and

$$C_{mn} = (-1)^{m+n} \frac{m-n}{m+n+1} \binom{-1/2}{m} \binom{-1/2}{n}. \qquad (80)$$

The states $\langle 0_b |$ and $| 0_d \rangle$ represent ground states of the b-oscillator and d-oscillator Fock spaces, respectively. Thus X_{FB}^a acts to the right on a bose state to give a fermi state and acts to the left on a fermi state to give a bose state. The operator X_{FB}^a is, among other things, a Dirac matrix. Dirac matrices enter via the d_0^i oscillators, which can be represented by Dirac matrices (see (39)). However, (76) only displays a single Dirac index a, corresponding to the right-hand side of the matrix, because the other index is contracted out when X_{FB}^a multiplies a fermion state on its left. In other words, matrix elements of X_{FB}^a only carry the index a, which is therefore properly regarded as a label of the operator. The ϑ dependences are introduced to keep track of mode numbers. They enable us to define $X_{n,FB}^a$ by

$$X_{FB}^a(\vartheta) = \sum_{n=-\infty}^{\infty} e^{-in\vartheta} X_{n,FB}^a. \qquad (81)$$

The factor $e^{i\vartheta/2}$ in (76) accounts for the fact that $c = \frac{1}{2}$ in (41), while there is no such constant in (69). Note that $X_{FB}^a(\vartheta)$ involves an odd number of b oscillators, and therefore the sum in (81) only involves integer values of n. (The vertices in [37–40] actually contain e^{iI} rather than $\sin I$. Removing the even part implements the even-G-parity projection of the boson sector.)

The X_{BF}^a operators satisfy two important identities. The first one is

$$D_m X_{BF}^a(\vartheta) - X_{BF}^a(\vartheta) B_m = e^{im\vartheta} \left(-i\frac{\partial}{\partial\vartheta} + \frac{m}{2} \right) X_{BF}^a(\vartheta), \qquad (82)$$

where D_m is given in (77) and

$$B_m = \frac{1}{2} \sum_{r=-\infty}^{\infty} \left(r - \frac{m}{2} \right) : b_{m-r}^i b_r^i : -\frac{1}{2}\delta_{m,0}. \qquad (83)$$

This formula can be proved using results given in [41]. The detailed reasoning is explained in Appendix B of [19]. The second fundamental identity involves the operator $X_{FB}^a(\vartheta)$, which is the adjoint of the X_{BF}^a. The formula is an operator identity in the boson sector

$$X_{m,BF}^a X_{n,FB}^b + \left(\gamma^0 X_{n,BF} \right)^b \left(X_{m,FB} \gamma^0 \right)^a = \delta^{ab}\delta_{m+n,0}. \qquad (84)$$

This remarkable identity has not been completely proved, but considerable evidence for its validity is presented in Appendix C of [19]. We prove there that the relation is satisfied when evaluated between boson ground states, and that it is true, in general, if one can show that the left-hand side of (84) commutes with an arbitrary bilinear of the form $b_r^i b_s^j$.

Physical states $|\psi\rangle$ in the fermi sector of the old formalism satisfy a generalized Dirac equation [2], $F_0|\psi\rangle = 0$, where

$$F_0 = \sqrt{\alpha'}\, \gamma \cdot p + \sum_{n=1}^{\infty} \left(\alpha_{-n}^i d_n^i + d_{-n}^i \alpha_n^i \right). \tag{85}$$

They also have definite handedness, which depends on the number of d oscillators involved. To incorporate these two properties, we define a projection operator

$$\Pi = \frac{F_0 \gamma^0}{2\sqrt{\alpha'}\, p_0} H, \tag{86}$$

where

$$H = \tfrac{1}{2}\left(1 \pm \gamma_{11}(-1)^{N_D}\right), \tag{87}$$

with

$$N_D = \sum_{n=1}^{\infty} d_{-n}^i d_n^i, \tag{88}$$

generalizes the Weyl projection $h = \tfrac{1}{2}(1 \pm \gamma_{11})$. Next we define modified X operators that incorporate the projection operator Π and some additional kinematical factors

$$S_{n,BF}^a = \left(\gamma_- X_{n,BF}\right)^a \Pi \left(p^0/p^+ \right)^{1/2}, \tag{89a}$$

$$S_{n,FB}^a = -\left(p^0/p^+ \right)^{1/2} \Pi \gamma^0 \left(X_{n,FB} \gamma^0 \gamma_+ \right)^a. \tag{89b}$$

These are then combined into a matrix

$$S_n^a = \begin{pmatrix} 0 & S_{n,BF}^a \\ S_{n,FB}^a & 0 \end{pmatrix}, \tag{90}$$

where the first row and column refer to the boson sector, and the second row and column to the fermion sector. The fundamental theorem is that these operators satisfy the algebra given in (61) and may, therefore, be identified as the S operators of the new formalism. The proof involves two steps. First one shows that the algebra is satisfied when evaluated between boson states as a consequence of (84). The second step is to argue that it must be true in the fermion sector as well because there are equal numbers of bose and fermi states at every level. The reader is referred to [19] for further details.

5. Discussion. The existence of the two formalisms described here raises a number of interesting mathematical questions. First, what is the physical or mathematical significance of the transformation described in §4 expressing the S operators in terms of the b and d operators? It appears reminiscent of the type of transformation that sometimes occurs in two-dimensional field theories where fermion operators are expressed in terms of boson ones. (32) and (49) cannot be considered to be equivalent field theories in this sense because the spectrum implied by (32) needs to be truncated before making the correspondence. Also, the proof of (84) needs to be completed before the correspondence can be considered to be fully established. One thing that is clear is that a deeper understanding of these connections should be sought.

Another peculiar feature of the correspondence between the two formalisms is that it is made after restriction to a physical gauge, rather than at the level of covariant gauge-invariant formulations. Put another way, no analogue of (31) has been found in the new formalism. The suspicion is that there is none, but this has not been proved, and the full significance of such a conclusion is not well understood.

A third area in which a deeper understanding of the mathematics of these models is called for concerns the uniqueness of the superstring theories. In particular, one wonders whether any theories exist in dimensions of spacetime other than ten. The theory of [4] has critical dimension $D = 2$, and therefore becomes almost trivial in a physical gauge (since there are no transverse coordinates). Certain uniqueness theorems for dual string models have been established [42, 43], but these predate the discovery of the new superstring formalism. Its structure raises the possibility of evading these theorems in the search for new theories.

REFERENCES

1. G. Veneziano, *Construction of a crossing-symmetric, Regge-behaved amplitude for linearly rising trajectories*, Nuovo Cimento A (11) **57** (1968), 190–197.

2. P. Ramond, *Dual theory for free fermions*, Phys. Rev. D (3) **3** (1971), 2415–2418.

3. A. Neveu and J. H. Schwarz, *Factorizable dual model of pions*, Nucl. Phys. B **31** (1971), 86–112; *Quark model of dual pions*, Phys. Rev. D (3) **4** (1971), 1109–1111.

4. M. Ademollo et al., *Supersymmetric strings and colour confinement*, Phys. Lett. B **62** (1976), 105–110; *Dual string with U*(1) *colour symmetry*, Nucl. Phys. B **111** (1976), 77–110.

5. Y. Nambu (Proc. Internat. Conf. on Symmetries and Quark Models, Wayne State University, 1969), Gordon & Breach, 1970, p. 269.

6. L. Susskind, *Dual-symmetric theory of hadrons.* I, Nuovo Cimento A (11) **69** (1970), 457–496.

7. J. Scherk and J. H. Schwarz, *Dual models for non-hadrons*, Nucl. Phys. B **81** (1974), 118–144.

8. A. Neveu and J. Scherk, *Connection between Yang-Mills fields and dual models*, Nucl. Phys. B **36** (1972), 155–161.

9. J. L. Gervais and A. Neveu, *Feynman rules for massive gauge fields with dual diagram topology*, Nucl. Phys. B **46** (1972), 381–401.

10. T. Yoneya, *Quantum gravity and the zero-slope limit of the generalized Virasoro model*, Nuovo Cimento Lett. **8** (1973), 951–955.

11. P. Goddard and C. B. Thorn, *Compatibility of the dual pomeron with unitarity and the absence of ghosts in the dual resonance model*, Phys. Lett. B **40** (1972), 235–238.

12. R. C. Brower, *Spectrum-generating algebra and no-ghost theorem for the dual model*, Phys. Rev. D **6** (1972), 1655–1662.

13. J. H. Schwarz, *Physical states and pomeron poles in the dual pion model*, Nucl. Phys. B **46** (1972), 61–74.

14. R. C. Brower and K. A. Friedman, *Spectrum-generating algebra and no-ghost theorem for the Neveu-Schwarz model*, Phys. Rev. D **7** (1973), 535–539.

15. Th. Kaluza, Sitzungsber. Preuss. Akad. Wiss. Berlin. Kl. Math. Phys. Tech. **1** (1921), 966.

16. O. Klein, Z. Phys. **37** (1926), 895.

17. F. Gliozzi, J. Scherk and D. I. Olive, *Supergravity and the spinor dual model*, Phys. Lett. B **65** (1976), 282–286; *Supersymmetry, supergravity theories and the dual spinor model*, Nucl. Phys. B **122** (1977), 253–290.

18. M. B. Green and J. H. Schwarz, *Supersymmetrical string theories*, Phys. Lett. B **109** (1982), 444–448.

19. _____, *Supersymmetrical dual string theory*, Nucl. Phys. B **181** (1981), 502–530.

20. J. H. Schwarz, *Superstring theory*, Phys. Rep. **89** (1982), 223.

21. J. Scherk, *An introduction to the theory of dual models and strings*, Rev. Modern Phys. **47** (1975), 123.

22. M. Jacob (ed.), *Dual theory*, North-Holland, Amsterdam, 1974.

23. P. Frampton, *Dual resonance models*, Benjamin, New York, 1974.

24. Y. Nambu, *Lectures at the Copenhagen Symposium*, 1970, unpublished.

25. O. Hara, *On origin and physical meaning of Ward-like identity in dual-resonance model*, Progr. Theoret. Phys. **46** (1971), 1549–1559.

26. T. Goto, *Relativistic quantum mechanics of one-dimensional mechanical continuum and subsidiary condition of dual resonance model*, Progr. Theoret. Phys. **46** (1971), 1560–1569.

27. P. Goddard, J. Goldstone, C. Rebbi and C. B. Thorn, *Quantum dynamics of a massless relativistic string*, Nucl. Phys. B **56** (1973), 109–135.

28. L. Brink, P. Di Vecchia and P. Howe, *A locally supersymmetric and reparametrization invariant action for the spinning string*, Phys. Lett. B **65** (1976), 471–474.

29. S. Deser and B. Zumino, *A complete action for the spinning string*, Phys. Lett. B **65** (1976), 369–373.

30. L. Brink and J. H. Schwarz, *Local complex supersymmetry in two dimensions*, Nucl. Phys. B **121** (1977), 285–295.

31. A. M. Polyakov, *Quantum geometry of bosonic strings*, Phys. Lett. B **103** (1981), 207–210; *Quantum geometry of fermionic strings*, Phys. Lett. B **103** (1981), 211–213.

32. S. Mandelstam, *Interacting-string picture of dual-resonance models*, Nucl. Phys. B **64** (1973), 205–235.

33. D. Z. Freedman, P. van Nieuwenhuizen and S. Ferrara, *Progress toward a theory of supergravity*, Phys. Rev. D (3) **13** (1976), 3214–3218.

34. S. Deser and B. Zumino, *Consistent supergravity*, Phys. Lett. B **62** (1976), 335–337.

35. P. van Nieuwenhuizen, *Supergravity*, Phys. Rep. **68** (1981), 189.

36. N. Marcus and J. H. Schwarz, (Caltech preprint CALT-68-910), *Field theories that have no manifestly Lorentz-invariant formulation*, Phys. Lett. B. **115** (1982), 111–114.

37. C. B. Thorn, *Embryonic dual model for pions and fermions*, Phys. Rev. D **4** (1971), 1112–1116.

38. J. H. Schwarz, *Dual quark-gluon model of hadrons*, Phys. Lett. B **37** (1971), 315–319.

39. E. Corrigan and D. I. Olive, *Fermion-meson vertices in dual theories*, Nuovo Cimento A (11) **11** (1972), 749–773.

40. E. Corrigan and P. Goddard, *The absence of ghosts in the dual fermion model*, Nucl. Phys. B **68** (1974), 189–202.

41. L. Brink, D. I. Olive, C. Rebbi and J. Scherk, *The missing gauge conditions for the dual fermion emission vertex and their consequences*, Phys. Lett. B **45** (1973), 379–383.

42. P. Ramond and J. H. Schwarz, *Classification of dual model gauge algebras*, Phys. Lett. B **64** (1976), 75–77.

43. C. Lovelace, *Systematic search for ghost-free string models*, Nucl. Phys. B **148** (1979), 253–282.

DEPARTMENT OF PHYSICS, CALIFORNIA INSTITUTE OF TECHNOLOGY, PASADENA, CALIFORNIA 91125

Lectures in Applied Mathematics
Volume 21, 1985

Gauging of Groups and Supergroups

P. van Nieuwenhuizen

ABSTRACT. We discuss various examples of free graded differential superalgebras which one encounters in the process of gauging (super)groups. We also discuss a few theorems on the general structure of such algebras. They determine the generalized curvatures and the purely group-theoretical part of the transformation rules which leave the action invariant.

Two main approaches to the gauging of groups are discussed: the group manifold approach and the R^2-approach. The latter only works for semisimple (super)groups, whereas the former can handle any (super)group. On the other hand, the R^2-approach has been the more successful for the superconformal groups.

1. Introduction. In these proceedings we will discuss the status of the gauging of groups (when we write groups we include supergroups). By "gauging of a group" one means a general method which associates to any given group a Lagrangian field theory. There exist many approaches to the gauging of groups, but in our opinion two methods have had more success than others: the R^2-method in Minkowski space-time and the group manifold approach. Neither approach is yet completely geometrical; the R^2-approach works only for semisimple groups, whereas the group manifold approach can handle any group. On the other hand, the R^2-method has reproduced, and even produced, some models which (so far) have not yet been reproduced in a fully satisfactory way by the group manifold approach, namely, the superconformal supergravities, so we will discuss both methods.

1980 *Mathematics Subject Classification.* Primary 81E20, 83C47.

139

When supergravity was first discovered [1–3] in 4-dimensional Minkowski space-time, group-theoretical aspects were not the main objects of interest. Of course, it was realized that the vielbein (viel = many in German, vier = four) field $V_\mu^m(x)$, the gravitino field $\psi_\mu^a(x)$ and the spin connection $\omega_{\mu n}^m(x)$ (= connection for local SO(3,1) transformations) did correspond to the generators P_m, Q_a and M_m^n of the super-Poincaré algebra [1]. However, the local algebra (i.e., the commutators of the local transformation rules which leave the action invariant) was more complicated than the global algebra [3]. For example, a typical commutator in the local algebra looks like

$$\left[\delta_s(\varepsilon_1(x)), \delta_s(\varepsilon_2(x))\right] V_\mu^m(x) = \left[\delta_{\text{diff}}(\xi^\mu(x)) + \delta_s(-\xi^\mu(x)\psi_\mu(x))\right.$$
$$\left. + \delta_L(-\xi^\mu(x)\omega_\mu^{mn}(x) + \cdots)\right] V_\mu^m(x)$$

$$(1.1)$$

where $\xi^\mu = \bar{\varepsilon}_2\gamma^m\varepsilon_1 V_m^\mu$ and V_m^μ is the inverse of V_μ^m. The symbol "diff" denotes a diffeomorphism, s denotes a local supersymmetry transformation, and L a local Lorentz (= SO(3,1)) transformation. In more detail:

(i) The commutator of *any* two local symmetries on a given field, say ϕ, which is part of a given representation Σ of the local algebra is again a sum of local symmetries acting on the *same* ϕ. (This defines a representation.) In the example in (1.1) the local symmetries are diffeomorphisms, local Lorentz and local supersymmetry transformations.

(ii) The structure constants can depend on all fields which constitute the representation; one speaks of structure *functions*. Hence, the representation in (1.1) is a particular kind of nonlinear representation. In (1.1) these structure functions appear between parentheses.

(iii) The same expression holds when one replaces V_μ^m by any other field in the representation; the structure functions are uniform within the multiplet. However, for another representation, the structure functions are different (but again uniform in that representation).

(iv) The global algebra is recovered by taking the parameters to be constant, and the flat-space limit in the structure functions

$$V_\mu^m(x) \to \delta_\mu^m, \quad \psi_\mu^a \to 0, \quad \omega_\mu^{mn} \to 0. \quad (1.2)$$

For example, (1.1) reduces to the well-known relation of global supersymmetry $[\bar{\varepsilon}_1 Q, \bar{\varepsilon}_2 Q] = \bar{\varepsilon}_2\gamma^m\varepsilon_1 P_m$, or, after extracting the parameters,

$$\{Q^a, Q^b\} = (\gamma^m)^{ab} P_m. \quad (1.3)$$

(v) A given multiplet Σ is usually larger than one might expect. For example, V_μ^m is part of a multiplet which contains, in addition to the expected ψ_μ^a and $\omega_{\mu n}^m$, a scalar S, a pseudoscalar P and an axial vector A_m [4]. (Actually, in this multiplet $\omega_{\mu n}^m$ is not an independent field, but a function of e_μ^m and ψ_μ^a.)

$$\Sigma = \left(V_\mu^m, \psi_\mu^a, S, P, A_m \right). \tag{1.4}$$

The dots in δ_L in (1.1) denote terms involving these fields S, P and A_m.

The local transformation rules, whose commutators define the local algebra, are relevant to the "gauging of groups" because they are to leave the action invariant. Thus, the program of gauging of groups splits into two parts: the construction of an action and the construction of the local transformation rules which leave the action invariant. Interestingly enough, the R^2-method starts with a geometric set of transformation rules, then constructs an action and then modifies the transformation rules (by imposing constraints on curvatures) such that the final transformation rules leave the final action (which is also modified by the constraints) invariant. The group manifold method, on the other hand, starts with the construction of the final action, and then constructs transformation rules which are on-shell (when the fields satisfy their Euler-Lagrange equations) correct, but whose final form (i.e., the off-shell rules) is, in general, only found by requiring that the action be invariant. Thus, both methods have unsatisfactory (i.e., nongeometric or, equivalently, non-group-theoretical) parts. The origin of the constraints is not entirely clear in the R^2-method, while the final fixing of the transformation rules in the group manifold approach is done "by hand" (by explicitly adding terms proportional to the equations of motion such that their sum leaves the given action invariant).

Only when the auxiliary fields (such as S, P and A_m in (1.4)) are part of the differential algebra (the "extended" group), something which is known to be the case in only one example, do the transformation rules follow in the group manifold approach in a purely geometrical way.

An immense amount of work in supergravity is currently being devoted to finding multiplets which represent the local gauge algebras of $\text{OSp}(N|4)$ and $\text{SU}(2,2|N)$. Technically this work is called "the auxiliary field problem for de Sitter (and Poincaré) and conformal supergravity" because $\text{Sp}(4) \sim \text{SO}(3,2)$ is the de Sitter group (whose contraction is the Poincaré algebra), while $\text{SU}(2,2)$ is the conformal group.

Due to the more complicated nature of the local (gauge) algebra, the construction of new supergravity models did not proceed by the "gauging of groups" but rather by choosing a multiplet of fields whose *states* did form a representation of the corresponding global algebra $\text{OSp}(N|4)$ or $\text{SU}(2,2|N)$. (The representations, in terms of states are not representations of the local algebra and do not contain auxiliary fields, which simplifies matters.) By subsequently using rather simple and direct, but non-group-theoretical methods (the so-called order-by-order-in-κ Noether method [5]), virtually all models known today were obtained.

The resulting models were, however, of a form which clearly hinted at a more central role for group theory. For example, naively gauging the supergroup $OSp(1|4)$ (whose generators are the above-mentioned P_m, Q_a, M_{nm}) reproduced all results of simple, ordinary supergravity except that the spin connection seemed to transform under local supersymmetry as $\delta_s \omega_{\mu n}^m = 0$, whereas, in reality, it is not inert when ω is considered an independent field (= first order formalism) [6]. (By naive gauging we mean: assuming that (2.5) is the complete set of transformation rules.) Later it was realized that if one expressed ω in terms of V_μ^m and ψ_μ^a by solving its algebraic field equation (second order formalism), it did not matter whether one computed the variation of the action using $\delta_s \omega_{\mu n}^m = 0$ or $\delta_s \omega_{\mu n}^m$ as obtained from the chain rule. In this sense $\delta \omega_{\mu n}^m = 0$ is correct [7]. Also, it was realized that this solution $\omega = \omega(V, \psi)$ of the ω field equation is the same as the solution of the constraint

$$R_{\mu\nu}^m(P) = 0, \qquad (1.5)$$

where $R_{\mu\nu}^m(P)$ is the curvature associated with the generator P_m (see §2). This constraint hinted, of course, again at group theory.

Similarly, naively gauging $SU(2,2|1)$ did reproduce all results of simple conformal supergravity, provided one imposed three constraints in this case [8]:

$$R_{\mu\nu}^m(P) = 0, \ (\gamma^m)^a_{\ b} R_{\mu\nu}^b(P) = 0, \text{ and the Einstein equations (!).}$$

$$(1.6)$$

Finally, naively gauging $OSp(2|4)$ reproduced all results of $N = 2$ extended supergravity (the theory which unifies electromagnetism with gravity [7]) except one result. Since this result is useful for understanding the difference between the philosophies of the R^2- and the group manifold approaches, we will quote it here in detail. In $N = 2$ supergravity there are the following fields: one field V_μ^m, two gravitino fields $\psi_\mu^{a,i}$ ($i = 1, 2$), and one photon field A_μ (with curl $F_{\mu\nu}$). The following transformation rule of the gravitinos leaves the action invariant:

$$\delta \psi_\mu^{a,i} = D_\mu \varepsilon^i + \varepsilon^{ij} F_{\rho\sigma} \left(\gamma^{\rho\sigma} \gamma_\mu \varepsilon^j \right). \qquad (1.7)$$

The first term on the right-hand side is the G-covariant derivative of the local supersymmetry parameter. This term is of pure group-theoretical origin (see §2) and would be the complete result by itself if one would gauge naively. The only thing of importance in the second term is the presense of a curvature. Apparently, the correct transformation law consists of two pieces: a purely group-theoretical piece and a curvature-dependent piece.

The basic achievement of the group manifold approach is to realize that the sum of these two terms is actually a diffeomorphism—not in Minkowski space-time, but in the group manifold. We explain this in §2.

In the R^2-approach one also can understand (and predict) the curvature term in (1.7); it is already present (when linearized in fields) in the representation of *global* supersymmetry in flat space. This, however, is a kind of hybrid explanation of (1.7). [For the linear representation of global supersymmetry where the carrier space consists of physical *states*, a complete mathematical theory, based on the method of induced representations, exists. But in the linear representation of the global algebra in terms of fields, auxiliary fields appear, and no complete mathematical theory exists. Indeed, (1.7) only leaves the action invariant and does not form a representation of the local or (in the linearized level) global algebra. For this one must find the extra auxiliary fields.]

A new development took place in the gauging of simple supergravity in $d = 11$ dimensions. In that theory there are an elevenbein V_μ^m, a gravitino ψ_μ^A ($A = 1, 32$ because the Clifford algebra in $d = 11$ is represented by 32×32 matrices, while $m, \mu = 0, 10$) and a 3-index antisymmetric tensor field $A_{\mu\nu\rho}$. In analogy with OSp(1|4) as the group of simple supergravity in $d = 4$, it seemed natural to take OSp(1|32) as the group of simple supergravity in $d = 11$. The generators of Sp(32) are products of 32×32 Dirac matrices, namely Γ_m, Γ_{mn} and Γ_{mnrstu}. Naive gauging of OSp(1|32) seemed to relate V_μ^m to Γ_m and ω_μ^{mn} to Γ_{mn}, while Γ_{mnrst} seemed to correspond to a field A_μ^{mnrst}. This field is not completely antisymmetric in all its six indices. It it would be totally antisymmetric, it could be interchanged with $A_{\mu\nu\rho}$ as far as a representation of global OSp(1|32) in terms of states would be concerned [9]. However, no corresponding Lagrangian field theory seemed to exist for the 6-index tensor, but only in the case of $A_{\mu\nu\rho}$. Thus OSp(1|32) did not seem to be the group of simple $d = 11$ supergravity.

Instead of a group, an "extended group" turned out to be the group-theoretical basis of the $d = 11$ model [10]. By "extended group" we mean a free graded differential superalgebra in which one finds not only 1-forms (as in the usual (dual) Lie algebras), but also p-forms with $p > 1$. In particular, $A_{\mu\nu\rho}$ was found to be a 3-form in such an algebra. Also, in $d = 10$ supergravity, and, more generally, in all the (many) models in which antisymmetric tensors appear, differential algebras are the basic group-theoretical structure. We shall discuss these algebras in some detail (§4). How one obtains actions from these differential algebras we shall not discuss, but refer to two sets of lectures on this subject [11].

2. Curvatures, covariant derivatives and diffeomorphisms. There are various approaches to the problem of gauging a given (super)group G, some of which use different definitions of curvatures than others. In the two approaches we will discuss, the R^2-approach in Minkowski space-time and the group manifold approach, the curvatures are G-curvature 2-forms

$$R^A = dh^A + \tfrac{1}{2}f^A_{BC}h^C \wedge h^B. \tag{2.1}$$

The h^A correspond in all approaches in a one-to-one way to the generators X_A of the (super)group

$$[X_A, X_B\} = X_C f^C_{AB}, \tag{2.2}$$

and in (2.1) the f^A_{BC} are the structure constants (ordinary real numbers) of G, and the contractions in (2.1) over C and B run over all generators of G. We will always contradict indices in a northeasterly fashion, without extra signs if the contracted indices are next to each other. Extra signs occur when contracted indices are not next to each other, but these signs are easily found as follows: first, write the contracted indices next to each other and make the result correct for the case that all uncontracted indices refer to commuting generators, then move indices, each time adding a minus sign when two anticommuting indices pass each other. In this way, one defines supersymmetric, supercyclic, etc.

The super-Jacobi identities

$$f^A_{BC}f^C_{PQ} + \text{supercyclic in } BPQ = 0 \tag{2.3}$$

are equivalent to the Bianchi identities

$$\nabla R^A \equiv dR^A + f^A_{BC}h^C \wedge R^B \equiv 0, \tag{2.4}$$

where ∇ is the G-covariant derivative in the adjoint representation.

In some other approaches one begins by defining curvatures in which the f^A_{BC} in (2.1) refer to a subgroup H of G. For example, in superspace supergravity one usually takes H equal to the Lorentz group. This we shall not do; however, in our discussion a subgroup H will also play a role (for example, in the *relative* cohomology theory; see §5).

The problem to solve is to find an action, constructed from the R^A and h^A, which is invariant under as many local symmetries as there are generators in G. The transformation rules under which the action is to be invariant are pure G-gauge transformations in the R^2-approach (to begin with) and diffeomorphisms in the group manifold approach (to begin with).

The G-gauge transformations are of the usual form: a G-covariant derivative of local parameters whose indices are in the adjoint representation.

$$\delta_{\text{gauge}}h^A = \nabla \varepsilon^A \equiv d\varepsilon^A + f^A_{BC}h^C\varepsilon^B. \tag{2.5}$$

Under an arbitrary variation of h^A into $h^A + \delta h^A$, where δh^A is infinitesimal (we assume that there exists a notion of "small" also for Grassmann variables), curvatures vary into a G-covariant derivative, $\delta R^A = \nabla(\delta h^A)$, but under the local gauge transformations in (2.5) curvatures transform homogeneously.

$$\delta_{\text{gauge}} R^A = \nabla \nabla \varepsilon^A = f^A_{BC} R^C \varepsilon^B. \tag{2.6}$$

This result expresses the well-known fact that the commutator of two covariant derivatives is a (sum of) curvatures.

The diffeomorphisms ($=$ general coordinate transformations "plus transport term" $=$ Lie derivatives) correspond to coordinate increases $z^\Lambda \to z^\Lambda + \xi^\Lambda(z)$. In the group manifold approach there are as many coordinates as there are generators, and there is a one-to-one correspondence, such that anticommuting generators correspond to anticommuting coordinates. (The group manifold is thus larger than superspace because it contains, in addition to x^μ and θ^α, Lorentz coordinates λ^{mn}. However, superspace is a station on the road from the group manifold to Minkowski space in the group manifold approach.) The diffeomorphisms act on the fields h^A_Π as follows (in the R^2-approach, replace Π by μ, where $\mu = 0, 3$):

$$\delta_{\text{diff}} h^A_\Pi = (-)^\Lambda \xi^\Lambda \partial_\Lambda h^A_\Pi + (-)^{\Lambda \Pi} h^A_\Lambda \partial_\Pi \xi^\Lambda. \tag{2.7}$$

The symbol Λ is 0 if Λ refers to a commuting generator and 1 if Λ refers to an anticommuting generator (see discussion under (2.2)).

One can rewrite (2.7) as a sum of a term as in (2.5) and a term as in (2.1)

$$\delta_{\text{diff}} h^A_\pi = (-)^{A\pi} \nabla_\pi \left(h^A_\Lambda \xi^\Lambda \right) + (-)^\Lambda \xi^\Lambda R^A_{\Lambda\pi}, \tag{2.8}$$

where from (2.1) we define that $R^A_{\Lambda\pi} = \partial_\Lambda h^A_\pi + \text{more}$ (with strength 2).

It is useful to employ forms because they take into account many of the ugly extra signs (while invariance of the action under diffeomorphisms will be automatic, even when the action is an integral over a curved hypersurface in the group manifold). In terms of forms $h^A \equiv h^A_\Lambda dz^\Lambda$, (2.8) becomes

$$\delta_{\text{diff}} h^A = \nabla \left(h^A_\Lambda \xi^\Lambda \right) + (-)^\Lambda \xi^\Lambda R^A_{\Lambda\pi} dz^\pi. \tag{2.9}$$

Let us introduce anholonomic local parameters by

$$\varepsilon^A \equiv h^A_\Lambda \xi^\Lambda. \tag{2.10}$$

In the case of the group manifold the h^A_Λ are square matrices which we assume to be invertible. In that case we can rewrite (2.9) entirely in terms of ε^A. The inverses h^Λ_A are defined by

$$h^\Lambda_A h^A_\pi = \delta^\Lambda_\pi \Rightarrow h^A_\Lambda h^\Lambda_B = \delta^A_B. \tag{2.11}$$

[For supermatrix algebra, see [5].] Inserting the completeness relation (2.11) twice into (2.9) one finds

$$\delta_{\text{diff}} h^A = \nabla \varepsilon^A + (-)^B \varepsilon^B R^A_{BC} h^C. \qquad (2.12)$$

In this form one sees that diffeomorphisms consist of a pure G-gauge transformation, plus extra curvature-dependent terms.

In the R^2-approach one begins with (2.5), but discovers subsequently that one needs constraints on certain of the curvatures. These constraints then modify (2.5) into a form like (2.8) (with Λ, Π equal to Minkowski indices μ, ν). To see how this mechanism works consider a constraint $C_A R^A_{\mu\nu} = 0$, where C_A may depend on fields. If this constraint is to hold identically, one must have

$$(\delta C_A) R^A_{\mu\nu} + C_A \delta R^A_{\mu\nu} \equiv 0. \qquad (2.13)$$

In general, the δh^A_μ which satisfy (2.13) are not given by (2.5) but need extra terms $\delta_{\text{extra}} h^A_\mu$:

$$\delta R^A_{\mu\nu} = f^A_{BC} R^C_{\mu\nu} \varepsilon^B + \nabla_\mu (\delta_{\text{extra}} h^A_\nu) - \mu \leftrightarrow \nu, \qquad (2.14)$$

and from (2.13) and (2.14) it is clear that these $\delta_{\text{extra}} h^A$ will, in general, contain curvatures, as in (2.12). Only if the constraints are gauge-invariant, i.e., if (2.13) holds when δh^A_μ is given by (2.5), no curvature terms are needed in δh^A_μ. In most cases this is the case, but in a few cases one finds extra curvature terms in the transformation rules.

In the group manifold approach one starts with (2.12) and has thus curvature terms in all transformation laws. In this approach *the field equations* in the group manifold relate the R^A_{BC} to each other, and many of the R^A_{BC} turn out to be zero. As a result, one ends up with (2.5) in many cases. However, (2.12) is not, in general, the correct law; it is only correct on-shell (when the Euler-Lagrange variational equations are satisfied), but off-shell one needs extra terms, proportional to the field equations in the x-space, and their form follows by requiring that the action be invariant. The action is obtained in a completely geometric way; the transformation rules, in general, are not (only when the aux-iliary fields are known, can they also be deduced in a geometric fashion). [This is a quite recent result, for which we refer the reader to [12].]

In the next section we will study an example of an R^2-theory, but there exist many more supergravity models which can be rewritten as R^2-theo-ries. The simplest case is simple ($N = 1$) supergravity, as well as ordinary Einstein gravity ($N = 0$ supergravity) [13]. The curvatures in this case are super-de Sitter curvatures, which are a sum of a super-Poincaré curvature and an extra term E involving only vierbeins. In the square $R_S R_S = (R_P + E)(R_P + E)$, the $R_P R_P$ terms are total derivatives, the cross terms $R_P E$ yield the super-Poincaré action while the EE terms yield the supercosmological term. Thus, one first obtains (super)gravity with a (super)cosmological term, and only subsequently can one reach the model without a (super)cosmological term by a (super)group contraction. Similar results hold for $N = 2$ supergravity [7].

3. An R^2-theory: Conformal supergravity. Although many (all)? supergravity models can be written as R^2-theories, in particular, $N = 1$ (simple) and $N = 2$ (the simplest extended) nonconformal supergravity, we shall only discuss here in some detail simple conformal supergravity because its properties are less well known.

In this case the supergroup we start from is the group SU(2, 2|1). The group SU(2, 2) is the conformal group (the group which leaves $\psi^\dagger \gamma_0 \chi$ invariant where ψ, χ are complex 4-component spinors). The 24 generators of SU(2, 2|1) form a graded Z_∞ superalgebra.

grade	−2	−1	0	+1	+2	
generator	K^m	S^a	D, A, M^{mn}	Q^a	P^m	(3.1)
field	f_μ^m	φ_μ^a	$b_\mu, A_\mu, \omega_\mu^{mn}$	ψ_μ^a	e_μ^m	

In the gradings −1 and +1 we find the odd generators $Q^a (= \sqrt{P^m})$ and $S^a (= \sqrt{K^m})$. This grading is useful to deduce how fields and curvatures transform. For example, the vierbein is S-inert (*Proof.* "S with what gives P"? *Answer.* Only with grade-3 generators, which do not exist). Other examples: in $\{Q^a, S^b\}$ one can only find D, A and M^{mn}, while $[Q^a, P^m] = 0$. The generators D and A generate scale and chiral transformations, respectively, while K^m generate conformal boosts, of which no satisfactory physical interpretation is known.

Since the action I is to be invariant under local scale transformations D, it should not contain dimensionful parameters. Moreover, I should be dimensionless. Let us assume that parity is preserved. This is a physical input which is mathematically (and perhaps physically) not necessary. Let us require that the action is *affine*, i.e., that it is of the form

$$I = \int d^4x Q_{AB} R^B \wedge R^A,$$

where Q_{AB} are constants. Then *the most general action reads*

$$I = \int d^4x \Big[\alpha R(M)^{mn} \wedge R(M)^{kl} \varepsilon_{mnkl}$$
$$+ \beta \bar{R}(Q) \gamma_5 R(S) + \gamma R(A) \wedge R(D) \Big], \qquad (3.2)$$

where $\bar{R}(Q)_a = R(Q)^b C_{ba}$ and α, β, γ are constants.

This action is invariant under local grade-0 (D, A, M^{mm})-gauge transformations, as one easily proves, using that curvatures rotate homogeneously. Let us now first consider local K, S transformations with parameters ξ_K^a and ε_S, respectively.

The action transforms under pure K-gauge transformations given by (2.5) as follows:

$$\delta_K I = \int d^4x \, \xi_K^m \Big[-4\alpha R(P)^n \wedge R(M)^{kl} \varepsilon_{mnkl}$$

$$- \beta \bar{R}(Q) \gamma_5 \gamma_m \wedge R(Q) - 2\gamma R(P)_m \wedge R(A) \Big]. \quad (3.3)$$

Snce these variations do not cancel, we need either to fix the parameters or to impose constraints on curvatures.

One of the basic lessons of supergravity is that it is simplest to eliminate as many independent fields as possible (the formulations of theories with some fields eliminated are generally called second-order theories). Let us consider the curvatures, and make a list of possible constraints on them which can, in principle, eliminate fields *algebraically*. (In superspace supergravity one also uses differential constraints, but we shall not consider such constraints here.)

TABLE 1. Table of constraints.

$$R(M)^{mn} = d\omega^{mn} + \omega^{mk} \wedge \omega_k^n + 2(e^m \wedge f^n - e^n \wedge f^m) + \frac{1}{2}\bar{\psi}\gamma^{mn}\varphi$$

$$R(D) = db - 2e^m \wedge f_m - \frac{1}{2}\bar{\psi} \wedge \varphi$$

$$R(A) = dA + \bar{\psi} \wedge i\gamma_5 \varphi$$

$$R(Q) = d\psi + \frac{1}{4}\omega^{mn} \wedge \gamma_{mn}\psi + \frac{1}{2}b \wedge \psi - \frac{3i}{4}A \wedge \gamma_5\psi - \gamma_m e^m \wedge \varphi$$

$$R(S) = d\varphi + \frac{1}{4}\omega^{mn} \wedge \gamma_{mn}\varphi - \frac{1}{2}b \wedge \varphi + \frac{3i}{4}A \wedge \gamma_5\varphi + \gamma_m f^m \wedge \psi$$

$$R(P)^m = de^m + \omega^{mn} \wedge e_n - \frac{1}{4}\bar{\psi} \wedge \gamma^m\psi - e^m \wedge b$$

$$R(K)^m = df^m + \omega^{mn} \wedge f_n + \frac{1}{4}\bar{\varphi} \wedge \gamma^m\varphi + f^m \wedge b$$

Looking at this table, we see that we can eliminate

$$b_\mu \text{ or } \omega_\mu^{mn} \text{ from } R(P); \quad f_\mu^m \text{ from } R(M)^{mn} \text{ or } R(D); \quad \varphi_\mu^a \text{ from } R(Q)^a.$$

$$(3.4)$$

However, we cannot put the whole $R_{\mu\nu}^{mn}(M) = 0$, because that would involve $6 \times 6 = 36$ constraints, whereas $f_{m\mu}$ has only $4 \times 4 = 16$ components. On the other hand, the contraction $R_{\mu\nu}^{mn}(M)e_n^\nu$ has 16 components and could be used. Similarly, $R_{\mu\nu}^a(Q)$ has too many components ($6 \times 4 = 24$) to eliminate φ_μ^a, but $\gamma^\mu R_{\mu\nu}^a(Q)$ has $4 \times 4 = 16$ components, just the right number to eliminate the 16 components of φ_μ^a. As to $R_{\mu\nu}^m(P)$, it has $4 \times 6 = 24$ components, the same number as $\omega_{\mu mn}$, so one could impose

$R_{\mu\nu}^m(P) = 0$. On the other hand, the weaker constraint $R(P)_{\mu\nu}^m e_m^\nu = 0$ also could be used to eliminate b_μ.

Looking back now at $\delta_K I$, we see that since all of $R(M)$ cannot be constrained to be zero, either $\alpha = 0$ or $R_{\mu\nu}^m(P) = 0$. As to the $R(Q)$ term in $\delta_K I$, again we have two choices: $\beta = 0$ or $R_{\mu\nu}(Q) = k\gamma_5 \varepsilon_{\mu\nu}^{\rho\sigma} R_{\rho\sigma}(Q)$, because $\bar{R}(Q)\gamma^m R(Q) = 0$ (since $C\gamma^m$ is symmetric and C antisymmetric). For consistency we must restrict $k^2 = 1/4$. Since we are interested in a nondegenerate theory, i.e., a theory with propagating gravitons and gravitinos, we must require that $\alpha \neq 0$, $\beta \neq 0$, hence we must impose the following constraints:

$$R(P)_{\mu\nu}^m = 0, \qquad R(Q)_{\mu\nu} = \pm\frac{1}{2}\gamma_5 \varepsilon_{\mu\nu}^{\rho\sigma} R_{\rho\sigma}(Q). \qquad (3.5)$$

The $R(Q)$ constraint eliminates the 12 components of $\varphi_\mu - \frac{1}{4}\gamma_\mu\gamma\cdot\varphi$. Since these constraints are invariant under K-transformations ("What with K gives P?" Nothing; *Idem*: "What with K gives Q?" Nothing), imposing these constraints does not lead to modifications in the K-gauge transformations of all fields, and hence, with these constraints, $\delta_K I$ is the same as without them.

Let us now analyze the local S-gauge transformations. Under pure group transformations

$$\delta_S I = \int d^4x \left[R(M)^{kl} \wedge \bar{\varepsilon}_S \gamma_{kl} R(Q)\left(2\alpha + \frac{1}{4}\beta\right) \right.$$

$$+ R(D) \wedge \bar{\varepsilon}_S \gamma_5 R(Q)\left(-\frac{1}{2}\beta + i\gamma\right)$$

$$+ R(A) \wedge \bar{\varepsilon}_S R(Q)\left(\frac{3i\beta}{4} - \frac{1}{2}\gamma\right) \qquad (3.6)$$

$$\left. - \beta\bar{R}(S)\gamma_5\gamma_m\varepsilon_S \wedge R(P)^m \right].$$

Since the constraints obtained in (3.5) are also S-invariant, the result for $\delta_S I$ does not acquire further terms due to modified transformations in $\delta\omega_\mu$ and $\delta\varphi_\mu$. Again, since not all of $R(M)^{cd}$ can be constrained to be zero, nor all of $R(Q)^a$, we must require

$$8\alpha + \beta = 0. \qquad (3.7)$$

The last term vanishes due to $R(P)^m = 0$, but the two remaining terms yield, using that $R(Q)$ is (anti-)self-dual,

$$\tilde{R}_{\mu\nu}(D)(-\beta/2 + i\gamma) + R_{\mu\nu}(A)(-3i\beta/2 + \gamma) = 0. \qquad (3.8)$$

This constraint contains $R_{\mu\nu}(D)$ and can thus be used to eliminate $f_{\mu\nu} - f_{\nu\mu}$.

On the other hand, the S-variation of this constraint is proportional to $aR_{\mu\nu}(Q) + \frac{1}{2}\gamma_5\varepsilon_{\mu\nu}{}^{\rho\sigma}R_{\rho\sigma}(Q)$ and, in order that $a = 1$, we must fix $\beta + 2i\gamma = 0$. Thus, all coefficients of the action are fixed at this point! [The K-variation of $R(D)$ is proportional to $R(P)$ while $R(A)$ is K-inert, hence the constraint in (3.8) is K-invariant and no extra $\delta_K h$ as in (2.14) are present.]

Finally, the local Q-transformations: Since $R(P)$ rotates under Q into $R(Q)$, the solved-for ω_μ^{mn} (a dependent field now) no longer transforms according to the group, but acquires an extra term

$$\delta\omega_{\mu mn} = -\frac{1}{4}\bar{\varepsilon}_Q\big(\gamma_\mu R(Q)_{mn} + \gamma_m R(Q)_{\mu n} - \gamma_n R(Q)_{\mu m}\big). \quad (3.9)$$

Similarly, the other constraints yield extra $\delta'f_{m\mu}$ and $\delta'\varphi_\mu$, which we will not need here. The Q-variation of the action now contains two terms proportional to $R(K)$: one term due to δ_Q (gauge) $R(S)$ and another term proportional to $\delta'\omega_\mu^{mn}$. Their sum cancels if and only if $\gamma^\mu\tilde{R}_{\mu\nu}(Q) = 0$. Together with the self-duality constraint of $R(Q)$ this implies the final Q-constraint

$$\gamma^\mu R_{\mu\nu}(Q) = 0. \quad (3.10)$$

With these constraints the action is invariant.

Note that the constraint $R_{\mu\nu}(A) \sim \tilde{R}_{\mu\nu}(D)$ converts the off-diagonal $R_{\mu\nu}(A)\tilde{R}_{\mu\nu}(D)$ term in the action into a kinetic term for A (or D)! Although we have sketched here in detail the derivation of the constraints using the action as guiding principle, one can also use kinematical arguments. The P-gauge transformations should be replaced by diffeomorphisms as invariances of the action. If all δ_Q (gauge) were unmodified, one would find that the action is invariant under the commutator of two δ_Q (gauge) transformations, hence under δ_P (gauge). Thus one must modify the Q-transformations of the fields which are eliminated (the exact law follows from the chain rule) or constrain curvatures. This explains that $R(P) = 0$ because $\delta_Q e_\mu^m = \frac{1}{2}\bar{\varepsilon}_Q\gamma^m\psi_\mu$ and $\delta_Q\psi_\mu$ is unchanged (ψ_μ cannot be eliminated, unlike φ_μ) so that the group commutator should coincide with a diffeomorphism: $R(P) = 0$! On the other hand, $\delta_Q\psi_\mu$ contains ω_μ^{mn}, and $\delta'\omega_\mu^{mn}$ must yield the extra $R(Q)$ term which modifies the P-gauge term into a diffeomorphism because we cannot put all of $R(Q)$ equal to zero. From $R(P) = 0$ alone one does not find the correct $\delta\omega'$; one needs, in addition, $\gamma^\mu\tilde{R}_{\mu\nu}(Q) = 0$. Finally, also A_μ and b_μ are not eliminated, and since δ_Q (gauge) $A_\mu \sim \bar{\varepsilon}_Q\gamma_5\varphi_\mu$, while δ_Q (gauge) $b_\mu \sim \bar{\varepsilon}_Q\varphi_\mu$, it is now $\delta'\varphi_\mu$ which must provide the required $R(A)$ and $R(D)$ terms which modify the P-gauge transformation into a diffeomorphism. The result is that the extra term $\delta'\varphi_\mu$ should be proportional to $R_{\mu\nu}(D)\varepsilon$ and $\gamma_5\varepsilon R_{\mu\nu}(A)$, and this result is the same as that when one imposes $\gamma^\mu R_{\mu\nu}(Q) = 0$ and $R_{\mu\nu}(D) + (i/4)\tilde{R}_{\mu\nu}(A) = 0$.

Summarizing: The superconformal action is quadratic in curvatures and purely affine, and the necessary constraints can either be deduced by requiring the action to be invariant under as many local symmetries as there are generators in the superalgebra, or by requiring that the commutator of two local Q-transformations yields a diffeomorphism instead of a local P-gauge transformation. The complete set of constraints reads

$$R(P)^m_{\mu\nu} = 0, \quad \gamma^\mu R_{\mu\nu}(Q) = 0, \quad \tilde{R}_{\mu\nu}(D) + iR_{\mu\nu}(A) = 0. \quad (3.11)$$

Using the Bianchi identity $D_{[\mu} R(P)^m_{\rho\sigma]} = 0$, one find that the antisymmetric part of the Ricci tensor $R_{\mu\nu}(M)$ is proportional to $R_{\mu\nu}(D)$, and thus one has a further constraint

$$R(M)_{\mu\nu} - R(M)_{\nu\mu} = (i/2)\tilde{R}_{\mu\nu}(A) + \bar{\psi}^\lambda \gamma_{[\mu} R(Q)_{\nu]\lambda}. \quad (3.12)$$

Actually, if one eliminates the nonpropagating field φ^m_μ from the action by solving its algebraic field equation, one gets the same result as imposing a constraint

$$R(M)_{\mu\nu} - \frac{1}{6} g_{\mu\nu} R(M) = \frac{i}{4} \tilde{R}_{\mu\nu}(A) + \frac{1}{2} \bar{\psi}^\lambda \gamma_\mu R(Q)_{\nu\lambda}. \quad (3.13)$$

(The terms on the right hand side make (3.13) S-invariant.)

Thus, this stronger constraint not only eliminates the antisymmetric part of $f_{\mu\nu}$ but also its symmetric part. However, it seems that this last elimination is a luxury and is not needed. After elimination of the full f^m_μ, ω^{mn}_μ and φ_μ, the remaining fields are e^m_μ, ψ_μ, b_μ and A_μ. Since only b_μ transforms under K-gauge transformations, while the action is K-invariant, the field b_μ must cancel in the action. It indeed does, as one can check explicitly.

A final word about why conformal and, especially, superconformal theories are useful in supergravity. Nonconformal supergravity can be considered as a conformal coupling of certain matter fields to the gauge fields of $SU(2,2|1)$ *in a particular gauge*. It simplifies matters enormously to work till the very end in a conformally invariant way and choose the gauges only at the very end.

Thus, all constraints are invariant under all local gauge invariances except under Q. This should be so, since the only (anti)commutator $\{A, B\} \sim P$ with $A, B \neq P$ is the $\{Q, Q\}$ anticommutator. If one uses (3.13), the local gauge algebra closes by accident: one does not need auxiliary fields in that case. If one uses (3.12), one finds in $\{Q, Q\}f^m_\mu$ a term proportional to the δ'_Q variation of the $(f_{\mu\nu} + f_{\nu\mu})$ field equation.

4. Free graded differential superalgebras. In supergravity models one often encounters antisymmetric tensor fields $A_{\mu_1 \cdots \mu_k}$. So far, no antisymmetric tensors with internal indices have been found ($A^B_{\mu_1 \cdots \mu_k}$). These antisymmetric tensors have no place in ordinary Lie (super)algebras where one only finds covariant vector fields h_μ^A, but they have a place in differential algebras.

The differential algebras we will consider have a finite number of generators h_p^A. The subscript p ($p \geq 1$) indicates that h is a p-form, while A labels the various p-form generators further. The h_p^A can be bosonic (even) or fermionic (odd). We have thus a $(Z_\infty \times Z_2)$-grading. The Z_∞-grading is indicated by "graded", while the Z_2-grading refers to the "super" in superalgebras. The "free" refers to the fact that all h_p^A are independent except that they satisfy the following properties:

$$h_p^A \wedge h_q^B = (-)^{pq+\sigma} h_q^B \wedge h_p^A, \tag{4.1}$$

where $\sigma = 0$, except when both h_p^A and h_q^B are fermionic, in which case $\sigma = 1$ since one gets an extra minus sign when one interchanges fermions. We will add even p-forms to even p-forms, and odd p-forms to odd p-forms, but multiply any two forms (multiplication is denoted by the wedge symbol $\wedge$) and consider the real numbers as the field over which the algebra is defined.

The differential operator d acts on a product of generators according to Leibnitz' rule as if it were a bosonic 1-form

$$d\left(h_p^A \wedge h_q^B\right) = \left(dh_p^A\right) \wedge h_q^B + (-)^p h_p^A \wedge dh_q^B. \tag{4.2}$$

This definition is in agreement with the usual representation $d = \partial_\Lambda dz^\Lambda$, whether Λ runs over the whole supergroup or only over Minkowski space-time.

The differential algebras we consider equate dh_p^A to a ($p + 1$)-form

$$dh_p^A = \sum \frac{1}{k} C\begin{pmatrix} A\, p_1 \cdots p_k \\ p\, A_1 \cdots A_k \end{pmatrix} h_{p_1}^{A_1} \wedge \cdots \wedge h_{p_k}^{A_k}, \tag{4.3}$$

and if h_p^A is bosonic (fermionic), so is dh_p^A. Here $p_1 + \cdots + p_k = p + 1$ and the C's are real numbers. The factor $1/k$ is inserted for later purposes and the C's have that (anti)symmetry in the index pairs (p_j, A_j) which is induced by the h's. The sum extends over all possible terms (including all partitions of $p + 1$) with this proviso: *the differential relations* (4.3) *must be consistent.* By this we mean that acting once more with d and using the Poincaré identity $dd = 0$ (again something which follows in the representation $d = \partial_\Lambda dz^\Lambda$), one finds an identity. Thus,

$$ddh_p^A = 0 = \sum C\begin{pmatrix} A\, p_1 \cdots p_k \\ p\, A_1 \cdots A_k \end{pmatrix} \sum \frac{1}{l} C\begin{pmatrix} A_1 q_1 \cdots q_l \\ p_1 B_1 \cdots B_l \end{pmatrix}$$
$$\cdot h_{q_1}^{B_1} \wedge \cdots \wedge h_{q_l}^{B_l} \wedge h_{p_2}^{A_2} \wedge \cdots \wedge h_{p_k}^{A_k}. \tag{4.4}$$

In other words, the generalized curvatures

$$R_p^A \equiv dh_p^A - \sum \frac{1}{k} C\begin{pmatrix} A\, p_1 \cdots p_k \\ p\, A_1 \cdots A_k \end{pmatrix} h_{p_1}^{A_1} \wedge \cdots \wedge h_{p_k}^{A_k} \tag{4.5}$$

vanish ($R_p^A = 0$) by definition and are consistent as well ($dR_p^A = 0$). The consistency conditions in (4.4) are generalized Jacobi identities. All this is standard in mathematics. These generalized curvatures R_p^A and the generators h_p^A are the building blocks for the action. There exists a complete set of rules to find an action for *any* group or supergroup, semisimple or not. We will not discuss here how to obtain the action, but refer to two sets of lectures on this subject [11]. Briefly, one has to find the general solution of a generalized cohomology equation

$$\nabla \nu_A^p + \left(\delta/\delta h_p^A \right)\Lambda = 0, \tag{4.6}$$

where Λ is a d-form (if the final Minkowski space-time has d dimensions) and where the Lagrangian density is the d-form $\Lambda + \nu_A^p \wedge R_p^A +$ terms with R^2, R^3,.... Clearly, ν_A^p is a $(d - (p + 1))$-form, and the covariant derivative ∇ is defined by the Leibnitz rule

$$\nabla\left(\nu_A^p \wedge S_p^A \right) = \left(\nabla \nu_A^p \right) \wedge S_p^A + \nu_A^p \wedge \nabla S_p^A = d\left(\nu_A^p \wedge S_p^A \right), \tag{4.7}$$

where ∇S_p^A is defined by

$$\nabla S_p^A = dS_p^A + \sum C \begin{pmatrix} A\, p_1 \cdots p_k \\ p\, A_1 \cdots A_k \end{pmatrix} S_{p_1}^{A_1} \wedge h_{p_2}^{A_2} \wedge \cdots \wedge h_{p_k}^{A_k}. \tag{4.8}$$

This definition of ∇ agrees with (4.5) where it yields Bianchi identities

$$\nabla R_p^A = 0. \tag{4.9}$$

We shall discuss examples of differential algebras as they are found in physics and enumerate open mathematical questions whose solution would be useful for physicists. We begin by quoting a theorem by D. Sullivan for nonsuperalgebras [14] whose extensions to superalgebras were discussed in [15].

THEOREM. *Every differential algebra splits into a contractible algebra and a minimal algebra.*

A contractible algebra consists of pairs of *generators* h_p^A, h_{p+1}^A such that

$$dh_p^A = h_{p+1}^A, \qquad dh_{p+1}^A = 0. \tag{4.10}$$

A minimal algebra is of the form $dh_p^A =$ sum of *products* (or zero). Thus, one need not consider mixed expressions such as $dh_p^A = h_{p+1}^A + h_p^A \wedge h_1^B + \cdots$. In particular, ordinary Lie superalgebras

$$dh_1^A = -\frac{1}{2} f_{BC}^A h_1^C \wedge h_1^B \tag{4.11}$$

are minimal algebras and consistency is equivalent here to the Jacobi identities.

Suppose we consider all p-form generators h_p^A, and collect all terms of the form $h_1^B \wedge h_p^C$ on the right-hand side of dh_p^A and bring them to the left-hand side. Then

$$(d - M_p) \wedge h_p^A = a_{p+1}^A, \qquad (4.12)$$

where M is a matrix 1-form ($M = M_B^A$) and a_{p+1}^A a $(p + 1)$-form, composed out of *products* of q-form generators with $q \leq p - 1$. Consistency implies two independent results since we are dealing with a free algebra

$$dM_p - M_p \wedge M_p = 0, \qquad (d - M_p)a_{p+1}^A = 0. \qquad (4.13)$$

These results imply the following:

(i) the matrices M_p form, for every p, a (trivial or nontrivial, reducible or not) representation of the superalgebra in (4.11) of the 1-forms;

(ii) the covariant derivatives $D_p = d - M_p$ (to be distinguished from the ∇) are nilpotent: $D_p D_p = 0$;

(iii) the a_{p+1}^A in (4.12) are closed under D_p. Denoting by A_k the algebra generated by all q-forms with $q \leq k$, the a_{p+1}^A are elements of A_{p-1} and are elements of a cohomology class H:

$$a_{p+1}^A \in H_{p+1}(M_p, A_{p-1}). \qquad (4.14)$$

A question of possible physical relevance is the following: Can one always trivialize a nontrivial cohomology class by adding extra 1-forms to be system? In other words, given (4.12), i.e., a $(p + 1)$-form constructed from generators with grade $p - 1$, which is closed under D_p but not exact, can one always add to (4.12) extra 1-form generators

$$dg_1^A = h_{BC}^A g_1^C \wedge g_1^B + k_{BC}^A g_1^C \wedge h_1^B + l_{BC}^A h_1^C \wedge h_1^B \qquad (4.15)$$

(without changing (4.12) and the original differential algebra) such that

$$a_{p+1}^A = D_p b_p^A, \qquad b_p^A \in A_1, \qquad (4.16)$$

in the extended system? D. Sullivan has given a counterexample: for a simple ordinary Lie algebra, the product of all 1-forms is closed under $D = d$ but can never be made exact. An open question is: In which cases is it possible to trivialize a given cohomology class?

The reason one requires b_p^A to lie in A_1 is that in that case one has a representation for h_p^A in terms of 1-forms, which means in terms of ordinary gauge fields. Mathematically, one can study the cases where $b_p^A \in A_{p-1}$; the physical relevance is not clear.

5. Differential algebras encountered in supergravity. In this section we assume familiarity with manipulations of elements of Clifford algebras (Dirac matrices). For a discussion see, for example, [16].

The differential algebras encountered in supergravity always contain the super-Poincaré algebra (or its extension, the super-de Sitter algebra; see below). In d dimensions it reads

$$R^{mn} \equiv d\omega^{mn} + \omega^{mk} \wedge \omega^n_k = 0, \tag{5.1a}$$

$$\mathcal{D}V^m + \bar\psi\gamma^m \wedge \psi = 0, \qquad \mathcal{D}V^m \equiv dV^m + \omega^{mn} \wedge V_n, \tag{5.1b}$$

$$\mathcal{D}\psi = 0, \qquad \mathcal{D}\psi \equiv d\psi + \frac{1}{4}\omega^{mn} \wedge \gamma_{mn}\psi. \tag{5.1c}$$

Here ψ^a is a fermionic 1-form ($a = 1, 4$), ω^{mn} and V^m are bosonic 1-forms, and $\bar\psi = \psi^T C$, where C is a matrix to be determined. (It is what physicists call the charge conjugation matrix.)

We will mostly be concerned with *relative* cohomologies. Given an algebra G and a subalgebra H, we consider a split $G = H \oplus K$. If, in the defining relations $D(M_p)h_p^A = a^A_{p+1}$, the a^A_{p+1} are constructed only from elements of K, while the C's in (4.3) are H-invariant constant tensors, then the a^A_{p+1} are elements of the relative cohomology class $H_{p+1}(M_p, A_{p-1}, H)$.

The relative cohomologies we are interested in have an H which always contains the Lorentz group. Thus, all relations in (5.1) must be good Lorentz relations. This means that, for example, $\bar\psi\gamma^m \wedge \psi$ should be a Lorentz vector, which means that (a) $C\gamma_{mn}C^{-1} = -(\gamma_{mn})^T$ (where $\gamma_{mn} \equiv \frac{1}{2}[\gamma_m, \gamma_n]$). Moreover, $\bar\psi\gamma^m \wedge \psi$ should not vanish identically, which means that (b) $C\gamma^m$ is a symmetric matrix. The condition (a) determines C up to a factor $\gamma_d \equiv \gamma_0\gamma_1 \cdots \gamma_{d-1}$. In odd d, $\gamma_d = I$, hence C is unique; but in even d there are thus two matrices C satisfying condition (a). In which dimensions does there exist a C satisfying condition (b) as well? The answer is [16]

$$d = 4, 8, 9, 10, 11 \mod 8. \tag{5.2}$$

We will now begin by investigating for which of the dimensions in (5.2) the relations in (5.1) are consistent. For completeness we recall that the matrices C satisfy, in any d,

$$C\gamma_m C^{-1} = -\gamma_m^T \text{ or } + \gamma_m^T \quad \text{for all } m, \tag{5.3}$$

and that C is the usual charge conjugation matrix, except that one can have a $+$ sign in (5.3) (because we are dealing with massless fields [16]).

The consistency of (5.1) is best studied by using the H-covariant derivative ($H = $ Lorentz group here) $\mathcal{D}$, instead of d, which satisfies

$$\mathcal{D}\mathcal{D} = R^{mn}L_{mn}, \qquad \mathcal{D}R^{mn} \equiv 0, \tag{5.4}$$

where L_{mn} are the Lorentz generators ($L_{mn} = \frac{1}{4}\gamma_{mn}$ in the spinor representation). The relation $\mathcal{D}R^{mn} \equiv 0$ is the Bianchi identity for the Lorentz group and holds whether or not $R_{mn} = 0$. It is at once clear that acting with $\mathcal{D}$ on (5.1) yields zero when (5.1) holds. Hence, in $d = 4, 8, 9, 10, 11$ mod 8, the system (5.1) is consistent. (In $d = 5, 6, 7$ one needs at least two fermionic 1-forms to obtain consistency [16].)

Let us now consider, as a first extension, the super-de Sitter algebra. It reads

$$R^{mn} + \alpha e^2 V^m \wedge V^n + \beta e \bar{\psi} \gamma^{mn} \wedge \psi = 0, \qquad (5.5a)$$

$$\mathcal{D} V^m + \bar{\psi} \gamma^m \wedge \psi = 0, \qquad (5.5b)$$

$$\mathcal{D} \psi + e \gamma^m V_m \wedge \psi = 0. \qquad (5.5c)$$

After rescaling V and ψ, the only free constants are α and β, but consistency of (5.5c) requires that

$$\mathcal{D}\mathcal{D}\psi = R^{mn} \frac{1}{4} \gamma_{mn} \psi = e\bar{\psi} \wedge \gamma_m \psi \wedge \gamma^m \psi - e^2 \gamma^m V_m \wedge \gamma^n V_n \wedge \psi. \qquad (5.6)$$

Hence, with (5.5a), $\alpha = 4$, while terms with three ψ cancel if

$$\frac{1}{4} \beta (\bar{\psi} \gamma^{mn} \wedge \psi) \wedge \gamma_{mn} \psi + (\bar{\psi} \gamma^m \wedge \psi) \wedge \gamma_m \psi = 0. \qquad (5.7)$$

Fierz rearranging $\bar{\psi}\psi\psi$ we find an identity

$$(\bar{\psi}\psi)\psi = -\frac{1}{\alpha} \left(\bar{\psi} \gamma^m \wedge \psi \gamma_m \psi - \frac{1}{2} \bar{\psi} \gamma^{mn} \psi \gamma_{mn} \psi + \frac{1}{5!} \bar{\psi} \gamma^{(5)} \psi \gamma^{(5)} \psi \right) = 0 \qquad (5.8)$$

[where $\gamma^{(5)}$ is a product of five Dirac matrices, totally antisymmetrized, and is absent from (5.8) when $d < 10$]. In $d = 4, 11$, $\bar{\psi}\psi = 0$ because $C^T = -C$ there, while in $d = 10$ one can choose a C with $C^T = -C$; however, in $d = 8, 9$ one has $C^T = +C$. Thus we conclude that we have consistency in $d = 4$ for $\beta = -2$, no consistency in $d = 8, 9$, while $d = 10, 11$ must be further investigated.

Actually, each term in (5.7) vanishes separately in $d = 4$ due to further Fierz identities, so β is still free in $d = 4$. Consistency of (5.5b) requires, however, $\beta = -2$ because

$$\mathcal{D}\mathcal{D} V^m = R^{mn} \wedge V_n = 2\bar{\psi} \gamma^m \wedge \mathcal{D}\psi = 2e\bar{\psi} \gamma^{mn} \wedge \psi \wedge V_n, \qquad (5.9)$$

while $R^{mn} \wedge V_n = -\beta \bar{\psi} \gamma^{mn} \psi \wedge V_n$. Consistency (of (5.5a)) follows, using $\mathcal{D} R^{mn} \equiv 0$, for $\alpha + 2\beta = 0$ in agreement with the previous results.

Can (5.7) be satisfied in $d = 10$ or $d = 11$? One should be careful; extra Fierz identities can sharpen the result in (5.8) as we saw. However, an analysis of Fierz identities, or a group-theoretical result [17], shows that, for no β, (5.7) can hold in $d = 10, 11$; hence the super-de Sitter algebra is only consistent in $d = 4$. This reflects the well-known fact that one cannot add a cosmological constant Λ to the action in $d = 11$, a result of [9] which is quite important for the recent renewed interest in Kaluza-Klein theories (in $d = 10$ one cannot add Λ because the necessary gravitino mass term vanishes, since ψ_μ is chiral).

Let us see whether the super-Poincaré and, in $d = 4$, the super-de Sitter algebras have nontrivial, relative or absolute (nonrelative) cohomologies. There are two theorems which may guide us [they have only been proven [17] by Chevalley and Eilenberg for nonsuperalgebras, but we assume their validity for superalgebras as well].

THEOREM I. *If G is a semisimple Lie algebra, the only nontrivial absolute cohomologies are in the identity representation. A fortiori this holds for relative cohomologies.*

THEOREM II. *If G is a semisimple Lie algebra, then in the identity representation one finds only nontrivial absolute cohomology elements if their grade is three or higher. At grade three one has $a_3 = f_{ABC}h^C \wedge h^B \wedge h^A$.*

Comment. Obviously this grade-three element is closed due to the super-Jacobi identities, but for superalgebras the super-Killing metric need not be nonsingular, even if the group is semisimple. Hence, $f_{ABC}h^C \wedge h^B \wedge h^A$ could vanish for simple superalgebras. Moreover, since in a_3, in principle, all 1-forms can appear, Theorem II does not hold for relative cohomologies.

Here are a few examples. For the Lorentz group, $a_3 = \text{tr}(\omega \wedge \omega \wedge \omega)$ is closed under d because $\text{tr}(\omega \wedge \omega \wedge \omega \wedge \omega) = 0$. For the Poincaré group, a_3 is the same since $g_{pp} = 0$, but for the de Sitter group $g_{pp} \neq 0$ and

$$a_3 = \text{tr}(\omega \wedge \omega \wedge \omega + 12V \wedge \omega \wedge V). \tag{5.10}$$

Now the supercases: For the $d = 4$ super-Poincaré algebra (which is, of course, nonsemisimple, so that the two theorems above do not apply), both $\text{tr}(\omega \wedge \omega \wedge \omega)$ and $\bar\psi\gamma^m \wedge \psi \wedge V_m$ are closed. But for the super-de Sitter algebra in $d = 4$,

$$a_3 = \text{tr}\Big(\omega \wedge \omega \wedge \omega + 12V \wedge \omega \wedge V$$
$$+ 24\bar\psi\gamma^m \wedge \psi \wedge V_m - 6\bar\psi\gamma^{mn} \wedge \psi \wedge \omega_{mn}\Big). \tag{5.11}$$

Let us now discuss relative cohomologies. Clearly, one can extend the $d = 4$ super-Poincaré algebra in $d = 4$ by a 3-form A or a 4-form B:

$$\begin{aligned} dA + \bar\psi\gamma^m \wedge \psi \wedge V_m &= 0, \\ dB + \bar\psi\gamma^{mn}\psi \wedge V_m \wedge V_n &= 0. \end{aligned} \tag{5.12}$$

Both extensions are consistent due to the identities

$$\bar\psi\gamma^m \wedge \psi \wedge \gamma_m\psi = 0, \qquad \bar\psi\gamma^{mn} \wedge \psi \wedge \gamma_{mn}\psi = 0. \tag{5.13}$$

Thus, in the super-Poincaré algebra there exist nontrivial relative cohomologies of grades three and four (they are in the identity representation). The A extension is not consistent in the super-de Sitter algebra (only (5.11) is consistent), but the B extension is consistent. These results are in agreement with Theorem II. The B extension has been found in the context of Kaluza-Klein compactification of 11 to 4 dimensions [18].

To find other relative cohomologies, we need further identities. One may show

$$\left. \begin{aligned} &\bar{\psi}\gamma^{mn} \wedge \psi \wedge \bar{\psi}\gamma_m\psi = 0, \\ &\bar{\psi}\gamma^{mn_1\cdots n_4} \wedge \psi \wedge \bar{\psi}\gamma_m \wedge \psi = 3\bar{\psi}\gamma^{[n_1n_2}\psi\bar{\psi}\gamma^{n_3n_4]}\psi = 0 \end{aligned} \right\} \quad \text{in } d = 11.$$

$$(5.14)$$

In $d = 10$ one finds stronger identities because the spinors ψ are chiral, meaning $\gamma^{10}\psi = \psi$.

$$\left. \begin{aligned} &\bar{\psi}\gamma^m \wedge \psi\bar{\psi} \wedge \gamma_m\psi = 0, \\ &\bar{\psi}\gamma^{mn} \wedge \psi = \bar{\psi}\gamma^{mnk} \wedge \psi = 0, \\ &\bar{\psi} \wedge \gamma^{ma_1\cdots a_4}\psi \wedge \bar{\psi} \wedge \gamma_m\psi = 0 \end{aligned} \right\} \quad \text{in } d = 10. \qquad (5.15)$$

Thus, one has the following extensions of the super-Poincaré algebra in $d = 11$:

$$dA + \bar{\psi}\gamma^{mn} \wedge \psi \wedge V_m \wedge V_n = 0,$$

or the same plus

$$dB + \bar{\psi} \wedge \gamma^{mn_1\cdots n_4}\psi \wedge V_m \wedge V_{n_1} \wedge \cdots \wedge V_{n_4}$$
$$+ \bar{\psi} \wedge \psi \wedge V_m \wedge V_n, \qquad (5.16)$$

and finally, in $d = 10$,

$$dA + \bar{\psi} \wedge \gamma^m\psi \wedge V_m = 0,$$
$$dB + \bar{\psi} \wedge \gamma^{m_1\cdots m_5}\psi \wedge V_{m_1} \wedge \cdots \wedge V_{m_5} = 0. \qquad (5.17)$$

The extension of the $d = 11$ super-Poincaré algebra with the 3-form A leads to the standard $d = 11$ supergravity theory. The extension with B shows that $H_7(I, A_3, \text{Lorentz})$ is nontrivial; however, one needs A: $H_7(I, A_1, \text{Lorentz})$ is trivial. Both extensions lead to the same field theory: B drops out of the action. The result that no extensions with B alone exist confirms a result, found by different methods [9], that no $d = 11$ supergravity exists with an antisymmetric 6-index tensor. In $d = 10$ one can have either B or A, and, indeed, both field theories have been constructed [19].

So far, only p-forms with $p > 1$ have been found in supergravity, which are in the identity representation of G. One can construct more general algebras in which there are general p-forms which are in a nontrivial representation of G. According to Theorem I we need a nonsemisimple (super)algebra, so we consider the super-Poincaré algebra. Its *defining representation* R_5 is given by 5×5 supermatrices

$$
\sigma = \left(
\begin{array}{c|c}
\frac{1}{4}\omega^{mn}\gamma_{mn} + \frac{1}{2}\gamma^m(1 + \gamma_5)V_m & \alpha(1 - \gamma_5)\psi \\
\hline
\beta\bar{\psi}\,(1 + \gamma_5) & 0
\end{array}
\right)
\tag{5.18}
$$

with $\alpha\beta = -2$. Indeed, the matrices σ satisfy the Cartan-Maurer equations $d\sigma + \sigma \wedge \sigma = 0$. The super-Poincaré algebra, being a contraction of the super-de Sitter algebra, is given by

$$
\sigma_{dS} = \left(
\begin{array}{c|c}
\frac{1}{4}\omega^{mn}\gamma_{mn} + \frac{e}{2}\gamma^m V_m & \alpha\psi \\
\hline
\beta\bar{\psi} & 0
\end{array}
\right).
\tag{5.19}
$$

One expects that there exists a matrix $U(e)$ such that $U(e)\sigma_{dS}U^{-1}(e) \to \sigma$ when $e \to 0$. Indeed, this is the case for $U(e) = \mathrm{diag}(1 + (e - 1)\gamma_5, 1)$.

We thus try to construct a pair (Λ^a, F), where Λ^a is a fermionic 4-component p-form and F a bosonic 1-component p-form, satisfying

$$
D(R_5)\binom{\Lambda}{F} = 0, \quad \binom{\Lambda}{F} \neq D(R_5)\binom{\lambda}{f}, \quad D(R_5) = d + \sigma.
\tag{5.20}
$$

Let us take (Λ^a, F) to be 3-forms (hence $p = 3$). Again, we will consider these cohomologies to be relative to the Lorentz group (meaning that all expressions are good Lorentz tensors: no bare ω^{mn} and the constants are Lorentz-invariant tensors, such as $(\gamma^m)_b^a$, etc.).

The most general form of (Λ^a, F) is (after an overall rescaling)

$$
\begin{aligned}
\Lambda &= \frac{1}{2}(a + b\gamma_5)\gamma^{mn}\psi \wedge V_m \wedge V_n, \\
F &= \bar{\psi}\gamma^m \wedge \psi \wedge V_m.
\end{aligned}
\tag{5.21}
$$

The covariant derivative in the defining representation follows from (5.18)

$$
\begin{aligned}
\mathcal{D}\Lambda + \frac{1}{2}\gamma^m(1 + \gamma_5)V_m \wedge \Lambda + \alpha(1 - \gamma_5)\psi \wedge F &= 0, \\
d(\alpha F) + \alpha\beta\bar{\psi}\,(1 + \gamma_5)\Lambda &= 0 \quad (\alpha\beta = -2).
\end{aligned}
\tag{5.22}
$$

We shall henceforth denote αF by F'.

Since $dF = 0$ according to (5.21), Λ must be proportional to $(1 - \gamma_5)$ according to (5.22). Hence $b = -a$. The equation for Λ is now satisfied if $a = -1$. Hence

$$\Lambda = -\frac{1}{2}(1 - \gamma_5)\gamma^{mn}\psi \wedge V_m \wedge V_n,$$
$$F = \bar{\psi}\gamma^m \wedge \psi \wedge V_m.$$
(5.23)

Thus we have found a set of 3-forms in a nontrivial representation of the super-Poincaré algebra which are closed. If they are not exact, we have a nontrivial relative cohomology class $H_3(R_5, A_1, \text{Lorentz})$. In that case we can extend the super-Poincaré algebra as follows:

$$D(R_5)\begin{pmatrix} \lambda \\ f \end{pmatrix} = \begin{pmatrix} \Lambda \\ F \end{pmatrix} \quad \text{with } (\lambda, f) \; new \; \text{2-forms}.$$
(5.24)

The covariant derivative was given in (5.22); thus we consider now the most general 2-forms (λ^a, f) constructed from the original 1-forms and show that (5.24) can never be satisfied for such (λ^a, f). The most general (λ, f) are

$$\lambda = (A + B\gamma_5)\gamma^m\psi \wedge V_m, \qquad f = 0,$$
(5.25)

and

$$\mathcal{D}\lambda + \frac{1}{2}\gamma^m(1 + \gamma_5)V_m \wedge \lambda + \frac{1}{2}(1 - \gamma_5)\psi \wedge f = \Lambda,$$
$$df' - 2\bar{\psi}\,(1 + \gamma_5)\lambda = F',$$
(5.26)

which cannot be solved when Λ, F are given by (5.23).

One can also contract the super-de Sitter algebra in a nonchiral way by first rescaling $\psi \to \sqrt{e}\,\psi$ and then choosing $U(e) = \text{diag}(e^{-1/2}, e^{-1})$. Going through the same steps, one finds another differential algebra [20]

$$\mathcal{D}X + \psi \wedge A + \frac{1}{2}\gamma^{mn}\psi \wedge V_m \wedge V_n = 0,$$
$$dA - \bar{\psi}\gamma^m \wedge \psi \wedge V_m = 0.$$
(5.27)

The corresponding field theories have not been constructed. Actually, this is an important topic, because it is generally believed that one cannot consistently couple antisymmetric tensors with internal indices.

Are there supergravities in $d > 11$? Usually one argues that they cannot appear, because they would contain interacting particles with spins $S > 2$, whereas such interactions seem forbidden in field theory, at least for spin $5/2$ [21]. However, this is a rather indirect argument, and the history of supergravity itself cautions against such categorical no-go theorems. Let us, therefore, see what the differential algebras have to offer on this question. One can find a set of gauge fields in $d = 12$ with equal numbers of bosonic and fermionic states:

vielbein V_μ^a	54	Majorana gravitino ψ_μ	288
scalar $A^{(0)}$	1	Majorana spinor λ	32
vector $A_\mu^{(1)}$	10		$\overline{320}$
2-index $A_{\mu\nu}^{(2)}$	45		
4-index $A_{\mu\rho\sigma\tau}^{(4)}$	$\underline{210}$		
	$\overline{320}$		

$$(5.28)$$

(Actually, Curtright has found more sets [22].) Let us try to extend the super-Poincaré algebra in $d = 12$ by [23]

$$dA^{(4)} + \left(\bar{\psi}\Gamma_{13}\Gamma^{mnr} \wedge \psi\right) \wedge V_m \wedge V_n \wedge V_r = 0. \qquad (5.29)$$

Consistency is violated because in $d = 12$,

$$\left(\bar{\psi}\Gamma_{13}\Gamma^{mnr} \wedge \psi\right) \wedge \left(\bar{\psi}\Gamma_m\psi\right) \neq 0. \qquad (5.30)$$

One should, in principle, also allow terms with $A^{(2)}$ and $A^{(1)}$ in $dA^{(4)}$. This has not been done, but one would expect that the outcome confirms the arguments about higher spin: no consistent extensions [involving $A^{(4)}$ *and* $A^{(2)}$, $A^{(1)}$] exist. However, one cannot exclude surprises.

We conclude with some examples whose consistency the reader may study (F_n = fermionic n-form, B_n = bosonic n-form):

(1) $dF_2 = F_1 \wedge F_1 \wedge F_1, dF_1 = 0$.

(2) $dF_2 = \alpha F_1 \wedge F_1 \wedge F_1 + \beta B_1 \wedge F_2$, $dB_1 = \gamma F_1 \wedge F_1$, $dF_1 = \zeta F_2 + \eta B_1 \wedge F_1$.

(3) $dB_1^i = \varepsilon^{ijk} B_1^j \wedge B_1^k, dA = \varepsilon^{ijk} B_1^i \wedge B_1^j \wedge B_1^k$.

(4) $dF_1^1 = 0, dF_1^2 = F_1^2 \wedge F_1^1$.

(5) $dB_2 = B_2 \wedge B_1 + F_1 \wedge F_1 \wedge B_1, dB_1 = 0, dF_1 = B_1 \wedge F_1$.

ADDED IN PROOF. Since the meeting, antisymmetric tensors in nontrivial representations of compact Lie algebras have been found in supergravity (P. K. Townsend, K. Pilch and P. van Nieuwenhuizen, Nucl. Phys. B **242** (1984), 377 and M. Pernia, K. Pilch and P. van Nieuwenhuizen, Phys. Lett. B **143** (1984), 103). This makes the study of nontrivial cohomologies advocated in the text even more important.

REFERENCES

1. D. Z. Freedman, P. van Nieuwenhuizen and S. Ferrara, Phys. Rev. D (3) **13** (1976), 3214.

2. S. Deser and B. Zumino, Phys. Lett. B **62** (1976), 335.

3. D. Z. Freedman and P. van Nieuwenhuizen, Phys. Rev. D (3) **14** (1976), 912.

4. S. Ferrara and P. van Nieuwenhuizen, Phys. Lett. B **74** (1978), 333; K. Stelle and P. C. West, Phys. Lett. B **74** (1978), 330.

5. P. van Nieuwenhuizen, Phys. Rep. **68** (1981), 223–229.

6. A. H. Chamseddine and P. C. West, Nucl. Phys. B **129** (1977), 39.

7. P. K. Townsend and P. van Nieuwenhuizen, Phys. Lett. B **67** (1977), 439.

8. M. Kaku, P. K. Townsend and P. van Nieuwenhuizen, Phys. Rev. D (3) **17** (1978), 3179; P. K. Townsend and P. van Nieuwenhuizen, Phys. Rev. D (3) **19** (1979), 3166.

9. H. Nicolai, P. K. Townsend and P. van Nieuwenhuizen, Lett. Nuovo Cimento (2) **30** (1981), 315; the $d = 11$ model was first constructed by E. Cremmer, B. Julia and J. Scherk, Phys. Lett. B **76** (1978), 409.

10. R. D'Auria and P. Fré, Nucl. Phys. B **201** (1982), 101.

11. R. D'Auria, P. Fré and T. Regge, *Supergravity '81* (S. Ferrara and J. G. Taylor, Eds.), Cambridge Univ. Press, London and New York, 1982; P. van Nieuwenhuizen, in Unified Field Theories of More Than 4 Dimensions, Including Exact Solutions (V. De Sabbata and E. Schnautzer, eds.), World Scientific, 1983.

12. R. D'Auria, P. Fré, P. K. Townsend and P. van Nieuwenhuizen, Ann. Physics (to appear).

13. S. MacDowell and F. Mansouri, Phys. Rev. Lett. **38** (1977), 739.

14. D. Sullivan, *Infinitesimal computations in topology*, IHES, Bures sur Yvette.

15. P. van Nieuwenhuizen, in Group Theoretical Methods in Physics (Proc. Istanbul Conf., 1982), Lecture Notes in Physics, vol. 180, Springer-Verlag, 1983.

16. _____ , *Supergravity '81* (S. Ferrara and J. G. Taylor, eds.), Cambridge Univ. Press, London and New York, 1982.

17. C. Chevalley and S. Eilenberg, *Cohomology theory of Lie groups and Lie algebras*, Trans. Amer. Math. Soc. **63** (1948), 85–24.

18. R. D'Auria and P. Fré, Phys. Lett. B **121** (1983), 141.

19. E. Bergshoeff, M. de Roo, B. de Wit and P. van Nieuwenhuizen, Nucl. Phys. B **195** (1982), 97; A. H. Chamseddine, Phys. Rev. D (3) **24** (1981), 3065.

20. L. Castellani, P. Fré, F. Giani, K. Pilch and P. van Nieuwenhuizen, Ann. Physics **146** (1983), 35.

21. F. A. Berends, B. de Wit, J. W. van Holten and P. van Nieuwenhuizen, Phys. Lett. B **83** (1979), 188; Nucl. Phys. B **154** (1979), 261; and J. Phys. A **13** (1980), 1643.

22. T. A. Curtright, Gainsville preprint UFTP–82–22, 1982.

23. L. Castellani, R. D'Auria, P. Fré, F. Giani, K. Pilch and P. van Nieuwenhuizen, Phys. Rev. D (3) **26** (1982), 1481.

CERN, GENEVA, SWITZERLAND

Current address: Institute of Theoretical Physics, State University of New York, Stony Brook, New York, 11794

II. SUPERSYMMETRY AND SUPERGRAVITY

Lectures in Applied Mathematics
Volume **21**, 1985

Semisimple Gauge Theories
and Conformal Gravity

C. Fronsdal

According to the special theory of relativity, we live in a $(3 + 1)$-dimensional Minkowski space; that is, a 4-dimensional, pseudo-Riemannian space with fixed pseudo-Euclidean metric, with signature 3, 1. The symmetry group of the metric is the Poincaré group, under which all physical laws are supposedly invariant. The generators of Poincaré transformations furnish the basic observables of physics: energy, momentum,.... General relativity replaces the fixed metric of the special theory by a more general metric with the same signature. The departure of this metric from flatness is interpreted as a new and independent physical excitation, susceptible to a physical interpretation in some ways similar to that of the electromagnetic field. The static features of general relativity account for gravitational phenomena in a very satisfactory way, but true dynamical manifestations of the new degrees of freedom have not yet been observed directly (Figure 1).

Unlike the special theory, general relativity does not have a metric isometry group. Consequently, there are no basic observables that can be identified experimentally by their property of being conserved. This poses very great problems of physical interpretation; it is necessary to suppose that a highly symmetric special metric intervenes somehow, either in an asymptotic region, or else as the point of departure of perturbation theory. The maximal dimension of the symmetry group of a metric in 4 dimensions is 10. With the signature 3, 1 the possible 10-parameter groups are: the Poincaré group of special relativity, and the two de Sitter groups. The latter are symmetries of spaces with constant

---

1980 *Mathematics Subject Classification*. Primary 22E46, 81E10, 83E99.

curvature. The value of the curvature constant is determined by the cosmological constant and is known experimentally to be exceedingly small.

Maxwell's equations for the free, classical, electromagnetic field are invariant under the 15-parameter group of conformal transformations of the metric. The conformal group seems to be intimately involved with the physics of massless particles.

The physics of these talks is the physics of massless particles. Space-time will be assumed to have a very small positive curvature, so that the symmetry group is SO(3, 2). The conformal group is locally isomorphic to SO(4, 2), so we shall be dealing with two simple groups. Mathematically, the subject is an introduction to the methods that have been used by physicists to construct, for their special needs, certain unitary representations of these groups. These methods are in some respects related to those discussed by J. Wolf and W. Schmid at this seminar. Physicists have to work with indecomposable representations in which the unitaries appear as subquotients. In fact, the appearance of indecomposable representations is a hallmark of gauge theories and of massless particles. Several examples will be discussed, along with some other aspects of massless physics. The problem, so far an open one, of quantum gravity, will be discussed at the end.

Phenomenon in Curved Space   Measurements in Flat Space

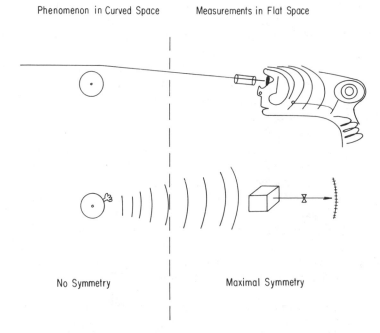

No Symmetry   Maximal Symmetry

FIGURE 1. Static and dynamic manifestations of the curved metric.

The Poincaré group is the semidirect product

$$\mathscr{P} = SO_0(3,1) \cdot T_4,$$

defined as the extension of the Lorentz group by its natural, 4-dimensional module. The de Sitter group $SO_0(3,2)$ is locally isomorphic to $Sp(2, R)$. The conformal group is locally isomorphic to $SO(4,2)$ and to $SU(2,2)$. The suffix "0" will be dropped.

Minkowski space is the homogeneous space

$$M = \mathscr{P}/SO(3,1),$$

and de Sitter space is (a covering of) $SO(3,2)/SO(3,1)$. The conformal group does not act globally on either; for this reason one replaces M by the compactification [1]

$$\overline{M} = SO(4,2)/W \times Z_2.$$

Here W is the 11-parameter Weyl group—the extension of the group of dilatations by $\mathscr{P}$. The double covering $SO(4,2)/W$ of $\overline{M}$ is homeomorphic to $S_1 \times S_3$. This space is also a double covering of compactified de Sitter space.

Positive energy representations. The translation subgroup T_4 of $\mathscr{P}$ contains the group of time translations. The infinitesimal generator of this group is identified with the Hamiltonian, and its (generalized) eigenvalues with the energy, the most basic of all physical observables. The interpretation requires that the operator associated with this generator be positive or at least bounded below. Our concern is thus exclusively with "positive energy representations." The de Sitter and conformal groups $SO(p,2)$ have maximal compact subgroups $SO(p) \times SO(2)$. In this case, the time translation group is the factor $SO(2)$ or its universal covering. The physically motivated emphasis on unitary representations with positive energy leads to representations with minimal weight; only such representations will be considered. Another limitation arises from the need to describe all representations in terms of fields on space-time, such as the vector field of electromagnetism. This limitation leads to indecomposable representations and to gauge theories.

Irreducible, positive energy representations of $SO(p,2)$ are easy to classify. Choose a maximal compact subgroup

$$K = SO(p) \times SO(2).$$

Let L_{05} denote the generator of the factor $SO(2)$, normalized so that the eigenvalues of $ad(L_{05})$ have unit spacing. A "minimal" weight is a pair (E_0, w), with E_0 real and w a dominant weight for $SO(p)$. A minimal weight (E_0, w) defines a minimal K-type for a unique, irreducible representation of $so(p,2)$, denoted $D(E_0, w)$. It is K-finite and the spectrum of L_{05} is $\{E_0 + k\}$, $k = 0, 1, 2, \ldots$. It can be integrated to an irreducible, projective representation of $SO(p, 2)$; this representation will also be denoted $D(E_0, w)$.

Necessary and sufficient conditions for the unitarity of the positive energy representations of $SO(3, 2)$ and $SO(4, 2)$ are well known. I recall the facts for $SO(3, 2)$. Choose a Cartan basis for $SO(3)$ such that w is a nonnegative half-integer s, then $D(E_0, s)$ is unitary if and only if [2]

$$E_0 \geqslant s + 1/2, \quad s = 0 \text{ or } 1/2,$$
$$E_0 \geqslant s + 1, \quad s = 1, 3/2, 2, \ldots.$$

The unitary limit is in each case the first reduction point encountered as E_0 is decreased. The limit points are the most interesting representations, for physical applications as well as for their special mathematical structure. They are all associated with gauge theories. Their intrinsic peculiarities are shown by their reduction on K.

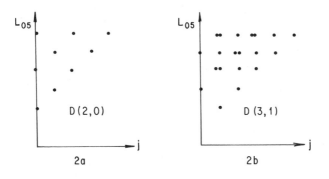

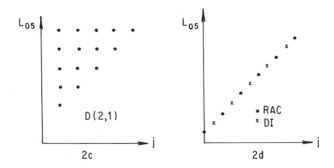

FIGURE 2. K-structures of some representations of $SO(3, 2)$.

We call "massive" those unitaries that satisfy the strict inequalities. Their K-structures are determined by s alone and have the typical wedge shape illustrated in Figures 2a and 2b. The limit representations $D(s + 1, s)$, $s \geqslant 1$, are called "massless"; their K-reductions are multiplicity-free and are illustrated in Figure 2c. The dramatically reduced K-structures of RAC $= D(\frac{1}{2}, 0)$ and DI $= D(1, \frac{1}{2})$ are shown in Figure 2d; these two are called Dirac singletons.

Simplest gauge theory, the RAC. The representations $D(E_0, 0)$ of $SO(3, 2)$ can be realized as follows [3]. The simplest fields on de Sitter space are sections of a complex line bundle. The universal covering of $SO(3, 2)$ acts naturally on such fields. The infinitesimal generators are vector fields and the second order Casimir element coincides with the Laplace-Beltrami operator $\square$. The eigenvalue equation for this operator is the wave equation

$$(\square + q\rho)\phi = 0, \qquad q \text{ real}, \rho = \text{curvature constant}.$$

If $q = E_0(E_0 - 3)$, then this equation admits solutions of definite K-type, spanning an $SO(3, 2)$ invariant space $\mathcal{V}(E_0)$ on which the K-spectrum is that of $D(E_0, 0)$, provided $E_0 > \frac{1}{2}$; it may be completed to a unitary $SO(3, 2)$ module (projective representation).

Considering $\mathcal{V}(E_0)$ as an $so(3, 2)$ module one can continue analytically to $E_0 = \frac{1}{2}$, at which point the module becomes indecomposable:

$$\lim_{E_0 \to 1/2} D(E_0, 0) = D(\tfrac{1}{2}, 0) \to D(\tfrac{5}{2}, 0).$$

The notation $A \to B$ denotes an extension of the representation A by the representation B. The unitary representation $D(\frac{5}{2}, 0)$ appears in an invariant subspace $\mathcal{V}_g$, while the interesting unitarizable representation $D(\frac{1}{2}, 0)$ is realized on a quotient.

The fields in $\mathcal{V}(\frac{1}{2})$ all belong to the same complex line bundle and can be interpreted as functions on the double covering of de Sitter space. The invariant subspace $\mathcal{V}_g \subset \mathcal{V}(\frac{1}{2})$, the subspace of "gauge fields", is characterized by rapid decrease at the boundary of a maximal Euclidean subvariety. The quotient $\mathcal{V}(\frac{1}{2})/\mathcal{V}_g$ can therefore be mapped bijectively to a space of functions on the 3-dimensional boundary of de Sitter space. In this way one can realize $D(\frac{1}{2}, 0)$ irreducibly in terms of functions on $S_1 \times S_2 = SO(3, 2)/W$, where W is the $(2 + 1)$-dimensional Weyl group [3]. This space is a double covering of a compactification of $(2 + 1)$-Minkowski space, and $SO(3, 2)$ acts on it as the conformal group of 3-dimensional space-time. Both RAC and DI remain irreducible when restricted to the 3-dimensional Poincaré group, and they are the only positive energy representations with this property. Hence, 3-dimensional space-time has only two types of massless particles—Dis and Racs [4];

they may be described by scalar and spinor fields, respectively. The direct sum DI + RAC has an extension to a very special representation (the metaplectic, or oscillator representation) of an orthosymplectic graded Lie algebra. The system described by one Rac field and one Di field has a unique supersymmetric Lagrangian (up to one real coupling constant) [5].

The nondecomposable representation encountered here is not directly utilizable in physics; it must be extended to a "Gupta-Bleuler triplet." I shall return to this problem in a wider context, after reviewing the role of massless representations in physics.

Massless representations. The unitary dual of $\mathcal{P}$ was calculated by Wigner in 1939. The massless, discrete helicity representations (massless representations for short) are distinguished by a whole series of very special properties; for example:

1. The associated representations of the enveloping algebra are characterized by an unusually large ideal, not generated by the center.

2. These representations are induced from a representation of $E(2) \cdot T_4$, trivial on the Abelian ideal of $E(2)$. This gives rise to imprimitivity systems and localization properties that are different from those of massive particles [6].

3. Construction of induced representations, in the sense of Mackey, leads us to a realization in terms of functions on the mass-hyperboloid $p_0^2 - p_1^2 - p_2^2 - p_3^2 = m^2$, an orbit of SO(3, 1) in the dual of T_4. In the massless case ($m = 0$) this becomes a cone and interesting things happen [7]. The space of differentiable vectors is not nuclear. The sharp momentum (generalized) eigenstates, widely used with relative impunity in the massive case, have paradoxical properties in the massless case, unless carefully defined as belonging to the dual of a space of sections of a nontrivial line bundle.

4. Realizations in terms of fields on M are always indecomposable (except in the case of helicity zero) and lead to interesting problems involving an indefinite metric.

5. Massless representations of $\mathcal{P}$ have unique extensions to the conformal group. This may be taken as the defining property of the massless representations.

Let us return to the description of the positive energy representations of SO(4, 2). The dominant weights of the compact subgroup SO(4) = SU(2) × SU(2) are parametrized by two half-integers, $w = (j_1, j_2)$. The irreducible representation $D(E_0, w) = D(E_0, j_1, j_2)$ is unitary if and only if [8]

$$E_0 \geq j_1 + j_2 + 1, \quad \text{when } j_1 j_2 = 0,$$
$$E_0 \geq j_1 + j_2 + 2, \quad \text{when } j_1 j_2 \neq 0.$$

The limit is the first reduction point encountered as E_0 is decreased. We call "massive" the representations for which the inequalities are strictly satisfied. Their K-structures depend on j_1, j_2 only. We call "massless" the representations $D(j + 1, j, 0)$ and $D(j + 1, 0, j)$; these representations have drastically reduced K-spectra; namely,

$$D(j + 1, j, 0)|_K = \sum_{n=0}^{\infty} D(j + n + 1) \times D\left(j + \frac{n}{2}, \frac{n}{2}\right).$$

Here $D(j + n + 1)$ is a one-dimensional representation of SO(2) and $D(j + \frac{n}{2}, \frac{n}{2})$ is an irreducible representation of SO(4).

A related property, which is behind all the physical manifestations of conformal invariance, is [9]

$$D(j + 1, j, 0)|_{\mathscr{P}} = D(0, j), \qquad D(j + 1, 0, j)|_{\mathscr{P}} = D(0, -j).$$

The two $\mathscr{P}$-representations are the massless representations with helicity j and $-j$.

De Sitter space also has its massless particles [6]. The associated massless representations of SO(3, 2) were defined already: $D(s + 1, s)$ for $s = 1, \frac{3}{2}, 2, \ldots$. One has

$$D(j + 1, j, 0)|_{SO(3,2)} = D(j + 1, 0, j)|_{SO(3,2)} = D(j + 1, j)$$

for $j = \frac{1}{2}, 1, \frac{3}{2}, \ldots$, and

$$D(1, 0, 0)|_{SO(3,2)} = D(1, 0) + D(2, 0).$$

These representations have direct physical applications; thus de Sitter electrodynamics uses two copies of $D(2, 1)$.

The singletons $RAC = D(\frac{1}{2}, 0)$ and $DI = D(1, \frac{1}{2})$ also play an important role in massless, de Sitter space physics. This is revealed by the following remarkable reduction formula [10]:

$$(RAC + DI) \times (RAC + DI) = D(1, 0) + D(2, 0) + 2\sum D(j + 1, j),$$

in which $j = \frac{1}{2}, 1, \frac{3}{2}, \ldots$. This shows that every massless particle can be looked upon as a composite system consisting of two singletons [10]. See also the contribution of R. Howe to these proceedings.

Quantizable representations. A common characteristic of the massless representations is that their realizations in terms of fields on space-time involve nondecomposable representations (extensions). This fact is of central importance in physics, since it imposes contraints (gauge invariance) that in turn are responsible for the vast predictive power of electrodynamics. The remainder of these lectures is devoted to a detailed study of these representations. It will be seen that field quantization imposes additional requirements leading to further extensions. The minimal extension, as in $D(\frac{1}{2}, 0) \rightarrow D(\frac{5}{2}, 0)$, is not sufficient. The following discussion is most directly concerned with de Sitter electrodynamics, but the structure that it reveals is very general in massless field theories.

We saw how $D(E_0, 0)$ can be realized in terms of scalar fields on de Sitter space. We next try to realize $D(E_0, s)$ in terms of symmetric tensor fields of integer rank s. This works as well for $s = 2, 3, \ldots$, as it does for $s = 1$, and we limit ourselves to the important case of vector fields. The group $SO(3, 2)$ acts in a natural way on vector fields, and we try to find an irreducible module spanned by fields of definite K-type and with the spectrum of L_{05} being $\{E_0 + k\}$, $k = 0, 1, 2, \ldots$. We also want to characterize this module as a space of solutions of a wave equation, if possible.

The second order Casimir element is fixed (at the correct value for $D(E_0, 1)$) by the wave equation

$$\Box A_\mu - \nabla_\mu \nabla \cdot A + \left(E_0^2 - 3E_0 - 1\right) \rho A_\mu = 0.$$

Here A_μ are the components of the vector field in some local chart, ∇ is the covariant derivative, $\nabla \cdot A$ the covariant divergence and $\Box$ the covariant d'Alembertian. Now it turns out that the center of the enveloping algebra is fixed if we require, in addition, the vanishing of the covariant divergence:

$$\nabla \cdot A = 0 \quad \text{(Lorentz condition)}.$$

The necessity for this condition can easily be understood. The field $\phi = \nabla \cdot A$ is a scalar field, likely to belong to a $D(E_0, 0)$; it must be excluded if $D(E_0, 1)$ is to appear irreducibly.

Now one easily verifies that an irreducible $D(E_0, 1)$ module, consisting of normalizable vector fields that satisfy these two equations, exists if $E_0 > 2$ [11]. This favorable situation does not remain, however, as one attempts to reach the massless representation $D(2, 1)$. Instead one finds

$$\lim_{E_0 \rightarrow 2} D(E_0, 1) = D(2, 1) \rightarrow D(3, 0).$$

For $2 > E_0 > 1$ one finds again the irreducible (nonunitary) representation $D(E_0, 1)$, but as $E_0 \rightarrow 1$,

$$\lim_{E_0 \rightarrow 1} D(E_0, 1) = D(2, 1) \rightarrow D(1, 1).$$

Instead of the desired irreducible representation $D(2, 1)$, one gets a choice between two indecomposables in which the massless unitary representation appears as a subquotient. Actually, de Sitter electrodynamics makes use of both of them.

Both invariant subspaces consist of exact vector fields. One verifies that exact vector fields solve our two conditions only if $E_0 = 1$ or 2. Note that the invariant subspace encountered above, in connection with the singleton $D(\frac{1}{2}, 0)$, has nothing to do with vector fields. The characteristic property of gauge theories is not the intervention of exact vector fields, but the realization of the relevant unitary representation as a subquotient of a nondecomposable one.

Let us turn to the problem of quantization. Canonical field quantization is based on a nondegenerate symplectic structure. In the unitary case, the imaginary part of the Hilbert metric provides an invariant, nondegenerate symplectic form. Quantization is therefore straightforward in the case of massive representations ($E_0 > 2$). The indecomposable representations encountered in the massless case admit only a degenerate, invariant symplectic form, which leads to quantization of the uninteresting component only.

Flat space electromagnetics solves this difficulty by giving up the "Lorentz condition" $\nabla \cdot A = 0$, and this is effective for de Sitter electrodynamics also. Instead of the nondecomposable limits of $D(E_0, 1)$ given above, one now finds the "Gupta-Bleuler triplets"

$$D(3, 0) \to D(2, 1) \to D(3, 0),$$

and

$$D(1, 1) \to D(2, 1) \to D(1, 1)$$
$$\quad\;\longrightarrow D(0, 0) \longrightarrow$$

Each has an invariant, nondegenerate (though indefinite) metric, and canonical (indefinite metric) quantization becomes possible. Of course, the indefiniteness of the metric poses new problems, the solution of which will be discussed very briefly at the end.

Returning to the RAC, one finds [3] that the minimal extension of $D(\frac{1}{2}, 0)$ by $D(\frac{5}{2}, 0)$ has to be imbedded in the triplet $D(\frac{5}{2}, 0) \to D(\frac{1}{2}, 0) \to D(\frac{5}{2}, 0)$. The extra field modes are found by going outside the bundle; they involve logarithms and are a-periodic. The periodic modes form an uncomplemented invariant subspace.

Conformal field theories. Conformal invariance is not an exact symmetry principle of physics, for most particles are massive. Nevertheless, massless particles play a very central role in most physical theories, including electrodynamics, gravity and modern gauge theories. It is believed that masses should be introduced as smooth perturbations on conformally invariant field theories, and that the primordial conformal symmetry is responsible for the fact that some of these theories make sense.

Conformal invariance is usually understood in the context of classical field theories, or as a property of the Green's functions of the Euclidean version of some quantum field theories [12]. It is possible to go further, and carry out conformally invariant quantization. However, it is not enough to notice that the massless representations have unique, irreducible extensions to the conformal group [13]; it is necessary to imbed the entire Gupta-Bleuler triplet in a representation of the conformal group: one must find the conformal triplets. We shall make use of Dirac's construction of $\overline{M}$ [14].

Consider the natural action of $SO(4, 2)$ in R^6, endowed with a pseudo-Euclidean metric and coordinates $\{y_\alpha\}$, $\alpha = 0, 1, \ldots, 5$. The metric δ is given by

$$y^2 \equiv \delta^{\alpha\beta} y_\alpha y_\beta = y_0^2 - y_1^2 - y_2^2 - y_3^2 - y_4^2 + y_5^2.$$

The invariant, projective cone $y^2 = 0$, $y \simeq \lambda y$ for $\lambda \neq 0$, is identified with $\overline{M}$ as a homogeneous space. Fields on $\overline{M}$ are homogeneous tensor fields on R^6, restricted to the cone. They are sections of a line bundle that is determined by the degree of homogeneity. Scalar fields of degree $N < 0$ carry the representation $D(-N, 0, 0)$. More precisely, this module is determined by its extraordinarily simple reproducing kernel $K(y, y') = (y \cdot y')^N$. This is a generalized function, defined as a limit of an analytic function of the pseudo-Euclidean inner product $y \cdot y'$. Its Fourier series is the standard expansion of the generating function for Gegenbauer polynomials. Each term has positive energy $\geq -N$ and is a finite sum of $SO(4)$ spherical functions. This expansion thus gives the K-reduction and the Hilbert metric.

To construct other positive energy representations, we reduce the direct products $D_n \times D(-N, 0, 0)$, where D_n is a finite-dimensional representation. The modules now consist of homogeneous fields on R^6, restricted to the cone, and taking values in a finite $SO(4, 2)$ module. Such direct products are usually reducible, but some are indecomposable. Conformal electrodynamics deals with vector fields on R^6, restricted to the cone, and its Gupta-Bleuler triplet is precisely $D_6 \times D(1, 0, 0)$, where D_6 is the natural, 6-dimensional representation. This direct product is algebraically equivalent to [15]:

$$D\left(1, \tfrac{1}{2}, \tfrac{1}{2}\right) \to \left(D(2, 1, 0) + D(2, 0, 1)\right) \to D\left(1, \tfrac{1}{2}, \tfrac{1}{2}\right)$$
$$\longrightarrow \quad D(0, 0, 0) \quad \longrightarrow$$

The reproducing kernel is simply $K(y, y') = \mathbf{1} \times (y \cdot y')^{-1}$, where $\mathbf{1}$ is the 6-dimensional unit matrix.

One recognizes the two massless representations in the center. There is no module of fields on $\overline{M}$ on which they appear by themselves. The invariant submodule $[D(2, 1, 0) + D(2, 0, 1)] \to D\left(1, \tfrac{1}{2}, \tfrac{1}{2}\right)$ is characterized by the condition

$$y \cdot A = 0 \quad \text{(conformal QED Lorentz condition)}.$$

The smallest, proper, invariant submodule $D(1, \frac{1}{2}, \frac{1}{2})$ consists of fields that may be called "gradients," or gauge fields. The operator of exterior differentiation, defined for forms on R^6, has no intrinsic meaning for fields defined only on the cone, since the partial derivatives are not tangential vector fields. It must be replaced by the second order differential operator defined by

$$\text{Grad}_\alpha = y_\alpha \partial^2 - (2\hat{N} + 4)\partial_\alpha,$$

where $y_\alpha = \delta_{\alpha\beta} y^\beta$, $\partial_\alpha = \partial/\partial y^\alpha$, $\partial^2 = \delta^{\alpha\beta} \partial_\alpha \partial_\beta$ and $\hat{N} = y^\alpha \partial_\alpha$. The gauge fields are vector fields of the form $A = \text{Grad}\,\Lambda$, where Λ is a scalar field of degree zero. Finally, it is important for the applications to have a wave equation; the entire Gupta-Bleuler module satisfies the equation $\partial^2 A = 0$, which makes sense because the degree of A is -1.

Formal calculations indicate that conformal QED may be equivalent to conventional quantum electrodynamics. Differences, if any, should show up in the renormalization program. A close examination may provide insight into the relationship between conformal invariance and renormalizability and is thus very much worthwhile. Of course, conformal invariance cannot be exact in a physical theory, for it is violated by the boundary conditions that define scattering. The space $\overline{M}$ is a homogeneous space for the conformal group. Massless particles interacting among themselves do not know where Minkowski infinity is and will never stop interacting. These problems, and the closely related difficulty that arises because $\overline{M}$ has no global causal structure, have been discussed at length elsewhere [16].

Conformal gravity [18]. Perhaps the most enticing of the unsolved problems of physics is the construction of a sensible quantum field theory of gravitational interactions. In spite of the very considerable progress that was achieved during the last ten years, the theory remains nonrenormalizable and hence ill-defined even as a perturbation theory.

Most attempts to quantize general relativity are heretical, in the sense that one abandons, at least provisionally, the geometrical interpretation. The field is viewed as an excitation in a flat, Minkowski space-time, differing from other fields such as electromagnetism, by its higher helicity. Of course, the higher helicity introduces very nontrivial and in fact unsolved difficulties. The same heretical attitude is taken here. Space-time is M or $\overline{M}$, and the problem is defined as that of constructing a quantum field theory in which the physical and propagating representations of the Poincaré group are the massless representations $D(0, \pm 2)$. The Poincaré invariant indefinite-metric quantization scheme is well known and leads to a nonrenormalizable field theory. We propose a conformally invariant quantization procedure, hoping for a better result.

The representations $D(0, \pm 2)$ are the irreducible restrictions to $\mathcal{P}$ of the massless representations $D(3, 2, 0)$ and $D(3, 0, 2)$ of SO(4, 2). One may begin by determining all those positive energy representations that can occur in association with these massless ones within an indecomposable module. The relevant ones will appear below. The next problem is to find all pairs (D_n, N), where D_n is a finite representation and N is real, such that the direct product $D_n \times D(-N, 0, 0)$ contains the massless representation $D(3, 2, 0)$ and/or $D(3, 0, 2)$. An answer to this question, as far as we know the simplest one, can be expressed as follows. One should take a module of mixed-symmetry transverse tensor fields of rank 3 and degree of homogeneity zero. The components satisfy

$$\psi_{\alpha\beta\gamma} = -\psi_{\beta\alpha\gamma}, \qquad \psi_{\alpha\beta\gamma} + \psi_{\beta\gamma\alpha} + \psi_{\gamma\alpha\beta} = 0,$$
$$y^\alpha \psi_{\alpha\beta\gamma} + y^\alpha \psi_{\alpha\gamma\beta} = 0.$$

To describe our results it is convenient to define an operator that is similar to exterior differentiation. It operates on antisymmetric tensors of rank 2 and on mixed tensor fields of rank 3:

$$(d\Lambda)_{\alpha\beta\gamma} = \mathbb{P} \; \mathrm{Grad}_\gamma \Lambda_{\alpha\beta}, \qquad (d\psi)_{\alpha\beta\gamma\delta} = \mathbb{P} \; \mathrm{Grad}_\delta \psi_{\alpha\beta\gamma}.$$

The Young tableaux are projection operators for the respective index symmetry types. One verifies that $d \circ d = 0$. There is a unique, irreducible, positive energy module of mixed symmetry, rank 3 tensor fields of degree zero. This is identified as $D(0, \frac{1}{2}, \frac{3}{2})$ and consists of fields in the image of d. [I overlook, for a moment, the fact that $D(0, \frac{3}{2}, \frac{1}{2})$ also appears here.] It is a limit of $D(\varepsilon, \frac{1}{2}, \frac{3}{2})$ as $\varepsilon \to 0$ in the unitary, projective dual. By examination of the same limit in the module, one first discovers that $D(1, 0, 2) \to D(0, \frac{1}{2}, \frac{3}{2})$ occurs in an invariant subspace. The known multiplicity of all these irreducible representations indicates that

$$D(3, 2, 0) \to D\left(0, \tfrac{1}{2}, \tfrac{3}{2}\right)$$

should also occur, and a long calculation confirms this. This submodule is the unique maximal module of positive energy tensor fields (rank 3, mixed symmetry, degree zero) in the kernel of d.

One has, therefore, the rather satisfying result that the massless representations appear as a cohomology quotient of the operator d: the space of closed fields modulo the space of exact fields carries precisely $D(3, 2, 0)$. The completion of the Gupta-Bleuler triplet can be carried out by a study of the reproducing kernel, or by investigating the approach of the module $D(\varepsilon, \frac{1}{2}, \frac{3}{2})$ to the limit $\varepsilon = 0$. The result is

$$
\begin{array}{c}
\phantom{D\left(0, \tfrac{1}{2}, \tfrac{3}{2}\right)} \xrightarrow{} D(3, 2, 0) \\[2pt]
D\left(0, \tfrac{1}{2}, \tfrac{3}{2}\right) \to D(-1, 0, 1) \to D\left(0, \tfrac{1}{2}, \tfrac{3}{2}\right) \\[2pt]
\phantom{D\left(0, \tfrac{1}{2}, \tfrac{3}{2}\right)} \xrightarrow{} D(1, 0, 2)
\end{array}
$$

The representation in the middle is finite; $D(1,0,2)$ is a nonunitary "ghost" and its elimination from the physical manifestations of the theory is a difficult problem.

Every $D(E_0, j_1, j_2)$ discussed here is accompanied, in the field module, by $D(E_0, j_2, j_1)$. The field module reduces to a direct sum of two indecomposables and the two massless, helicity ± 2 representations each have their own Gupta-Bleuler triplets. That this must be so follows from the fact that the minimal weights $(3,2,0)$ and $(3,0,2)$ are not Weyl equivalent. [The minimal weights $(2,1,0)$ and $(2,0,1)$ of the unitary representations of electrodynamics *are* Weyl equivalent.]

To clarify the physical content of this construction it helps to express the fields in terms of intrinsic geometrical objects. The 3-tensor ψ combines an intrinsic 3-tensor Γ of conformal degree zero and an intrinsic symmetric 2-tensor h of conformal degree -1. The equation

$$\boxplus \ \Gamma_{\mu\nu\lambda,\rho} = \boxplus \ \delta_{\mu\lambda}h_{\nu\rho} \qquad (\mu, \nu, \lambda, \rho = 0, 1, 2, 3),$$

fixes Γ in terms of h, up to a gauge field, and h satisfies Einstein's linearized field equation. We have thus the rather unexpected result that Einstein's linearized field equations are imbedded in a system of equations that is conformally invariant.

The table summarizes the information about Gupta-Bleuler triplets. Notice the wide variety of "Lorentz conditions." Notice also the unwanted and unexpected appearance of finite representations, and the nonunitary $D(1,0,2)$. Is their inclusion required by the group structure (as I suspect) or do they appear because of the choice of the representation space?

TABLE 1

	Scalar	Transverse	Gauge	Lorentz condition
Singletons	$D(\frac{5}{2},0)$ ⟶	$D(\frac{1}{2},0)$ ⟶	$D(\frac{5}{2},0)$	periodic
de Sitter QED	$D(3,0)$ ⟶	$D(2,1)$ ⟶	$D(3,0)$	
	$D(1,1)$ ⟶	$D(2,1)$ ⟶	$D(1,1)$	$\nabla \cdot A = 0$
		$D(0,0)$		
Conformal QED	$D(1,\frac{1}{2},\frac{1}{2})$ ⟶	$D(2,1,0) + D(2,0,1)$ ⟶	$D(1,\frac{1}{2},\frac{1}{2})$	$y \cdot A = 0$
		$D(0,0,0)$		
Conformal Gravity	$D(0,\frac{1}{2},\frac{3}{2})$ ⟶	$D(3,2,0)$	$D(0,\frac{1}{2},\frac{3}{2})$	$d\psi = 0$
		$D(1,0,2)$		
		$D(-1,0,1)$		

Finally, a few words about the indefinite metric. The fields, in all the theories that have been thoroughly analyzed, interact with "currents." Currents play two roles: they are the sources from which the fields radiate, and they are the measurement devices by means of which the fields are observed. Mathematically, currents are functionals (linear in the simplest cases) of the fields. These functionals vanish on the minimal, proper subspace of gauge fields (see table), a property that is usually called current conservation. This property of the currents is effective in decoupling the two outer members of the Gupta-Bleuler triplets. The finite-dimensional representations in the middle of the triplets are believed to cause no serious problems, but the "ghost" $D(1, 0, 2)$ of conformal gravity is an ill omen for the success of this theory.

REFERENCES

1. R. Penrose, Proc. Roy. Soc. London Ser. A **284** (1965), 159.
2. N. T. Evans, J. Math. Phys. **8** (1967), 170.
3. M. Flato and C. Fronsdal, J. Math. Phys. **22** (1981), 1100.
4. B. Binegar, J. Math. Phys. **23** (1982), 1511.
5. C. Fronsdal, *The Dirac supermultiplet*, Phys. Rev. D (3) **26** (1982), 1988.
6. E. Angelopoulos, M. Flato, C. Fronsdal and D. Sternheimer, Phys. Rev. D (3) **23** (1981), 1278.
7. M. Flato, C. Fronsdal and D. Sternheimer, Comm. Math. Phys. **90** (1983), 563.
8. G. Mack, Comm. Math. Phys. **55** (1977), 1.
9. G. Mack and I. T. Todorov, J. Math. Phys. **10** (1969), 2078.
10. M. Flato and C. Fronsdal, Lett. Math. Phys. **2** (1978), 421; Phys. Lett. B **97** (1980), 236.
11. C. Fronsdal, Phys. Rev. D (3) **12** (1975), 3819.
12. I. T. Todorov et al., *Conformal invariance in quantum field theory*, Scuola Normale Superiore, Pisa, 1978; S. L. Adler, Phys. Rev. D (3) **6** (1972), 3445 and **8** (1973), 2400.
13. E. Angelopoulos and M. Flato, Lett. Math. Phys. **2** (1978), 405.
14. P. A. M. Dirac, Ann. of Math. (2) **37** (1936), 429.
15. B. Binegar, C. Fronsdal and W. Heidenreich, *Conformal QED*, J. Math. Phys. (to appear).
16. B. Binegar, M. Flato, C. Fronsdal and S. Salamó, Czechoslovak J. Phys. B **32** (1982), 429.
17. E. Angelopoulos, University of Dijon preprint, 1981.
18. B. Binegar, C. Fronsdal and W. Heidenreich, Phys. Rev. D (3) **27** (1983), 2249.

DEPARTMENT OF PHYSICS, UNIVERSITY OF CALIFORNIA, LOS ANGELES, CALIFORNIA 90024

Lectures in Applied Mathematics
Volume **21**, 1985

Dual Pairs in Physics: Harmonic Oscillators, Photons, Electrons, and Singletons

Roger Howe

1. Introduction. Jacobi invented θ-functions in the 1820s. Since then they have been used in many investigations by generations of number theorists. They are involved in many fascinating identities of number-theoretical and combinatorial import, and they provide one of the most effective ways to construct automorphic forms. In the early 1960s André Weil, inspired especially by the work of C. L. Siegel, provided a representation-theoretic foundation for the theory of θ-functions [**WA1, WA2**]. The details of Weil's discoveries are beyond the scope of this discussion, but, in brief, he found that θ-functions were intimately connected with a most singular projective unitary representation ω of the symplectic group. This representation ω arises by virtue of the existence of an action by automorphisms of the symplectic group on a certain two-step nilpotent group, commonly called the Heisenberg group.

Weil was of course working in the adelic formalism of modern number theory, so he considered symplectic groups and Heisenberg groups with coefficients in a general local field. When the field is **R**, the real numbers, both groups are Lie groups and so have associated Lie algebras. The commutation relations of the Heisenberg Lie algebra are, as its name suggests, essentially Heisenberg's Canonical Commutation Relations (CCR) which form the foundation of quantum mechanics. Furthermore, motivated by quantum field theory, David Shale, a student of Irving Segal, had several years earlier constructed the representation ω of the real symplectic group [**Sh**]. The remarkable coincidence that ω is involved

---

1980 *Mathematics Subject Classification.* Primary 81-02, 22E50, 81G20, 35-02.

in a fundamental way both in number theory and physics was discovered very soon, even before the publication of Weil's papers, and is one of the most fascinating of the many interesting properties of the representation ω.

Somewhat later, in order to systematize certain investigations in the theory of θ-series by Shalika, Tanaka, Gelbart, Rallis, Schiffman and others, I formulated the notion of a *reductive dual pair* of subgroups of the symplectic group. As the reader will soon see, a dual pair is a fairly highly structured object, and it was, indeed, conceived of to clarify a particular phenomenon in representation theory and automorphic forms. Therefore I was surprised to gradually realize that this notion, too, seems to be, on a formal level at least, quite closely related to physics. In fact, it is related to physics on a much broader scale than the representation ω originally appeared to be. The points of contact between dual pairs and physics seem to occur on several fronts. They include at least the following:

Fundamental formalism	(1.1i)
Massless particles	(1.1ii)
Classical equations	(1.1iii)
Supersymmetry	(1.1iv)

Although these seem, to this author at least, fairly different kinds of topics within physics, the mathematical theory of dual pairs to be discussed is both highly structured and fairly compact. At the very least, this would seem to imply something about the kind of mathematical structures physicists like to create. I would like to pose as a question to the reader whether there is anything more fundamental going on: can one formulate a principle which explains and/or unifies these several levels of interaction, and are there implications for the future?

2. Definition of dual pairs. Before getting into more specific topics, I want to set out the concept which serves to organize the discussion. It is very simple and makes sense in a general, abstract context. Let S be a group, and let G, G' be a pair of subgroups of S. We say (G, G') form a *dual pair* of subgroups of S if G' is the (full) centralizer of G in S (that is, G' consists of all elements of S commuting with G), and vice versa.

The notion of dual pair is the obvious group theoretical analogue of the concept of mutual commutants, familiar from the theory of associative algebras, and especially important for von Neumann algebras [**Dx**] (and hence to C^*-physics). We briefly recall that if A is an associative algebra and $X \subseteq A$ is a set, then X', the *commutant* of X, is the set of all elements of A which commute with all elements of X. Clearly X' is an associative subalgebra of A. If X is also the commutant of X', then X and X' are called *mutual commutants*.

Besides the clear analogy between them, the notions of dual pair and mutual commutants are connected via representation theory. Let π: $S \to \text{End } V$ be a representation of the group S on the vector space V, i.e., a homomorphism from S into the algebra of linear operators on V. Then if (G, G') is a dual pair of subgroups of S, obviously $\pi(G)$ and $\pi(G')$ will commute with each other. In other words, if $A(\pi(G'))$ denotes the algebra spanned by the operators $\pi(g')$ for g' in G', we have

$$A(\pi(G')) \subseteq (A(\pi(G)))' \tag{2.1}$$

and vice versa. Of course in general one would expect $A(\pi(G'))$ to be a rather small subalgebra of $(A(\pi(G)))'$. It would, therefore be an interesting property of the pair (G, G') and the representation π if the inclusion (2.1) were actually an equality. We will see below some examples of this.

There is also an obvious version of the notion of dual pairs for Lie algebras. If $\mathfrak{s}$ is a Lie algebra, and $X \subseteq \mathfrak{s}$ is a subset, then set

$$X' = \{s \in \mathfrak{s}: [s, x] = 0 \text{ for all } x \in X\}.$$

Here $[\,,\,]$ is the bracket operation in $\overline{\mathfrak{s}}$. Call X' the *centralizer* of X in $\mathfrak{s}$. Then a pair $(\mathfrak{g}, \mathfrak{g}')$ of Lie subalgebras of $\mathfrak{s}$ is a *dual pair* in $\mathfrak{s}$ if $\mathfrak{g}'$ is the centralizer of X in $\mathfrak{s}$, and vice versa. If S is a Lie group, and (G, G') are a dual pair of connected subgroups of S, then the pair $(\mathfrak{g}, \mathfrak{g}')$ of Lie algebras of (G, G') are a dual pair in the Lie algebra $\mathfrak{s}$ of S.

3. Basic formalism. The formal structure within which dual pairs have proved of interest is the *lingua franca* of modern physics, the Canonical Commutation Relations and, more explicitly, creation and annihilation operators. (This is point (i) in the list (1.1).) Many of the objects defined below will consequently already be quite familiar to physicists, more so even than to mathematicians. Consider an associative algebra generated by $2n$ symbols a_i (the annihilation operators) and a_i^+ (the creation operators), for $1 \leq i \leq n$, subject to the commutation relations

$$[a_i, a_j] = 0 = [a_i^+, a_j^+], \qquad [a_i, a_j^+] = \delta_{ij} = \begin{cases} 0 & \text{if } i \neq j, \\ 1 & \text{if } i = j. \end{cases} \tag{3.1}$$

(We have normalized Planck's constant so as to make equations (3.1), to which we will refer as the CCR, as simple as possible.) This algebra is frequently called the *Weyl algebra* and will be denoted $\mathfrak{W}_n$, or often simply $\mathfrak{W}$.

Concretely, the algebra $\mathfrak{W}$ consists of all possible linear combinations (sums with complex coefficients) of all possible products of the a_i and a_i^+. Let $\mathfrak{W}^{(l)}$ be the finite-dimensional subspace consisting of linear combinations of at most l-fold products of the a_i and a_i^+. If $\mathfrak{W}^{(l)}\mathfrak{W}^{(k)}$ denotes the set of all possible sums of all possible products, one from

$\mathfrak{W}^{(l)}$ and one from $\mathfrak{W}^{(k)}$, then clearly

$$\mathfrak{W}^{(l)}\mathfrak{W}^{(k)} = \mathfrak{W}^{(l+k)}. \tag{3.2}$$

The space $\mathfrak{W}^{(0)}$ simply reduces to **C**, the scalars. Let W be the linear span of the a_i and a_i^+, i.e., the set of all linear combinations

$$\sum r_i a_i + s_i a_i^+, \qquad r_i, s_i \in \mathbf{C}.$$

Then

$$\mathfrak{W}^{(1)} = W \oplus \mathbf{C}. \tag{3.3}$$

Although $\mathfrak{W}^{(1)}$ generates W as an associative algebra, the CCR says $\mathfrak{W}^{(1)}$ is closed under taking commutators; in other words $\mathfrak{W}^{(1)}$ is a Lie subalgebra of $\mathfrak{W}$. More precisely, the CCR may be reformulated to say

$$[\mathfrak{W}^{(1)}, \mathfrak{W}^{(1)}] \subseteq \mathfrak{W}^{(0)} \cong \mathbf{C}. \tag{3.4}$$

Still more precisely, the commutator mapping

$$w_1, w_2 \to [w_1, w_2] \in \mathbf{C}$$

defines a complex-valued, bilinear, skew-symmetric form on W. We will denote the commutator form on W by $\langle\ ,\ \rangle$. It is easy to see from the CCR that the commutator form $\langle\ ,\ \rangle$ is nondegenerate in the sense that if w_1 in W is nonzero, then there is w_2 in W such that $\langle w_1, w_2 \rangle \neq 0$. Such a (bilinear, skew-symmetric, nondegenerate) form is called *symplectic*. In this new language, the CCR, formula (3.1), say that the a_i and a_i^+ form a basis for the vector space W such that

$$\langle a_i, a_i \rangle = 0 = \langle a_i^+, a_i^+ \rangle, \qquad \langle a_i, a_j^+ \rangle = \delta_{ij}. \tag{3.5}$$

A basis for W which can be grouped into two subsets such that the analogues of relations (3.5) hold is called a *symplectic basis* for W. It is a standard fact (cf. [A]) that any symplectic form allows a symplectic basis. In other words, up to choice of coordinates, there is only one symplectic form on a vector space of a given dimension. Thus, to say that $\mathfrak{W}$ is generated by W, and that (3.3) and (3.4) hold, and the commutator form on W is symplectic is an equivalent, more geometric, way of formulating the CCR. It brings out the fact that $\mathfrak{W}$ is built in a canonical way from a simple kind of geometric object, a symplectic form.

The equations (3.3) and (3.4) say that $\mathfrak{W}^{(1)} = W \oplus \mathbf{C}$ is a Lie algebra. Equation (3.4) implies that any 3-fold commutator $[w_1[w_2, w_3]]$ is zero. A Lie algebra in which this holds is called *two-step nilpotent*. Since the Lie algebra structure of $\mathfrak{W}^1$ embodies the CCR, it is frequently called the *Heisenberg Lie algebra*.

Besides the algebra of the CCR, physics requires a notion of unitarity, the infinitesimal analogue of which is skew-symmetry. Thus we define a complex conjugate-linear involution $^+: \mathfrak{W} \to \mathfrak{W}$ by the recipe

$$(a_i)^+ = a_i^+, \qquad (a_i^+)^+ = a_i, \qquad (3.6\text{i})$$

$$s^+ = \bar{s}, \qquad s \in \mathbf{C}, \qquad (3.6\text{ii})$$

$$(uv)^+ = v^+ u^+, \qquad u, v \in \mathfrak{W}. \qquad (3.6\text{iii})$$

In (3.6ii), the symbol $\bar{s}$ denotes the usual complex conjugate of the complex number s. An element $u \in \mathfrak{W}$ is called skew-symmetric (resp. symmetric) if $u^+ = -u$ (resp. $u^+ = u$). Since the map $^+$ is complex conjugate linear, the set of skew-symmetric elements $\mathfrak{W}^-$ of $\mathfrak{W}$ forms a real subspace of $\mathfrak{W}$, i.e., $\mathfrak{W}^-$ is preserved by multiplication by real scalars but not by complex scalars. One has the direct sum of real vector spaces $\mathfrak{W} = \mathfrak{W}^- \oplus i\mathfrak{W}^-$.

In particular, if we set

$$\mathfrak{h} = \mathfrak{W}^{(1)} \cap \mathfrak{W}^-, \qquad (3.7)$$

then $\mathfrak{h}$ is a real Lie algebra whose complex span is $\mathfrak{W}^{(1)}$. We call $\mathfrak{h}$ the real Heisenberg Lie algebra. If we set $W_{\mathbf{R}} = W \cap \mathfrak{W}^-$, then obviously from (3.3) we have

$$\mathfrak{h} = W_{\mathbf{R}} \oplus i\mathbf{R}. \qquad (3.8)$$

Here $i\mathbf{R} \subseteq \mathbf{C}$ is the imaginary axis. Note that $\langle \, , \, \rangle$ will take pure imaginary values on $W_{\mathbf{R}}$.

Associated to $\mathfrak{h}$ by the standard facts of Lie theory [**Se**] is a simply connected Lie group H. Connecting $\mathfrak{h}$ and H there is the exponential map

$$\exp: \mathfrak{h} \to H. \qquad (3.9)$$

The group multiplication in H can then be expressed via the Campbell-Hausdorff formula [**Se**] which in this case is quite simple:

$$\exp(x_1)\exp(x_2) = \exp(x_1 + x_2 + \tfrac{1}{2}[x_1, x_2]), \qquad x_i \in \mathfrak{h}. \ (3.10)$$

In terms of H we may express the Stone-von Neumann Theorem, the basic statement of the essential uniqueness of the CCR, as follows.

THEOREM 3.1 (STONE-VON NEUMANN). *There is an irreducible unitary representation ρ of H with the property that if ρ is defined on $\mathfrak{h}$ in the usual way, i.e.,*

$$\rho(x) = \lim_{t \to 0} \frac{\rho(\exp(tx)) - 1}{t}, \qquad x \in \mathfrak{h},$$

then the CCR are verified, i.e.,

$$[\rho(w_1), \rho(w_2)] = \langle w_1, w_2 \rangle 1, \qquad w_1, w_2 \in W_{\mathbf{R}}. \qquad (3.11)$$

Here 1 is the identity operator. The representation ρ is unique up to unitary equivalence.

Observe that by virtue of formula (3.11), the algebra of operators generated by $\rho(\mathfrak{h})$ will be canonically isomorphic to $\mathfrak{W}$. In other words the representation ρ of H gives rise to a representation, which we will also call ρ, of $\mathfrak{W}$. It is possible to give very explicit realizations of ρ, and we will later consider one standard realization; but for now ρ acts on an "abstract quantum-mechanical Hilbert space".

A useful feature of formulating the CCR in terms of the symplectic form $\langle \, , \, \rangle$ is that it makes clear the full symmetry inherent in the CCR. Indeed, let $g: W \to W$ be a linear map which is an isometry of $\langle \, , \, \rangle$ i.e., the map g satisfies

$$\langle g(w_1), g(w_2) \rangle = \langle w_1, w_2 \rangle. \tag{3.12}$$

Such a map g is called a symplectic map, and the group of all symplectic maps is called the (complex) symplectic group, and denoted $Sp_{\mathbb{C}}$. Since the structure of $\mathfrak{W}$ is completely determined by the symplectic form $\langle \, , \, \rangle$, a symplectic map g on W immediately and uniquely extends to an automorphism of $\mathfrak{W}$, satisfying and defined by

$$g(uv) = g(u)g(v), \qquad u, v \in \mathfrak{W}.$$

The symplectic group $Sp_{\mathbb{C}}$ is of course a Lie group and so has a Lie algebra $\mathfrak{sp}_{\mathbb{C}}$, the symplectic Lie algebra of "infinitesimal isometries" of the form $\langle \, , \, \rangle$. For a linear map $T: W \to W$, define $\exp T$ by the usual formula

$$\exp T = \sum_{n=0}^{\infty} \frac{T^n}{n!}. \tag{3.13}$$

Then $T \in \mathfrak{sp}_{\mathbb{C}}$ if $\exp(sT) \in Sp_{\mathbb{C}}$ for all $s \in \mathbb{R}$. Differentiating the relation

$$\langle \exp sT(w_1), \exp sT(w_2) \rangle = \langle w_1, w_2 \rangle$$

at $s = 0$ yields the criterion

$$\langle T(w_1), w_2 \rangle + \langle w_1, T(w_2) \rangle = 0 \tag{3.14}$$

for T to belong to $\mathfrak{sp}_{\mathbb{C}}$. Similarly, differentiating the relation

$$(\exp sT(u))(\exp sT(v)) = \exp sT(uv), \qquad u, v \in \mathfrak{W},$$

yields an extension of T to $\mathfrak{W}$ satisfying

$$T(u)v + uT(v) = T(uv), \qquad u, v \in \mathfrak{W}. \tag{3.15}$$

A map satisfying (3.15) is called a *derivation* of $\mathfrak{W}$. Thus we have an action of $\mathfrak{sp}_{\mathbb{C}}$ on $\mathfrak{W}$ by derivations.

A crucial property of $\mathfrak{W}$ is that the action of $Sp_{\mathbb{C}}$, or more properly $\mathfrak{sp}_{\mathbb{C}}$, on $\mathfrak{W}$ is already implicit in the structure of $\mathfrak{W}$ itself. Consider the commutator bracket operation not simply on W, but on all of $\mathfrak{W}$. By leaving the first element in $[u, v]$ fixed and varying the second we obtain a linear map from $\mathfrak{W}$ to itself. This is called adu. Briefly, the identity

$$\mathrm{adu}(v) = [u, v] = uv - vu, \qquad u, v \in \mathfrak{W}, \qquad (3.16)$$

defines adu. It is trivial to check that

$$\mathrm{adu}(vw) = \mathrm{adu}(v)w + v\,\mathrm{adu}(w);$$

that is, the map adu is a derivation.

By an easy induction from relation (3.4) one finds that

$$[\mathfrak{W}^{(k)}, \mathfrak{W}^{(l)}] \subseteq \mathfrak{W}^{(k+l-2)}. \qquad (3.17)$$

This implies that if $u \in \mathfrak{W}^{(2)}$, then $\mathrm{adu}(\mathfrak{W}^{(l)}) \subseteq \mathfrak{W}^{(l)}$. Since the $\mathfrak{W}^l$ are finite-dimensional, there is no problem about defining exp(adu) by formula (3.13). Then a formal calculation shows that since adu is a derivation, its exponential exp(adu) is an automorphism of $\mathfrak{W}$.

Consider the subspace of $\mathfrak{W}^{(2)}$ spanned by the quadratic expressions

$$a_i a_j, \qquad \frac{1}{2}\left(a_i a_j^+ + a_j^+ a_i\right), \qquad a_i^+ a_j^+ . \qquad (3.18)$$

It is easy to check that for any element u of this space one has $\mathrm{adu}(W) \subseteq W$ and furthermore the map

$$u \to \mathrm{adu} \,|\, W$$

defines an isomorphism between this space and the Lie algebra $\mathfrak{sp}_{\mathbf{C}}$ of infinitesimal isometries of $\langle\ ,\ \rangle$. For this reason, we denote the span of the operators (3.18) by $\mathfrak{sp}_{\mathbf{C}}$. From our discussion we see that the exponentials exp(adu), $u \in \mathfrak{sp}_{\mathbf{C}}$, realize the action of the symplectic group on $\mathfrak{W}$ described above.

Let $\mathfrak{sp}_{\mathbf{R}} = \mathfrak{sp} \subseteq \mathfrak{sp}_{\mathbf{C}}$ denote the real Lie subalgebra of elements which are antisymmetric with respect to the involution $^+$ of $\mathfrak{W}$. Then ad $\mathfrak{sp}$ will preserve the real subspace $W_{\mathbf{R}} \subseteq W$ and will act on the real Heisenberg algebra $\mathfrak{h}$. Also the automorphisms exp(adu), for $u \in \mathfrak{sp}$, will commute with the involution $^+$. They will generate the Lie group $Sp_{\mathbf{R}} = Sp$ of isometries of $W_{\mathbf{R}}$; this is the *real* symplectic group. Using formula (3.10) we can make Sp act as a group of automorphisms of H in such a way that exp: $\mathfrak{h} \to H$ commutes with the actions of Sp on $\mathfrak{h}$ and H.

Recall the representation ρ of H, and the resulting representation ρ of W. The space $\rho(\mathfrak{sp})$ is a Lie algebra of skew-adjoint operators. It is a natural to wonder if there is a unitary representation of Sp of which $\rho(\mathfrak{sp})$ is the infinitesimal version. The general problem of this sort was studied by Mackey [**My**] and this particular case was settled by Shale [**Sh**] and Weil [**WA1**]. The answer is slightly subtle: there is not a representation of Sp, but rather of the 2-fold cover $Sp^\sim$ of Sp. This group $Sp^\sim$ looks as follows. It contains a central subgroup C_2 of order 2 such that

$Sp^\sim / C_2 \cong Sp$; and $Sp^\sim$ is a nontrivial extension of Sp by C_2 in the sense that it is *not* the direct product $Sp \times C_2$.

THEOREM 3.2 (SHALE-WEIL). *There is a unitary representation ω of $Sp^\sim$ on the space of the representation ρ of H (cf. Theorem 3.1) such that*

$$\omega(g)\rho(h)\omega(g)^{-1} = \rho(g(h)), \qquad g \in Sp^\sim, h \in H. \qquad (3.19)$$

Here $g \in Sp^\sim$ acts on H via its image in Sp. In particular, one has

$$\omega(\mathfrak{sp}) = \rho(\mathfrak{sp}).$$

We call ω the *oscillator representation* of $Sp^\sim$, for reasons given in [H1].

We finish this section on formalism with some further comments on $\mathfrak{sp}_\mathbf{C}$. First, let us note that $\mathfrak{sp}_\mathbf{C}$ is a complement of $\mathfrak{W}^{(1)}$ in $\mathfrak{W}^{(2)}$, so that we have the decompositions

$$\mathfrak{W}^{(2)} = \mathfrak{sp}_\mathbf{C} \oplus \mathfrak{W}^{(1)} = \mathfrak{sp}_\mathbf{C} \oplus W_\mathbf{C} \oplus \mathbf{C}. \qquad (3.20)$$

Second, it is useful to consider some extra structure on $\mathfrak{sp}_\mathbf{C}$. Let the linear span of the $a_i a_j$ be denoted $\mathfrak{sp}_\mathbf{C}^{(0,2)}$; the linear span of the $a_i^+ a_j^+$ by $\mathfrak{sp}_\mathbf{C}^{(2,0)}$; and the linear span of the $(\frac{1}{2})(a_i a_j^+ + a_j^+ a_i)$ by $\mathfrak{sp}_\mathbf{C}^{(1,1)}$. Then we may check the following facts about commutators in $\mathfrak{sp}_\mathbf{C}$.

$$\left[\mathfrak{sp}_\mathbf{C}^{(0,2)}, \mathfrak{sp}_\mathbf{C}^{(0,2)}\right] = 0 = \left[\mathfrak{sp}_\mathbf{C}^{(2,0)}, \mathfrak{sp}_\mathbf{C}^{(2,0)}\right],$$

$$\left[\mathfrak{sp}_\mathbf{C}^{(1,1)}, \mathfrak{sp}_\mathbf{C}^{(0,2)}\right] = \mathfrak{sp}_\mathbf{C}^{(0,2)}, \qquad \left[\mathfrak{sp}_\mathbf{C}^{(1,1)}, \mathfrak{sp}_\mathbf{C}^{(2,0)}\right] = \mathfrak{sp}_\mathbf{C}^{(2,0)},$$

$$\left[\mathfrak{sp}_\mathbf{C}^{(1,1)}, \mathfrak{sp}_\mathbf{C}^{(1,1)}\right] = \mathfrak{sp}_\mathbf{C}^{(1,1)} = \left[\mathfrak{sp}_\mathbf{C}^{(0,2)}, \mathfrak{sp}_\mathbf{C}^{(2,0)}\right]. \qquad (3.21)$$

In particular each of the $\mathfrak{sp}_\mathbf{C}^{(i,j)}$ is a Lie subalgebra of $\mathfrak{sp}_\mathbf{C}$. The subalgebras $\mathfrak{sp}_\mathbf{C}^{(0,2)}$ and $\mathfrak{sp}_\mathbf{C}^{(2,0)}$ are abelian. The subalgebra $\mathfrak{sp}_\mathbf{C}^{(1,1)}$ is isomorphic to $\mathfrak{gl}_\mathbf{C}$, the Lie algebra of all $n \times n$ matrices. It has a one-dimensional center, spanned by the element

$$\sum_{i=1}^n \frac{(a_i a_i^+ + a_i^+ a_i)}{2}. \qquad (3.22)$$

This element will be recognized by many as the Hamiltonian for a system of n independent quantum harmonic oscillators.

The abelian subalgebras $\mathfrak{sp}_\mathbf{C}^{(2,0)}$ and $\mathfrak{sp}_\mathbf{C}^{(0,2)}$ are interchanged by the involution $^+$, so that individually they do not intersect $\mathfrak{sp}_\mathbf{R} = \mathfrak{sp}$. However $\mathfrak{sp}_\mathbf{C}^{(1,1)}$ is stabilized by $^+$, and one has

$$\mathfrak{sp}_\mathbf{C}^{(1,1)} \cap \mathfrak{sp} \cong \mathfrak{u}_n = \mathfrak{u} \qquad (3.23)$$

the Lie algebra of $n \times n$ skew-Hermitian matrices, which is the Lie algebra of the $n \times n$ unitary group U_n. This is the maximal compact subgroup of Sp.

Finally, we should remark on the fact that the elements of $\mathfrak{sp}_{\mathbf{C}}^{(1,1)}$ are symmetrized products of the a_i and a_j^+. This is necessitated purely from the internal structure of $\mathfrak{W}$. In order to make $\mathfrak{sp}_{\mathbf{C}}$ to be closed under taking commutators, i.e., to make it be a Lie algebra, it is necessary to symmetrize in the a_i and a_j^+ as we have done. However this simple, formally required procedure has substantial consequences. In the first place, it is responsible for the necessity to replace Sp by its 2-fold cover $Sp^\sim$ in Theorem 4.2. Second, in physics it affects the form of the Hamiltonian for the quantum harmonic oscillator, resulting in the half-integers in the spectrum and the existence of the zero-point energy. Later, when the same formalism is used to construct a free boson field, it leads to the first example of renormalization. Finally, again in mathematics, Weil [WA1] showed it was intimately related to the law of Quadratic Reciprocity.

4. Basic results on dual pairs in Sp. It is inside Sp that we wish to consider dual pairs, and we wish to consider how they (more precisely, their inverse images in $Sp^\sim$) act under the oscillator representation.

The group Sp is not a general abstract group; it acts in a concrete way on a given vector space. We use the extra structure available in the situation to put an extra condition on the dual pairs we will consider. A group $G \subseteq Sp$ is called *reductive* if whenever $U \subseteq W_{\mathbf{R}}$ is a subspace invariant by G, there is also a complementary G-invariant subspace V, i.e., V is G-invariant and we have a direct sum decomposition $W_{\mathbf{R}} = U \oplus V$. A *reductive dual pair* (G, G') in Sp is then a dual pair in which each member is reductive.

The collection of reductive dual pairs in Sp is a manageable but interesting set. We can give a complete description of such pairs. First we can define a notion of an irreducible pair. If $U \subseteq W$ is a subspace, define

$$U^\perp \{w \in W: \langle w, u \rangle = 0 \text{ for all } u \in U\}.$$

Let $(G, G') \subseteq Sp$ be a reductive dual pair. If we can find a subspace $U \subseteq W$ such that $W = U \oplus U^\perp$, and such that U is invariant by G and by G', then we call (G, G') *reducible*. If we cannot find such a U, then (G, G') is *irreducible*. An easy argument shows that all dual pairs are built in a simple way out of irreducible ones. Furthermore, we can give a complete list of the irreducible pairs. They fall into two classes, type I, and type II. The type I pairs fall into families which correspond to division algebras over $\mathbf{R}$ with involution. There are four of these: $\mathbf{R}$ itself; $\mathbf{C}$ with the identity involution; $\mathbf{C}$ with complex conjugation; and Hamilton's quaternions $\mathbf{H}$, with quaternionic conjugation. The corresponding families of irreducible pairs are as follows.

TABLE 4.1. Type I irreducible dual pairs in Sp.

Division algebra with involution	Dual pair family
$(\mathbf{R}, 1)$	$(O_{p,q}, Sp_{2m}(\mathbf{R})) \subseteq Sp_{2m(p+q)}(\mathbf{R})$
$(\mathbf{C}, 1)$	$(O_p(\mathbf{C}), Sp_{2m}(\mathbf{C})) \subseteq Sp_{4mp}(\mathbf{R})$
$(\mathbf{C}, -)$	$(U_{p,q}, U_{r,s}) \subseteq Sp_{2(p+q)(r+s)}(\mathbf{R})$
$(\mathbf{H}, -)$	$(Sp_{p,q}, O_{2m}^*) \subseteq Sp_{4m(p+q)}(\mathbf{R})$

In Table 4.1, the notations for the various groups are those in Helgason [He], except our $Sp_{2n}(\mathbf{R})$ and $Sp_{2n}(\mathbf{C})$ are his $Sp(n, \mathbf{R})$ and $Sp(n, \mathbf{C})$. The $-$ in $(\mathbf{C}, -)$, and $(\mathbf{H}, -)$ indicates the appropriate conjugation.

The type II pairs are simpler to describe. They fall into families corresponding to division algebras over $\mathbf{R}$, namely $\mathbf{R}$, $\mathbf{C}$, and $\mathbf{H}$. They are as follows.

TABLE 4.2. Type II irreducible dual pairs in Sp.

Division algebra	Dual pair family
$\mathbf{R}$	$(GL_n(\mathbf{R}), GL_m(\mathbf{R})) \subseteq Sp_{2nm}(\mathbf{R})$
$\mathbf{C}$	$(GL_n(\mathbf{C}), GL_m(\mathbf{C})) \subseteq Sp_{4nm}(\mathbf{R})$
$\mathbf{H}$	$(GL_n(\mathbf{H}), GL_m(\mathbf{H})) \subseteq Sp_{8nm}(\mathbf{R})$

Observe that, ignoring centers and connected components, the Tables 4.1 and 4.2 contain every classical group and no exceptional groups.

Irreducible dual pairs in Sp typically arise via some sort of tensor product decomposition of the symplectic vector space. To give the idea, we will give the representative example $(O_n, Sp_{2m}(\mathbf{R}))$. Let $W_{\mathbf{R}} = M_{n,2m}(\mathbf{R})$, the vector space of real $n \times 2m$ matrices. Then O_n, the orthogonal group in n variables acts on $M_{n,2m}(\mathbf{R})$ by multiplication on the left. Similarly $Sp_{2m}(\mathbf{R})$, the real symplectic group in $2m$ variables, can act on $M_{n,2m}(\mathbf{R})$ by multiplication on the right. Obviously these two actions commute with one another. It remains to define on $W_{\mathbf{R}}$ a symplectic form which is invariant under these two actions. We recall O_n is the group of real $n \times n$ matrices A such that $A^t A = I_n$ where A^t is the transpose of A and I_n is the $n \times n$ identity matrix. Similarly, $Sp_{2m}(\mathbf{R})$ is the group of real $2m \times 2m$ matrices B such that $B^t J B = J$, where

$$J = \begin{bmatrix} 0 & I_m \\ -I_m & 0 \end{bmatrix}.$$

Define a symplectic form on $M_{n,2m}(\mathbf{R})$ by the formula

$$\langle S, T \rangle = \operatorname{tr}(SJT^t - TJS^t).$$

Here tr is the trace function on the $n \times n$ matrices.

It is straightforward to check that this form is O_n- and $Sp_{2m}(\mathbf{R})$-invariant. Hence $(O_n, Sp_{2m}(\mathbf{R})) \subseteq Sp(W_{\mathbf{R}})$, and it can be shown they form a reductive dual pair in this large symplectic group.

There is one main fact about reductive dual pairs in Sp, and like the rest of the theory it comes in a number of versions. We will state the algebraic and Hilbert space versions.

We have seen that $Sp_\mathbf{C}$, hence $Sp_\mathbf{R}$ acts on $\mathfrak{W}$ as a group of automorphisms. If $G \subseteq Sp_\mathbf{R}$ is a subgroup, let $\mathfrak{W}^G$ be the set of elements of $\mathfrak{W}$ invariant by G. Since G acts by automorphisms, one sees that $\mathfrak{W}^G$ will be a subalgebra of $\mathfrak{W}$. Suppose $(G, G') \subseteq Sp$ is a reductive dual pair. Then if $\mathfrak{g}$ and $\mathfrak{g}'$ are the Lie algebras of G and G', then $(\mathfrak{g}, \mathfrak{g}')$ form a reductive dual pair in $\mathfrak{sp}$. Since $\mathfrak{sp} \subseteq \mathfrak{sp}_\mathbf{C} \subseteq \mathfrak{W}^{(2)}$, and the action of Sp on $\mathfrak{sp}$ is simply by conjugation, we see that

$$\mathfrak{g}' = \mathfrak{W}^G \cap \mathfrak{sp}.$$

Thus $\mathfrak{g}'$ is sort of the set of quadratic invariants for G in $\mathfrak{W}$.

THEOREM 4.1 [H1]. *For a reductive dual pair* $(G, G') \subseteq Sp$, *the algebra* $\mathfrak{W}^G$ *is generated as associative algebra by* $\mathfrak{g}'$, *the Lie algebra of* G'.

Theorem 4.1 summarizes a considerable portion of Weyl's book, *The classical groups*, and can be used to perform a variety of well-known mathematical computations (see [H1]). We will see later its connection with physics (§§6, 7), but for the moment we are more interested in its Hilbert space analogue.

To formulate this analogue we need the notion of the *von Neumann algebra* generated by a set of operators. This concept is slightly involved technically, so we make no attempt to define it here, and simply refer the reader to one of the standard references on the subject [Dx].

Let (G, G') again be a reductive dual pair in Sp. Let $\tilde{G}$ and $\tilde{G}'$ be the inverse images of $\tilde{G}$ and $\tilde{G}'$ in $Sp^\sim$. It is not difficult to verify that $\tilde{G}$ and $\tilde{G}'$ commute with one another, so they form a dual pair in $Sp^\sim$. Hence the images $\omega(\tilde{G})$ and $\omega(\tilde{G}')$ under the oscillator representation will certainly also commute.

THEOREM 4.2 [H5]. *Let* (G, G') *be a reductive dual pair in* Sp. *Consider the oscillator representation* ω *of the 2-fold cover* $Sp^\sim$ *of* Sp. *Let* $\mathfrak{A}(\tilde{G})$ *and* $\mathfrak{A}(\tilde{G}')$ *be the von Neumann algebras generated by* $\omega(\tilde{G})$ *and* $\omega(\tilde{G}')$ *respectively. Then* $\mathfrak{A}(\tilde{G})$ *and* $\mathfrak{A}(\tilde{G}')$ *are mutual commutants in the algebra of all bounded operators on the Hilbert space where* ω *is realized.*

There is a close connection between Theorems 4.1 and 4.2. Let $\mathfrak{H}$ be the Hilbert space on which ω is realized. From Theorems 3.1 and 3.2 we know that the Weyl algebra $\mathfrak{W}$ is represented as an algebra of operators on (some dense subspace of) $\mathfrak{H}$. Further when $\mathfrak{W}$ is so realized, the action of Sp on $\mathfrak{W}$ comes from (is the restriction to $\mathfrak{W}$ of) the action by conjugation of $\omega(Sp^\sim)$ on operators on $\mathfrak{H}$. Therefore, for $G \subseteq Sp$, the algebra $\mathfrak{W}^G$ is simply the subalgebra of $\mathfrak{W}$ of operators which commute with $\omega(\tilde{G})$.

Since $\mathfrak{W}$ consists of unbounded operators and is an algebraic rather than an analytic object, in general neither of Theorems 4.1 nor 4.2 implies the other. However, when either G or G' is compact, Theorem 4.2 follows directly from Theorem 4.1.

Also if G or G' is compact the true significance of Theorem 4.2 can be made tangible. The point is that there results from Theorem 4.2, a correspondence between representations of G and representations of G'. In general, this must be formulated in a roundabout technical way, but if G or G' is compact, it can be stated quite directly.

So let us assume from now on that one of the groups, say G', is compact. Then we can decompose the Hilbert space $\mathcal{H}$ on which ω is defined into isotypic components for $\tilde{G}'$. That is, we may write

$$\mathcal{H} = \sum_{\sigma' \in (\tilde{G}')^{\wedge}} \mathcal{H}_{\sigma'} \tag{4.1}$$

where $(\tilde{G}')^{\wedge}$ denotes the set of equivalence classes of irreducible unitary representations of $\tilde{G}'$, and for $\sigma' \in (\tilde{G}')^{\wedge}$, the space $\mathcal{H}_{\sigma'}$ is the sum of all the $\omega(\tilde{G}')$-invariant, irreducible subspaces of $\mathcal{H}$ on which $\tilde{G}'$ acts by a representation of class σ'. Of course, some $\mathcal{H}_{\sigma'}$ may be zero.

The next result follows immediately from Theorem 4.2.

THEOREM 4.3. *Let $(G, G') \subseteq Sp$ be a reductive dual pair, with G' compact. Let $\mathcal{H}$ be the Hilbert space on which the oscillator representation ω is realized, and let* (4.1) *be the decomposition of $\mathcal{H}$ into isotypic subspace for $\tilde{G}'$. Then each $\mathcal{H}_{\sigma'}$ is also preserved by $\omega(\tilde{G})$, and is irreducible as a joint $\tilde{G} \times \tilde{G}'$ representation. One has a factorization*

$$\mathcal{H}_{\sigma'} \cong \sigma \otimes \sigma' \tag{4.2}$$

where σ is an irreducible unitary representation of $\tilde{G}$. Furthermore, the correspondence $\sigma \leftrightarrow \sigma'$ is a bijection between the representations of $\tilde{G}$ and $\tilde{G}'$ involved. (That is, σ determines σ' and vice versa.)

REMARK. Special cases of Theorem 4.2 have been discovered both by physicists [**GS, MQ**], and by mathematicians [**Ge, GK, KV, R, Sa**].

It turns out that the representations σ of G which arise in Theorem 4.3 are of a special kind, called *holomorphic representations* (or are the contragredients of holomorphic representations). To fully explicate this terminology would take too long, but we can explain the structure of these σ.

Let $\mathfrak{g}$ and $\mathfrak{g}'$ be the Lie algebras of G and G'. Recall that $\mathfrak{sp}_{\mathbb{C}}^{(1,1)} \cap \mathfrak{sp}$ is the Lie algebra of a maximal compact subgroup of Sp. Since G' is compact, we may assume that $\mathfrak{g}'_C \subseteq \mathfrak{sp}_C^{(1,1)}$. Then we will have a decomposition

$$\mathfrak{g}_C = \left(\mathfrak{g}_C \cap \mathfrak{sp}_C^{(2,0)}\right) \oplus \left(\mathfrak{g}_C \cap \mathfrak{sp}_C^{(1,1)}\right) \oplus \left(\mathfrak{g}_C \cap \mathfrak{sp}_C^{(0,2)}\right)$$

$$\overset{\text{def}}{=} \mathfrak{g}_C^{(2,0)} \oplus \mathfrak{g}_C^{(1,1)} \oplus \mathfrak{g}_C^{(0,2)}. \tag{4.3}$$

The Lie algebra $\mathfrak{g}_C^{(1,1)} \cap \mathfrak{g} = \mathfrak{g} \cap \mathfrak{sp}_C^{(1,1)}$ will be the Lie algebra of a maximal compact subgroup of G, and will be isomorphic to a product of unitary Lie algebras. The Lie algebras $\mathfrak{g}_C^{(2,0)}$ and $\mathfrak{g}_C^{(0,2)}$ will of course be abelian.

A holomorphic representation σ of $\tilde{G}$ has the following structure.

> (i) Let σ_0 be the space of vectors in σ which are annihilated by all operators $\sigma(\mathfrak{g}_C^{(0,2)})$. Then σ_0 is nonzero, is invariant under $\sigma(\mathfrak{g}_C^{(1,1)})$, and defines an irreducible representation for $\mathfrak{g}_C^{(1,1)}$.
>
> (ii) The whole space of σ is generated by the algebra generated by $\mathfrak{g}_C^{(2,0)}$ applied to σ_0. $\tag{4.4}$

It turns out that Theorem 4.3 provides a systematic way to realize the holomorphic representations of the groups to which it applies. (Referring to Table 4.1, we see these are the groups $Sp_{2n}(\mathbf{R})$, $U_{p,q}$, and O_{2m}^*.) In fact, one has the following classification theorem. Partial versions of it were given in [**EP, H6, J, P**], and the complete result is proved in [**EHW**].

THEOREM 4.4. *Let G be a group which appears in an irreducible reductive dual pair (G, G') in Sp with G' compact. Let G' vary through all possible compact members of the pair (G, G'). Then the representations σ of $\tilde{G}$ arising as in Theorem 4.3, when restricted to the commutator subgroup $\tilde{G}^{(2)}$ of $\tilde{G}$, exhaust all holomorphic and antiholomorphic representations of $\tilde{G}^{(2)}$.*

Beyond classification, the theory of dual pairs is an effective device for studying various features of holomorphic representations. We will discuss some properties of these representations that intersect the physics literature.

For ease of formulation, we will restrict ourselves to the examples of the pairs $(U_{p,q}, U_n)$. However, completely analogous results hold for the pairs (O_{2m}^*, Sp_p) and $(Sp_{2m}(\mathbf{R}), O_p)$. We will convene in what follows that $p \geq q$. The group $U_{p,q}$ acts on $\mathbf{C}^{p+q}$, preserving a certain Hermitian pseudometric. We can take this metric to be

$$z_1\bar{z}_{p+1} + z_2\bar{z}_{p+2} + \cdots + z_q\bar{z}_{p+q} + z_{q+1}\bar{z}_{q+1} + \cdots + z_p\bar{z}_p \tag{4.5}$$

for $z = (z_1, z_2, \ldots, z_{p+q}) \in \mathbf{C}^{p+q}$. In other words, the matrix of the associated Hermitian-bilinear form is

$$\begin{bmatrix} 0 & 0 & I_q \\ 0 & I_{p-q} & 0 \\ I_q & 0 & 0 \end{bmatrix}$$

where I_m is the $m \times m$ identity matrix.

Let $\mathbf{C}^k \subseteq \mathbf{C}^{p+q}$ be the subspace of z such that $z_i = 0$ for $i > k$. For $k \leq q$, the metric (4.5) is identically zero on $\mathbf{C}^k$. Let P_k be the subgroup of $U_{p,q}$ which maps $\mathbf{C}^k$ into itself. The group P_k looks as follows. There is a decomposition (the Levi decomposition)

$$P_k = M_k N_k \tag{4.6a}$$

where N_k is normalized by M_k. The group M_k consists of matrices

$$\begin{bmatrix} A & 0 & 0 \\ 0 & B & 0 \\ 0 & 0 & A^{*-1} \end{bmatrix} \tag{4.6b}$$

where $A \in GL_k(\mathbf{C})$, $B \in U_{p-k, q-k}$, and A^* is the conjugate transpose of A. The group N_k consists of matrices

$$\begin{bmatrix} I_k & S & T - \dfrac{1}{2}SS^* \\ 0 & I_{p+q-2k} & -S^* \\ 0 & 0 & I_k \end{bmatrix} \tag{4.6c}$$

where T is a skew-Hermitian $k \times k$ matrix. The matrices of the form

$$\begin{bmatrix} I_k & 0 & T \\ 0 & I_{p+q-2k} & 0 \\ 0 & 0 & I_k \end{bmatrix}$$

form the center Z_k of N_k, which as can easily be seen is isomorphic to the additive group of $k \times k$ skew-Hermitian matrices. The group Z_k is normal in P_k, and the action of the group M_k on Z_k can be seen to be just the usual action of $GL_k(\mathbf{C})$ on the $k \times k$ skew-Hermitian matrices. Thus

$$\begin{bmatrix} A & 0 & 0 \\ 0 & B & 0 \\ 0 & 0 & A^{*-1} \end{bmatrix} \begin{bmatrix} I & 0 & T \\ 0 & I & 0 \\ 0 & 0 & I \end{bmatrix} \begin{bmatrix} A & 0 & 0 \\ 0 & B & 0 \\ 0 & 0 & A^{*-1} \end{bmatrix}^{-1} = \begin{bmatrix} I & 0 & ATA^* \\ 0 & I & 0 \\ 0 & 0 & I \end{bmatrix}.$$

Let σ be a unitary representation of $U_{p,q}$, and consider the restriction of σ to Z_q. Since Z_q is abelian, the restricted representation will have a spectral decomposition involving certain characters of Z_q. These characters all have the form

$$\chi_S: \begin{bmatrix} I & 0 & T \\ 0 & I & 0 \\ 0 & 0 & I \end{bmatrix} \rightarrow e^{\operatorname{tr}(TS)}$$

where S is a $k \times k$ Hermitian matrix and tr denotes the trace of a matrix. Since this representation is a restriction from $U_{p,q}$, the set of characters involved in the spectral decomposition of $\sigma \mid Z_q$ will be invariant under conjugation by M_q. In other words, if χ_S occurs in $\sigma \mid Z_q$, then so does χ_{ASA^*} for $A \in GL_q(\mathbf{C})$. It is well known that every $q \times q$ Hermitian matrix S is of the form ADA^* where D is diagonal, with only 0 and ± 1 for diagonal entries. More precisely, we may take $D = D_{r,s}$ with r ($+1$)'s and s (-1)'s, and $q - r - s$ 0's on the diagonal. The numbers r and s completely characterize the $GL_q(\mathbf{C})$ orbit of S. The number $r + s$ is called the *rank* of S. We will say the representation σ of $U_{p,q}$ has N_q-rank k if all the characters χ_S involved in the spectral decomposition of $\sigma \mid Z_q$ have rank $S \leqslant k$, and some have rank $S = k$.

Clearly the N_q-rank is some measure of the size of a representation. Another well-known measure of the size of a representation is the Gelfand-Kirillov dimension. To give a precise definition of it is beyond the scope of the present discussion; we refer the reader to [**GeKi**]. However, roughly speaking, if a representation σ is realized on a space of vector-valued functions on a manifold X, the Gelfand-Kirillov dimension of σ will be the dimension of X.

For holomorphic representations, the notions of N_q-rank and Gelfand-Kirillov dimension are closely related and are tied in nicely with the theory of dual pairs.

THEOREM 4.5. *Let σ be a holomorphic irreducible unitary representation of $\tilde{U}_{p,q}$.*

(a) *If σ has N_q-rank $k \leqslant q$, then the Gelfand-Kirillov dimension of σ is $k(p + q - k)$.*

(b) *If $k < q$, then σ is of the form $\psi \otimes \sigma_1$ where ψ is a one-dimensional unitary character of $\tilde{U}_{p,q}$, and σ_1 is associated to a representation σ_1' of $\tilde{U}_k$ as in Theorem 4.4.*

(c) *If S is $q \times q$ Hermitian, and χ_S is involved in the spectral decomposition of $\sigma \mid Z_q$, then S is positive semidefinite (and of rank $\leqslant k$).*

(d) *If $l \geqslant k < q$, then $\sigma \mid \tilde{P}_l$ is irreducible, and the restriction $\sigma \mid \tilde{P}_l$ determines σ as a unitary representation of $\tilde{U}_{p,q}$.*

Theorem 4.5 does not appear elsewhere in the literature although parts of it do. It may be approached by the methods of [**H6**].

Finally we will make some remarks about tensor products. These are of interest in physics in considering combined states of two or more particles. The dual pair formalism allows one to reduce many questions about Clebsch-Gordan coefficients of holomorphic representations to analogous questions about finite-dimensional representations of compact groups. We will again use $U_{p,q}$ for purposes of illustration.

The dual pair $(U_{p,q}, U_m)$ lives inside $Sp_{2(p+q)m}(\mathbf{R})$. Denote by τ_m the restriction to $\tilde{U}_{p,q}$ of the oscillator representation of $Sp_{2(p+q)m}(\mathbf{R})$. The basis for this discussion is the relation

$$\tau_m \otimes \tau_n \cong \tau_{m+n}. \tag{4.7}$$

This follows from the general formalism of the oscillator representation. Consider now the holomorphic irreducible representations σ_1 and σ_2 of $\tilde{U}_{p,q}$, such that σ_1 occurs in τ_m and σ_2 occurs in τ_n. Then from Theorem 4.3, we know that the σ_1-isotypic subspace of τ_n defines a $\tilde{U}_{p,q} \times \tilde{U}_m$ representation of the form $\sigma_1 \otimes \sigma_1'$ where $\sigma_1' \in (\tilde{U}_m)\hat{\ }$. Similarly the σ_2-isotypic subspace of τ_n has the form $\sigma_2 \otimes \sigma_2'$ for some $\sigma_2' \in (\tilde{U}_n)\hat{\ }$. Therefore the tensor product $\sigma_1 \otimes \sigma_2$ occurs in the subspace

$$(\sigma_1 \otimes \sigma_2) \otimes (\sigma_1' \otimes \sigma_2')$$

inside τ_{m+n}. This subspace may be identified as the $(\sigma_1' \otimes \sigma_2')$-isotypic subspace for $\tilde{U}_m \times \tilde{U}_n \subseteq \tilde{U}_{n+m}$. From this relation we easily get the following result (cf. [H7]).

THEOREM 4.6. *Let σ_1 and σ_2 be holomorphic representations of $\tilde{U}_{p,q}$. Suppose σ_1 is paired as in Theorem 4.3 with the representation σ_1' of $\tilde{U}_m$, and σ_2 is paired with the representation σ_2' of $\tilde{U}_n$. Then $\sigma_1 \otimes \sigma_2$ decomposes as a discrete direct sum of holomorphic representations. A holomorphic representation σ_3 of $\tilde{U}_{p,q}$ will occur in $\sigma_1 \otimes \sigma_2$ only if σ_3 is matched with a representation σ_3' of $\tilde{U}_{m+n}$ via Theorem 4.3. If this holds, then the multiplicity with which σ_3 occurs in $\sigma_1 \otimes \sigma_2$ is the same as the multiplicity with which $\sigma_1' \otimes \sigma_2'$ occurs in the restriction of σ_3' to $\tilde{U}_m \times \tilde{U}_n$.*

5. Examples: massless particles and singletons. A simple but interesting dual pair is $(U_{p,q}, U_1)$ in $Sp_{2(p+q)}(\mathbf{R})$. It has been the subject of a special study in [SW]. If both p and q are positive, each character of U_1 occurs in the restriction of the oscillator representation, so we obtain a family of holomorphic representations of $\tilde{U}_{p,q}$ indexed by the integers. By Theorem 4.5 these representations have N_q-rank 1 and Gelfand-Kirillov dimension $p + q - 1$. If p and q are both two or more, Theorem 4.5 says the restrictions of these representations to $(\tilde{U}_{p,q})^{(2)} \cong SU_{p,q}$ yields all holomorphic unitary representations of Gelfand-Kirillov dimension $p + q - 1$. In fact, it can be shown that these representations exhaust *all* unitary representations of $SU_{p,q}$ of Gelfand-Kirillov dimension $p + q - 1$. Also, this is the smallest possible Gelfand-Kirillov dimension for nontrivial representations of $SU_{p,q}$.

The case $p = q = 2$ is relevant to physics, for $SU_{2,2}$ is a 2-fold cover of the identity component of $SO_{4,2}$, known in physics as the conformal group. In this case the representations coming from the pair $(U_{2,2}, U_1)$ were studied by other methods by Mack, Salam and Todorov [MS, MT] and were called *ladder representations*, because of the structure of their restrictions to the maximal compact subgroup $U_2 \times U_2$ of $U_{2,2}$. The ladder representations are significant because of their connection with massless particles. The Poincaré group $(SO_{3,1} \times \mathbf{R}^x) \times\!\!\!| \, \mathbf{R}^4$ is essentially the subgroup P_2 of $SU_{2,2}$, as defined by formulas (4.6). In this case, since $p = q = k = 2$, the subgroup N_2 equals its center Z_2. It is isomorphic to 2×2 Hermitian matrices, which is well known to be isomorphic to Minkowski space. (Since we are dealing with 2×2 matrices, the determinant function is quadratic, and can be identified to the Lorentz metric.) Under this isomorphism, the rank 1 matrices (which are the same as the matrices of determinant 0) are identified to the light cone [Jo].

According to Theorem 4.5, the restriction of a ladder representation to P_2 is irreducible, and determines the representation from which it comes. Since these representations are rank 1, they will, if regarded as representations of (a cover of) the Poincaré group, appear to have Minkowski-space spectrum concentrated on the (forward) light cone. Hence by the yoga of matching particles to representations, they correspond to mass zero particles. One can see that particles of all possible (integer or half-integer) spins arise this way. From another point of view, one can say that each massless particle automatically allows the full conformal group as symmetry group, and in a unique way.

Another dual pair that is quite simple but also of considerable interest is the pair $(Sp_{2n}(\mathbf{R}), O_1)$. Here $Sp_{2n}(\mathbf{R})$ is the whole symplectic group, and $O_1 \cong \{\pm 1\}$ is its center. According to Theorem 4.3, the representation ω decomposes into 2 pieces, ω^+ and ω^-, corresponding to the two characters, trivial and signum, of O_1. (Actually $\tilde{O}_1$ is cyclic of order 4 and its element of order 2 acts as minus the identity, so Theorem 4.3 would have us deal with the two characters of $\tilde{O}_1$ which are -1 on the element of order 2, and it is not immediately clear how to identify these with characters of O_1; but in practice there is no problem.) An analogue of Theorem 4.5 tells us that ω^+, ω^-, and their contragredients are the entire set of unitary representations of $\tilde{Sp}_{2n}$ (or any of its covering groups) of Gelfand-Kirillov dimension n; further this is the minimum Gelfand-Kirillov dimension of any nontrivial representation of $\tilde{Sp}_{2n}$ (or any of its covering groups). These facts underscore the very special nature of ω.

For $n = 2$, the representations $\omega^{\pm}$ have occurred in the physics literature. In this case Sp_4 is a 2-fold cover of (the identity component) of

$SO_{3,2}$, a group intermediate between the Lorentz group and the conformal group. The representations $\omega^\pm$ were discovered as (Lie algebra) representations of $SO_{3,2}$ by Dirac [**Dr**], who called them singletons. Recently the theory of singletons has been developed by Fronsdal, Flato and others [**FF1**, **FF2**]. The two singleton particles have been called "Di" and "Rac", and the possible 2-particle states formable from Dis and Racs have been analyzed. It was found that each tensor product $\omega^\pm \otimes \omega^\pm$ decomposed into a multiplicity free sum of irreducible unitary representations, labeled by positive integers (plus two labeled by zero). It turns out that each of these representations (except the two labeled by zero) extends (in exactly two ways) to an irreducible representation of the conformal group (actually $SU_{2,2}$). Moreover, the resulting representations of $SU_{2,2}$ are the ladder representations discussed earlier. This remarkable coincidence allows one to interpret each massless particle as a bound state of two singletons.

The above facts have a natural interpretation in terms of dual pairs. As just discussed, the representations $\omega^\pm$ are the components of ω, and come from the pair (Sp_{2n}, O_1) by means of Theorem 4.3. In analogy with the discussion preceding Theorem 4.6, the tensor products $\omega^\pm \otimes \omega^\pm$ are certain subspaces of the representations of Sp_{2n} associated to the pair (Sp_{2n}, O_2); these subspaces are characterized by how they transform under $O_1 \times O_1$. How they decompose is described in analogy with Theorem 4.6. The group O_2 has a very simple structure: the subgroup SO_2 is normal of index 2, and is abelian—it is just the unit circle. Hence all the representations of O_2 are two dimensional, and contain the characters $\theta \to e^{in\theta}$ of SO_2, except when $n = 0$, when there are two one-dimensional representations of O_2 trivial on SO_2. From these facts, one obtains in analogy with Theorem 4.6 the decomposition of $\omega^\pm \otimes \omega^\pm$.

The connection of $\omega^\pm \otimes \omega^\pm$ with the ladder representations may be seen as follows. The group $SO_2 \subseteq O_2$, is, as noted, simply the unit circle, and if $\mathbf{R}^2$ is identified to $\mathbf{C}$ by the usual recipe, the group SO_2 becomes U_1. Thus we have inside Sp_{4n} the two dual pairs (Sp_{2n}, O_2) and $(U_{n,n}, U_1)$, and since $U_1 \subseteq O_2$, we have $Sp_{2n} \subseteq U_{n,n}$. Such a system of dual pairs related by inclusions has been dubbed by Kudla [**Ku**] a pair of *see-saw pairs*. Since, as just described, each nontrivial representation of U_1 is contained exactly once in exactly one representation of O_2, the fact that the Sp_{2n} representations coming from (Sp_{2n}, O_2) extend (in two ways) to ladder representations follows from a general duality phenomenon for see-saw pairs (cf. [**H7**]). (Theorem 4.6 is also an example of this duality.) When $n = 2$, we have the diagram

$$
\begin{array}{ccc}
Sp_4 & \subseteq & SU_{2,2} \\
\downarrow & & \downarrow \\
SO_{3,2} & \subseteq & SO_{4,2}
\end{array}
$$

and we recover the results of [**FF1**] concerning 2-particle states for singletons.

Since the basic formalism of creation and annihilation operators is so intimately connected with the theory of the harmonic oscillator, we should explain how dual pairs relate to this ubiquitous object of physics. As noted above, in the formalism of creation and annihilation operators, the Hamiltonian for a system of n independent quantum harmonic oscillators is given by expression (3.21). But as stated just before (3.21), that operator is also the center of $\mathfrak{u}_n$, the $n \times n$ unitary Lie algebra, the Lie algebra of the maximal compact subgroup U_n of Sp. Moreover, as we can now recognize, the group U_n together with its center U_1 form a dual pair in Sp. Therefore another application of Theorem 4.3 tells us that the U_1 eigenspaces of the oscillator representation are precisely the irreducible subspaces for U_n. (Since $U_1 \subseteq U_n$, the joint $U_n \times U_1$ action reduces to just the U_n action. And since both members of the pair are compact, the spaces are finite dimensional.) But since the Schrödinger-Hermite operator (3.21) is the infinitesimal generator for U_1, the U_1 eigenspaces are the same as the eigenspaces for the harmonic oscillator Hamiltonian, i.e., are the energy eigenstates of n harmonic oscillators. These states are well known to be degenerate. However, Theorem 4.3 may in this context be seen to assert that the energy degeneracies of n harmonic oscillators are completely accounted for by the symmetries of the Hamiltonian (3.21) embodied in U_n.

Looking ahead slightly to the next section, we note that the Hamiltonian (3.21) is often written as the differential operator

$$\Delta - r^2 \tag{5.1}$$

where Δ is the Laplace operator on $\mathbf{R}^n$ and r is the usual Euclidean length of a vector. This operator is obviously invariant under the Euclidean rotation group O_n, which is a subgroup of the operator's full symmetry group, U_n. The group O_n will account for some but not all of the energy degeneracies of (4.1). The full symmetry group U_n is not clearly in evidence here. That the creation-annihilation formalism does exhibit the (dual pair) symmetry structure of the harmonic oscillator is an advantage to its use in treating this system.

6. Concrete realizations and classical equations. So far in this discussion we have not given any specific realization of the representation ρ of the Heisenberg group asserted to exist in Theorem 3.1, or of the associated representation ρ of the Weyl algebra $\mathfrak{W}$, or of the oscillator representation ω of $\tilde{Sp}$; and for many purposes a concrete model of ρ is irrelevant. However, models for ρ can be quite useful, and in this section we will discuss the most common one. It will turn out that several of the best known equations of physics are closely associated to dual pairs as written in this way. (We mention in passing that a considerable number of topics in pure mathematics are also closely related to this embodiment of dual pairs (see [H1]).)

Specifically, consider the following equations.

The wave equation	(6.1i)
Laplace's equation	(6.1ii)
Maxwell's equation	(6.1iii)
Dirac's equation	(6.1iv)

Each of these equations has been the focus of study for an era of mathematical physics, from the mid-eighteenth century for the wave equation through the mid-twentieth for Dirac's equation. It turns out that for each of these equations, there is a dual pair to which it is closely related and which can be used to explain important aspects of its structure. Furthermore, the relation between an equation and its associated dual pair is essentially the same for all these equations. Thus the structure of dual pairs provides a fairly tight formal link between these equations of different times and different physical phenomena.

We take the Hilbert space $\mathcal{H}$ on which ρ and ω are realized to be $L^2(\mathbf{R}^n)$. The Lie algebra $\mathfrak{h}$ is the span of the operators $\frac{\partial}{\partial x_j}$, ix_j, i. The creation and annihilation operators may be taken to be

$$a_j = \frac{1}{\sqrt{2}}\left(\frac{\partial}{\partial x_j} + x_j\right), \qquad a_j^+ = \frac{1}{\sqrt{2}}\left(-\frac{\partial}{\partial x_j} + x_j\right).$$

Thus $\mathfrak{W}$ is simply the algebra of all polynomial coefficient differential operators. And $\mathfrak{sp}_{\mathbf{C}}$ is the space of all differential operators of total degree 2, and symmetrized in the x's and $\frac{\partial}{\partial x}$'s. More explicitly it is the span of the operators

$$x_i x_j, \qquad \frac{1}{2}\left(x_i\frac{\partial}{\partial x_j} + \frac{\partial}{\partial x_j}x_i\right) = x_i\frac{\partial}{\partial x_j} + \frac{1}{2}\delta_{ij}, \qquad \frac{\partial^2}{\partial x_i \partial x_j}.$$

To explain the connection between the equations of the list (6.1) and dual pairs in Sp, it is probably simplest to start with Laplace's equation (6.1ii). This of course is just

$$\Delta\varphi = 0 \tag{6.2}$$

where φ is a smooth function on $\mathbf{R}^n$ and

$$\Delta = \sum_{i=1}^{n} \frac{\partial^2}{\partial x_i^2} \tag{6.3}$$

is the Laplace operator. The connection with dual pairs begins to emerge with the observations that (i) the Laplacian is a second order operator, and so belongs to $\mathfrak{sp}_C$, and (ii) it has a large symmetry group, namely O_n. If we look for other operators which commute with O_n, we immediately find multiplication by r^2, where

$$r^2 = \sum_{i=1}^{n} x_i^2 \tag{6.4}$$

is the square of the usual Euclidean length of a vector. We know the second order operators commuting with O_n will be a Lie algebra. Taking the commutator of Δ and r^2 we find

$$[\Delta, r^2] = 4\left(\sum_{i=1}^{n} x_i \frac{\partial}{\partial x_i} \right) + 2n. \tag{6.5}$$

The main part of this expression, namely

$$E = \sum_{i=1}^{n} x_i \frac{\partial}{\partial x_i} \tag{6.6}$$

is just the Euler degree operator: applied to a polynomial of degree d, it multiplies the polynomial by d. One can check that taking further commutators produces no new operators. The 3 operators, Δ, r^2, and $4E + 2n$, span a Lie algebra, which can be seen to be isomorphic to $\mathfrak{sp}_2$. In fact, the operators

$$\frac{\Delta}{2}, \quad -\frac{r^2}{2} \quad \text{and} \quad E + \frac{n}{2}$$

have the same commutation relations as the standard basis

$$\begin{bmatrix} 0 & 0 \\ 1 & 0 \end{bmatrix}, \quad \begin{bmatrix} 0 & 1 \\ 0 & 0 \end{bmatrix}, \quad \text{and} \quad \begin{bmatrix} 1 & 0 \\ 0 & -1 \end{bmatrix}$$

of $\mathfrak{sl}_2$. Thus we find ourselves dealing with the dual pair (O_n, SL_2).

We observe that in the current realization of $\mathfrak{W}$, the space $\mathcal{P}$ of polynomials is invariant by $\mathfrak{W}$. A complete analogue of Theorem 4.3 holds for the action of $\mathfrak{sp}_C$ on $\mathcal{P}$. We can apply the analogue (to which we will refer still as Theorem 4.3) to learn simultaneously about polynomial solutions of Laplace's equation and the structure of the polynomials as an O_n-module. On general principles we can write $\mathcal{P} = \sum_{j=1}^{\infty} \mathcal{G}_j$ where the $\mathcal{G}_j$ are the isotypic subspaces for the action of O_n on $\mathcal{P}$. Since the operators from $\mathfrak{sl}_2$ commute with O_n, they will preserve the spaces $\mathcal{G}_j$. Thus the $\mathcal{G}_j$ are joint modules for O_n and for $\mathfrak{sl}_2$. A first consequence of Theorem 4.3 is that each $\mathcal{G}_j$ is irreducible under the joint action of O_n and $\mathfrak{sl}_2$. Let $\mathcal{H}$ be the space of harmonic polynomials, the solutions of Laplace's equation. Again since O_n commutes with Δ we see that $\mathcal{H}$ will be a direct sum of its intersections with the $\mathcal{G}_j$. Set $\mathcal{H} \cap \mathcal{G}_j = \mathcal{H}_j$, so that $\mathcal{H} = \Sigma \, \mathcal{H}_j$. In the terminology of representation theory $\mathcal{H}_j$ is the space of "lowest weight vectors" for $\mathcal{G}_j$ considered as an $\mathfrak{sl}_2$ module. Further reasoning from Theorem 4.3 yields the following facts.

PROPOSITION 6.1. (a) *Each space* $\mathcal{H}_j$ *is irreducible under the action of* O_n. *It consists of the elements of lowest degree in* $\mathcal{G}_j$.

(b) *Each subspace of* $\mathcal{P}$ *which is invariant and irreducible under the action of* O_n *defines an* O_n-*module isomorphic to a unique* $\mathcal{H}_j$.

(c) *Each element of* $\mathcal{G}_j$ *has a unique expression as a sum*

$$\sum_{l \geqslant 0} r^l h_l, \qquad h_l \in \mathcal{H}_j.$$

These statements will be recognized as the essential results of the theory of spherical harmonics. Focusing on the Laplace operator and its solutions, we may summarize the situation as follows.

(a) The Laplace operator allows a large group of symmetries (namely O_n). This symmetry group is one member of a reductive dual pair in Sp. The Laplace operator is then an element of a "standard basis" for the Lie algebra of the second member of the dual pair (namely $\mathfrak{sl}_2$).

(b) The space of solutions of Laplace's equations are characterized as being vectors of a certain type relative to the action of $\mathfrak{sl}_2$ (the second member of the dual pair).

(c) Theorem 4.3 yields a precise description of the space of solutions of the Laplace's equation as module for O_n (the first member of the dual pair). In particular, this module is multiplicity-free.

We should remark also on the mathematical point that from the perspective of this example, Theorem 4.3 becomes a generalization of the classical theory of spherical harmonics.

The relation of the wave equation to dual pairs is very similar to that of Laplace's equation. To get the wave equation one replaces the Euclidean metric $\Sigma \, x_i^2$ on $\mathbf{R}^n$ with the Lorentz metric $\Sigma \, x_i^2 - t^2$ on $\mathbf{R}^{n+1}$. On the algebraic level, i.e., when dealing with polynomials, the theory of the two equations is identical. On the transcendental level, however, the two equations have different behavior, reflecting the fact that the usual orthogonal group O_n, the symmetry group of Δ, is compact, whereas the Lorentz group $O_{n,1}$, the symmetry group of the wave operator $\square = \Delta - \partial^2/\partial t^2$, is noncompact. The transcendental version of the theory of dual pairs can describe the more general solutions of the wave equation. In particular, it can account for Huyghens' Principle, and the fact that it holds when the number n of space dimensions is odd, but not when n is even [H3, H4].

7. Maxwell's equation and supersymmetry. To account for Maxwell's equations and Dirac's equation we must resort to a more sophisticated construction. However, the critical features of dual pairs carry over to the fancier set-up which, indeed, is already described in [H1] and [Ti].

We will give a detailed account only for Maxwell's equation. First, recall them in their classical form [Jc]. They involve an electric field E and a magnetic field H. Each of these is a 3-vector:

$$E = (E_1, E_2, E_3), \qquad H = (H_1, H_2, H_3),$$

and each of the components of E and H is a function of the space variables x, y, z and time t. Maxwell's equations are

$$\nabla \cdot E = \rho, \tag{7.1a}$$

$$\nabla \cdot H = 0, \tag{7.1b}$$

$$\nabla \times E + \partial H / \partial t = 0, \tag{7.1c}$$

$$\nabla \times H - \partial E / \partial t = J. \tag{7.1d}$$

Here ∇ is the standard formal differential operator of classical vector analysis:

$$\nabla = \left(\frac{\partial}{\partial x}, \frac{\partial}{\partial y}, \frac{\partial}{\partial z} \right).$$

The quantity ρ is the electrical charge density, and J is the current density. To keep the discussion simple we will consider only Maxwell's equation in empty space, so that $\rho = J = 0$.

To make contact with dual pairs, we must put equations (3.1) in a different form.[1] Define an *electromagnetic 2-form* on $\mathbf{R}^4$ by

$$\mathscr{E} = E_1 \, dx \wedge dt + E_2 \, dy \wedge dt + E_3 \, dz \wedge dt$$

$$+ H_1 \, dy \wedge dz - H_2 \, dx \wedge dz + H_3 \, dx \wedge dy. \tag{7.2}$$

Then equations (4.1b) and (4.1d) may be directly verified to be equivalent to the mathematically natural equation $d\mathscr{E} = 0$, i.e., the form $\mathscr{E}$ should be closed.

To account for the other two of Maxwell's equations, we recall that electromagnetism was responsible for the discovery of the Lorentz transformations, and of the Lorentz group $O_{3,1}$, the isometry group of the Minkowski metric $x^2 + y^2 + z^2 - t^2$ on $\mathbf{R}^4$. Maxwell's equations are well known to be invariant under $O_{3,1}$. Associated to an inner product on $\mathbf{R}^n$ such as the Minkowski metric is a *Hodge star map* [St, p. 23]

$$*: \Lambda^P(\mathbf{R}^n) \to \Lambda^{n-p}(\mathbf{R}^n). \tag{7.3}$$

---

[1] The earliest occurrence known to me of this form of Maxwell's equations is in [WH]. I thank C. H. Wilcox for this reference.

We will describe $*$ very briefly. Let $(\,,\,)$ be the inner product on $\mathbf{R}^n$. Then $(\,,\,)$ induces on each $\Lambda^p(\mathbf{R}^n)$ an inner product $(\,,\,)_p$. Choose an element $\nu \in \Lambda^n(\mathbf{R}^n)$ such that $(\nu, \nu)_n = 1$. Then $*$ is defined by the condition

$$\omega \wedge *(\omega') = (\omega, \omega')_p \nu, \qquad \omega, \omega' \in \Lambda^p(\mathbf{R}^n).$$

Note that $*$ (for the Minkowski metric) will map $\Lambda^2(\mathbf{R}^4)$ to itself. We can apply $*$ pointwise to differential forms. It turns out, and is easy to check, that equations (4.1a) and (4.1d) amount to the condition that $*\mathcal{E}$ is a closed form. Hence Maxwell's equations in empty space can be written

$$d\mathcal{E} = 0 = *^{-1}d*\mathcal{E}. \qquad (7.5)$$

In this formulation of Maxwell's equations, the electromagnetic field is not a scalar quantity, but an exterior form. Thus to bring these equations into the fold of dual pairs, we must extend the scope of these objects so that they also encompass exterior forms. This in fact was already done in [H1]. We sketch the essential features of the broader formalism.

Consider the exterior algebra $\Lambda^*(\mathbf{R}^m)$. This algebra has natural analogues of the operators of multiplication by x_i and of differentiation $\partial/\partial x_i$ on the polynomial algebra $\mathcal{P}(\mathbf{R}^m)$. Let dx_i, $1 \leq i \leq m$, form a basis for $\Lambda^1(\mathbf{R}^m)$. Then

$$\wedge_i: \omega \to dx_i \wedge \omega, \qquad \omega \in \Lambda^*(\mathbf{R}^m), \qquad (7.6a)$$

is clearly parallel to multiplication by x_j. The analogue of differentiation is known as *inner multiplication*, and denoted $\lrcorner_i$. Write a general exterior form as $\omega_1 + dx_i \wedge \omega_2$ where both ω_1 and ω_2 do not involve dx_i, that is, are formed from the dx_j with $j \neq i$. Then

$$\lrcorner_i(\omega_1 + dx_i \wedge \omega_2) = \omega_2. \qquad (7.6b)$$

In analogy with the CCR, formulas (2.4), the operators $\wedge_i$ and $\lrcorner_j$ satisfy *anticommutation relations* (CAR)

$$\{\lrcorner_i, \wedge_j\} = \lrcorner_i \wedge_j + \wedge_j \lrcorner_i = \delta_{ij},$$

$$\{\lrcorner_i, \lrcorner_j\} = 0 = \{\wedge_i, \wedge_j\}. \qquad (7.7)$$

It is furthermore not hard to check that the full matrix algebra $\mathrm{End}(\Lambda^*(\mathbf{R}^m))$ is generated by the $\wedge_i$ and $\lrcorner_j$, and the CAR form a full set of relations for the $\wedge_i$ and $\lrcorner_j$ as generators of $\mathrm{End}(\Lambda^*(\mathbf{R}^m))$. We will denote $\mathrm{End}(\Lambda^*(\mathbf{R}^m))$ viewed as an algebra generated by the $\wedge_i$ and $\lrcorner_j$ subject to the CAR (4.7) by $\mathcal{C}_m = \mathcal{C}$. The notation stems from the fact that this description of $\mathcal{C}$ is in fact the standard description of the *Clifford algebra* for the totally split inner product $(\,,\,)_s$ on $\mathbf{R}^{2m}$ given by

$$((x, y), (x', y'))_s = x \cdot y' + y \cdot x', \qquad x, x', y, y' \in \mathbf{R}^m, \qquad (7.8)$$

where $x \cdot y$ is as above the usual inner product on $\mathbf{R}^m$. Physicists will recognize the $\wedge_i$ and $\lrcorner_i$ as the creation and annihilation operators for fermions.

Thus the situation for the "antisymmetric variables" embodied in $\Lambda^*(\mathbf{R}^m)$ is entirely parallel to that for the more usual symmetric variables living in the polynomial algebra $\mathcal{P}(\mathbf{R}^n)$. Moreover we may combine the two types of variables. This may be done straightforwardly by simply splicing the two together in a tensor product $\mathcal{P}(\mathbf{R}^n) \otimes \Lambda^*(\mathbf{R}^m)$, or in a more organic way by a formalism of "graded inner product spaces" [Ti]. In either case one considers objects which are sums of products $p\omega$ where p is a polynomial on $\mathbf{R}^n$ and ω is an exterior form on $\mathbf{R}^m$. There is no need for m and n to be equal, but if they are, the space $\mathcal{P}(\mathbf{R}^n) \otimes \Lambda^*(\mathbf{R}^n)$ may be interpreted as the space of "polynomial coefficient differential forms" on $\mathbf{R}^n$.

The combined space $\mathcal{P}(\mathbf{R}^n) \otimes \Lambda^*(\mathbf{R}^m)$ admits as algebra of endomorphisms the combined algebra $\mathfrak{W}_n \otimes \mathcal{C}_m$. This is of course generated by the operators $\{x_i, \partial/\partial x_j\}_{i,j \leq n}$ and $\{\wedge_k, \lrcorner_l\}_{k,l \leq m}$, acting in the obvious way on the appropriate variables. By analogy with the theory described above for $\mathfrak{W}_n$, we consider the elements of $\mathfrak{W}_n \otimes \mathcal{C}_m$ which are of second total order in all these generators. We may distinguish 3 subspaces of this system:

(i) The operators of order two in the x_i and $\partial/\partial x_j$. This is simply the Lie algebra $\mathfrak{sp}_\mathbf{C}$ discussed above.

(ii) The operators of order two in the $\wedge_k$ and $\lrcorner_l$. By considerations very similar to those producing $\mathfrak{sp}_\mathbf{C}[J]$, one sees that these also form a Lie agebra, namely the orthogonal Lie algebra of the inner product $(,)_s$ defined in equation (7.8). We denote this Lie algebra by $\mathfrak{o}$.

(iii) The operators that involve one x_i or $\partial/\partial x_y$ and one $\wedge_k$ or $\lrcorner_l$. There are 4 types of these, namely

$$x_i \wedge_k, \quad x_i \lrcorner_l, \quad \frac{\partial}{\partial x_y} \wedge_k, \quad \text{and} \quad \frac{\partial}{\partial x_j} \lrcorner_l.$$

We have denoted the span of the ix_j and $\partial/\partial x_j$ by W. Let us denote the span of the $\wedge_k$ and $\lrcorner_l$ by U. Then it is clear that this collection of operators is naturally isomorphic to $W_\mathbf{C} \otimes U$, and we will identify the two spaces.

What sort of algebraic system do the above components form? The spaces $\mathfrak{sp}_\mathbf{C}$ and $\mathfrak{o}$, as already stated, form Lie algebras under the commutator law derived from $\mathfrak{W}$ and $\mathcal{C}$ respectively. Moreover, ad $\mathfrak{sp}$ and ad $\mathfrak{o}$ both preserve $W \otimes U$. The question is what multiplication law to assign to $W \otimes U$. It turns out (and again is a simple computation based on the CCR and CAR) that the anticommutator of two elements in $W_\mathbf{C} \otimes U$ lies in $\mathfrak{sp}_\mathbf{C} \oplus \mathfrak{o}$. Thus $\mathfrak{os} = \mathfrak{sp}_\mathbf{C} \oplus \mathfrak{o} \oplus (W_\mathbf{C} \otimes U)$ forms what is known as a *graded Lie algebra* or a *Lie superalgebra* [**Kc, FK**]. The objects have been of interest to physicists recently in connection with the theory of supergravity [**NF**]. In fact $\mathfrak{os}_{2m,\,2n} = \mathfrak{os}$ is one of the simple Lie superalgebras usually called the *orthosymplectic algebra*.

REMARK. One of the best-known elements of $\mathfrak{os}_{2n,2n}$ is

$$d = \sum_{i=1}^{n} \wedge_i \frac{\partial}{\partial x_i}$$

the familiar operator of exterior differentiation; it lies in $W_{\mathbf{C}} \otimes U$. (This shows that in this formalism the operator d is second order, rather than first order as it is in the usual sense, and increases the similarity between d and the second order operators discussed in §3.) So also does the formal adjoint

$$\delta = \sum_{i=1}^{n} x_i \lrcorner_i.$$

The operators d and δ satisfy the anticommutation relation

$$d\delta + \delta d = \deg_{\mathcal{P}} \oplus \deg_{\Lambda}$$

where $\deg_{\mathcal{P}}$ is the usual Euler degree operator on $\mathcal{P}$, and $\deg_{\Lambda}$ is its analogue on Λ^*. This relation is significant in the proof of the Hilbert Syzygy Theorem [**J**, Chapter 6].

One can define dual pairs in $\mathfrak{os}$ in analogy with pairs in an ordinary Lie algebra. It is to such a dual pair that Maxwell's equations are related. Specifically, let the Lorentz group $O_{3,1}$ act on $\mathbf{R}^4$ in the usual way, and consider the induced action on $\mathcal{P}(\mathbf{R}^4) \otimes \Lambda^*(\mathbf{R}^4)$, the polynomial coefficient differential forms. Then the fixed points for $O_{3,1}$ acting on $\mathfrak{os}_{8,8}$ is also an orthosymplectic Lie superalgebra, namely $\mathfrak{os}_{2,2} \cong \mathfrak{sl}_2 \oplus \mathfrak{o}_2 \otimes (\mathbf{R}^2 \otimes \mathbf{R}^2)$ and $(\mathfrak{o}_{3,1}, \mathfrak{os}_{2,2})$ form a dual pair in $\mathfrak{os}_{8,8}$. The operators spanning $\mathfrak{os}_{2,2}$ are as follows.

$\mathfrak{sl}_2$:   the Minkowski metric, the wave operator or D'Alembertian $\Box$, the (modified) Euler degree operator $E + 2$.   (7.9a)

$\mathfrak{o}_2$:   the (modified) exterior degree operator   (7.9b)

$\mathbf{R}^2 \otimes \mathbf{R}^2$:   the exterior differential d, its natural adjoint δ, and the conjugates of these by the Lorentz Hodge star, $*^{-1}d*$ and $*^{-1}\delta*$.   (7.9c)

Thus the two operators d and $*^{-1}d*$ which intervene in Maxwell's equations (4.5) are members of the natural basis for the "super" part of the Lie superalgebra $\mathfrak{os}_{2,2}$, the other member of the dual pair defined by the natural section of the Lorentz group on differential forms on $\mathbf{R}^4$. The parallel with the Laplace and wave equations as discussed in §3 is clear. Also, the well-known connection between Maxwell's equations and the wave equation is provided naturally by the multiplication in $o\mathfrak{s}_{2,2}$; one has the identity

$$\{d, *^{-1}d*\} = d(*^{-1}d*) + (*^{-1}d*)d = \square. \tag{7.10}$$

Thus although physicists became explicitly interested in Lie superalgebras only several years ago, these algebras actually entered physics in a disguised form when Maxwell formulated his theory of electromagnetism over 100 years ago.

8. Concluding remarks. To deal with Dirac's equation, the procedure is similar to that for Maxwell's equations. Instead of differential forms on $\mathbf{R}^4$, one deals with spinors. The associated algebra is $\mathfrak{W}_4 \otimes \tilde{\mathcal{C}}$ where $\tilde{\mathcal{C}}$ is the Clifford algebra of $\mathbf{R}^4$ equipped with the Lorentz metric. The dual pair is $(o_{3,1}, o\mathfrak{s}_{2,1})$. Here $o_{3,1}$ is again the Lorentz Lie algebra, and $o\mathfrak{s}_{2,1} = \mathfrak{sl}_2 \oplus (\mathbf{R}^2 \otimes \mathbf{R}^1)$. The $\mathfrak{sl}_2$ is the same as in (4.9a); the o_1 is actually trivial; and the "super part" $\mathbf{R}^2 \otimes \mathbf{R}^1$ has a basis of two elements, one of which is the Dirac operator.

The list (6.1) by no means exhausts the set of equations on which light can be shed by the oscillator representation and the theory of dual pairs. We saw in §5 that the energy degeneracies of the harmonic oscillator Hamiltonian were accounted for by the pair (U_n, U_1). Also the heat equation is quite closely related to the oscillator representation, but in a rather different way from those of the list (6.1). Also, if we consider nonzero eigenvalue problems, instead of just the kernels of the operators of (6.1), then we are talking about the Helmboltz and Klein-Gordan equations. The Dirac equation is actually usually formulated as a nonzero eigenvalue problem. In the terminology of representation theory, the distinction between the Laplace and Helmboltz equation, or the wave and Klein-Gordan equations, is the difference between N-fixed vectors and Whittaker vectors. Both are well-studied objects.

A main point of the discussion of §§6–7 is that all the equations of list (6.1) are both very important in physics and related to dual pairs in essentially the same way. The theory of dual pairs provides a common link between these equations, and differences between the equations may be accounted for by differences between the dual pairs to which they are associated.

Another equation which is much less well known than those of list (6.1), but which relates to dual pairs in basically the same way, is Weyl's equation for the neutrino [**Jo**]. This is essentially a refinement of the zero-mass Dirac's equation. For Weyl's equation, a relatively subtle property of dual pairs in the Pin group, the 2-fold cover of the orthogonal group, comes into play. Here the disconnectedness of the orthogonal group plays an important role.

Finally, I should remark that there appear to be other connections between dual pairs and physics than the ones discussed here (cf. [E, I, JV]), but I do not know enough about them to formulate them in detail.

REFERENCES

[A] E. Artin, *Geometric algebra*, Interscience, New York, 1957.

[Dr] P. Dirac, *A remarkable representation of the 3 + 2 de Sitter group*, J. Math. Phys. **4** (1963), 901–909.

[Dx] J. Dixmier, *Von Neumann algebras*, North-Holland, New York, 1981.

[E] M. Englefield, *Group theory and the Coulomb problem*, Interscience, New York, 1972.

[EHW] T. Enright, R. Howe, and N. Wallach, *A classification of unitary highest weight modules*, (Proc. 1982 Conf. on Reductive Groups, Park City, Utah) (to appear).

[EP] T. Enright and R. Parthasarathy, *A proof of a conjecture of Kashiwara and Vergne*, Non-Commutative Harmonic Analysis (Proceedings, Marseilles, 1980), J. Carmona and M. Vergne, eds., Lecture Notes in Math., vol. 880, Springer-Verlag, Berlin and New York, 1981, pp. 74–90.

[FF1] M. Flato and C. Fronsdal, *One massless particle equals two Dirac singletons*, Lectures in Math. Phys., vol. 2, Riedel, Dordrecht, Holland, 1978, pp. 421–426.

[FF2] _____, *On Dis and Racs*, Phys. Let. **97B** (1980), 236–240.

[FK] P. Freund and I. Kaplansky, *Simple supersymmetries*, J. Math. Phys. **17** (1976), 228–231.

[Ge] S. Gelbart, *Examples of dual reductive pairs*, Automorphic Forms, Representations and L-functions. Part I, A. Borel and W. Casselman, eds., Proc. Sympos. Pure Math., vol. 33, Amer. Math. Soc., Providence, R.I., 1979, pp. 287–296.

[GEKi] I. Gelfand and A. Kirillov, *Sur les corps liés aux algèbres envellopantes des algèbres de Lie*, Inst. Hautes Études Sci. Publ. Math. **31** (1966), 5–20.

[GK] K. Gross and R. Kunze, *Bessel functions and representation theory*. II, J. Funct. Anal. **25** (1977), 1–49.

[Gu] M. Günaydin, *Unitary realizations of the non-compact symmetry groups of supergravity*, Ref. Th. 332—CERN, preprint, 1981.

[GS] M. Günaydin and C. Saglioğlu, *Oscillator like unitary representations of non-compact groups with a Jordan structure and the non-compact groups of supergravity*, Ref. Th. 3209—CERN, preprint, 1918.

[He] S. Helgason, *Differential geometry and symmetric spaces*, Academic Press, New York, 1962.

[H1] R. Howe, *Remarks on classical invariant theory*, preprint.

[H2] _____, *θ-series and invariant theory*, Automorphic Forms, Representations, and L-Functions. Part I, Proc. Sympos. Pure Math., vol. 33, Amer. Math. Soc., Providence, R.I., 1979, pp. 275–285.

[H3] _____, *On Huyghens' Principle*, preprint.

[H4] _____, *On the role of the Heisenberg group in harmonic analysis*, Bull. Amer. Math. Soc. (N.S.) **3** (1980), 821–843.

[H5] _____, *Transcending classical invariant theory*, preprint.

[H6] _____, *On a notion of rank for unitary representations of the classical groups*, Harmonic Analysis and Group Representations (Proc. C.I.M.E. II ciclo 1980), A. Figà-Talamanca, ed., Liguori Editore, Naples, 1980, 223–332.

[H7] _____, *Reciprocity laws in the theory of dual pairs* (Proc. 1982 Conf. on Reductive Groups, Park City, Utah) (to appear).

[HW] L. Hugheston and R. Ward., eds., *Advances in twistor theory*, Pitman, San Francisco, 1974.

[I] F. Iachello, *Dynamical symmetries in nuclei*, Lecture, 1978 Group Theory Conf., Austin, Texas, preprint.

[Jc] J. Jackson, *Classical electrodynamics*, Wiley, New York, 1962.

[J] N. Jacobson, *Basic algebra*. II, W. H. Freeman, San Francisco, 1979.

[Jk] H. Jakobsen, *On singular holomorphic representations*, Invent. Math. **62** (1980), 67-78.

[JV] H. Jakobsen and M. Vergne, *Wave and Dirac operators and representation of the conformal group*, J. Funct. Anal. **24** (1977), 52-106.

[Jo] R. Jost, *General theory of quantized fields*, Lectures in Appl. Math., vol. 4, Amer. Math. Soc., Providence, R.I., 1965.

[Kc] V. Kac, *Lie superalgebras*, Adv. in Math. **26** (1977), 8-96.

[KV] M. Kashiwara and M. Vergne, *On the Segal-Shale-Weil representations and harmonic polynomials*, Invent. Math. **44** (1978), 1-48.

[Ku] S. Kudla, *Holomorphic Siegel modular forms attached to SO(n, 1)*, preprint.

[MS] G. Mack and A. Salam, *Finite-component field representations of the conformal group*, Ann. of Phys. **53** (1969), 174-202.

[MT] G. Mack and I. Todorov, *Irreducibility of the ladder representations of U(2,2) when restricted to the Poincare subgroup*, J. Math. Phys. **101** (1969), 2078-2085.

[My] G. Mackey, *Unitary representations of group extensions*. I, Acta Math. **99** (1958), 265-311.

[MQ] M. Moshinsky and C. Quesne, *Linear canonical transformations and their unitary representations*, J. Math. Phys. **12** (1971), 1772-1780.

[NF] P. Van Nieuwenhuizen and D. Freedman, eds., *Supergravity*, North Holland, Amsterdam, New York, 1971.

[O] E. Onofri, *Dynamical quantization of the Kepler manifold*, J. Math. Phys. **17** (1976), 401-408.

[P] R. Parthasarathy, *Criteria for the unitarizability of some highest weight modules*, Proc. Indian Acad. Sci. **89** (1980), 1-24.

[R] W. Rossman, *Some Weil representations*, preprint.

[Sa] M. Saito, *Représentations unitaires des groupes symplectiques*, J. Math. Soc. Japan **24** (1972), 232-251.

[Se] J. P. Serre, *Lie algebras and Lie groups*, Benjamin, New York, 1965.

[Sh] D. Shale, *Linear symmetries of free boson fields*, Trans. Amer. Math. Soc. **103** (1962), 149-167.

[St] S. Sternberg, *Lectures on differential geometry*, Prentice-Hall, Englewood Cliffs, N.J., 1964.

[SW] S. Sternberg and J. Wolf, *Hermitian Lie algebras and metaplectic representations*. I, Trans. Amer. Math. Soc. **238** (1978), 1-43.

[Ti] H. Tilgner, *Graded generalizations of Weyl and Clifford algebras*, J. Pure Appl. Algebra **10** (1977), 163-168.

[WA1] A. Weil, *Sur certains groupes d'operateurs unitaires*, Acta Math. **111** (1964), 143-211.

[WA2] _____, *Sur la formule de Siegel dans la théorie des groupes classiques*, Acta Math. **113** (1965), 1-87.

[WH] H. Weyl, *Die naturlichen Randwerteaufgaben in Aussenraum für Strahlungsfelder beliebiger Dimension und beliebigen Ranges*, Math. Z. **56** (1952), 105-119.

DEPARTMENT OF MATHEMATICS, YALE UNIVERSITY, NEW HAVEN, CONNECTICUT 06520

Lectures in Applied Mathematics
Volume **21**, 1985

Langlands Classification and
Unitary Dual of $SU(2,2)$

A. W. Knapp[1]

This paper continues a theme addressed by some seminar participants: investigation of the irreducible unitary representations of semisimple Lie groups. For the group

$$G = SU(2,2) = \left\{ g \in SL(4,\mathbf{C}) \mid g^* \begin{pmatrix} 1 & & & \\ & 1 & & \\ & & -1 & \\ & & & -1 \end{pmatrix} g = \begin{pmatrix} 1 & & & \\ & 1 & & \\ & & -1 & \\ & & & -1 \end{pmatrix} \right\},$$

which is locally isomorphic to the conformal group of space-time, the classification problem has been completely solved in joint work with B. Speh [8].

The detailed answer appears in [8], and some parts of that answer will be reproduced presently. Qualitatively there are no surprises, and moreover the argument shows that all the unitary representations are unitary for simple reasons. This is so even for the ladder representations that have been studied by a number of mathematical physicists!

Instead of concentrating here on the answer to the classification problem, we shall emphasize the approach to such a problem. In particular, the Langlands classification (of irreducible "admissible" representations) and its method of proof will be of special interest because the classification and proof may well have independent applications in mathematical physics and they are available in full only as unpublished

1980 *Mathematics Subject Classification.* Primary 22E45.

[1] Supported by NSF Grant MCS 80-01854. The research on the unitary dual of $SU(2,2)$ was joint work with B. Speh [8].

notes [**6, 7, 9, 3, 14, 11**]. Partial proofs appear in published form in Warner [**15**], Borel and Wallach [**2**] and Miličić [**12**].

Historically, approaches to classification of the irreducible unitary representations of semisimple groups have always proceeded in two steps:

(1) Get control of some larger class of representations (such as the irreducible "admissible" representations defined below in §1).

(2) Decide which ones are unitary.

In fact, in some groups, seemingly only a little more complicated than $SU(2, 2)$, the pattern of unitary parameters is much more complicated than in $SU(2, 2)$. Thus either this two-step approach is forced on us as a nontrivial subdivision of the problem, or else people are completely off track in the current thinking of what kinds of variables to use as parameters.

Step (1) above is nowadays solved in considerable generality—completely in connected semisimple groups G having faithful matrix representations. The relevant theorem is the Langlands classification [**9**] obtained in 1973. This result is not simply a generalization to G of earlier techniques, and we shall contrast the earlier techniques with the Langlands approach in §1.

Step (2) is the relatively new part for $SU(2, 2)$. For $SU(2, 2)$ this step involves only a little more than an exercise with known techniques. The techniques succeed because of the small size of the group. Aspects of this step are the subject of §2.

1. Admissible representations of $SU(2, 2)$. Let K be the maximal compact subgroup of $G = SU(2, 2)$ given by

$$K = S(U(2) \times U(2)) = SU(2, 2) \cap U(4) = \left\{ \begin{pmatrix} * & 0 \\ 0 & * \end{pmatrix} \right\}.$$

A (continuous) representation π of G on a (complex) Hilbert space V, say with K acting unitarily, is said to be *admissible* if each irreducible representation of K occurs only finitely often in the restriction $\pi|_K$. It is known [**5**] that irredicible unitary representations are admissible.

In this case we can pass to the Lie algebra $\mathfrak{g}$ of G, which acts on

$$V_K = \{ v \in V | \pi(K)v \text{ spans a finite-dimensional space} \}.$$

The vector space V_K is called the space of *K-finite vectors* in V, and the representation of $\mathfrak{g}$ on V_K provides an instance of an *admissible representation of* $(\mathfrak{g}, K)$ (= a complex vector space on which $\mathfrak{g}$ and K both have representations, in compatible fashion, with every vector K-finite and with each irreducible representation of K occurring only finitely often).

A number of properties of admissible representations were established by Harish-Chandra [5]. The representation of G on V is irreducible if and only if the representation of $(\mathfrak{g}, K)$ on V_K is algebraically irreducible. An irreducible admissible representation of $(\mathfrak{g}, K)$ comes from a unitary representation of G if and only if the representation of $(\mathfrak{g}, K)$ admits an inner product such that $\mathfrak{g}$ acts by skew-Hermitian operators, and any two such irreducible unitary representations of G leading to the same irreducible representation of $(\mathfrak{g}, K)$ are unitarily equivalent. (We summarize this condition on the representation of $(\mathfrak{g}, K)$ by saying it is *infinitesimally unitary*.)

It is convenient to define equivalence of admissible representations of G on the Lie algebra level, saying that two admissible representations of G are *infinitesimally equivalent* if the corresponding representations of $(\mathfrak{g}, K)$ are algebraically equivalent. Every irreducible admissible representation of $(\mathfrak{g}, K)$ arises from an admissible representation of G, according to a theorem of Lepowsky [10]. Let $\mathfrak{k}$ be the Lie algebra of K.

The traditional approach to Step (1) of the introduction would be as follows: Write $\mathfrak{g} = \mathfrak{k} \oplus \mathfrak{p}$, where $\mathfrak{p}$ is the set of Hermitian members of $\mathfrak{g}$, and complexify to get $\mathfrak{g}^C = \mathfrak{k}^C \oplus \mathfrak{p}^C$. We can assume that we understand the action of $\mathfrak{k}^C$ in an irreducible representation, and we want to understand the action of $\mathfrak{p}^C$. In the first place, this action has to be consistent with the bracket relations in $\mathfrak{g}^C$. (Recall that $[\mathfrak{k}^C, \mathfrak{p}^C] \subseteq \mathfrak{p}^C$ and $[\mathfrak{p}^C, \mathfrak{p}^C] \subseteq \mathfrak{k}^C$.) But also it must satisfy some further relations because the center $Z(\mathfrak{g}^C)$ of the universal enveloping algebra (the generalized Casimir elements) must act as scalars. In the case of $SU(2,2)$, $Z(\mathfrak{g}^C)$ is a full polynomial algebra (with no relations other than commutativity) in three operators that one could write down explicitly if necessary; so we essentially have three additional relations beyond the bracket relations. The idea is to play these relations off against each other and see what happens.

This calculation seems to be just barely possible for $SU(2,2)$ (cf. Angelopoulos [1]), but the result that is obtained does not obviously fall into any general pattern.

The Langlands approach to Step (1) is quite different. One begins by constructing models for some irreducible admissible representations; these will be denoted $J(P, \sigma, \nu)$. To construct these representations, let $G = KA_{\min}N_{\min}$ be an Iwasawa decomposition of G. Here $A_{\min}$ and $N_{\min}$ have Lie algebras $\mathfrak{a}_{\min}$ and $\mathfrak{n}_{\min}$, respectively, where

$$\mathfrak{a}_{\min} = \left\{ \begin{pmatrix} 0 & 0 & s & 0 \\ 0 & 0 & 0 & t \\ s & 0 & 0 & 0 \\ 0 & t & 0 & 0 \end{pmatrix} \right\} \tag{1.1}$$

and where $\mathfrak{n}_{\min}$ is a certain nilpotent subalgebra of $\mathfrak{g}$ of dimension 6. We define $M_{\min}$ to be the centralizer of $A_{\min}$ in K; this is the group generated by the scalar fourth roots of unity and the circle group

$$T = \left\{ \begin{pmatrix} e^{i\theta} & & & \\ & e^{-i\theta} & & \\ & & e^{i\theta} & \\ & & & e^{-i\theta} \end{pmatrix} \right\}. \tag{1.2}$$

Then $P_{\min} = M_{\min} A_{\min} N_{\min}$ is a closed subgroup of G that plays the same role for $SU(2,2)$ that the upper triangular group plays for $SL(n,\mathbf{R})$ or $SL(n,\mathbf{C})$.

There is a standard series of admissible representations $U(P_{\min}, \sigma, \nu)$ induced from $P_{\min}$ known as the *nonunitary principal series*. The parameter σ is a unitary character of the compact abelian group $M_{\min}$, acting in the one-dimensional complex vector space $V^\sigma \cong \mathbf{C}$. The parameter ν is a complex-valued real-linear functional on $\mathfrak{a}_{\min}$, which we may write in coordinates as

$$\nu = cf_1 + df_2, \tag{1.3}$$

where f_1 and f_2 on the matrix (1.1) are s and t, respectively. Then $\sigma \otimes e^\nu \otimes 1$ is a representation of $P_{\min}$ (nonunitary unless ν is imaginary), and we let

$$U(P_{\min}, \sigma, \nu) = \operatorname{ind}_{P_{\min}}^G (\sigma \otimes e^\nu \otimes 1).$$

The conventions in the definition of $U(P_{\min}, \sigma, \nu)$ are that G is to act on the left on the representation space in the form

$$\left\{ f \in L^2(K, V^\sigma) \mid f(km) = \sigma(m)^{-1} f(k) \text{ for } k \in K, m \in M_{\min} \right\}, \tag{1.4}$$

and the parameters are arranged so that unitary data lead to unitary representations U ("Mackey induction").

In the notation of (1.3), let $\rho = 3f_1 + f_2$. If ν in (1.3) satisfies $\operatorname{Re} c > \operatorname{Re} d > 0$ and f and g are continuous, then one has the limit formula

$$\lim_{a \to +\infty} e^{-(\nu - \rho)\log a} \big(U(P_{\min}, \sigma, \nu, ma) f, g \big) = \left(\int_{N_{\min}^*} \sigma(m) f(n^*) \, dn^* \right) \overline{g(1)} \tag{1.5}$$

for m in $M_{\min}$. Here "*" refers to adjoints, and $a \to +\infty$ means that $(f_1 - f_2)(\log a) \to +\infty$ and $f_2(\log a) \to +\infty$. Equation (1.5) is easy to see on a formal level by changing variables from K to $N_{\min}^*$ and passing to the limit, provided one is content to omit the justification of the interchange of limit and integral. We shall rewrite (1.5) a little. We define an operator by

$$A(P_{\min}^* : P_{\min} : \sigma : \nu)f(x) = \int_{N_{\min}^*} f(xn^*) dn^*. \tag{1.6}$$

Under our assumption on ν that Re $c >$ Re $d > 0$, one can show that this is a bounded operator on the space (1.4) that satisfies

$$U(P_{\min}^*, \sigma, \nu)A(P_{\min}^* : P_{\min} : \sigma : \nu) = A(P_{\min}^* : P_{\min} : \sigma : \nu)U(P_{\min}, \sigma, \nu). \tag{1.7}$$

We can then rewrite (1.5) as

$$\lim_{a \to +\infty} e^{-(\nu - \rho)\log a}\big(U(P_{\min}, \sigma, \nu, ma)f, g\big)_{L^2(K, V^\sigma)}$$
$$= \big(\sigma(m)\big(A(P_{\min}^* : P_{\min} : \sigma : \nu)f\big)(1), g(1)\big)_{V^\sigma}. \tag{1.8}$$

It is easy to see from (1.8) that if f is not in the kernel of $A(P_{\min}^* : P_{\min} : \sigma : \nu)$, then f is cyclic for $U(P_{\min}, \sigma, \nu)$. It follows readily that $U(P_{\min}, \sigma, \nu)$ has a unique irreducible quotient $J(P_{\min}, \sigma, \nu)$, known as the *Langlands quotient*, and that $J(P_{\min}, \sigma, \nu)$ is isomorphic to the image of $A(P_{\min}^* : P_{\min} : \sigma : \nu)$. In view of (1.7), we can therefore regard $J(P_{\min}, \sigma, \nu)$ as operating in a subspace of $U(P_{\min}^*, \sigma, \nu)$.

A version of this construction works when $P_{\min}$ is replaced by any larger closed subgroup of G. There are four such subgroups P (including $P_{\min}$ and G) in the case of $SU(2, 2)$, and they are listed explicitly in [8, p. 44]. Each can be written as $P = MAN$ with $M \supseteq M_{\min}$, $A \subseteq A_{\min}$ and $N \subseteq N_{\min}$. Except in the case of $P_{\min}$, M will be noncompact. We form

$$U(P, \sigma, \nu) = \text{ind}_P^G(\sigma \otimes e^\nu \otimes 1),$$

where σ is an irreducible unitary representation of M whose $(K \cap M)$-finite matrix coefficients are in $L^{2+\varepsilon}(M)$ for every $\varepsilon > 0$ (i.e., σ is *irreducible tempered*) and where ν is a complex-valued real-linear functional on the Lie algebra of A such that Re ν satisfies a suitable positivity condition. Then (1.6), with the subscripts "min" erased, is a convergent integral for K-finite f, (1.7) holds at least on the Lie algebra level when the subscripts "min" are erased, and (1.8) is valid for f and g K-finite when the subscripts "min" are erased. In the same way it then follows that $U(P, \sigma, \nu)$ has a unique irreducible quotient $J(P, \sigma, \nu)$, the "Langlands quotient," and $J(P, \sigma, \nu)$ is isomorphic to the image of $A(P^* : P : \sigma : \nu)$.

Langlands showed conversely that all irreducible admissible representations of G are obtained this way, and each arises from this construction only once. A precise statement follows. We shall refer to the relevant triple (P, σ, ν) as the "Langlands parameters" of an irreducible admissible representation.

THEOREM (LANGLANDS CLASSIFICATION FOR $SU(2, 2)$). *The equivalence classes of irreducible admissible representations of $SU(2, 2)$, under infinitesimal equivalence, stand in one-one correspondence with triples (P, σ, ν), where*

$P = MAN$ is a closed subgroup of $SU(2, 2)$ containing $P_{\min}$,

σ is an irreducible tempered representation of M, two such representations being regarded as the same if they are unitarily equivalent,

ν is a complex-valued real-linear functional on the Lie algebra of A with $\operatorname{Re} \nu$ in the open positive Weyl chamber.

The correspondence is that (P, σ, ν) corresponds to the class of $J(P, \sigma, \nu)$.

We shall give the idea behind the proof of completeness. Let π be irreducible admissible on a Hilbert space V, and consider a finite K-stable block of matrix coefficients, which we write as $E_2\pi(x)E_1$, where E_1 and E_2 are orthogonal projections. This function has known behavior on the left and right under K. Also, if we regard each member X of the universal enveloping algebra of $\mathfrak{g}^{\mathbf{C}}$ as a left-invariant differential operator on G, then we have

$$X(E_1\pi(x)E_2) = E_1\pi(x)\pi(X)E_2.$$

For X in the center $Z(\mathfrak{g}^{\mathbf{C}})$, $\pi(X)$ is a scalar, and it follows that

$$Z(E_1\pi(x)E_2) = c(Z)E_1\pi(x)E_2 \quad \text{for } Z \in Z(\mathfrak{g}^{\mathbf{C}}). \qquad (1.9)$$

Now $G = K\overline{A_{\min}^+}K$, and it turns out that one can use the transformation laws under K to rewrite the system (1.9) as a system of differential equations on $A_{\min}^+$ with variable coefficients. Solutions of the rewritten system will be functions on $A_{\min}^+$ whose value at each point is in the space W of linear maps from $\operatorname{image}(E_1)$ to $\operatorname{image}(E_2)$ commuting with the action of $M_{\min}$. One can arrange that the coefficient functions in the rewritten system have values in the space of linear maps from W into W.

The rewritten system behaves as if it has a regular singular point at $+\infty$ in $A_{\min}$, except that the domain is two-dimensional. All solutions have series expansions (about $+\infty$) with coefficients in W, and the space of solutions is finite-dimensional.

Any leading term of a solution leads in a natural way to an imbedding of the K-finite vectors of π as a subrepresentation in some $U(P^*, \sigma, \nu)$, hence to a realization as $J(P, \sigma, \nu)$. In more detail, there are only finitely many candidates for growth/decay rates on $A_{\min}$ of leading terms, independently of E_1 and E_2. Thus fix v' in V_K. For any v in V_K choose E_1

and E_2 with $E_1 v = v$ and $E_2 v' = v'$. Then the analysis above enables us to expand $(\pi(x)v, v')$ in series and to pick off the coefficient of the leading term we are studying. The result is a linear functional l on V_K with good behavior relative to $\mathfrak{a}_{\min}$ and with $l(\pi(\mathfrak{n}^*_{\min})V_K) = 0$. Formally we complete the argument by mapping v in V_K to the function f_v on G given by $f_v(x) = l(\pi(x)^{-1}v)$ to obtain an imbedding of V_K in an induced representation from $A_{\min}N^*_{\min}$ to G. Bringing in the $M_{\min}$ behavior, we extract a $\mathfrak{g}$-commuting mapping of V_K into the space of some $U(P^*_{\min}, \sigma_0, \nu_0)$ with $\text{Re } \nu_0$ in the closed positive Weyl chamber. Separating out directions in which $\text{Re } \nu_0 = 0$ and building a bigger M subgroup from them, we obtain a $\mathfrak{g}$-commuting mapping of V_K into the space of some $U(P^*, \sigma, \nu)$ with $\text{Re } \nu$ in the open positive Weyl chamber and with σ irreducible tempered, provided the leading term that we use is suitably extremal.

2. Irreducible unitary representations of $SU(2,2)$. In view of the results mentioned in §1, the problem of classifying the irreducible unitary representations of $SU(2,2)$ comes down to deciding which Langlands quotients $J(P, \sigma, \nu)$ are infinitesimally unitary. This decision is simplified by the following general facts:

(1) If $J(P, \sigma, \nu)$ is infinitesimally unitary, then *a fortiori* $J(P, \sigma, \nu)$ has a nonzero invariant Hermitian form. The existence of such a form is equivalent with the condition that $J(P, \sigma, \nu)$ be infinitesimally equivalent with its complex contragredient, which can be realized as $J(P^*, \sigma, -\bar{\nu})$. This equivalence forces a conjugacy of (P, σ, ν) with $(P^*, \sigma, -\bar{\nu})$, and we conclude that if $J(P, \sigma, \nu)$ is infinitesimally unitary, then there exists w in the normalizer of A in K with $wPw^{-1} = P^*$, $w\sigma \cong \sigma$ and $w\nu = -\bar{\nu}$.

(2) If $J(P, \sigma, \nu)$ has a nonzero invariant Hermitian form, then the form is unique up to a scalar and is given by a simple modification of $(A(P^* : P : \sigma : \nu)f, g)$. The question of unitarity is then whether this specific form is semidefinite.

(3) There are some standard techniques to make the decision in many cases whether the form in (2) is semidefinite. We list two of them.

(a) Continuity: If the operator appearing in the form is definite for one value of ν (e.g., if it is the identity), then it must remain definite on any connected set where the symmetry conditions of (1) hold as long as the operator remains invertible on the K-finite vectors. And the operator is invertible until some reducibility occurs for $U(P, \sigma, \nu)$.

(b) Boundedness of matrix coefficients: The K-finite matrix coefficients are unaffected by changing the inner product for an admissible representation, and they must be bounded for a unitary representation. Hence the K-finite matrix coefficients of $J(P, \sigma, \nu)$ are bounded if $J(P, \sigma, \nu)$ is infinitesimally unitary. The limit relation (1.5), rewritten with P in place of $P_{\min}$, enables us to conclude that $J(P, \sigma, \nu)$ cannot be infinitesimally unitary if Re ν lies outside a certain bounded region.

Now let us return to $G = SU(2, 2)$. We defined a circle subgroup T in (1.2), and we let γ be the diagonal matrix

$$\gamma = \mathrm{diag}(1, -1, 1, -1).$$

Then

$$M_{\min} = T \oplus \{1, \gamma\}, \tag{2.1}$$

and the dual group is correspondingly parametrized by

$$\hat{M}_{\min} \leftrightarrow \{(n, \pm)\}. \tag{2.2}$$

There are four closed subgroups P containing $P_{\min}$, and we shall discuss unitarity of some of the representations attached to two of them. (For a full discussion of unitarity in all cases, see [8, Main Theorem and §5].)

For one of them, $P = MAN$ has dimension 10 with

$$M = SL(2, \mathbf{R}) \oplus T,$$

with A one-dimensional, and with N equal to a Heisenberg group of dimension 5. We discuss only the representations σ of M given by

$$\sigma \leftrightarrow (D_k^{\pm}, n),$$

where $D_k^{\pm}$ is a discrete series ($k \geq 2$) or limit of discrete series ($k = 1$) of $SL(2, \mathbf{R})$ and where n refers to a character of the circle group T. Here dim $A = 1$, and general fact (1) at the start of this section says we may take ν to be real-valued (and equal to a positive multiple of f_1). The unitary points among such ν are given by $\nu = cf_1$ for the following values of c:

$$
\begin{array}{ll}
0 < c \leqslant 1 & \text{if } k \equiv n \bmod 2, \\
0 < c \leqslant 2 & \text{if } |n| = k - 1, \\
\text{no } c & \text{in the remaining cases.}
\end{array}
$$

The Langlands quotients that occur at the endpoints are of special interest. For $c = 2$ and $|n| = k - 1$, $J(P, \sigma, \nu)$ is a ladder representation, and all ladder representations (except the trivial representation) are of this form. The usual proofs (e.g., [4]) that the ladder representations are unitary use complex variable theory and/or the Fourier transform. Here we obtain the unitarity as a consequence of general fact (3a) after seeing that the induced representation at $\nu = 0$ is irreducible.

For $c = 1$ and $k \equiv n \bmod 2$, some of the representations $J(P, \sigma, \nu)$ are highest weight representations (but not of ladder type), and others are not.

Now let us consider $P_{\min}$. The group $M_{\min}$ and its characters are given in (2.1) and (2.2). Let us suppose $\nu = cf_1 + df_2$ is real with $c > d > 0$. For the character $\sigma \leftrightarrow (0, +)$, the unitary points are $(c, d) = (3, 1)$, where the trivial representation of G occurs, and the points with $c \leqslant 1$; those with $c = 1$ arise from degenerate complementary series induced from the 10-dimensional MAN with the trivial representation on M. For the character $\sigma \leftrightarrow (0, -)$, the unitary points are the ones with $c + d \leqslant 2$; those with $c + d = 2$ arise from degenerate complementary series induced from the 11-dimensional MAN with the signum character on the two-component group M.

For the character $\sigma \leftrightarrow (2n, +)$ with $n \neq 0$, the unitary points are the points with $c \leqslant 1$; those with $c = 1$ arise from degenerate complementary series induced from the 10-dimensional MAN with a unitary character on M. In the limiting case $d = 1$ along the line $c = 1$, the Langlands quotient is of the type constructed explicitly by Strichartz [13] for the analysis of the discrete spectrum of $L^2(SO(4, 2)/SO(3, 2))$.

Note added in proof. Since this paper was written, some of the material in [3] and [11] has been published as [16].

REFERENCES

1. E. Angelopoulos, *Sur les représentations unitaires irréductibles de $\overline{SO}_0(p, 2)$*, C. R. Acad. Sci. Paris Ser. I **292** (1981), 469–471.

2. A. Borel and N. Wallach, *Continuous cohomology, discrete subgroups, and representations of reductive groups*, Princeton Univ. Press, Princeton, N. J., 1980.

3. W. Casselman, *Systems of analytic partial differential equations of finite codimension*, manuscript, 1975.

4. L. Gross, *Norm invariance of mass-zero equations under the conformal group*, J. Math. Phys. **5** (1964), 687–695.

5. Harish-Chandra, *Representations of semisimple Lie groups on a Banach space. I*, Trans. Amer. Math. Soc. **75** (1953), 185–243.

6. _____, *Some results on differential equations*, manuscript, 1960.

7. _____, *Differential equations and semisimple Lie groups*, manuscript, 1960.

8. A. W. Knapp and B. Speh, *Irreducible unitary representations of $SU(2, 2)$*, J. Funct. Anal. **45** (1982), 41–73.

9. R. P. Langlands, *On the classification of irreducible representations of real algebraic groups*, mimeographed notes, Institute for Advanced Study, Princeton, N. J., 1973.

10. J. Lepowsky, *Algebraic results on representations of semisimple Lie groups*, Trans. Amer. Math. Soc. **176** (1973), 1–44.

11. D. Miličić, *Notes on asymptotics of admissible representations of semisimple Lie groups*, manuscript, 1976.

12. _____, *Asymptotic behaviour of matrix coefficients of the discrete series*, Duke Math. J. **44** (1977), 59–88.

13. R. Strichartz, *Harmonic analysis on hyperboloids*, J. Funct. Anal. **12** (1973), 341–383.

14. N. R. Wallach, *On regular singularities in several variables*, manuscript, 1975.

15. G. Warner, *Harmonic analysis on semi-simple Lie groups*. Vol. II, Springer-Verlag, Berlin and New York, 1972.

16. W. Casselman and D. Miličić, *Asymptotic behavior of matrix coefficients of admissible representations*, Duke Math. J. **49** (1982), 869–930.

DEPARTMENT OF MATHEMATICS, CORNELL UNIVERSITY, ITHACA, NEW YORK 14853

Lectures in Applied Mathematics
Volume 21, 1985

Quantum Mechanics from the Point of View of the Theory of Group Representations[1]

George W. Mackey

Introduction. The purpose of my talk today is to give mathematicians familiar with the theory of group representations some idea of how miraculously well that theory fits the needs of modern elementary particle physics. It does so in two quite different ways. It is first of all a powerful tool for getting numerical answers to concrete problems. Secondly, it contributes to the conceptual foundations of the subject by providing an elegant and more or less complete mathematical model for the underlying physical theory. The first way is rather better known than the second and may be regarded as a "noncommutative" analogue of Fourier's method for solving the linear partial differential equations of heat flow, wave propagation, etc. It was discovered by Wigner in 1927 only a year or so after the discovery of quantum mechanics itself by Heisenberg and Schrödinger. The second way was pointed out by Hermann Weyl, also in 1927, but apart from some mysterious remarks by Weyl himself, was little developed for a quarter of a century or so.

I shall concern myself more or less exclusively with the second way, which one may, with some justice, call "Weyl's program" although Weyl himself did not carry it very far. Moreover I shall combine my presentation with an attempt to explain the nature of quantum mechanics to mathematicians in their own language.

In meetings such as this one designed to bring mathematicians and physicists together, it seems to be customary for the physicists to attempt

---

1980 *Mathematics Subject Classification*. Primary 81G20, 81E99.
[1] A substantial part of this paper was written while the author was a member of the Research Institute for Mathematical Sciences at Berkeley, California.

to explain their subject and its problems to the mathematician while the latter attempt to teach the physicists about branches of mathematics which they do not know well and which might be useful to them. Thus it might seem strange that I, a "pure" mathematician, have decided to talk about physics and attempt to explain it to my colleagues. Would I not do better to leave this to the physicists? I do not think so. There is a very real sense in which mathematicians and physicists speak different languages (with misleadingly similar vocabularies) and just as the best translations of literature are done by mature speakers of the *target* language, so I believe that mathematics is best explained to physicists by other physicists who have gone to considerable trouble to learn it and physics is best explained to mathematicians in corresponding fashion.

I have used the phrases "quantum mechanics" and "elementary particle physics" almost as though they were interchangeable. While this is far from being the case, the two phrases denote entities that are so closely connected that I had no hesitation in using them as I did. The point is this: It has long been a goal of physics to show that physical phenomena can be qualitatively and quantitatively accounted for as logical consequences of the hypothesis that all matter is made of very small indecomposable "atoms" interacting with one another through known laws. Since the time of Newton, it had been thought that these laws would be analogous to those governing the motion of the planets and, after the chemical revolutions led by Lavosier in the late 18th century and by Dalton in the early 19th, one had reason to hope that Dalton's atoms were the hoped for indecomposable objects. If this were so, the basic problem would be to discover the interaction law between these atoms and then apply Newtonian mechanics to deduce the existence and properties of molecules and bulk matter. This program failed for two reasons: (1) The chemists' atoms turned out not to be indecomposable, and the interaction laws between them to be very complex mathematical consequences of much simpler interaction laws between their more fundamental constituents; (2) Newtonian mechanics turned out to be only a limiting case of a truer and more subtle mechanics and to give grossly wrong answers for particles as small as the lighter constituents of atoms. This refinement of Newtonian mechanics is now called quantum mechanics. Its chief application is to the study of the structure of matter and of its ultimate constituents, and without it such a study could not proceed very far, thus the close relationship between "quantum mechanics" on the one hand and "elementary particle physics" on the other.

Before proceeding to details I think it will be useful if I present an overall summary of what I propose to say. The numbered items below are, in part, statements which I propose to explain and justify and, in part, topics which I shall discuss.

(1) Let $\mathfrak{S}$ be the group of isometries of physical space S. Let T be the group of time translations. Then a particle "is" a projective unitary representation W of $\mathfrak{S} \times T$ together with a system of imprimitivity based on S for the restriction $W \upharpoonright \mathfrak{S}$ of W to $\mathfrak{S}$.

(2) It follows from (1) that $W \upharpoonright \mathfrak{S}$ is induced by the subgroup of $\mathfrak{S}$, leaving a point of S fixed. The inducing representation "is" the spin of the particle.

(3) An "interaction" between two particles W^1, P^1 and W^2, P^2 "is" a selfadjoint operator J in the space of $W^1 \times W^2$ which commutes with all $W^1_{x,0} \times W^2_{x,0}$ and all $P^1_E \times P^2_F$.

(4) The bound state mechanism and the concepts of "composite" and "elementary" particles.

(5) "Identity" of particles and the boson-fermion dichotomy.

(6) Maxwell's equations as the infinitesimal generator of a unitary representation V of T and the reinterpretation of V as defining the dynamics of a quantum mechanism "particle".

(7) Dirac's quantization of the field dynamics defined by Maxwell's equations and how it may be reinterpreted as the quantum mechanics of an indefinite number of bosons.

(8) The idea that all elementary particles arise from fields just as photons arise from the Maxwell field. Electrons and "second quantization".

(9) Difficulties with (8).

For a considerably more detailed treatment of this material the reader is referred to [4, pp. 159–271].

1. The nature of quantum mechanics. Let me begin by recalling and commenting upon the conceptual framework of quantum mechanics in the rigorous form given it by von Neumann. For each physical system one has a one-to-one correspondence between the "observables" of the system and the selfadjoint operators in some separable Hilbert space $\mathcal{H}$. In addition, one has a one-to-one correspondence between the "pure states" of the system and the one-dimensional subspaces of this same Hilbert space $\mathcal{H}$. The significance of this correspondence is that given the selfadjoint operator A_θ corresponding to an observable θ, and the one-dimensional subspace L_s corresponding to a pure state s, one can compute the probability distribution which results when one measures θ in the pure state s as follows: Let P denote the projection-valued measure on the real line assigned to A_θ by the spectral theorem and let ψ be any unit vector in L_s. Then the probability that a measurement of θ in the state s will lead to a result in the set E is $(P_E(\psi) \cdot \psi)$. In addition to these two correspondences one has also a unitary representation $t \to V_t$ of the real line in the Hilbert space $\mathcal{H}$ which describes the dynamics of the physical system. Specifically, if at time t_0 the system is in the pure state s_0, then at time t_1 it is the state s_1 where s_1 is the unique pure state such that $V_{t_1-t_0}(L_{s_0}) = L_{s_1}$. Given the two correspondences and the unitary representation V, one is in a position to answer questions about the system by solving well-defined mathematical problems.

At first glance it is rather difficult to see what the above has to do with classical mechanics of which it is, in fact, both a refinement and a close analogue. I shall try to explain. In the classical mechanics of a finite number of interacting particles, a pure state is a set of simultaneous positions and velocities. Let Ω denote the collection of all possible such sets. If n "unconstrained" particles move in Euclidean three-space, then Ω is an open subset of R^{6n}. Modulo the refinement of replacing velocities by "momenta", it is the so-called *phase space*. An *observable* in classical mechanics is a real-valued function defined on Ω. Given an observable f and a pure state ω in Ω, one can compute $f(\omega)$ the *value* of the observable f when the system is in the pure state ω. The *number* $f(\omega)$ is the analogue of the *probability distribution* $E \to (P_E(\psi) \cdot \psi)$, and indeed one of the characteristic features of quantum mechanics is that one is forced to give up the idea that there are states in which all observables have definite values.

The classical analogue of the one-parameter unitary group $t \to V_t$ which defines the dynamics of a quantum mechanical system is a one-parameter group $t \to \alpha_t$ of one-to-one transformations of Ω onto itself. It has the property that if ω_0 is the state of the system at the time t_0 then $\alpha(t_1 - t_0)(\omega_0)$ is the state at the time t_1. Of course one is seldom, if ever, given $t \to \alpha_t$ directly. This one-parameter family usually preserves the differentiable structure of Ω and is differentiable in t so that its "trajectories", i.e. the curves $t \to \alpha_t(\omega_0)$, are smooth enough to have tangent vectors at every point. By the theory of differential equations the resulting *vector field* on Ω uniquely determines $t \to \alpha_t$, and it is usually the vector field that is directly given. Indeed, in a sense the basic insight of Issac Newton is that the laws of physics can be much more simply stated in "infinitesimal" form. More precisely, in most cases they can *only* be stated in infinitesimal form. To get them in global form involves integrating differential equations: the so-called equations of motion. The situation in quantum mechanics is analogous but technically simpler. By a well-known theorem of M. H. Stone, every unitary representation $t \to V_t$ of the additive group of the real line may be written uniquely in the form $V_t = e^{iHt}$ where H is a selfadjoint operator. What one is usually given is the selfadjoint operator H. The skew-adjoint operator iH is evidently the analogue of the vector field defining $t \to \alpha_t$. Since $d(V_t\psi_0)/dt = iHV_t\psi_0$, one can regard $V_t(\psi_0)$ as obtained by integrating the differential equation $d\psi(t)/dt = iH\psi(t)$. This is the celebrated

Schrödinger equation in abstract form. In many examples $\mathcal{H}$ is a function space and H is a differential operator. Under these circumstances the Schrödinger equation becomes a linear partial differential equation.

The analogy between classical and quantum mechanics becomes much closer if we change our point of view toward the former in a simple way which seems at first to be a bit perverse. First of all, let us generalize the notion of pure state in classical mechanics as one does in statistical mechanics—by defining a (not *necessarily* pure) state to be a probability measure defined on the Borel subsets of Ω. (The Borel subsets are the members of the smallest σ-field containing the open sets.) The pure states may then be identified with those particular states having one-point supports. The nonpure states are sometimes called *mixed states*. If f is the (Borel) function defining an observable and μ is the probability measure on Ω defining a state, then one cannot necessarily assign a definite value to the observable f in the state μ. When μ is a mixed state we have (just as in quantum mechanics) only a probability measure on the real line; that is, the probability measure $E \to \mu(f^{-1}(E))$. There are also mixed states in quantum mechanics. Notice that each unit vector ψ in the Hilbert space $\mathcal{H}$ defines a function on all projection operators via $P \to (P(\psi) \cdot \psi)$ and this function is used in assigning a probability measure to each pair consisting of a pure state and an observable. It has the following important formal properties:

(1) If $P = I$, $(P(\psi) \cdot \psi) = 1$.

(2) If $P = 0$, $(P(\psi) \cdot \psi) = 0$.

(3) If $P_1, P_2, \ldots$ are mutually orthogonal and $P = P_1 + P_2 + \cdots$, then $(P(\psi) \cdot \psi) = \sum_{j=1}^{\infty} (P_j(\psi) \cdot \psi)$.

It is easy to see that *any* nonnegative real-valued function on the projections with properties (1), (2) and (3) can be used in the same way as $P \to (P(\psi) \cdot \psi)$ to assign a probability measure on the line to each selfadjoint operator and hence to each quantum mechanical observable. The mixed states of quantum mechanics are all such functions which are not of the form $P \to (P(\psi) \cdot \psi)$. That mixed states can exist can be seen at once from the obvious fact that any convex combination of nonnegative functions on projections with properties (1)–(3) is again such a function. Moreover, as shown by Gleason [1], the particular mixed states defined by von Neumann (and Weyl) are the only ones that exist. These are the (possibly countably infinite) convex combinations of pure states or equivalently those of the form $P \to \text{Trace}(AP)$ where A is a nonnegative selfadjoint operator whose trace exists and is one. Their existence makes quantum statistical mechanics possible and his observation that this is so may be von Neumann's most important contribution to theoretical physics. It is significant that in both classical mechanics and quantum mechanics the pure states are precisely those states which are *not* convex combinations of other states. A key difference between classical mechanics and quantum mechanics lies in the fact that in the latter even pure states do not assign definite values to most observables.

The change in viewpoint alluded to above consists in replacing the point set Ω by the Boolean algebra $\mathcal{L}_\Omega$ of all of its Borel subsets. It is obvious that one can define states without mentioning the points of Ω. That observables may also be so defined is less obvious but nevertheless is a fact. For each real-valued Borel function f on Ω let f^{-1} denote the set function which maps each Borel subset E of the real line into the set $f^{-1}(E)$ of all ω in Ω such that $f(\omega)$ is in E. Then f^{-1} is a mapping from the Borel subsets of the line into $\mathcal{L}_\Omega$ which completely determines f and has the following properties:

(1) $f^{-1}(-\infty, \infty) = \Omega$ and $f^{-1}(\varnothing) = \varnothing$ where $\varnothing$ is the empty set.

(2) If $E_1, E_2, \ldots$ are disjoint Borel subsets of $-\infty, \infty$ and $E = E_1 \cup E_2 \cup \cdots$, then $f^{-1}(E) = f^{-1}(E_1) \cup f^{-1}(E) \cup \cdots$ and $f^{-1}(E_i)$ is disjoint from $f^{-1}(E_j)$ whenever E_i and E_j are disjoint.

Conversely, it is not difficult to see that any mapping from Borel subsets of the real line to $\mathcal{L}_\Omega$ which has properties (1) and (2) is f^{-1} for a unique real-valued Borel function on Ω. Let us call such mappings $\mathcal{L}_\Omega$-*valued measures on the line*. Then from the $\mathcal{L}_\Omega$-valued measure on the line associated with a given observable $\mathcal{O}$ and the probability measure μ on $\mathcal{L}_\Omega$ associated with a state s, we get the probability measure on the line $E \to \mu(f^{-1}(E))$, describing what happens when one measures observable $\mathcal{O}$ in the state s. Note that the points in Ω need not be mentioned. Everything may be expressed in terms of the Boolean algebra $\mathcal{L}_\Omega$. Moreover, in doing so we do *not* use the fact that $\mathcal{L}_\Omega$ is a Boolean algebra. Only the fact that it is an orthocomplemented partially ordered set (with certain extra properties) plays a role.

To pass to quantum mechanics it is only necessary to replace the orthocomplemented partially ordered set $\mathcal{L}_\Omega$ by the orthocomplemented partially ordered set $\mathcal{L}_\mathcal{H}$ where $\mathcal{L}_\mathcal{H}$ is the set of all projection operators in a separable Hilbert space $\mathcal{H}$. Here $P_1 \leqslant P_2$ if and only if $P_1 P_2 = P_2 P_1 = P_1$ and the complement of P is $I - P$. The whole of the von Neumann formulation of quantum mechanics emerges if one simply expresses states and observables in terms of $\mathcal{L}_\Omega$ and then passes to the corresponding notions for $\mathcal{L}_\mathcal{H}$; that is, if one identifies observables in quantum mechanics with $\mathcal{L}_\mathcal{H}$-valued measures on the line and states in quantum mechanics with probability measures on $\mathcal{L}_\mathcal{H}$. $\mathcal{L}_\mathcal{H}$-valued measures on the line are readily seen to coincide with what are more commonly called projection-valued measures or spectral measures, and by the spectral theorem they correspond one-to-one in a natural way with the selfadjoint operators. The connection between states in quantum mechanics and probability measures on $\mathcal{L}_\mathcal{H}$ has been discussed above. As far as dynamics is concerned it is only necessary to observe that the most general one-parameter group of automorphisms of $\mathcal{L}_{\mathcal{H}}$ is defined (in the obvious way) by a continuous one-parameter group $t \to W_t$ of unitary operators, and that $t \to W_t$ and $t \to V_t$ define the *same* automorphism group if and only if $W_t = e^{-iat}V_t$ for some real number a. Writing $W_t = e^{iHt}$, by Stone's theorem one sees that $V_t = e^{i(H+aI)t}$. Thus the observable which defines the dynamics in quantum mechanics, is determined only up to an additive constant just as it is in quantum mechanics.

In addition to the close formal analogy between classical and quantum mechanics just described, there is a precise sense in which classical mechanics is a limiting special case of quantum mechanics. If one tries to apply quantum mechanics to the motions of particles of masses of everyday magnitudes one finds that the limitations on finding states in which all interesting observables have precise values became very slight. While one still cannot have absolutely precise values, one can have very highly concentrated probability distributions which persist in time for all relevant observables simultaneously. When these change in time according to the laws of quantum mechanics, the points in phase space which they approximate move (approximately) according to the laws of classical mechanics.

2. The particle concept in quantum mechanics. Let $\mathcal{H}$ be the Hilbert space associated with a quantum mechanical system and let $t \to V_t$ be the unitary representation of the line describing the time evolution of the system. Let S denote physical space and suppose that the group $\mathcal{E}$ of all isometries of space is actually an automorphism group for the physics of our system, that is, that there exists a homomorphism of $\mathcal{E}$ into the automorphisms of $\mathcal{L}_{\mathcal{H}}$ which transforms the observables of the system in a covariant manner to be spelled out below. Assuming, as we shall, that every element of $\mathcal{E}$ is x^2 for some x in $\mathcal{E}$ and that suitable regularity conditions hold, one can show that this homomorphism may be implemented by a unitary projective representation U of $\mathcal{E}$ with $\mathcal{H}$ the space $\mathcal{H}(U)$ of U. Given a "particle" in our system and given a Borel subset E of space S, let P_E be the selfadjoint operator corresponding to the observable which is 1 when the particle is in E and 0 when it is not. Then P_E must have its spectrum in the set $\{0, 1\}$ and so must be a projection operator. To be given the P_E is to be given all possible coordinate observables for the particles. Indeed, if g is any real-valued Borel function on S, that is, any generalized coordinate for a particle, then $E \to P_{g^{-1}(E)}$ is the projection-valued measure on the line defining the selfadjoint operator associated with the corresponding quantum observable. It is not hard to convince oneself that the mapping $E \to P_E$ must have the properties of a projection-valued measure on S; that is,

(1) $P_\emptyset = 0, P_s = I,$

(2) $P_E P_F = P_{E \cap F}$ for all E and F,

(3) $P_{E_1 \cup E_2 \cup \cdots} = P_{E_1} + P_{E_2} + \cdots$ when $E_i \cap E_j = 0$ for $i \neq j$.

Now given the projection-valued measure on S, $E \to P_E$ and the unitary representation $t \to V_t$ one can not only compute the selfadjoint operator corresponding to any generalized coordinate but also the selfadjoint operator corresponding to any generalized velocity. The latter, when it exists, is

$$\frac{d}{dt} V_t^{-1} A V_t \Big|_{t=0}$$

where A is the selfadjoint operator corresponding to the generalized coordinate. The particle, as far as its position is concerned, is completely described by the pair P, V and until we invoke invariance under $\mathfrak{S}$ nothing more can be said. *Any* pair P, V consisting of a projection-valued measure in space and a unitary representation of the line could describe a conceivable particle. However, invariance under the group $\mathfrak{S}$ of spatial isometries turns out to imply drastic restrictions on both P and V. If S is ordinary Euclidean three space so that $\mathfrak{S}$ is a semidirect product of the subgroup of all translations and the group of rotations about some origin O then it is easy to see that invariance under translations in the x direction must mean that $U_\alpha^{-1} X U_\alpha = X + \lambda I$ whenever X is the selfadjoint operator corresponding to the x coordinate and α is the element of $\mathfrak{S}$ corresponding to translation by λ units in the x direction. Applying the same idea in a slightly less transparent situation, one finds quite easily that for general S

$$U_\alpha^{-1} P_E U_\alpha = P_{[E]\alpha}$$

for all Borel subsets E of S and all α in $\mathfrak{S}$. Notice that if $\mathfrak{S}$ acted trivially on S (which of course it does not) then this condition would amount to the statement that the P_E and the U_α must commute. Now $\mathfrak{S}$ does act trivially on time and it is not difficult to see that invariance under $\mathfrak{S}$ implies that the U_α and V_t must commute. (Actually the most immediate conclusion is that the U_α and V_t must commute up to a factor which depends upon α and t. However, this factor can be eliminated in all the usual models for physical space and we shall ignore it here.)

It follows then that a particle is described by a triple P, U, V whose components are related by the two identities

(a) $U_\alpha^{-1} P_E U_\alpha = P_{[E]\alpha}$,

(b) $U_\alpha V_t = V_t U_\alpha$

for all real t, all $\alpha \in \mathfrak{S}$ and all Borel subsets E of space S. Now (b) implies that if we define $W_{\alpha,t} = U_\alpha V_t$ for all $\alpha, t \in \mathfrak{S} \times T$ then W is a unitary projective representation of $\mathfrak{S} \times T$. Here T is the additive group of the real line regarded as the "time translation group". Conversely, if

W is any unitary projective representation of $\mathfrak{S} \times T$ (whose projective multiplier is the product of a multiplier for $\mathfrak{S}$ and the trivial multiplier for T as is here the case), then W is uniquely the product of a projective unitary representation U of $\mathfrak{S}$ and a unitary representation V of T as above. In other words we may replace the triple P, U, V by the pair P, W, eliminate condition (b), and think of condition (a) as applying to P and the restriction $W \upharpoonright \mathfrak{S}$ of W to $\mathfrak{S} (= \mathfrak{S} \times e)$.

The fact that a particle is described by a pair P, W such that P and $W \upharpoonright \mathfrak{S}$ satisfy (a) is an explanation of item (1) in the list at the end of the Introduction except for the identification of condition (a) with P's being a "system of imprimitivity" for U based on S. From one point of view this is a matter of definition, but the use of this particular terminology may be worth explaining. Let U be any unitary representation of a group G and suppose that its space $\mathcal{H}(U)$ admits a direct sum decomposition $\mathcal{H}(U) = \mathcal{H}_1 \oplus \mathcal{H}_2 \oplus \cdots$ into subspaces which are not necessarily invariant under the U_x but have the weaker property that, for each x in G and each index j, there exists an index j' depending on x and j such that $U_x(\mathcal{H}_j) = \mathcal{H}_{j'}$. Suppose, in other words, that the $\mathcal{H}_j$, while not necessarily invariant under the U_x, are merely permuted among themselves by these operators. By analogy with a well-known concept in the theory of permutation groups, such a system of subspaces has long been referred to as a "system of imprimitivity" for the representation U. the notion defined above is a generalization which allows the direct sum decomposition to be "continuous" or "a direct integral" and is suggested by the spectral theorem. To see the connection let P_j denote the projection on the closed subspace $\mathcal{H}_j$. Evidently the condition $U_x(\mathcal{H}_j) = \mathcal{H}_{j'}$ is equivalent to the condition $U_x P_j U_x^{-1} = P_{j'}$. Now let S denote the set of indices which label the $\mathcal{H}_j$ and for each subset E of S let P_E denote the sum of all the P_j with $j \in E$. Then $E \to P_E$ is clearly a projection-valued measure defined on S, and the condition $U_x P_j U_x^{-1} = P_{j'}$ implies that $U_x P_E U_x^{-1} = P_{[E]x^{-1}}$ where $[E]x^{-1}$ refers to the obvious action of G on S. Evidently the notion of system of imprimitivity for a group representation U as just defined is totally equivalent to the notion of a countable G-space S together with a projection-valued measure P on S such that $U_x P_E U_x^{-1} = P_{[E]x^{-1}}$ for all x in G and all subsets E of S. The generalization that we need here is that in which the countable G-space S is replaced by a G-space S together with a G-invariant σ-field of "Borel sets", i.e. a "Borel space". Since it often happens that $P_{\{s\}} = 0$, for all s in S one has indeed a more general notion.

3. Analysis of the particle concept: The imprimitivity theorem, induced representations and the "spin" of a particle. Let us return for the moment to the discrete case of a system of imprimitivity for a unitary representation U of a group G, and let us specialize further by assuming that the action of G on the index space S is transitive; i.e. that there is just one G orbit in S. Pick s_0 arbitrarily and let K_{s_0} denote the subgroup of all x in G such that $[s_0]x = s_0$. Then the representation U when restricted to K_{s_0} evidently leaves the subspace $\mathfrak{H}_{s_0}$ fixed. Hence it defines a unitary representation L^{s_0} of K_{s_0} in $\mathfrak{H}_{s_0}$. The reader should now have no difficulty in verifying that the representation U of G can be reconstructed knowing only the representation L^{s_0} of K_{s_0} and the action of G on S, and, in addition, that the latter is isomorphic in an obvious sense to the action of G on the space G/K_{s_0} of all cosets $K_{s_0}y$ defined by the formula $(K_{s_0}y)x = K_{s_0}(yx)$.

One can reverse the procedure outlined in the preceding paragraph. Let K be an arbitrary closed subgroup of a separable locally compact group G such that the right coset space G/K is countable. Let L be an arbitrary continuous unitary representation of the subgroup K. One may define a continuous unitary representation U^L of G as follows. Its space is the set of all functions f from G to the space $\mathfrak{H}(L)$ of L which satisfy the identity

(1) $$f(kx) = L_k f(x)$$

for all k in K and all x in G and, in addition, the boundedness condition

(2) $$\sum_{G/K} (f(x) \cdot f(x)) < \infty.$$

Condition (2) finds its meaning in the fact that $(f(kx) \cdot f(kx)) = (L_k f(x) \cdot L_k f(x)) = (f(x) \cdot f(x))$ is constant on the right cosets Kx and its sum over these cosets is a well-defined positive number or ∞. The set of all such functions is a Hilbert space in which $\|f\|^2 = \Sigma_{G/K}(f(x) \cdot f(x))$ and this Hilbert space is invariant under right translations. The operators U_x^L are defined by the formula

(3) $$(U_x^L(f))(y) = f(yx),$$

that is, by right translation. The unitary representation U^L of G just defined is called the representation of G induced by the unitary representation L of K.

Now consider the subspace of the space of U^L consisting of all functions f which vanish outside of a particular right coset Kx_0. If we denote it by $\mathfrak{H}_{Kx_0}$ we see at once that the whole space $\mathfrak{H}(U^L)$ is a direct sum of the $\mathfrak{H}_{Kx_0}$ for all right cosets in G/K and that $U_y(\mathfrak{H}_{Kx_0}) = \mathfrak{H}_{Kx_0y^{-1}}$. Moreover, the subgroup leaving $\mathfrak{H}_K$ fixed is precisely K and the restriction of U^L to K leaves $\mathfrak{H}_K$ invariant and reduces there to L.

In other words, every induced representation U_L in which L is a unitary representation of a closed subgroup K of countable index in G has a canonically associated system of imprimitivity P^L based on the coset space G/K. Conversely, if V is any unitary representation of G and

P is any system of imprimitivity for V based on a countable coset space G/K, then there exists a unitary representation L of K such that the pair V, P is unitarily equivalent to the pair U^L, P^L, and L is uniquely determined up to equivalence. This theorem is the discrete case of what we shall refer to as the *imprimitivity theorem*. For finite groups it was known to Frobenius.

The relevance of these considerations to the quantum mechanics of one-particle systems lies in the fact that the imprimitivity theorem reduces the classification of pairs P, U to the classification of the unitary representations of the subgroup K_{s_0} of G, and that pairs P, U are part of the definition of a particle in quantum mechanics. There is a difficulty in that the quantum mechanical S is physical space and is *not* countable. However, in 1949 the author found and sketched a proof of a generalization of the discrete imprimitivity theorem [2] which applies whenever the system of imprimitivity P is defined on a coset space G/K where K is any closed subgroup of any separable locally compact group G. One defines U^L when G/K is not countable by replacing summation over G/K by integration with respect to a "quasi invariant" measure and insisting that functions are identified whenever they are equal almost everywhere. It is no longer so straightforward to find L where P and U are given but it can be done by indirect means. A number of different proofs of this (generalized) imprimitivity theorem have now been given. The reader will find references to some of these in [3, p. 235].

Having thus described and motivated the imprimitivity theorem, let us examine its implications for the analysis of the particle concept begun in §2. Recall that a particle is completely described by a triple P, U, V where P is a projection-valued measure defined on space, U is a unitary projective representation of the group $\mathfrak{S}$ of isometries of S, and V is a unitary representation of the line, these three objects being restricted by the identities

(a) $U_\alpha^{-1} P_E U_\alpha = P_{[E]\alpha}$,

(b) $U_\alpha V_t = V_t U_\alpha$ for all $\alpha \in \mathfrak{S}$, all Borel subsets E of S and all real t. For simplicity we replace $\mathfrak{S}$ by its simply connected double covering $\tilde{\mathfrak{S}}$ which enables us to deal with ordinary representations rather than projective ones. $\tilde{\mathfrak{S}}$ is a semidirect product of the three-dimensional vector space R^3 and the simply connected double cover SU(2) of the rotation group. Now $\tilde{\mathfrak{S}}$ acts transitively on S so that the latter may be identified with the coset space $\tilde{\mathfrak{S}} | K_{s_0}$ where K_{s_0} is the subgroup of $\tilde{\mathfrak{S}}$ leaving an origin s_0 fixed. K_{s_0} is, of course, isomorphic to the compact group SU(2).

Since (a) is nothing more or less than the assertion that P is a system of imprimitivity for U, the imprimitivity theorem has the following far-reaching consequence. It tells us that (to within unitary equivalence) there is just one pair P, U satisfying (a) for each unitary representation L of SU(2), and that we may take it to be the pair P^L, U^L where U^L is the unitary representation of $\tilde{\mathfrak{S}}$ induced by L and P^L is the canonically associated system of imprimitivity. Thus, to determine all possible triples P, U, V, one need only determine all possible unitary representations L of SU(2), and for each L determine the most general unitary representation V of the additive group of the real line such that each V_t commutes with all U_α^L.

The solution of the first problem is well known and this solution carries with it a fundamental classification of particles. One of the most basic facts about any particle is the particular representation L of SU(2) that occurs in the above analysis, and when this representation is irreducible one speaks of a particle with a "spin" which depends upon the equivalence class of L. When L is reducible it usually turns out that the particle can be thought of as existing in as many different forms as there are constituents and that this form can change into another. In particular, one can think of the neutron and the proton as the two possible different forms of a single particle whose associated L is a direct sum of two equivalent irreducible representations of SU(2). The fact that they are equivalent reflects the fact that the neutron and the proton have the *same* spin.

Before explaining the appropriateness of the term "spin" and assigning a numerical value to it, it will be necessary to discuss some of the consequences of the fact that P and U may be realized in the form P^L, U^L for some L. Straightforward arguments based on the definition of U^L lead to the conclusion that the Hilbert space of our particle may be concretely realized as $\mathfrak{L}^2(E^3, \mathfrak{K}(L))$ (i.e. the space of all square summable functions on physical space *with values* in $\mathfrak{K}(L)$) in such a manner that

(1) $$\left(P_E f(x, y, z) \right) \equiv \phi_E(x, y, z) f(x, y, z)$$

where ϕ_E is the characteristic function of E, and

(2) $$U_\alpha(f)(x, y, z) = f(x + x_0, y + y_0, z + z_0)$$

when α is the translation

$$x, y, z \mapsto x + x_0, \quad y + y_0, \quad z + z_0,$$

$U_\alpha(f)(x, y, z) = L_\alpha((x, y, z)\alpha)$ when α is a rotation.

These facts about P and U have the following more or less immediate consequences:

(A) If g is any real-valued measurable function of the coordinates x, y, z of the particle, then the selfadjoint operator representing the associated quantum mechanical observable is that which takes f into gf. In particular,

(i) the operators $f \to xf$, $f \to yf$ and $f \to zf$ correspond, respectively, to the x, y and z coordinates of the particle;

(ii) the probability that the particle will be found in the subset S of E^3 when in the state defined by f is $\iiint_S \|f(x, y, z)\|^2 \, dx \, dy \, dz$.

(B) Let $x_0 \to U_{x_0,0,0}$ denote the one-parameter unitary group obtained by restricting U to translations in the x direction and let M_x denote the selfadjoint operator such that $U_x = e^{itM_x}$ (Stone's theorem). Then $M_x(f) = i\, \partial f / \partial x$. Similar statements hold for translations in the y and z directions.

(C) Let $\theta \to U_\theta^z$ denote the one-parameter subgroup of all rotations about the z-axis, and let Ω_z denote the unique selfadjoint operator such that $e^{i\Omega_z \theta} = U_\theta$. Then Ω_z is the sum of two operators. One of these is the differential operator $i(x\, \partial/\partial y - y\, \partial/\partial x)$. The other is an operator of the form $f \to A^z f$ where A^z is a selfadjoint operator in the space $\mathcal{H}(L)$ of the representation L. Analogous statements are true for the group of rotations about the x- and y-axes. In particular, if L is the identity representation, then $A^z = A^x = A^y = 0$.

To explain the significance of consequences (B) and (C) we note that an observable defined by a selfadjoint operator T will have a constant probability distribution in every state if and only if $V_t T = T V_t$ for all t. Such observables are the quantum analogues of "integrals of the motion" in classical mechanics. Since the U_α and the V_t commute, it follows that the M's and the Ω's defined above must all commute with all V_t and, hence, that the observables defined by these operators are "integrals of the motion". A more detailed analysis shows that they reduce in the classical limit to constant multiples of the linear and angular momentum observables. Thus (B) and (C) allow us to specify which operators define the linear and angular momentum observables just as (A) allows us to specify which operators define the coordinate observables.

Note that the coordinate and linear momentum observables necessarily satisfy the celebrated Heisenberg commutation relations which have not been assumed as such. Actually they are implicit in the invariance principles on which our discussion has been based, and this way of introducing them seems to the writer to have considerably more a priori plausibility than Heisenberg's reliance on analogy with the classical Poisson brackets.

While the form of the coordinate and linear momentum operators is independent of the representation L of K, this is not so for the angular momentum operator. Whenever the representation L is other than the identity, the differential operators $x\, \partial/\partial y - y\, \partial/\partial x$, etc. must be changed by the addition of an operator which is defined by an operator on $\mathcal{H}(L)$ and has only a finite number of distinct eigenvalues whenever $\mathcal{H}(L)$ is finite dimensional. It is as though the angular momentum were a sum of two components, one of which depends upon L and can fail to exist. The physicists like to think of this extra angular momentum as caused by a rotation or "spinning" of the particle about an axis through itself rather than its motion through space and speak of the "spin angular momentum".

The eigenvalues of the operator $x\,\partial/\partial y - y\,\partial/\partial x$ and its analogues are precisely the integers. Thus the (orbital) angular momentum of a particle necessarily takes on only those values which are integer multiples of some unique least positive value. When L is irreducible and has dimension $2j + 1$ where j is a positive integer or half-integer, it turns out that the eigenvalues of the spin angular momentum operators are $-jh$, $(-j + 1)h, \ldots, (j + 1)h$ where h is the minimum positive value of the orbit angular momentum operator. One says, accordingly, that the particle "has spin j". For the electron and the proton, for example, $2j + 1$ is 2 or $j = \frac{1}{2}$, and one says that these are particles of spin $\frac{1}{2}$.

4. Possibilities for V and the dynamics of a particle. Having seen that the pair P, U is completely determined by choosing an arbitrary unitary representation L of K, we now have to find the most general unitary representation ($t \to V_t$ of the additive group of the real line) which commutes with a given $U = U^L$, that is, such that $U_\alpha V_t = V_t U_\alpha$ for all α in $\tilde{\mathscr{E}}$ and all t in T. Since, by Stone's theorem, $V_t = e^{itH}$ for some selfadjoint operator H and $V_t U_\alpha = U_\alpha V_t$ for all α and t if and only if $U_\alpha H U_\alpha^{-1} = H$ for all α, our problem reduces to that of finding the most general selfadjoint operator H which commutes with all U_α^L. The solution of this problem is an immediate consequence of general theorems about unitary group representations which, in particular, permit us to find the decomposition of U^L into irreducible constituents. The answer is simplest in the case of a particle without spin or different "forms", that is, a particle for which L is the one-dimensional identity. In that case, it follows that H must be some real-valued "function" of the Laplace operator $-(\partial^2/\partial x^2 + \partial^2/\partial y^2 + \partial^2/\partial z^2)$; in other words, that there exists a real-valued Borel function ϕ defined on the positive reals such that $H = \phi(H_0)$ where H_0 is the selfadjoint operator associated with $-(\partial^2/\partial x + \partial^2/\partial y^2 + \partial^2/\partial z)$ and $\phi(H_0)$ is defined using the spectral theorem. (If H has an orthonormal basis of eigenvectors $\phi_1, \phi_2, \ldots$ with $H_0(\phi_j) = \lambda_j \phi_j$ then $\phi(H_0)$ is defined by the fact that $\phi(H_0)(\phi_j) = \phi(\lambda_j)\phi_j$.) One can define $\phi(H_0)$ in this particular case without using the spectral theorem by exploiting the fact that the Fourier transform defines a unitary operator which carries differential operators into multiplication operators. Indeed, let

$$\hat{f}(u, v, w) = \int\int\int e^{2\pi i(ux + vy + wz)} f(x, y, z)\, dx\, dy\, dz$$

be the Fourier transform of the function f which we assume for the moment to be both differentiable and square summable. Then a simple calculation using integration by parts shows that

$$\widehat{\frac{\partial f}{\partial x}}(u, v, w) = -2\pi i u \hat{f}(u, v, w)$$

with analogous formulas for $\widehat{\partial f/\partial y}$ and $\widehat{\partial f/\partial z}$. Thus, if

$$g(x, y, z) = -\left(\partial^2 f/\partial x^2 + \partial^2 f/\partial y^2 + \partial^2 f/\partial z^2\right)(x, y, z),$$

then

$$\hat{g}(u, v, w) = +4\pi^2(u^2 + v^2 + w^2)\hat{f}(u, v, w),$$

and we define $\phi(-(\partial^2/\partial x^2 + \partial^2/\partial y^2 + \partial^2/\partial z^2))$ as the operator which takes f into the inverse Fourier transform of the function

$$u, v, w \mapsto \phi\left(4\pi(u^2 + v^2 + w^2)\right)\hat{f}(u, v, w).$$

Since H evidently commutes with all V_t, the observable corresponding to this operator is an integral of the motion in the sense defined above. As with the quantum mechanics observables associated with the infinitesimal generators of one-parameter subgroups of translations and rotations, it is a constant multiple of the quantum mechanical analogue of a well-known and basic integral of the motion in classical mechanics. It is the so-called *energy* observable.

It is time to say a word about the multiplicative constants mentioned in connection with the linear and angular momentum observables and with the energy observable. These "constants" are, in fact, all the same and coincide in value with $\hbar = h/2\pi$ where h is the fundamental constant introduced by Planck in 1900 when he made the discovery initiating the so-called "old quantum theory". From our present point of view it would be natural to define energy and momentum so that the associated operators coincide with the related infinitesimal generators. However, physics is conservative and abandons outdated definitions slowly if at all. One needs a conversion factor to adjust the old definitions to the "natural" ones suggested by quantum mechanics. In the end the celebrated constant of Planck can be best understood as just 2π multiplied by this conversion factor. If one begins with an axiomatic treatment of quantum mechanics and deduces classical mechanics from it as a limiting case, which holds when masses and energies are large, one finds that there is no need to introduce Planck's constant at all. Its value can always be made equal to 2π by choosing a unit of mass suitably. Unlike classical mechanics quantum mechanics provides a "natural" mass unit and no arbitrary choice is necessary.

Let

$$\Omega = \left(\frac{\hbar}{i}\frac{\partial}{\partial x}\right)^2 + \left(\frac{\hbar}{i}\frac{\partial}{\partial y}\right) + \left(\frac{\hbar}{i}\frac{\partial}{\partial z}\right)^2.$$

Then the corresponding observable is, on one hand, the quantum analogue of the square of the total angular momentum and, on the other
hand, a constant—$\hbar^2$ times the Laplacian. Since the energy operator $H/\hbar$
is a function of the Laplacian

$$H = \hbar\phi\big(-\left(\partial^2/\partial x^2 + \partial^2/\partial y^2 + \partial^2/\partial z^2\right)\big),$$

it is also a function of Ω, $H = \hbar\phi(\Omega/\hbar^2)$. In other words, the function ϕ
introduced above determines and is determined by the function relating
the energy to the square of the total linear momentum.

Now in classical mechanics the total energy of a free particle of mass m
and velocity v is $mv^2/2 = (mv)^2/2m$ and is the constant $1/2m$ times the
square of the total linear momentum. If we assume that this relationship
persists in quantum mechanics we are led to the conclusion that $\phi(v) = kv$
for some constant k, and hence that the operator H for a free spinless
particle is just $-k(\partial^2/\partial x^2 + \partial^2/\partial y^2 + \partial^2/\partial z^2)$ where the constant k is
inversely proportional to the mass of the particle. However, it is not
necessary to appeal to classical mechanics to arrive at this conclusion.
The group theoretical analysis given in §2 and §3 can be extended to
study the consequences of assuming spacetime invariance—either that of
Galileo and Newton or that of Einstein. In the first case one shows that,
for a spinless particle, $\phi(v)$ necessarilly has the form $\phi(v) = kv$. In the
second case one finds instead that $\phi(v) = c[m_0^2 c^2 + r^2]^{1/2}$ where c is the
velocity of light and m_0 is a parameter depending on the particle and
which is usually referred to as the rest mass. In either case the quantum
mechanics of a single spinless particle is determined by specifying a
single real parameter. More generally, that is, if nothing is assumed about
spacetime invariance, a whole function (the energy momentum relationship) has to be specified before the quantum mechanical properties of the
particle are known.

If ϕ is a unit vector in Hilbert space which describes the state of the
particle at any particular time $t = 0$, then t time units later it will be
described by $\phi_t = V_t(\phi) = e^{itH}\phi$. Hence when ϕ is in the dense domain of
H, one has

$$\frac{d}{dt}\phi_t = iH\phi_t,$$

and this is an abstract form of the celebrated Schrödinger equation.
When the particle is spinless, ϕ is a complex-valued function of x, y, z,
and ϕ_t is a complex-valued function of x, y, z, t. If this is denoted by ψ
then $d\phi_t/dt$ becomes $\partial\phi(x, y, z, t)/\partial t$; similarly when $\phi(v) = kv$ so that

$$H = -k\big(\partial^2/\partial x^2 + \partial^2/\partial y^2 + \partial^2/\partial z^2\big),$$

then $iH\phi_t$ becomes

$$-ki\left(\frac{\partial^2\psi}{\partial x^2} + \frac{\partial^2\psi}{\partial y^2} + \frac{\partial^2\psi}{\partial z^2}\right)$$

and the Schrödinger equation becomes

$$\frac{\partial\psi}{\partial t} = -ki\left(\frac{\partial^2\psi}{\partial x^2} + \frac{\partial^2\psi}{\partial y^2} + \frac{\partial^2\psi}{\partial z^2}\right)$$

which reduces to the usual nonrelativistic Schrödinger equation for a particle of mass m and spin 0, if we take $k = \hbar/2m$.

When L is no longer the identity, the analysis is somewhat more complicated but is easily handled by the standard theory of group representations. We shall not give further details here but content ourselves instead with a few remarks. First of all, the function $\phi(r)$ has to be replaced by several independent such functions, but as long as L is finite dimensional there will only be a finite number of them. Secondly, the assumption of either Galilean relativity or Einstein's relativity eliminates all functions of r and replaces them by a simple real parameter—the particle mass—provided that L is irreducible. Thirdly, whether one assumes any kind of relativistic invariance or not, one can seek operators H of the form

$$H_0 + iH_i\frac{\partial}{\partial x} + iH_2\frac{\partial}{\partial y} + iH_3\frac{\partial}{\partial z}$$

where H_0, H_1, H_2, H_3 are all "matrix" operators $\psi_1, \psi_2,\ldots,\psi_n \to \Sigma\, a_{ij}\psi_j$ and the a_j are complex constants. Moreover, one can prove that they exist (with the postulated invariance properties) for certain choices of L. The celebrated Dirac equations for the electron are of this form with L a direct sum of two replicas of the two-dimensional irreducible representation of SU(2), and they are relativistically invariant. The fact that a relativistically invariant H *of this form* does not exist for particles of spin zero is the basis of Dirac's celebrated argument that spin is a consequence of special relativity.... However, this argument is no longer considered to be valid. The π-meson has a relativistically invariant H *and* spin zero. The H for the π-meson is *not* defined by a differential operator but by the square root of such an operator. By using matrices on vector-valued functions as above, one can find square roots of constants added to the Laplacian which are first order differential operators. However, contrary to Dirac's belief, nature does not insist on such elegance.

5. Particle interactions. Let U^1, P^1 and U^2, P^2 define two particles. Let us form the tensor product $\mathcal{H}(U^1) \times \mathcal{H}(U^2)$ of the two Hilbert spaces

$\mathcal{H}(U^1)$ and $\mathcal{H}(U^2)$. Let $U^1 \otimes U^2$ denote the representation α, $t \to U^1_{\alpha,t}$ $\times U^2_{\alpha,t}$ of $\tilde{\mathfrak{E}} \times T$, and let $P^1 \times P^2$ denote the unique projection-valued measure on $S \times S$ such that $(P^1 \times P^2)_{E \times F} = P^1_E \times P^2_F$. Then the pair $U^1 \otimes U^2$, $P^1 \times P^2$ can be interpreted as a quantum mechanical system with two particles each of which behaves as though the other did not exist. If one wants a mathematical model for the more realistic case in which the particles do influence one another, it is not unreasonable to seek it amongst models which are small modifications of the above. Of course the restriction V of $U^1 \otimes U^2$ must change and it proves possible to find a partially axiomatic justification for supposing that this is all that changes. More explicitly, one is led to assume that the interacting system is described by $P^1 \times P^2$ and a representation U of $\tilde{\mathfrak{E}} \times T$ whose restriction to $\tilde{\mathfrak{E}} \times 0$ is just $U^1 \otimes U^2$. Such a representation is, of course, uniquely determined by its restriction $\tilde{V}$ to $e \times T$. By Stone's theorem this restriction may be written in the form $\tilde{V}_t = e^{itH}$ where H is some selfadjoint operator. If $V_t = e^{itH_0}$ then $J = H - H_0$ is what must be added to the "noninteracting" dynamical operator H_0 to obtain the actual one. We shall call it the *interaction* between the particles in question. A priori J could be an arbitrary selfadjoint operator but the partially axiomatic treatment mentioned above leads to the conclusion that J "must" lie in the commuting algebra of $U^1 \times U^2$ restricted to $\tilde{\mathfrak{E}}$ and also must commute with all projection operators $(P^1 \times P^2)_E$ where E is a Borel subset of $S \times S$.

The problem of finding the most general possible J satisfying the conditions just described turns out to be solvable by the methods of the general theory of unitary group representations. We shall not be able to give details here but shall content ourselves with a few remarks about the results. The simplest case is that in which both particles have spin zero so that the Hilbert space of the system is $\mathcal{L}^2(E^6)$, the space of all square summable functions of the six variables $x_1, y_1, z_1, x_2, y_2, z_2$. In this case one can show that J must be of the form

$$J(f)(x_1, y_1, z_1, x_2, y_2, z_2)$$
$$= \rho\left(\left[(x_1 - x_2)^2 + (y_1 - y_2)^2 + (z_1 - z_2)^2\right]^{1/2}\right)$$
$$\times f(x_1, y_1, z_1, x_2, y_2, z_2)$$

where ρ is some real-valued Borel function defined on the positive real axis. Thus the interaction between two spinless particles is defined by a certain function of the distance between them just as in classical mechanics. The determination of the function is, of course, a matter for experiment.

Now suppose that our particles have spins defined by representations L^1 and L^2 of SU(2) so that the Hilbert space of our system consists of square summable functions from E^6 to the tensor product $\mathcal{K}(L^1) \otimes \mathcal{K}(L^2)$. In that case J may be shown to take the form

$$Jf(x_1, y_1, z_1, x_2, y_2, z_2) = A(x_1, y_1, z_1, x_2, y_2, z_2)$$
$$\times f(x_1, y_1, z_1, x_2, y_2, z_2)$$

where A is a Borel function from the positive real axis to the selfadjoint operators in the Hilbert space $\mathcal{K}(L^1) \times \mathcal{K}(L^2)$. This A is subject to further conditions which reduce in the spinless case to the assumption that A depends only on $[(x_1 - x_2)^2 + (y_1 - y_2)^2 + (z_1 - z_2)^2]^{1/2}$. In the general case they cannot be stated so simply. We shall content ourselves here with the following remarks:

(a) The set $\mathcal{C}$ of all functions A which satisfy the conditions in question is a real vector space which is closed under multiplication by functions of the form

$$\rho\left(\left[(x_1 - x_2)^2 + (y_1 - y_2)^2 + (z_1 - z_2)^2\right]^{1/2}\right)$$

where ρ is as above.

(b) When $\mathcal{K}(L^1)$ and $\mathcal{K}(L^2)$ are finite dimensional then $\mathcal{C}$ admits a finite basis $A_1, \ldots, A_r$ such that every member is uniquely of the form

$$\sum_{j=1}^{r} \rho_j\left(\left[(x_1 - x_2)^2 + (y_1 - y_2)^2 + (z_1 - z_2)^2\right]^{1/2}\right)$$
$$\times A_j(x_1, y_1, z_1, x_2, y_2, z_2).$$

(c) The integer r under (b) depends upon L^1 and L^2 and is 6 in the special case in which L^1 and L^2 are both two-dimensional and irreducible, that is, when both particles have spin $\frac{1}{2}$.

(d) One of the A_j may always be taken to be the function which is the identity operator for all $x_1, y_1, z_1, x_2, y_2, z_2$.

Whenever A is $\rho([(x_1 - x_2)^2 + (y_1 - y_2)^2 + (z_1 - z_2)^2]^{1/2})$ multiplied by the identity operator, one says that interaction is "independent of the spin". In more general cases one thinks of the interaction as dependent both on the distance between the particles and of the orientations of their "spin angular momentum" both with respect to one another and with respect to the "orbital angular momentum".

6. Interactions when there are more than two particles. Consider three quantum mechanical particles defined by U^1, P^1, U^2, P^2 and U^3, P^3. Let J_{12}, J_{23} and J_{31} be the selfadjoint operators in $\mathcal{K}(U^1) \times \mathcal{K}(U^2), \mathcal{K}(U^2) \times \mathcal{K}(U^3)$ and $\mathcal{K}(U^3) \times \mathcal{K}(U^1)$, respectively, which define the interactions between these particles taken in pairs. Form the Hilbert space

$\mathcal{K}(U^1) \times \mathcal{K}(U^2) \times \mathcal{K}(U^3)$, remembering that this Hilbert space is naturally isomorphic to $\mathcal{K}(U^2) \times \mathcal{K}(U^1) \times \mathcal{K}(U^3)$, etc. Form $J_{12} \times I$, $I \times J_{23}$ and the corresponding operator involving J_{13}. Denote these by $\tilde{J}_{12}$, $\tilde{J}_{23}$ and $\tilde{J}_{31}$. Finally let $J = \tilde{J}_{12} + \tilde{J}_{23} + \tilde{J}_{31}$. Then the dynamics of the three-particle system due to the interactions of the three particles in pairs is given by the one-parameter group $t \to e^{iHt}$ where $H = J + H_0$ and e^{iH_0t} is $U^1 \otimes U^2 \otimes U^3$ restricted to $e \times T$. The generalization to n particles should be obvious. When the dynamics of an n-particle system can be described in terms of two-particle interactions as we have just done, one says that the particles interact through "two body forces". It is possible to envisage more complicated interactions involving "k body forces" where $k > 2$. However, these seem to occur mainly, if not exclusively, when special relativity produces other complications as well and we shall not discuss them here. We warn the reader at this point that the mathematical model for interacting particles presented in this and the preceding section is only valid when all of the particles described are different. If two or more of the particles are "identical" a rather profound correction has to be made. We shall discuss this in §8.

7. The bound state concept and the distinction between elementary and composite particles. It is a well-known theorem in classical mechanics that a system of interacting particles such as the sun and the planets may be "factored" into a product of two independent systems. Specifically, one finds that the motion of the particles relative to their common center of gravity is independent of the motion of the latter and, vice versa, the motion of the center of gravity is independent of the relative motion of the particles. In more technical terms one can introduce new variables and divide them into two subsets in such a manner that the differential equations governing the changes of one set do not involve the other. Somewhat more abstractly, the phase space of the system may be decomposed as a product space in such a way that the one-parameter group of transformations describing the time evolution of the system is a product of such groups in the two factors. A parallel theorem exists in quantum mechanics and turns out to be of fundamental importance in describing how matter is constructed out of the basic particles known as protons, neutrons and electrons. Given a system of interacting particles as described above and given in addition that the system has suitable spacetime invariance properties in the Galilean sense, one can write the underlying Hilbert space $\mathcal{K}$ as a tensor product $\mathcal{K}_0 \times \mathcal{K}_R$ so that the one-parameter group $t \to V_t$ defining the dynamics is a tensor product $t \to V_t^0 \times V_t^R$ of one-parameter subgroups in $\mathcal{K}_0$ and $\mathcal{K}_R$. Moreover, one can interpret these factors as describing the motion of the center of gravity and motion relative to the center of gravity, respectively. Actually one can do better than this and factor the restriction of W to $SU(2) \times T$ in a corresponding fashion. Consider the $\mathcal{K}_R$ part of the factorization of this restriction. Its restriction to $e \times T$ is just V^R and its restriction to $SU(2)$ will be a unitary representation M of $SU(2)$ such that the M_β for β in $SU(2)$ commute with all V_t^R. Of course, $V_t^R = e^{iH^R t}$ where H^R is a selfadjoint operator. Let us look at the spectrum of H^R. It may have a discrete part or it may be purely continuous. In the latter event one says that the given system of particles has no bound states. Supposing then that H^R has a discrete part, let λ be any eigenvalue and let $\mathcal{K}_R^\lambda$ denote the λ eigenspace of H^R, i.e. the null space of $H^R - \lambda I$. Since the M_β commute with the V_t^R they commute with H^R and leave $\mathcal{K}_R^\lambda$ invariant. The restriction of M to $\mathcal{K}_R^\lambda$ may be decomposed as a direct sum of irreducible subrepresentations each of which acts in a well-defined subspace $\mathcal{K}_R^{\lambda,L}$ of $\mathcal{K}_R^\lambda$. Here L denotes an irreducible subrepresentation of M. Of course $\mathcal{K}_R^{\lambda,L}$ is an invariant subspace of H^R and one in which H^R reduces to λI where I is the identity operator. Consider, finally, the tensor product $\mathcal{K}_0 \times \mathcal{K}_R^\lambda$. This is a subspace of the Hilbert space $\mathcal{K}_0 \times \mathcal{K}_R$ which is the Hilbert space for the original system of particles. It is not difficult to show that this subspace is an invariant subspace for the representation W of $\widetilde{\mathcal{E}} \times T$. If the state vector for the particle system is in the subspace $\mathcal{K}_0 \times \mathcal{K}_R^{\lambda,L}$ at any time t_0, it remains there forever. Moreover, the position observables for the center of gravity of the system are defined by selfadjoint operators which also leave the subspace $\mathcal{K}_0 \times \mathcal{K}_R^{\lambda,L}$ invariant. If we simply ignore the rest of the Hilbert space and look at W and the center of gravity position observable operators restricted to $\mathcal{K}_0 \times \mathcal{K}_R^{\lambda,L}$, we find ourselves looking at the model for a single free particle whose spin is described by the irreducible representation L of $SU(2)$. In fact, when the system is in a state described by a vector in $\mathcal{K}_0 \times \mathcal{K}_R^{\lambda,L}$, its behavior cannot be distinguished from that of a single free particle whose mass is the sum of the masses of the original particles and whose spin is $(d - 1)/2$ where d is the dimension of $\mathcal{K}(L)$. All other degrees of freedom of the system are "frozen". One speaks of the system of particles as being in a "bound state" although the use of the word *state* in this connection is inconsistent with its earlier use. A bound state is neither a pure or a mixed state as defined earlier. It is a *collection* of pure states which behave as the set of all possible pure states of a free particle.

It is clear that finding the possible bound states of a given system of interacting particles reduces mathematically to finding the eigenvectors of the selfadjoint operator H^R and sorting them out according to how they behave under a certain representation of $SU(2)$. Actually one uses the commutativity of H^R with this representation as a tool in breaking the eigenvector—eigenvalue problem down into easier subproblems—one for each equivalence class of irreducible unitary representations of SU(2). From a conceptual point of view it is worth noting that the postulated Galilean invariance may be used to show that the representation W of $\tilde{6} \times T$ has a canonical extension to a projective representation of the ten-dimensional Galilean group, and that the bound states defined above correspond exactly to the discrete irreducible constituents of this representation.

An actual physical particle is said to be *composite* if it can be shown to be a bound state, in the above sense, of particles of smaller mass. Otherwise it is said to be an *elementary particle*. For example, an atom is a bound state composed of a "nucleus" and a finite number of particles called "electrons". Electrons are thought to be elementary particles but nuclei are bound states composed of a finite number of "neutrons" and "protons". While protons and neutrons were thought for a time to be elementary it is now believed that they are bound states composed of "quarks". Now a composite particle can only reveal its composite character by interacting with other particles and need not do so unless sufficient energy is available to be passed on to the composite particle. One speaks of some composite particles as being more tightly bound than the others. In more precise terms this is a question of the difference between the eigenvalue λ of H^R and the expected value of H^R in various states not in the collection defined by the bound state. A consequence is that one can never be sure that an apparently elementary particle is not, in fact, composite. There is a further complication in that when the binding is sufficiently "tight" the energies involved are so great that Galilean invariance is no longer an adequate approximation. In that case the very concept of a stable bound state loses its rigorous meaning. Morever, the Einsteinian relativity that replaces Galilean relativity has implications about the introconvertability of mass and energy that cast doubt in the distinction between elementary and composite particles in other ways. It turns out that most atomic nuclei are sufficiently tightly bound so that their failing to be elementary cannot be detected without going to great lengths to produce extraordinarily high energies. Correspondingly, most of chemistry may be understood by applying the bound state concept of quantum mechanics to systems of a finite number of electrons and atomic nuclei; both the electrons and the atomic nuclei being regarded as elementary. Moreover, the interaction between these may be taken to be a sum of two body interactions as described in §6, and the two body interactions may be taken to be spin independent and to be defined by functions ρ of the form cQ_1Q_2/r where c is a universal constant, and Q_1 and Q_2 are positive or negative numbers characteristic of the particle and called its charge. All electrons have charge -1 and all atomic nuclei have charge equal to a positive integer. An *atom* of atomic number n is a bound state of a nucleus of charge n and n electrons. A *molecule* is a bound state of a finite number of nuclei of charges $n_1, n_2, \ldots, n_k$ and $n_1 + n_2 + \cdots + n_k$ electrons. To decide whether a finite collection of atoms whose nuclei have charges $n_1, n_2, \ldots, n_k$ can combine to form a molecule, one considers the system consisting of these nuclei and $n_1 + n_2 + \cdots + n_k$ electrons and studies the operator H^R. If it has a discrete component to its spectrum the atoms can combine, otherwise they cannot. The so-called binding energy of the molecule is determined by the relevant eigenvalue of H^R. Usually the system will have many bound states. Then the one of lowest energy is called the *ground state* and the others are called *excited states*. When interaction with the outside world causes an atom or molecule to pass from an excited state to one of lower energy, the energy difference is emitted as a "quantum of electromagnetic radiation" (see below) whose frequency is the energy difference divided by h. In this way the mathematics of the bound state mechanism produces a well-defined algorithm for explaining and calculating the sharp lines that one finds in the spectra of the light and other electromagnetic radiation emitted by atoms and molecules. It was the search for such an algorithm that provided one of the chief motivations for the discovery of quantum mechanics in the middle 1920s.

The fact that quantum mechanics together with the bound state mechanism explains not only atomic and molecular spectra but also the periodic table, and the binding of atoms into molecules is the inspiration for a celebrated statement by Dirac. In the introduction to a paper published in 1929 in Volume 123 of Series A of the Proceedings of the Royal Society of London, he says

> The general theory of quantum mechanics is now almost complete, the imperfections that still remain being in connection with the exact fitting in of the theory with relativity ideas. These give rise to difficulties only when high speed particles are involved, and are therefore of no importance in the consideration of atomic and molecular structure and ordinary chemical reactions in which it is, indeed, usually sufficiently accurate if one neglects relativity variation of mass with velocity and assumes only Coulomb forces between the various electrons and atomic nuclei. The underlying physical laws necessary for the mathematical theory of a large part of physics and the whole of chemistry are thus completely known and the difficulty is only that the exact application of these laws leads to equations much too complicated to be soluble.

We close this section with the remark that while it is indeed difficult to find exact solutions to the equations in question, enormous qualitative insight and useful approximation techniques have been obtained by applying the theory of group representations in a manner that cannot be adequately described here.

8. Identity of particles, the boson-fermion dichotomy and the Pauli exclusion principle. If one forgets the warning at the end of §6 and proceeds to study the bound state question for a system of nuclei and electrons as outlined in §5–7, one discovers that our model is incorrect in that it predicts far too many bound states. In particular, if one is trying to predict the spectrum of an atom, one predicts very many lines that do not exist in nature. Some, but by no means all, of these omissions are explained by a theory that calculates the intensity of spectral lines and sometimes gives a zero answer. It seems that many bound states predicted by the unmodified theory simply do not exist in nature. This circumstance raises two questions: (a) How can the theory be modified so as to eliminate the unwanted states? (b) Must this modification be simply accepted ad hoc or is there some plausible rationale for it?

As far as question (a) is concerned there is a simple and elegant answer. We present it initially in the special case of the sytem consisting of a nucleus of charge n and n electrons. The underlying Hilbert space is then the tensor product $\mathcal{H}_N \times \mathcal{H}_e \times \mathcal{H}_e \times \cdots \times \mathcal{H}_e$ where $\mathcal{H}_N$ is the Hilbert space of the nucleus N considered as a particle and each $\mathcal{H}_e$ is a replica of the Hilbert space of a single electron. Since the factors $\mathcal{H}_e$ are all identical there is a natural unitary representation B of the symmetric group of n objects S_n in the n-fold tensor product $\mathcal{H}_e \times \mathcal{H}_e \times \cdots \times \mathcal{H}_e$. This representation carries the permutation π into the unique unitary operator B_π which carries $\phi_1 \times \phi_2 \times \cdots \times \phi_n$ into $\phi_{\pi(1)} \times \phi_{\pi(2)} \times \cdots \times \phi_{\pi(n)}$. If we decompose this representation as a direct sum of subrepresentations, each of which is a direct sum of equivalent irreducibles, we obtain a direct sum decomposition of $\mathcal{H}_e \times \mathcal{H}_e \times \cdots \times \mathcal{H}_e$ whose constituents are parametrized by the equivalence classes of irreducible unitary representations of S_n. These include precisely two one-dimensional representations. One of these is just the identity representation I. The other is the so-called alternating representation A which is the one-dimensional identity operator for all even permutations and the negative of the one-dimensional identity operator for all odd permutations. Let $\mathcal{H}^s$ and $\mathcal{H}^A$ denote the subspaces of $\mathcal{H}_e \times \mathcal{H}_e \times \cdots \times \mathcal{H}_e$ defined as above by I and A, respectively. They are known (respectively) as the symmetric and antisymmetric subspaces of $\mathcal{H}_e \times \mathcal{H}_e \times \cdots \times \mathcal{H}_e$. Of course, $\mathcal{H}_N \times \mathcal{H}^S$ and $\mathcal{H}_N \times \mathcal{H}^A$ are subspaces of the system Hilbert space $\mathcal{H}_N \times \mathcal{H}_e \times \mathcal{H}_e \times \cdots \times \mathcal{H}_e$. The answer to question (a) may now be stated as follows. The system Hilbert space $\mathcal{H}_N \times \mathcal{H}_e \times \cdots \times \mathcal{H}_e$ must be modified by being replaced by the subspace $\mathcal{H}_N \times \mathcal{H}^A$. States corresponding to unit vectors not contained in this subspace simply do not exist in nature. This modification in the Hilbert space has profound consequences and it is not immediately obvious that it may be done without upsetting other parts of the theory. The chief apparent difficulty of this sort is that the selfadjoint operators defining the position and momentum observables of the individual electrons do not leave the subspace $\mathcal{H}_N \times \mathcal{H}^A$ invariant and hence do not define observables for the modified system in any evident way. On the other hand, algebraic combinations of these operators *which are symmetric* in all the electrons do leave $\mathcal{H}_N \times \mathcal{H}^A$ invariant and so define observables in the modified model for the system. Consider, for example, the case in which $n = 3$ and let X_1, X_2 and X_3 denote the selfadjoint operators in the original Hilbert space $\mathcal{H}_N \times \mathcal{H}_e \times \mathcal{H}_e \times \mathcal{H}_e$ which correspond to the x coordinates of the three electrons. Then $X_1 + X_2 + X_3$, $X_1^2 + X_2^2 + X_3^2$ and $X_1^3 + X_2^3 + X_3^3$ all leave $\mathcal{H}_N \times \mathcal{H}^A$ invariant and have well-defined counterparts in the modified Hilbert space. Notice, however, that if a, b and c are any three real numbers and we know $a + b + c$, $a^2 + b^2 + c^2$ and $a^3 + b^3 + c^3$, then we can compute abc and $ab + bc + ac$ by the elementary theory of symmetric functions. Knowing $a + b + c$, $ab + bc + ac$ and abc we know the coefficients of a cubic equation whose roots are a, b and c. Solving this equation we may compute the three numbers a, b and c. However, *we have no way of telling which root is a, which root is b and which root is c*. Put, otherwise, our modified theory lets us talk about the coordinates of our three electrons and discuss their variation in time, provided we do this entirely by means of symmetric functions of them or, equivalently, provided we never ask which coordinate corresponds to which electron. This suggests a fundamental change in our conception of an electron which makes it even less like a billiard ball than things like the Heisenberg uncertainty principle might already imply. To explain what I have in mind it is useful to consider an analogy. Let a string be stretched between two supports and set in motion by deforming it as shown in Figure 1. Here the two deformations are supposed to be identical in size and shape, equidistant from the supports and traveling toward one another with equal and opposite velocities. According to the classical laws of motion of the linearized string the two deformations will pass through one another unchanged and after a while the string will be in the state illustrated in Figure 2. According to these same laws the two deformations will be reflected at the supports and after a time the string will be as shown in Figure 3.

FIGURE 1

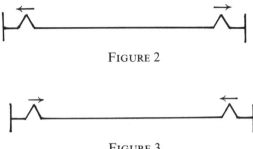

FIGURE 2

FIGURE 3

Notice that the string as a string is in precisely the same state when described in Figure 3 as it was when it was described in Figure 1. Nothing has changed. On the other hand, if one thinks of the two deformations as moving particle-like entities, these two "particles" seem to have changed places. The point of this discussion is that electrons are more like deformations of a string than like billiard balls. When two of them change places nothing has happened physically. Thus it *cannot make sense* to ask which electron has x-coordinate x_1 and which has x-coordinate x_2. Our answer to question (b) is that the replacement of $\mathfrak{IC}_N \times \mathfrak{IC}_e \times \mathfrak{IC}_e \times \cdots \times \mathfrak{IC}_e$ by $\mathfrak{IC}_N \times H^A$ simply takes account of this fundamental indistinguishability of electrons.

More generally, where one studies systems of interacting particles in quantum mechanics and some of the particles are "mutually identical", one must replace the corresponding n-fold tensor product of replicas of the "same" Hilbert space by either its antisymmetric or its symmetric subspace. Which of these one must do is a property of the particle which turns out (for poorly understood reasons) to be a function of the spin of the particle. When the particle has spin $\frac{1}{2}$, $\frac{3}{2}$, $\frac{5}{2}$,... it is the antisymmetric subspace that must be used, and when the spin is 0, 1, 2,... the symmetric subspace is the relevant one. In the first case one says that the particle is a fermion and in the second that it is a boson. This terminology stems from the fact that in quantum statistical mechanics the identity of the particles has far reaching consequences which are quite different for fermions and bosons. The first case was studied by Fermi and Dirac and the second by Bose and Einstein. One refers accordingly to "Fermi-Dirac statistics" and to "Bose-Einstein statistics". The facts about the spins of fermions and bosons are referred to as "the connection between spin and statistics". This connection is a theorem in certain approaches to axiomatic field theory.

Let φ_1, φ_2,... be an orthonormal basis in the Hilbert of spaces of states for a single fermion. Let this Hilbert space be $\mathfrak{IC}_0$. Then the vectors $\varphi_{v_1} \times \varphi_{v_2} \times \cdots \times \varphi_{v_n}$ constitute an orthonormal basis for the tensor product $\mathcal{H}_0 \times \mathcal{H}_0 \times \cdots \times \mathcal{H}_0$ (n factors). Let $\mathfrak{M}$ be the subspace spanned by all the permutations of this product of vectors, i.e. by $\varphi_{v_2} \times \varphi_{v_1} \times \varphi_{v_3} \times \cdots$, etc. One can prove that $\mathfrak{M}$ intersected with the antisymmetric subspace is either $\{0\}$ or is one dimensional and that the latter occurs if and only if $\varphi_{v_1}, \varphi_{v_2}, \ldots, \varphi_{v_n}$ are all distinct. This fact is known as the Pauli exclusion principle—especially when applied to electrons. It is sometimes loosely stated in the form "Two identical fermions cannot be in the same state at the same time".

9. Quantum mechanics applied to electromagnetic radiation: the photon concept. In the development of the classical theory of interactions between charged particles, between magnets, and between magnets and charged particles, an important advance was made by Faraday in emphasizing a "field" rather than an "action at a distance" point of view. One thinks of a charged particle as experiencing an acceleration when at a particular point of space not directly because of the existence of distant charged particles, but indirectly because the existence of these distant particles has modified the space surrounding them. The modification of space produced by the charged particles is described by a vector-valued function defined throughout space and called the electric field $x, y, z \rightarrow \mathbf{E}(x, y, z)$. In a similar spirit one has a magnetic field described by another vector-valued function $x, y, z \rightarrow \mathbf{H}(x, y, z)$. Faraday's work was conceived and written up in a rather intuitive manner avoiding formal mathematics as much as possible. It was continued by Maxwell (born forty years later) whose first paper was published at about the same time as Faraday's last. Maxwell was much more mathematically inclined and began by translating Faraday's ideas into mathematical form. Specifically, he found that many of the discoveries of Faraday and his predecessors could be summed up in the assertion that electric and magnetic fields change in time in such a manner that the functions describing these changes, i.e. $x, y, z, t \rightarrow \mathbf{E}(x, y, z, t)$ and $x, y, z, t \rightarrow \mathbf{H}(x, y, z, t)$ satisfy the following partial differential equations.

$$\operatorname{curl} \mathbf{E} = -\frac{1}{c}\frac{\partial \mathbf{H}}{\partial t}, \qquad \operatorname{div} \mathbf{E} = -4\pi\rho,$$

$$\operatorname{curl} \mathbf{H} = \frac{4\pi i}{c}, \qquad \operatorname{div} \mathbf{H} = 0$$

where c is a constant having the dimensions of a velocity and ρ and i are, respectively, the charge and current densities in space. Maxwell was disturbed by a certain asymmetry in these equations and conjectured that the equation $\operatorname{curl} \mathbf{H} = 4\pi i/c$ should be modified by adding the term $(1/c)\partial \mathbf{E}/\partial t$ to the right-hand side. When one does this, the equations in empty space where $\rho \equiv 0$ and $i \equiv 0$ take the beautifully simple and symmetric form

$$\frac{1}{c}\frac{\partial \mathbf{H}}{\partial t} = -\text{curl}\,\mathbf{E}, \qquad \frac{1}{c}\frac{\partial \mathbf{E}}{\partial t} = \text{curl}\,\mathbf{H},$$

$$\text{div}\,\mathbf{H} = 0, \qquad\qquad \text{div}\,\mathbf{E} = 0.$$

Moreover, if one differentiates the top two equations with respect to t and performs some straightforward eliminations, one deduces the two equations

$$\frac{1}{c^2}\frac{\partial^2 \mathbf{E}}{\partial t^2} = \frac{\partial^2 \mathbf{E}}{\partial x^2} + \frac{\partial^2 \mathbf{E}}{\partial y^2} + \frac{\partial^2 \mathbf{E}}{\partial z^2},$$

$$\frac{1}{c^2}\frac{\partial^2 \mathbf{H}}{\partial t^2} = \frac{\partial^2 \mathbf{H}}{\partial x^2} + \frac{\partial^2 \mathbf{H}}{\partial y^2} + \frac{\partial^2 \mathbf{H}}{\partial z^2}.$$

In other words, both the electric and magnetic fields satisfy the so-called wave equation, and this implies that these fields are propagated through space as waves travelling with velocity c. Moreover, the equations $\text{div}\,\mathbf{H} = 0$ and $\text{div}\,\mathbf{E} = 0$ imply that these waves are so-called "transverse waves". This was all rather exciting because it was already known that the constant c was approximately equal to the velocity of light and that light consisted in some sense of transverse waves. On the other hand, no one had any idea what light waves, in fact, consisted of; that is, what it was whose wavelike changes manifested themselves as light. Maxwell's proposal that light consisted in the wavelike changes of an electric field and a magnetic field, interrelated by the equations above, was verified a decade or so later by experiments of Hertz.

One now speaks of the electromagnetic field, of Maxwell's equations, and of electromagnetic radiation. Ordinary light consists of electromagnetic radiation whose frequency of vibration lies between certain narrow limits to which the human retina is sensitive. The notion of frequency of vibration is given a precise mathematical meaning by applying Fourier analyses to the solutions of Maxwell's equations. Every solution with finite total energy (see below) is a "continuous superposition" of solutions, each of which has a sharply defined frequency, wave length and propagation direction.

Electromagnetic radiation is produced whenever charges are accelerated and, in particular, when they are made to oscillate rapidly back and forth. When this happens mechanical energy disappears and apparently enters the field because it may be recovered at distant points by letting the radiation act on suitable devices. Experiments are consistent with the hypothesis that the energy in the field is distributed throughout space, the energy density at the point x, y, z at time t being

$$\frac{1}{8\pi}\left(\left|\mathbf{E}(x, y, z, t)\right|^2 + \left|\mathbf{H}(x, y, z, t)\right|^2\right),$$

and, in fact, we can recover the law of conservation of energy by thinking of the electromagnetic field as a dynamical system which interacts with more conventional dynamical systems and interchanges energy with them.

This nonmaterial "honorary" dynamical system is most simply and elegantly described in the idealized special case in which space is free of matter and the electromagnetic field is all that exists. Recall, first, that in a conventional dynamical system of n interacting particles one studies the changes in time of the $3n$-tuple of particle coordinates. Labeling these q_1, $q_2, \ldots, q_{3n}$, one finds that in many important cases these time changes are governed by differential equations of the form $m_j d^2 q_j / dt^2 = -\partial V / \partial q_j$ where the m_j are positive real numbers which are equal in triples and V is a function of $q_1, \ldots, q_{3n}$, which describes the interaction. Introducing the auxiliary variables $p_j = m_j dq_1 / dt$, these equations may be written as $6n$ first order equations

$$\frac{dp_j}{dt} = -\frac{\partial V}{\partial q_j}, \quad \frac{dq_j}{dt} = \frac{p_j}{m},$$

and (when V is reasonably well behaved) it follows from the general theory of differential equations that there is exactly one solution having given values for the q_j and the p_j at some fixed time t_0. Since the p_j and the dq_j / dt determine one another, one can say that the future of the system is determined once one knows the *configuration* $q_1, q_2, \ldots, q_{3n}$, and the *rate of change of that configuration* $dq_1 / dt, \ldots, dq_{3n} / dt$ at some fixed time t_0. The pair consisting of the configuration and its rate of change is called the state of the system. One checks easily that the function

$$q_1, \ldots, q_{3n}, p_1, \ldots, p_{3n} \rightarrow \sum_{j=1}^{3n} \frac{p_j^2}{2m} + V(q_1, q_2, \ldots, q_{3n})$$

$$= \tilde{H}(q_1, \ldots, q_{3n}, p_1, \ldots, p_{3n})$$

has the property that it remains constant during the motion of the system. It is called the total energy function. Noticing that $\partial \tilde{H} / \partial q_j = \partial V / \partial q_j$ and $\partial \tilde{H} / \partial p_j = p_j / m$, one sees that the equations of motion may be rewritten in the symmetrical and elegant form

$$\frac{dq_j}{dt} = \frac{\partial \tilde{H}}{\partial p_j}, \quad \frac{dp_j}{dt} = -\frac{\partial \tilde{H}}{\partial q_j}.$$

This is called the *Hamiltonian* form.

To think of Maxwell's equations as being the equations of motion of a dynamical system, one thinks of the electric field in empty space as a "configuration" whose changes in time are to be studied. The analogue of the function V is the part of the total energy in the electromagnetic field which is due to the electric field $\mathbf{E}$, i.e.

$$\frac{1}{8\pi} \iiint |\mathbf{E}(x, y, z)|^2 \, dx \, dy \, dz.$$

The analogue of the equations $dq_j/dt = p_j/m$ is the Maxwell equation $d\mathbf{E}/dt = c \operatorname{curl} \mathbf{H}$. Since the vector operator curl is one-to-one on square summable vector-valued functions with zero divergence, H can be "defined" as $(1/c)\operatorname{curl}^{-1}(\partial \mathbf{E}/\partial t)$ and so regarded as an auxiliary variable analogous to the $3n$-tuple $p_1, \ldots, p_{3n}$. Maxwell's equations

$$\frac{\partial \mathbf{E}}{\partial t} = c \operatorname{curl} H \quad \text{and} \quad \frac{\partial \mathbf{H}}{\partial t} = -c \operatorname{curl} E$$

now become the analogues of

$$\frac{dq_j}{dt} = \frac{\partial \tilde{H}}{\partial p_j} \quad \text{and} \quad \frac{dp_j}{dt} = -\frac{\partial \tilde{H}}{\partial q_j},$$

once we see how to think of $c \operatorname{curl} \mathbf{H}$ and $-c \operatorname{curl} \mathbf{E}$ as "partial derivatives" of the total energy

$$\frac{1}{8\pi} \iiint |\mathbf{E}(x, y, z)|^2 \, dx \, dy \, dz + \frac{1}{8\pi} \iiint |\mathbf{H}(x, y, z)|^2 \, dx \, dy \, dz.$$

This can be done using the concept of "functional differentiation" which first arose in the calculus of variations and is a substitute for partial differentiation in dealing with functions whose arguments are other functions. However, we shall omit the details in the interest of brevity. In the end one finds that one can regard the electromagnetic field as differing in no essential respect from a classical dynamical system except in its infinite dimensionality, and even that difference disappears if one augments the classical systems by those describing vibrating strings and other continua. The two Maxwell equations containing the time are just the equations of motion of this system and the other two restrict the vector fields being considered.

Our next remark is that the dynamical system defined by Maxwell's equations is simpler than most in that the equations of motion are linear. This makes them easy to solve and implies that the one-parameter group of automorphisms of the "phase space" that one obtains by integrating the equations is a one-parameter group of linear transformations. This statement makes sense because the phase space itself is linear. It is simply the set of all suitably differentiable pairs of vector-valued functions on space x, y, $z \to \mathbf{E}(x, y, z)$, $\mathbf{H}(x, y, z)$ such that $\operatorname{div} \mathbf{E} = \operatorname{div} \mathbf{H} = 0$ and both are square integrable. Using the square integrability one has a natural norm in the phase space and this norm is preserved under the one-parameter group $t \to A_t$ of linear transformations describing the motion. This is, of course, just a reflection of the constancy of the total energy and the fact that the norm is defined in terms of the energy. It is, of course, possible to form the completion of this space and it is easy to see that it is a real Hilbert space $\mathcal{H}$ to which the linear operators A_t have unique continuous extensions $\tilde{A}_t$. Now there is a canonical way, which we shall not stop to describe, of introducing a multiplication by $i = \sqrt{-1}$ in the real Hilbert space $\mathcal{H}$ which is such that it becomes a complex Hilbert space in which the operators $\tilde{A}_t$ are complex linear and, in fact, unitary. We remark in passing that the fact that the phase space of the electromagnetic field, regarded as a dynamical system, can be completed to a complex Hilbert space in which the dynamics is described by a one-parameter unitary group is not a peculiarity of the electromagnetic field. The same thing is true for linear continuum mechanics both for the linearized vibrating string and the linearized elastic solid.

The invariance of Maxwell's equations under translations and rotations leads easily to the conclusion that our (now) complex Hilbert space $\mathcal{H}$ admits a natural representation U of the simply connected covering group $\tilde{\mathfrak{E}}$ of the group $\mathfrak{E}$ generated by rotations and translations. Moreover, the U_d and the $\tilde{A}_t$ commute so that we arrive at a unitary representation of $\tilde{\mathfrak{E}} \times T$. Thus the phase space of the dynamical system defining the electromagnetic field (slightly augmented by certain limit points) together with its dynamics is the same kind of mathematical object as the quantum mechanical model for a single free particle. One can well ask whether there exists a particle in nature whose quantum mechanical behavior is described by the representation of $\tilde{\mathfrak{E}} \times T$ which we have just associated with the *classical* electromagnetic field: There is—at least if one generalizes the particle concept slightly—and it is known as a *photon*.

Now all this is very suggestive and hints at a resolution of the celebrated paradox of the "old quantum theory" of 1900–1925 in which light was mysteriously conceived as consisting simultaneously of waves and particles. However, as formulated above, the reinterpretation of the classical mechanics of the electromagnetic field raises more questions than it answers and is little more than a curious fact. The real meaning of this curiosity becomes apparent when we follow Dirac and seek to quantize the electromagnetic field, that is, seek a quantum mechanical system whose observables are the observables of the electromagnetic field and whose changes in time in the classical limit are the same as those predicted by Maxwell's equations.

While there are serious obstacles in extending standard quantization procedures from the finite-dimensional dynamical systems consisting of interacting particles to the infinite-dimensional systems defined by continua, these difficulties are largely mitigated when the equations of motion are linear. This is essentially because a linear system can be factored into independent finite-dimensional systems—although this factorization may be into continuum-many parts. However, without explicitly carrying out this factorization one can demonstrate one of its most important consequences. This is that there is a very simple and general relationship between the unitary representations $t \to V_t$ of the real line which describes the *classical* motion of a *linear* system and the unitary representation $t \to \tilde{V}_t$ of the real line which describes the motion of the corresponding quantized system. This relationship may be summed up by the equation

$$\tilde{V}_t = I \oplus V_t \oplus_s (V \otimes_s V)_t \oplus (V \otimes_s V \otimes_s V) \oplus \cdots$$

where I denotes the (one-dimensional) identity representation and $(V \otimes_s V \otimes_s \cdots \otimes_s V)$ denotes the *symmetrized nth power* of the representation V. To define symmetrized nth power of a representation, one takes $\mathcal{H}(V) \times \mathcal{H}(V) \times \cdots \times \mathcal{H}(V)$ and defines the symmetric subspace precisely as was done in §8 in discussing the identity of particles. This is an invariant subspace for the n-fold tensor product $V \otimes \cdots \otimes V$ (n factors) and restricting to this subspace defines the symmetrized nth power of V.

It remains to show how this relationship between the classical phase space and the quantum mechanical Hilbert space for the dynamics of the electromagnetic field leads to the photon concept, and makes sense of the fact that electromagnetic radiation is at the same time a system of particles and a field whose disturbances are propagated in a wavelike manner. On the one hand, it is not difficult to write down an explicit selfadjoint operator in the Hilbert space

$$\mathcal{H}(I) \oplus \mathcal{H}(V_t) \oplus \mathcal{H}(V) \otimes_s \mathcal{H}(V) \oplus \cdots$$

corresponding to each field observable; a typical field observable being the average value of the x component of the magnetic field in some specific small volume in space. One can then study the interplay between the one-parameter unitary group $\tilde{V}_t$ and these selfadjoint operators, and so study the quantum mechanical behavior of the classical dynamical system defined by Maxwell's equations. On the other hand, one can ignore the selfadjoint operators defining the field observables and look instead at the observables suggested by the interpretation of $\mathcal{H}(V_t)$ as the Hilbert spaces of states for a single free particle. The subspace $\mathcal{H}(V \otimes_s V \otimes_s \cdots \otimes_s V) = \mathcal{H}(V) \otimes_s \mathcal{H}(V) \otimes_s \cdots \otimes_s \mathcal{H}(V)$ can then be interpreted as the Hilbert space of states for a system of n identical copies of this particle under the assumption (consistent with its spin) that the particle is a boson. In other words, the whole Hilbert space can be interpreted as describing the free (noninteracting) motion of a system consisting of an indefinite number of identical bosons. Although it is one system it can be viewed in two quite different ways. Depending upon which observables are scrutinized, it can appear alternatively as a (quantized) system of waves or as a quantized system of particles called photons. In somewhat different language and from a somewhat different point of view, this reconciliation of the 250-year-old controversy as to whether light consists of waves or particles was found by P.A.M. Dirac and published in a famous paper "The quantum theory of the emission and absorption of radiation", Proc. Roy. Soc. (1927), 243–265.

10. Electrons as field quanta: second quantization and the Dirac field. If one can find particles such as photons arising in a natural or automatic way without being postulated by "quantizing" the dynamical system represented by the electromagnetic field, can one not do likewise with other particles such as electrons? Could one not go in the reverse direction and discover or invent a field whose quantization would lead to electrons as Dirac's quantization of the electromagnetic field led to photons? One might even hope to see the fermion nature of electrons built into this quantization as was the case for the boson nature of photons. There is even an obvious way to look for the appropriate field. This is to start with the quantum mechanical model for a single free electron and, reversing what was done in §9, try to reinterpret it as a classical continuum. Since the Hilbert space of a free electron consists of the square summable functions from three-space to a two-dimensional complex Hilbert space, this appears to be a feasible program. However, to carry it out and arrive at particles which are fermions, it seems to be necessary to alter the usual quantization procedure in such a way that the generalized coordinates and momenta are described by selfadjoint operators which do not satisfy the Heisenberg commutation rules but a variant of them in which commutators $PQ - QP$ are replaced by anticommutators $PQ + QP$. The writer does not know of a plausible a priori justification for this procedure and has always felt uneasy about it. In any event, the program was successfully carried out by Jordan and Wigner in 1928 and in the course of time it became standard procedure to postulate an underlying field for every particle thought to be elementary. Since one "quantizes" this field and the field itself is obtained by quantizing a free particle motion, one speaks of *second quantization*. One speaks of the electron field as the Dirac field because it was Dirac who first described the way in which the Schrödinger equation for the free electron must be changed in order to be relativistically invariant (see §4).

Regarding particles as the quanta of suitably defined fields is of major importance in the still incompletely realized goal of constructing a logically coherent synthesis of quantum mechanics and special relativity. It seems to be difficult to make a theory of interacting particles which is also relativistically invariant (even in classical mechanics) but rather easy to do so for interacting fields. One then seeks to extend Dirac's quantization of the electromagnetic field to the system of interacting fields which (hopefully) describe the particles of nature. This done, one can deduce the behavior of the particles from that of the fields whose quanta they are. While this program is a plausible one and has been the goal of elementary particle physics for half a century, it presents great difficulties which so far have been only very partially overcome.

From one point of view, the key difficulty is that the interactions, which it seems reasonable to postulate between fields, lead to nonlinear equations of motion and this prevents factoring the whole system into finite-dimensional components. On the other hand, the standard methods for quantizing finite-dimensional nonlinear dynamical systems, when looked at rigorously rather than formally, lose much of their meaning when one drops the finite-dimensionality assumption.

The physicists have struggled with the difficulties with great resourcefulness and ingenuity and have had a number of striking partial successes. At the moment there is considerable optimism among a large contingent concerning the future of certain recent ideas. The writer must confess, however, that he is unconvinced that the key to the problem has yet been found.

We remark, in conclusion, that in the face of the stubborn difficulties presenting a complete reconciliation of quantum mechanics with special relativity, it is all too easy to forget the enormous success of nonrelativistic quantum mechanics in explaining low energy phenomena, in fact, almost everything that one can observe outside of highly specialized laboratories. It is only when one wants to understand the structure of nuclei rather than accept them as "elementary" that one is forced to consider energies at which relativity becomes important and is led into experiments which produce the current proliferation of new (but very short-lived) particles. Even the structure of nuclei can be understood to a fair degree of approximation by applying the theory of nonrelativistic bound states to collections of neutrons and protons assumed to interact in definite but rather complicated ways of the sort indicated in §5.

Bibliography

1. A. M. Gleason, *Measures on the closed subspaces of a Hilbert space*, J. Math. Mech. **6** (1957), 885–893.

2. G. W. Mackey, *Imprimitivity for representations of locally compact groups*, Proc. Nat. Acad. Sci. **35** (1949), 537–545.

3. _____, *The theory of unitary group representations*, Univ. of Chicago Press, Chicago, Ill., 1976.

4. _____, *Unitary group representations in physics, probability and number theory*, Benjamin/Cummings, Reading, Ma., 1978.

DEPARTMENT OF MATHEMATICS, HARVARD UNIVERSITY, CAMBRIDGE, MASSACHUSETTS 02138

Lectures in Applied Mathematics
Volume **21**, 1985

Phase-Space Representations

Daniel Sternheimer

ABSTRACT. Lie groups and algebras can be represented alternatively
(nonoperatorially) by means of deformed products of the associative
algebra of functions over some symplectic manifolds (coadjoint orbits
in general). A presentation of the main tools and notions of this new
theory is given here, and a summary of the results obtained is
presented.

Introduction and summary. The development of quantum physical
theories has contributed in a significant manner to—and has often
initiated—the development of the theory of Lie group representations.
Either as symmetries of physical states or as covariance groups of
observables, Lie groups act through their (in general unitary) operatorial
representations on Hilbert spaces. However, quantum theories are usually
"quantizations" of their "classical limit" obtained when $\hbar \to 0$, and
classical theories are related to some phase space. The operatorial formulation is lost at the classical level, but the symmetries (especially the
covariance groups) often persist.

A number of (more or less rigorous) expressions of the classical limit
have been given, in particular via the notion of coherent states (an
overdetermined complete set of generalized vectors "in" the Hilbert
space). Recently it has been shown [1] that quantum mechanics can be
treated as a deformation, in the sense of the mathematical theory [2] of
deformations of Lie or associative algebras, of classical mechanics. The
fact that there is a connection between deformation theory and quantization, or rather between contractions and classical limit, was certainly

1980 *Mathematics Subject Classification.* Primary 22E99, 16A58, 53C15, 81A78.
Key words and phrases. Coadjoint orbits, deformation theory, group representations.

suspected by many mathematical physicists; this fact was rigorously established. But the crucial point was to show that quantum mechanics can be developed in an *autonomous* manner on the same observables as classical mechanics (functions over phase-space), using a deformed algebraic composition law—a fact which has obvious conceptual advantages, and also some practical ones (especially when the phase-space is not vectorial, due to constraints).

This way of looking at some level of a theory as a deformation (or as its inverse, a contraction) of another level, when both are properly formulated, is quite general in physics (cf. e.g. [3]). In more or less the same way as quantization as indicated above, it should be possible to look at statistical mechanics as a deformed theory with deformation parameter $\beta \neq 0$. The thermodynamic limit in spin systems has been associated [4] with the contraction of spin $j \to \infty$ rotation group representations to the Euclidean limit—and this can be translated in the above-mentioned algebraic framework. Similarly, the notion of "large N limit" (cf. e.g. [5]) can be translated into the limit $N \to \infty$ of an $SO(N)$-quantization (cf. §2a), also formulated in the same algebraic framework.

We are thus led naturally to look for a realization of group representations in terms of deformed products over some phase-space. This need for a "representation theory without operators" is also suggested by a number of other facts. For instance, some of the classical observables play a preferred role in the quantization procedure—and they generate a finite-dimensional subgroup of the canonical transformations [1]; the orbits in the coadjoint representation play a central role in the so-called geometric quantization approach [6]—and are useful in representation theory also [7]; etc.

The purpose of this contribution is to sketch the progress which has been made (in the past 6 years) in this new branch of Lie group representation theory: the representations of Lie algebras and Lie groups by means of deformed products of functions on suitable phase-spaces (typically, orbits in the coadjoint representation). This development is completely *autonomous* but effort will be made to relate these representations (defined by the phase-space and a deformed product on it) to the corresponding (in a sense to be made precise) operatorial representations.

We shall start with some basic facts and results on quantization as a deformation, including the paradigm of the Weyl quantization realized with Moyal products and brackets on $\mathbf{R}^{2l}$—and the metaplectic algebra of preferred observables associated with it. Then (§2) we shall present the main definitions and (at least on the formal level) tools of the autonomous phase-space representation theory, and of its operatorial correspondence. Finally we shall present a number of examples of low-dimensional groups or of classes of groups, for which the theory has been developed. There is still a lot to be done, and I shall be glad if this contribution brings some new contributors to this promising aspect of group representation theory.

1. Deformations, quantization and preferred observables.

(a) Let W be a phase-space, i.e. a symplectic manifold (of dimension $2l$) with a closed 2-form F such that F^l is a volume element ($\neq 0$ everywhere). Denote by $N = C^\infty(W)$ and by $E(N, \nu)$ the space $N((\nu))$ of formal series in some parameter ν, with coefficients in N. If i denotes the interior product, an isomorphism between the tangent and cotangent bundles is realized by $\mu: TW \ni X \mapsto -i(X)F \in T^*W$ and it extends to higher tensors, so that $\Lambda = \mu^{-1}(F)$ is a 2-tensor. The Poisson bracket on N (or $E(N, \nu)$) can then be written

$$P(u, v) = \{u, v\} = i(\Lambda)(du \wedge dv), \qquad u, v \in N, \qquad (1)$$

and its powers (as bidifferential operator) can be written, on a local chart $U \subset W$ with coordinates $(x_i)_{i=1}^{2l}$ (and $\partial_i = \partial/\partial x_i$, etc.)

$$P^r(u, v) = \Lambda^{i_1 j_1} \cdots \Lambda^{i_r j_r} \partial_{i_1 \cdots i_r} u \partial_{j_1 \cdots j_r} v. \qquad (2)$$

We consider deformations of the associative algebra N, and of the Lie algebra (N, P), in the sense of the general theory of deformations of associative or Lie algebras [2], of the form (the cochains C_r being bidifferential operators)

$$u *_\nu v = uv + \sum_{r=1}^\infty \nu^r C_r(u, v) \in E(N, \nu), \tag{3}$$
$$u, v \in N \text{ or } E(N, \nu),$$

$$[u, v]_\lambda = P(u, v) + \sum_{r=1}^\infty \lambda^r C_{2r+1}(u, v) \in E(N, \lambda), \tag{4}$$
$$u, v \in N \text{ or } E(N, \lambda),$$

where $C_1(u, v) = P(u, v)$ and $C_r(u, v) = (-1)^r C_r(v, u)$. These conditions ensure that (3) gives (4) with $\lambda = \nu^2$. The typical example is $W = \mathbf{R}^{2l}$ and $C_r = (1/r!)P^r$ (Moyal product [8] or bracket [9]): in the Weyl quantization procedure [10], the product $u *_\nu v = \exp(\nu P)(u, v)$ is mapped into the product of the operators corresponding to u and v, with $\nu = \frac{1}{2}i\hbar$; the usual quantum mechanics can then be developed entirely in $E(N, \nu)$, and several examples (including of course the harmonic oscillator, the universal paradigm, and the hydrogen atom) have been explicitly computed [1] in this autonomous framework.

(b) The existence problem of $*$-products is a more delicate question. When $b_3 = \dim H^3(W, \mathbf{R}) = 0$, J. Vey [11] showed that there exists a bracket (4), and Neroslavsky and Vlasov [12] and Lichnerowicz [13] proved the existence of $*_\nu$-products where, in addition, the cochains C_r are null on the constants and have the same principal symbol as $(1/r!)P^r$ (these products are called Vey products). The condition $b_3 = 0$ is by no means necessary; in particular, there exist $*_\nu$-products on cotangent bundles to a large class of homogeneous spaces, [1, 14] and on $W = T^*M$ when M is parallelizable [15] (e.g. a Lie group); this result has been extended, using the same technique, to any cotangent bundle by M. de Wilde and P. Lecomte; in fact these authors have proved (May 1983) the existence of $*$-products on *any* symplectic manifold..

The first terms (up to C_3) of a Vey product can be given a closed form, using a symplectic connection Γ (the expression $\exp(\nu P)(u, v)$ with covariant derivatives instead of ordinary ones in (2) gives an associative product iff Γ has no curvature). Therefore only those (ν-independent) symplectomorphisms of W which preserve Γ can preserve the Vey products of brackets: they form a finite-dimensional subgroup (for $\mathbf{R}^{2l}$, the corresponding Hamiltonians belong to $\mathfrak{sp}(l, \mathbf{R}) \cdot \mathfrak{h}_l$, where $\mathfrak{h}_l$ is the Heisenberg algebra, i.e. they are all polynomials of degree ≤ 2). Their generators (in N) form an algebra of preferred observables, for which the Poisson and Moyal brackets with any other observable in N coincide— and therefore they are those Hamiltonians for which the classical and quantum trajectories in N coincide. Another choice of quantum ordering (such as the standard ordering, which is chosen for the usual pseudodifferential operators in mathematics) will give an equivalent $*$-product, but will, in general, restrict the algebra of (ν-independent) preferred observables to $\mathfrak{h}_l$: an equivalence (in the sense of the theory of deformations) may be a nontrivial transformation from a group theoretical point of view. This is not so surprising since, if $b_2 = 0$, all nontrivial deformations of (N, P) can be transformed [1] into Moyal brackets (by equivalence and possibly a polynomial change in the deformation parameter), which is a kind of uniqueness result for quantum mechanics, while the UIR (unitary irreducible representations) of the Heisenberg group H_l are only projectively equivalent one to another.

(c) Let $\mathcal{Q}$ be a real Lie algebra of dimension m, $\mathcal{Q}^*$ its dual, the duality being written $\langle \cdot, \cdot \rangle$, G a connected group with Lie algebra $\mathcal{Q}$. A 2-tensor on $\mathcal{Q}^*$ can be defined [7] by $\Lambda_\xi(a, b) = \langle \xi, [a, b] \rangle$, $\xi \in \mathcal{Q}^*$, $a, b \in \mathcal{Q}$; it is degenerate. G acts on $\mathcal{Q}^*$ via the coadjoint representation $G \ni g \mapsto \mathrm{Ad}^* g = {}^{tr}\mathrm{Ad}(g^{-1})$, and on each orbit of G in $\mathcal{Q}^*$ the restriction of Λ defines [7] a symplectic structure and a Poisson bracket by (1). We can therefore, on each orbit, look for deformations of the Lie and associative algebras N. [For $\mathcal{Q} = \mathfrak{h}_l$, the orbits are $\mathbf{R}^{2l}$ with the usual symplectic structure, and are parametrized by the value of the center of $\mathfrak{h}_l$ on them: the Moyal product on any of them will be the autonomous version of the

UIR's of H_l (the value of the deformation parameter being linearly related to that of the center of $\mathfrak{h}_l$ in the UIR's), the correspondence between both aspects being realized by the Weyl quantization rule.] To be consistent, we shall have to require that, on $\mathcal{C}$, the deformed bracket coincides with the Poisson bracket of the corresponding functions (on $\mathcal{C}^*$). In many cases (but not always) $\mathcal{C}$ will also be an algebra of preferred observables in the above sense.

2. Some basic facts for group representations by ∗-products on phase spaces.

(a) *Invariance and covariance of ∗-products.* Let G be a connected Lie group, acting by symplectomorphisms on a symplectic manifold W. Denote by $\mathcal{C}$ its Lie algebra; we shall suppose that the elements $X \in \mathcal{C}$, acting on W, are globally Hamiltonian vector fields, the Hamiltonians u_X (defining $\mathcal{C}$) of which satisfy the commutation relations of $\mathcal{C}$ with respect to the Poisson bracket on W. (This is always true when W is an orbit for the coadjoint representation in $\mathcal{C}^*$.) A ∗-product on $E(N, \nu)$ is said to be $\mathcal{C}$-*covariant* if the associated bracket coincides on $\mathcal{C}$ with the Poisson bracket $\{\cdot, \cdot\}$:

$$\{u_X, v_Y\} = (2\nu)^{-1}(u_X * v_Y - v_Y * u_X) \equiv [u_X, v_Y], \qquad X, Y \in \mathcal{C}.$$

(5)

Then one can prove [17] that there exists a representation τ of $\tilde{G}$ (the universal covering of G) by automorphisms of $(E(N, \nu); *)$ such that

$$\tau_g(u) = \left(I + \sum_{s \geqslant 1} \nu^s \tau_g^s\right)(g \cdot u) \qquad \forall g \in \tilde{G}, u \in N \text{ or } E(N, \nu) \quad (6)$$

where τ_g^s is a differential operator on W and $(g \cdot u)(\xi) = u(g^{-1} \cdot \xi)$. This justifies the definition. The ∗-product is said to be *G-invariant* if the τ_g^s can be taken 0, i.e. if $g \cdot (v * w) = (g \cdot v) * (g \cdot w), \forall g \in G$, or equivalently if

$$\{u_X, v * w\} = \{u_X, v\} * w + v * \{u_X, w\},$$

$$\forall v, w \in N, X \in \mathcal{C}. \quad (7)$$

If $\mathcal{C}$ is "large enough" in the sense that the vector fields $X \in \mathcal{C}$ generate the tangent space at each point $\xi \in W$, then it is *equivalent* [17] to assume that the ∗-product is $\mathcal{C}$-*covariant and G-invariant* and that $\mathcal{C}$ is an *algebra of preferred observables*: $\{u, v\} = [u, v] \forall u \in \mathcal{C}, v \in N$. (In [17], (5) defines what is called an "$\mathcal{C}$-quantization" and (6) the G-covariance, and it is shown that (5) implies the existence of τ satisfying (6); we have chosen to reduce the number of definitions). If the ∗-product is $\mathcal{C}$-covariant, it is enough [17] to assume (7) for v and w in the associative ∗-algebra $\mathcal{P}$ generated by $\mathcal{C}$ (the ∗-representation of the enveloping algebra of $\mathcal{Q}$) in order to show that $\mathcal{Q}$ is an algebra of preferred observables. Many examples of *-products are G-invariant, but such *-products do not exist for all Lie groups or all "representations" of Lie groups. For instance, some nilpotent groups [18], or the Poincaré group orbit of zero mass and helicity [19], do not possess G-invariant Vey products: only $\mathcal{Q}$-covariance can be obtained, but this is sufficient for a reasonable definition of *-representation. Note that (5) shows that the map $\mathcal{Q} \ni X \mapsto (2\nu)^{-1} u_X \in N$ is a Lie algebra morphism from $\mathcal{Q}$ to $(N, [\cdot, \cdot])$.

(b) *-representations of groups. Let us keep the same definitions and notation as before, and consider a *-product which is $\mathcal{Q}$-covariant. Let us introduce [1] the function Exp (the *-exponential) from $\mathcal{Q}$ into $E(\mathcal{P}, \nu^{-1})$, where $\mathcal{P}$ is the *-algebra generated by $\mathcal{Q}$ (or into $E(N; \nu, \nu^{-1})$) defined by

$$\mathrm{Exp}(X) = \sum_{n=0}^{\infty} \frac{(2\nu)^{-n}}{n!} (u_X *)^n,$$

$$X \in \mathcal{Q}, (u_X *)^n = u_X * \cdots * u_X \ (n \text{ factors}). \tag{8}$$

Then if $Z = Z(X, Y)$ is defined by the Campbell-Hausdorff formula (in $\tilde{G}$): $\exp(X) \cdot \exp(Y) = \exp(Z)$, for $X, Y \in \mathcal{Q}$, the function Exp satisfies

$$\mathrm{Exp}(X) * \mathrm{Exp}(Y) = \mathrm{Exp}(Z). \tag{9}$$

If $e^X = \exp X$ is the element of $\tilde{G}$ corresponding to $X \in \mathcal{Q}$ in the exponential map, we can now define $E(e^X) = \mathrm{Exp}(X)$ and, in view of (9), a group homomorphism E from $\tilde{G}$ into $(E(\mathcal{P}, \nu^{-1}); *)$ (or $(E(N; \nu, \nu^{-1}); *)$) by extension to all of $\tilde{G}$. The introduction of the parameter ν^{-1} follows from the fact that, due to (5), the elements $X \in \mathcal{Q}$ have to be represented by $(2\nu)^{-1} u_X$, and thus each term $(2\nu)^{-n}(u_X *)^n \in (2\nu)^{-n} E(N, \nu)$. Therefore the $\mathrm{Exp}(X)$ given by (8), and their *-products (in particular the elements $E(g)$, $g \in \tilde{G}$) have to be treated as formal series of two (a priori independent) parameters ν and ν^{-1} with coefficients in N. A closer study is necessary in order to make sure that we get finite terms when ν^{-1} is taken as the inverse of ν (cf. [18] for the nilpotent case). The introduction of the *-algebra $\mathcal{P}$ and of $E(\mathcal{P}, \nu^{-1})$ obviously circumvents this problem.

We are now in position to define a map $\mathcal{E}$ from a space D, stable under convolution, of test functions on $\tilde{G}$, into $E(\mathcal{P}, \nu^{-1})$ by

$$\mathcal{E}(f) = \int_{\tilde{G}} f(g) E(g^{-1}) \, dg, \quad f \in D. \tag{10}$$

Invariance of the *-product means that $\mathcal{E} \circ \mathrm{ad}_{\mathcal{Q}}(X) = \mathrm{ad}^*_{\mathcal{Q}}(X) \circ \mathcal{E}$. We shall thus [1, 20–23] define such a distribution (valued in $E(N; \nu, \nu^{-1})$ or $E(\mathcal{P}, \nu^{-1})$) on a Lie group G as a star representation of G. By integration of $\mathfrak{S}$ over W with respect to a suitable quasi-invariant measure we get a (formal series of terms, each of which is a) scalar-valued distribution on G. We may call it the *character* associated with the *-representation $\mathfrak{S}$. The terminology is indeed appropriate since it is in fact an invariant eigendistribution on G (as follows from formula (12) below). Integration over W corresponds to taking the trace of the operators related to the functions on W in a suitable Weyl correspondence. The definition of a *-representation in terms of the distribution $\mathfrak{S}$ rather than E or Exp may be more suitable in a number of cases; indeed the function $E(g, \xi)$, $\xi \in W$, turns out to be analytic almost everywhere on G, but need not be analytic at the identity.

(c) The *correspondence* between *-representations and operatorial representations can be established by several methods (which may be used simultaneously) such as:

(i) Comparison of the characters.

(ii) Introduction [21] of an invariant *Weyl correspondence* Ω between functions on W and operators on a suitable Hilbert space (typically, a space of functions on a manifold of half the dimension of W, given by some polarization, as in the geometric quantization approach). This will be a generalization of the quantization rules adapted to the case of H_j:

$$ u \mapsto \Omega(u) = \int \tilde{u}(\xi, \eta) \exp((\xi \cdot P + \eta \cdot Q)/i\hbar)\omega(\xi, \eta)\, d\xi\, d\eta $$

where $\tilde{u}$ is the inverse Fourier transform of u, P and Q are selfadjoint operators satisfying the usual canonical commutation relations (those of $\mathfrak{h}_l$) and ω is some weight function depending on the quantization rule which is chosen ($\omega = 1$ for the Weyl rule [10], $\omega = \exp(-\frac{1}{4}(\xi^2 + \eta^2))$ for the normal ordering, etc; see e.g. [24]). Here u is a function (or distribution) on $\mathbf{R}^{2l}$ and $\Omega(u)$ an operator (possibly unbounded) on $L^2(\mathbf{R}^l)$; the analytical properties and various expressions of this correspondence and its inverse (sometimes called the Wigner mapping, since E. Wigner [25] was the first to give the "distribution function" on phase-space $\mathbf{R}^{2l}$ corresponding to the projector on a vector of $L^2(\mathbf{R}^l)$) have been studied (and are still being studied) by many authors (see e.g. [24 and 26] and references quoted therein).

This (generalized) Weyl correspondence Ω should intertwine between the *-representation $\mathfrak{S}$ and the usual representation U, i.e. $U = \Omega \circ \mathfrak{S}$. (The characters should be the same for $\mathfrak{S}$ and U.)

(iii) Introduction of a *-polarization*: this is a subspace S of $E(N, \nu^{-1})$ stable and irreducible under the representation $(g, f) \mapsto E(g) * f, g \in G$, $f \in S$. In practice it will be built [18, 20, 21] using a subalgebra $\mathfrak{N}$ of $\mathcal{Q}$ and a character ξ_0 on $\mathfrak{N}$ as the space of solutions of the equation $f * \mathrm{u}_X = \xi_0(X)f, \forall x \in \mathfrak{N}$. (Typically, ξ_0 will be a point in an orbit W of

G in $\mathcal{Q}^*$ and $\mathfrak{N}$ a subalgebra of $\mathcal{Q}$ subordinate to ξ_0, i.e. such that $\langle \xi_0, [\mathfrak{N}, \mathfrak{N}] \rangle = 0$.) Then the representation $g \mapsto E(g) *$ will be equivalent to a (unitary) irreducible representation of G. (This method is of course well adapted to the nilpotent case [18].)

(iv) Spectral theory for the center of the enveloping algebra and for compact generators. The spectrum of a generator H in $\mathcal{Q}$ (or in the enveloping algebra) can be defined by solutions to the $*$-eigenvalue equations $H * \varphi = \lambda \varphi = \varphi * H$; when $H \in \mathcal{Q}$, it is often more convenient to consider the $*$-exponential $\mathrm{Exp}(itH)$ as a distribution in t (and on W): the spectrum of H will then [1] be the support of the Fourier transform (in t) of this distribution. In a Weyl transformation, the spectrum so defined will correspond to the operatoral spectrum [1].

(d) To *construct* $*$-representations, many methods can be used. The starting point will, in general, be an orbit W for the coadjoint representation in $\mathcal{Q}^*$ (or a collection of orbits, such as $\mathcal{Q}^*$ itself). This is still another manifestation of the effectiveness of the orbit method (see e.g. W. Schmid's lecture [27]) in dealing with representation theory. On the formal level, the quantization of the orbits corresponding to unitary representations (admissible orbits, cf. Kirillov's conjecture [28]) will manifest here mainly by the greater simplicity of the formulas obtained (and thus by a better chance of giving a precise meaning to the formal calculations); but there is here (at least formally) a greater flexibility for the choice of the orbits in their class.

Another possibility is to study a "$*$-regular representation", i.e. a G-invariant $*$-product on T^*G, which has very recently been shown to exist [15]. In particular, one may try to define and find a "$*$-decomposition" of this $*$-regular representation.

Given a symplectic (or Poisson [1]) manifold W on which G acts, a first step may be to check that there *exists* a $*$-product. If $b_3(W) \neq 0$, one has to check that at each level of the deformation one passes through the zero class of the Hochschild 3-cohomology of the associative algebra N. If we look for G-invariant $*$-products, it is enough [19] to compute the G-invariant de Rham 3-cohomology of W: its vanishing will show the existence of a G-invariant linear connection on W. These existence results will in general not give an explicit formula for the $*$-product (except for the first 3 or 4 terms).

A more explicit construction for a $*$-representation may be (especially in the semisimple case) to build the $*$-*exponential* as a solution of differential equations given by the center of the enveloping algebra. Let $Q = p(X_i)$ be such a central element, the X_i being a basis of $\mathcal{Q}$, i.e. coordinate functions on $\mathcal{Q}^*$; its restriction to W will be a constant (still denoted by Q), and the same will be true for $\hat{Q} = p(* X_i)$, where we replace ordinary product by $*$-product in the polynomial expression of Q.

From (9) we get, by differentiation,

$$X * (\text{Exp}) = -2\nu l(X)\,\text{Exp}, \qquad (\text{Exp}) * X = 2\nu r(X)\text{Exp} \qquad (11)$$

where $l(X)$ and $r(X)$ are the left and right (respectively) invariant differential operators on G defined by $X \in \mathcal{G}$, and therefore, if p is homogeneous of degree k, we get

$$\left[l(Q) - (-2\nu)^{-k}\hat{Q} \right] \text{Exp} = 0 \qquad (12)$$

a differential equation which permits [1, 20–23] us to find explicitly the function Exp in a number of cases. No fixed relation is assumed between $\hat{Q}$ (the "Casimir" of the representation) and Q (the parameter of the orbit), though some restrictions may be needed. It also turns out that the *-exponential and the operators $X *$ may have a simpler form when the value of Q is adapted to the orbit. In fact, in the examples that have been computed (cf., in particular, [20] for the rank-1 groups), accomodation of the character to the orbit and selection of some families of orbits by some quantization condition (such as the integrability of a de Rham cohomology class) permit us to obtain the function E in a rather simple and closed form; but this accomodation is by no means necessary, and several computations may be carried through without it. Note also that, except for the nilpotent case and the harmonic oscillator representations (which are closely related to the Heisenberg groups; cf. also [29]), the operators $X *$ will have a pseudodifferential expression, which may require some care in handling.

Finally, once a *-representation has been constructed, one may start looking for a *-polarization and a (generalized) Weyl application. The mapping $\mathcal{E}$ can provide it [20]. Indeed, if G_0 is a closed subgroup of G, and ζ_0 is some character of G_0, the subspace D_0 of test functions f on G satisfying $f(g_0 g) = \zeta_0(g_0)f(g)$, $g_0 \in G_0$, $g \in G$, carries an induced representation T_0 of G. Then $\mathcal{E}(D_0)$ can be taken as a *-polarization subspace S, and $\mathcal{E}$ will map $T_0(g)f$ into $E(g)*\mathcal{E}(f)$ for f in D_0: it intertwines T_0 and the representation given by $E(g) *$ on D_0. Now if we consider the operator (for some function or distribution $\tilde{u}$ on G),

$$\hat{u} = T_0(\tilde{u}) = \int_G T_0(g^{-1})\tilde{u}(g)\,dg$$

then $\mathcal{E}(\hat{u}f) = u * f$, where u is the function on W defined by

$$u = \int_G E(g^{-1})\tilde{u}(g)\,dg.$$

Therefore $\mathcal{E}$ (or E) defines a generalized Wigner mapping $\hat{u} \mapsto \Omega^{-1}(\hat{u}) = u$.

3. Examples (a short review).

(a) The first example treated has been $\mathcal{C} = \mathfrak{sp}(l, \mathbf{R}) \cdot \mathfrak{h}_l$, represented (the oscillator $*$-representation) with polynomials of degree ≤ 2 in p_α, q_α ($\alpha = 1, \ldots, l$) and the Moyal product on $\mathbf{R}^{2l}$: $u*v = \exp(\nu P)(u, v)$, P being the usual Poisson bracket. This representation has been used in various physical problems. It has also the remarkable property of being a representation of the supersymmetry generated by the polynomials of order 2 (the Lie algebra $\mathfrak{sp}(l, \mathbf{R})$) and 1 (the p_α and q_α). The correspondence with usual representations is given by Weyl quantization. For $l = 1$, the restriction to $\mathfrak{sl}(2, \mathbf{R})$ corresponds to a representation of the twofold covering of $SL(2, \mathbf{R})$ (the metaplectic group $Mp(1, \mathbf{R})$), which is the sum of the two representations $D^+(\frac{1}{4})$ and $D^+(\frac{3}{4})$ for which the spectrum of the compact generator $H = \frac{1}{4}(p^2 + q^2)$ is $\{\frac{1}{4} + n\}$ and $\{\frac{3}{4} + n\}$, $n \in \mathbf{N}$.

(b) Using essentially the method mentioned at the end of §2(d), several examples have been treated, with various degress of completeness. They are the 3-dimensional algebras $\mathfrak{so}(3)$, $\mathfrak{so}(2, 1)$ and $\mathfrak{e}(2)$ (of the Euclidean group of the plane), for which the treatment is essentially complete. A number of results are known for higher groups, based on this method.

For instance [20] in the case of $G = SU(2)$ realized as the 3-sphere S^3 (an element $g \in G$ being realized by $x \in \mathbf{R}^4$ with $x_0^2 + \vec{x}^2 = 1$), denoting by $\vec{L}$ the usual basis of $\mathcal{C} = \mathfrak{su}(2)$ considered as functions on the orbit W defined by $Q = \vec{L} \cdot \vec{L} = (2l\hbar)^2$, the adapted $*$-representation with Casimir value $\hat{Q} = 4\hbar^2 l(l + 1)$ (with $2l - 1 \in \mathbf{N}$) is defined by the function $E(g) = (x_0 + \vec{x} \cdot \vec{L}/2i\hbar l)^{2l}$.

(c) The case of the Poincaré group has been studied somewhat extensively. The coadjoint orbits are known. The existence of invariant $*$-products in the positive mass case has been proved [19, 30] (by the vanishing of the invariant de Rham 3-cohomology), but the natural construction on the orbits [19, 22] (by restriction from Moyal on $\mathbf{R}^8$ to the cotangent bundle of the mass hyperboloid, for the spin 0 case, and symplectic product with a product on S^2 for spin $\neq 0$) gives only a covariant, Lorentz invariant, $*$-product. In the massless, zero-helicity case, a covariant $*$-product has been built [31] but there are no invariant Vey products [19] (since there are no invariant linear connections; the orbit is the cotangent bundle to the vertexless light-cone); for nonzero helicity, the orbits do not have a simple description (in this case, the differentiable vectors for the unitary representations are [32] sections of a nontrivial complex line bundle over the cone, and a similar complication arises for these coadjoint orbits); this makes the explicit expression of an associated covariant $*$-product more complicated.

(d) The nilpotent groups have recently been exhaustively studied [18, 33]. The case of the Heisenberg groups H_l is of course solved by the Moyal product on $\mathbf{R}^{2l}$, and it is invariant. For general nilpotent groups (when the central series has length $\geqslant 3$) only covariant products can be expected. Each orbit of the coadjoint representation being symplectomorphic to some $\mathbf{R}^{2l}$, the Moyal product there will give a (covariant) ∗-product on the orbit, for which the ∗-exponential can be handled rather easily (due to the nilpotency); this construction is (up to automorphism) independent of the choice of the symplectic global chart for W. A ∗-polarization can then be introduced, giving a linear representation equivalent to the unitary irreducible representation associated with the orbit, which establishes correspondence with the usual theory. [Note, however, that in all the ∗-representation approach the construction is completely autonomous, correspondence being made only for convenience of the conservative reader!]

(e) For compact groups coadjoint orbits, the dimensions of the de Rham cohomology (in degree 1, 2 and 3) have been studied (it is enough to consider the case of simple groups, for which $b_1 = 0$ if they are simply connected, b_2 is the dimension of a maximal torus of the stabilizer of a point in the orbit, and $b_3 = 0$). Then there are ∗-products, the invariance of which is being studied. [Since, when we restrict ourselves to group invariant forms, $b_2^{\mathrm{inv}} = 1$ and $b_3^{\mathrm{inv}} = 0$, there is in this case a unique invariant ∗-product: this result has just been obtained by S. Gutt.]

On the cotangent bundle to any Lie group it has very recently been shown [15, 34] that there exists a G-invariant ∗-product. This follows from an explicit study of the Hochschild 3-cocycles occurring in the step by step construction of a ∗-product on a parallelizable manifold, and from the form of the 2-cochains C_r thus obtained. This ∗-product seems to play the role of a ∗-regular representation of G. For a compact Lie group G, it has been suggested [35] that this ∗-regular representation should be decomposable in a way similar to the Peter-Weyl decomposition of the (linear) regular representation, giving at once all the ∗-representations. In the general case, a ∗-Plancherel formula might be a useful notion.

(f) Finally, it may be worthwhile to stress one point. The general theory (§2) was presented at a rather heuristic level; in particular, the questions of convergence of the formal series introduced, the precise domains of test-functions needed, etc., were not touched. However, in the examples that were treated, the derivations were mathematically rigorous —and in many instances closed formulas could be obtained.

REFERENCES

1. F. Bayen, M. Flato, C. Fronsdal, A. Lichnerowicz and D. Sternheimer, *Quantum mechanics as a deformation of classical mechanics*, Lett. Math. Phys. **1** (1977), 521–530; *Deformation theory and quantization*, Ann. Physics **111** (1978), 61–151, see also D. Sternheimer, *Deformation theory applied to quantization and group representations*, Lecture Notes in Phys., vol. 153, Springer-Verlag, Berlin and New York, 1982, pp. 314–318.

2. M. Gerstenhaber, *On the deformation of rings and algebras*, Ann. of Math. (2) **79** (1964), 59–103.

3. M. Flato, *Deformation view of physical theories*, Czechoslavak J. Phys. B **32** (1982), 472–475.

4. D. Arnal and J. C. Cortet, *A group theoretical approach of thermodynamic limits in spin systems*, Lett. Math. Phys. **1** (1977), 505–512; *Geometrical theory of contractions of groups and representations*, J. Math. Phys. **20** (1979), 556–563.

5. L. G. Yaffe, *Large N limits as classical mechanics*, Rev. Modern Phys. **54** (1982), 407–435.

6. B. Kostant, *Quantization and unitary representations*, Lecture Notes in Math., vol. 170, Springer-Verlag, Berlin and New York, 1970, 87–208; J. M. Souriau, *Structure des systèmes dynamiques*, Dunod, Paris, 1970; D. Simms, *Symplectic geometry and quantization*, Lectures Notes in Phys., vol. 153, Springer-Verlag, Berlin and New York, 1982, pp. 184–189.

7. A. K. Kirillov, *Eléments de la théorie des representations*, "Mir", Moscow, 1974.

8. A. Groenewold, *On the principles of elementary quantum mechanics*, Physica **12** (1946), 405–460.

9. J. Moyal, *Quantum mechanics as a statistical theory*, Proc. Cambridge Philos. Soc. **45** (1949), 99–124.

10. H. Weyl, *Gruppentheorie und Quantenmechanik*, Hirzel Verlag, Leipzig, 1928.

11. J. Vey, *Déformation du crochet de Poisson sur une variété symplectique*, Comment. Math. Helv. **50** (1975), 421–454.

12. O. M. Neroslavskii and A. T. Vlassov, *Sur les déformations de l'algèbre des fonctions d'une variété symplectique*, C. R. Acad. Sci. Paris Sér. A **292** (1981), 71.

13. A. Lichnerowicz, *Existence and equivalence of twisted products on a symplectic manifold*, Lett. Math. Phys. **3** (1979), 435–502.

14. _____, *Construction of twisted products for contangent bundles of classical groups and Stiefel manifolds*, Lett. Math. Phys. **2** (1977), 133–143.

15. M. Cahen and S. Gutt, *Regular star-representations of Lie algebras*, Lett. Math. Phys. **6** (1982) pp. 395–404.

16. S. Gutt, *Déformations formelles de l'algèbre des fonctions différentiables sur une variété symplectique*, Thesis, Bruxelles, 1980; *Equivalence of deformations and associated star-products*, Lett. Math. Phys. **3** (1979), 297–309.

17. D. Arnal, J. C. Cortet, P. Molin and G. Pinczon, *Covariance and geometrical invariance in star-quantization*, J. Math. Phys. **24** (1983), 276–283.

18. D. Arnal, **-products and representations of nilpotent groups*, Pacific J. Math. (1984) (to appear).

19. P. Molin, *Invariance et covariance de structures une variété symplectique*, Thesis, Université de Dijon, 1981.

20. C. Fronsdal, *Some ideas about quantization*, Rep. Math. Phys. **15** (1978), 111–145.

21. _____, *Invariant star-product quantization of the one-dimensional Kepler probelm* J. Math. Phys. **20** (1979), 2226–2232.

22. D. Arnal, J. C. Cortet, M. Flato and D. Sternheimer, *Star-products: quantization and representations without operators*, Field theory, Quantization and Statistical Physics (E. Tirapegui, Ed.), D. Reidel, 1981, pp. 85–111.

23. F. Bayen and C. Fronsdal, *Quantization on the sphere*, J. Math. Phys. **22** (1981), 1345–1349.

24. G. S. Agarwal and E. Wolf, *Calculus for functions of noncommuting operators and general phase-space methods in quantum mechanics*, Phys. Rev. D (3) **2** (1970), 2161–2225.

25. E. Wigner, *Quantum correction for thermodynamic equilibrium*, Phys. Rev. **40** (1932), 749–759.

26. A. Voros, *Sur les développements semi-classiques*, Thèse, Orsay no. 1843, 1977 and related papers; I. Daubechies, *Representation of quantum mechanical operators by kernels on Hilbert spaces of analytic functions*, Thesis, Brussels, 1979 and related papers; C. Moreno and Ortega–Navarro, *Deformations of the algebra of functions on Hermitian symmetric spaces resulting from quantization*, Preprints, Collège de France (to be published); J. M. Maillard, private communication, 1982.

27. W. Schmid, Lecture, AMS-SIAM Summer Sem. Applications of Group Theory in Physics and Math. Physics, 1982.

28. A. A. Kirillov, *Characters of unitary representations of Lie groups*, Functional Anal. Appl. **2** (1968), 40–55.

29. R. Howe, *On the role of the Heisenberg group in harmonic analysis*, Bull. Amer. Math. Soc. (N. S.) **3** (1980), 821–843.

30. P. Molin, *Existence de star-produits fortement invariants sur les orbites massives de la coadjointe du groupe de Poincaré*, C. R. Acad. Sci. Paris Sér. A **293** (1981), 309–312.

31. P. Molin, D. Arnal and J. C. Cortet, *Star-produit et représentation de masse nulle du groupe de Poincaré*. C. R. Acad. Sci. Paris Sér. A **291** (1980), 327–330.

32. M. Flato, C. Fronsdal and D. Sternheimer, *Difficulties with massless particles?*, Comm. Math. Phys. (1983) (to appear).

33. V. Lugo, *An associative algebra of functions on the orbits of nilpotent Lie groups*, Lett. Math. Phys. **5** (1981), 509–516.

34. S. Gutt, *An explicit star-product on the cotangent bundle of a Lie group*, Lett. Math. Phys. **7** (1983), 249–258.

35. M. Flato, private communication, 1982.

CENTRE NATIONAL DE LA RECHERCHE SCIENTIFIQUE (CNRS), PARIS, FRANCE

Mailing address: Physique-Mathématique, Faculté des Sciences-Mirande, Université de Dijon, BP 138, 21004 Dijon Cedex, France

Lectures in Applied Mathematics
Volume 21, 1985

Classifying Representations by Lowest K-types

David A. Vogan, Jr.[1]

0. Introduction. Let G be a connected semisimple Lie group with finite center and K a maximal compact subgroup. The group K can be made to play roughly the same role in the representation theory of G that a maximal torus does for a compact group. The point of these notes is to explain that assertion. As motivation, and to recall some technical background, I begin with a summary of the Cartan-Weyl theory for compact groups in §1. §2 describes some of Harish-Chandra's basic theory of infinite-dimensional representations. In §3, I define "lowest K-types" in analogy with highest weights for compact groups. Theorem 3.3 says that these K-types always have multiplicity one. §4 is perhaps the most technical part: it is devoted to the construction of representations with specified lowest K-types. The main result is Proposition 4.11. §5 explains how these representations can be used in a classification of all representations (Theorem 5.12). There are no proofs included, but all the structural results and parametrizations have been made fairly explicit for $SL(2n, \mathbf{R})$. Proofs of the results stated may be found in [1 and 2], and [3] is a good reference for §2. It is always dangerous to be brief about credit, but the general shape of the results of §§3–5 is largely due to Harish-Chandra, Knapp and Langlands.

1. Compact groups. Let K be a compact connected Lie group. Since the unitary groups are so well known, I will use an orthogonal group as a running example:

$$SO(2n) = \{2n \times 2n \text{ real orthogonal matrices of determinant } 1\}.$$

$$(1.1)$$

1980 *Mathematics Subject Classification.* Primary 22E46.
[1] Supported in part by NSF grant MCS 82-02127.

Put

$$\hat{K} = \{\text{equivalence classes of irreducible}$$
$$\text{unitary representations of } K\}; \qquad (1.2)$$

the goal is to describe the set $\hat{K}$. The first step is to choose a *maximal torus* T inside K; this is a maximal connected abelian subgroup of K. To describe such a group for $SO(2n)$, write

$$r(\theta) = \begin{pmatrix} \cos\theta & \sin\theta \\ -\sin\theta & \cos\theta \end{pmatrix},$$

the matrix of the rotation of the plane through an angle θ. Now if $\theta_1, \ldots, \theta_n$ are all real numbers, define

$$r(\theta_1, \ldots, \theta_n) = 2n \times n \text{ matrix with } 2 \times 2 \text{ blocks,}$$
$$r(\theta_1), r(\theta_2), \ldots, r(\theta_n) \text{ along diagonal.}$$

This matrix performs rotations in each of n orthogonal planes (in $\mathbf{R}^{2n}$) separately. Finally, put

$$B = \{r(\theta_1, \ldots, \theta_n) \mid \theta_i \in \mathbf{R}\} \subseteq SO(2n). \qquad (1.3)$$

This is a connected abelian group: it is obviously just a product of n copies of the rotation group $SO(2)$ of the plane. It is not hard to check that any orthogonal matrix commuting with B lies in B; so B is a maximal torus in $SO(2n)$.

Returning to the general case, write t_0 for the Lie algebra of T, and t_0^* for its dual vector space. Any irreducible representation $\chi \in \hat{T}$ is necessarily one dimensional and therefore, maps T into the set of complex numbers of absolute value 1. The differential $d\chi$ of χ is a map of t_0 into $\mathbf{C}$, defined by

$$\chi(\exp(tZ)) = e^{t(d\chi(Z))} \qquad (1.4)$$

for $Z \in t_0$ and t any real number. Since $\chi(\exp(tZ))$ has absolute value 1, $d\chi$ must take purely imaginary values:

$$d\chi: t_0 \to i\mathbf{R}. \qquad (1.5a)$$

I will also write

$$d\chi \in i(t_0)^* \qquad (1.5b)$$

to mean the same thing. Since $d\chi$ determines χ by (1.4), this proves

LEMMA 1.6. *Suppose T is a compact connected abelian Lie group. Then $\hat{T}$ may be identified with a subset of $i(t_0)^*$ by identifying a representation with its differential.*

Henceforth we will often use the letter χ alone for $d\chi$ as well, confusing representations of T and their differentials. Notice that this proposition gives a pretty good solution to the original problem of describing $\hat{K}$, in the case when K is abelian. (The problem of deciding what subset of $i(t_0)^*$ corresponds to $\hat{T}$ can be given a nice abstract answer, but from a concrete point of view it is perhaps better simply to check when a possible $d\chi$ gives a well-defined χ by plugging into (1.4).)

For the group B of (1.3), $\mathfrak{b}_0$ can be identified with $\mathbf{R}^n$ by requiring

$$\exp(\theta_1,\ldots,\theta_n) = r(\theta_1,\ldots,\theta_n).$$

The elements of $i(\mathfrak{b}_0)^*$ corresponding to elements of $\hat{B}$ are the linear functionals $\chi(m_1,\ldots,m_n)$ defined by

$$\chi(m_1,\ldots,m_n)(\theta_1,\ldots,\theta_n) = \sum_{j=1}^{n} im_j\theta_j;$$

here each m_j is an integer. The corresponding element of $\hat{B}$ is defined by

$$\chi(m_1,\ldots,m_n)(r(\theta_1,\ldots,\theta_n)) = \prod_{j=1}^{n} e^{im_j\theta_j}. \tag{1.7}$$

Returning to the general case, let π be an irreducible representation of K on a complex vector space V. For each $\chi \in \hat{T}$, the χ-weight space of V is

$$V(\chi) = \{ v \in V \mid \pi(t)v = \chi(t)v, \text{ all } t \in T \}$$
$$= \text{largest subspace of } V \text{ which is invariant under } T,$$
$$\text{and isomorphic to a direct sum of copies of } \chi. \tag{1.8}$$

Obviously

$$V = \bigoplus_{\chi \in \hat{T}} V(\chi);$$

this is called the *weight space decomposition* of V (or π). If $V(\chi)$ is nonzero, we call χ a *weight* of π. The *multiplicity* of χ is the dimension of $V(\chi)$.

As an example, let π be the natural representation of $SO(2n)$ on $\mathbf{C}^{2n}$. The group B acts in the usual basis by

$$\pi(r(\theta_1,\ldots,\theta_n))e_{2j+1} = \cos\theta_j e_{2j+1} - \sin\theta_j e_{2j+2},$$
$$\pi(r(\theta_1,\ldots,\theta_n))e_{2j+2} = \sin\theta_j e_{2j+1} + \cos\theta_j e_{2j+2}.$$

If we put

$$u_j = e_{2j+1} + ie_{2j+2}, \qquad v_j = e_{2j+1} - ie_{2j+2},$$

then

$$\pi(r(\theta_1,\ldots,\theta_n))u_j = e^{i\theta_j}u_j, \qquad \pi(r(\theta_1,\ldots,\theta_n))v_j = e^{-i\theta_j}v_j.$$

The weights of π are therefore the $2n$ elements

$$\{\chi(0,\ldots,0,\pm 1,0,\ldots,0)\} \subseteq \hat{B}$$

(notation (1.7)), each having multiplicity one.

It is not too hard to show that the weights of any $\pi \in \hat{K}$ (and their multiplicities) determine π. A good way to think of π is as this set, arrayed in the vector space $i(t_0)^*$ like the vertices of a crystal. In the example of SO(4) acting on $\mathbf{C}^4$, the weights are $(\pm 1, 0)$, $(0, \pm 1)$; the picture is presented in the figure.

For SO(6) acting on $\mathbf{C}^6$, we get the six vertices of an octahedron in space. More general representations lead to far more complicated pictures, however. Although it is sometimes necessary to understand these pictures in great detail, the representations can be identified with much less data. To explain this, we need to choose a linear ordering of the vector space $i(t_0)^*$. This means picking a basis $\{e_1, \ldots, e_n\}$ of $i(t_0)^*$, and defining

$$\sum a_i e_i < \sum b_i e_i \quad \text{if and only if}$$

$$a_1 < b_1; \quad \text{or} \quad a_1 = b_1 \text{ and } a_2 < b_2;$$

$$\text{or} \quad a_2 = b_2 \text{ and } a_3 < b_3, \text{ etc.} \tag{1.9}$$

(For obvious reasons, such an ordering is called *lexicographic*.) If π is an irreducible representation of K, the *highest weight* of π is the largest weight of π with respect to the ordering (1.9). The first main result is

PROPOSITION 1.10 (CARTAN AND WEYL). *Two irreducible representations of K having the same highest weight are equivalent. The highest weight of any irreducible representation has multiplicity one.*

The set $\hat{K}$, which is defined in terms of Hilbert spaces and operators in a very abstract way, may therefore be identified with a subset of $i(t_0^*)$, which is very concretely constructed from K. To complete this description of $\hat{K}$, we need to specify which elements of $i(t_0^*)$ can be highest weights of irreducible representations.

Write $N(T)$ for the normalizer of T in K. This group acts on T by conjugation; and this induces actions on $\hat{T}$, t_0 and $i(t_0)^*$. Now $N(T)$ contains T, which (as an abelian group) acts trivially on itself by conjugation. The quotient group

$$W = W(K, T) = N(T)/T \tag{1.11}$$

therefore on all of these groups as well. A weight $\chi \in i(t_0)^*$ is called *dominant* if for every $w \in W$, $w\chi \leqslant \chi$; here the order is that of (1.9).

PROPOSITION 1.12 (CARTAN AND WEYL). *An element $\chi \in i(t_0)^*$ is the highest weight of an irreducible representation of K if and only if*
(a) χ *is dominant, and*
(b) χ *is the differential of a representation of T.*

Computing $N(B)$ when $K = SO(2n)$ (see (1.3)) is a little messy. Fix i and j distinct integers between 1 and n, and define elements of $SO(2n)$ (acting on the standard basis $e_1, \ldots, e_{2n}$) by

$$\sigma_{ij}^{\pm}(e_{2i+1}) = e_{2j+1}, \qquad \sigma_{ij}^{\pm}(e_{2j+1}) = e_{2i+1},$$

$$\sigma_{ij}^{\pm}(e_{2i+2}) = \pm e_{2j+2}, \qquad \sigma_{ij}^{\pm}(e_{2j+2}) = \pm e_{2i+2},$$

$$\sigma_{ij}^{\pm}(e_k) = e_k, \qquad k \notin \{2i+1, 2i+2, 2j+1, 2j+2\}.$$

It is easy to check that these elements belong to $N(B)$; in fact,

$$\sigma_{ij}^{\pm}\big(r(\theta_1, \ldots, \theta_n)\big)\big(\sigma_{ij}^{\pm}\big)^{-1} = r(\phi_1, \ldots, \phi_n).$$

Here

$$\phi_i = \pm\theta_j, \quad \phi_j = \pm\theta_i, \quad \phi_k = \theta_k, \qquad k \notin \{i, j\}.$$

That is, $\sigma_{ij}^{\pm}$ interchanges the i and j coordinates, possibly changing the signs of both. Using a little theory, or a messy calculation, one can show that the elements $\sigma_{ij}^{\pm}$ (together with B itself) generate $N(B)$. It follows that W acts on B by permuting the θ_i coordinates arbitrarily and changing an even number of signs. This description carries over directly to the action on the elements $\chi(m_1, \ldots, m_n)$ of $\hat{B}$ (see (1.7)). We can order $i(b_0)^*$ using the basis $\chi(1, 0, \ldots, 0)$, $\chi(0, 1, 0, \ldots, 0)$, etc.; the result (following (1.9)) is that $\chi(m_1, \ldots, m_n) < \chi(p_1, \ldots, p_n)$ if and only if $m_1 < p_1$; or $m_1 = p_1$ and $m_2 < p_2$, etc. Putting these two descriptions (of W and the ordering) together, we find that

$$\chi(m_1, \ldots, m_n) \text{ is dominant for } SO(2n) \text{ if and only if}$$

$$m_1 \geqslant m_2 \geqslant \cdots \geqslant m_{n-1} \geqslant |m_n|. \tag{1.13}$$

Such decreasing strings of integers therefore parametrize the representations of $SO(2n)$.

There is another characterization of the highest weight which may serve to motivate the material on noncompact groups. Fix a positive definite inner product

$$\langle \,, \rangle \text{ on } i(t_0)^*, \tag{1.14}$$

invariant under $W(K, T)$.

PROPOSITION 1.15 (CARTAN AND WEYL). *Suppose* (π, V) *is an irreducible representation of* K, *of highest weight* χ. *If* λ *is any other weight of* π, *then*

$$\langle \lambda, \lambda \rangle \leq \langle \chi, \chi \rangle.$$

Equality holds if and only if there is an element $w \in W(K, T)$ *such that* $\lambda = w\chi$.

The set of weights of π therefore sits inside a sphere in $i(\mathfrak{t}_0)^*$, touching it exactly at the W-translates of the highest weight. This proposition may be regarded as the key to all the deeper results about $\hat{K}$: the Weyl character formula, for example, can be proved using only this fact and formal arguments.

COROLLARY 1.16. *Write* $\hat{T}/W(K, T)$ *for the set of orbits of* $W(K, T)$ *on* $\hat{T}$. *There is a one-to-one correspondence from* $\hat{T}/W(K, T)$ *onto* $\hat{K}$; *the orbit* $W(K, T) \cdot \lambda$ *corresponds to the unique representation having* λ *as one of its longest weights.*

2. Noncompact groups: Harish-Chandra's theory. Let G be a connected semisimple Lie group with finite center. The running example will be

$$\mathrm{SL}(2n, \mathbf{R}) = \{2n \times 2n \text{ real matrices of determinant } 1\}. \quad (2.1)$$

It turns out to be easier to study representations of G which are not necessarily unitary.

DEFINITION 2.2. An *irreducible representation* of G is a pair (π, V), with V a complex Banach space, and π a homomorphism of G into the group of bounded invertible operators on V. We assume that

(a) for every $v \in V$, the map $g \to \pi(g)v$ from G to V is continuous; and

(b) if $V_0 \subseteq V$ is a closed subspace such that $\pi(g)V_0 \subseteq V_0$ for all $g \in G$, then $V_0 = \{0\}$ or V.

We say that (π, V) is *quasisimple* if all the higher order Casimir operators (that is, the center of the enveloping algebra) act by scalars on V.

Irreducible unitary representations of G are automatically quasisimple; and, in fact, there are no known examples of irreducible representations which are not.

When the group was compact, it was possible to understand representations by considering their restriction to a smaller, simpler group (the maximal torus). Exactly the same method applies here. A *maximal compact subgroup* of G is a compact subgroup of largest possible dimension; we fix such a subgroup K. When $G = \mathrm{SL}(2n, \mathbf{R})$, we can take

$$K = \mathrm{SO}(2n). \quad (2.3)$$

If (π, V) is any representation of G, and $\mu \in \hat{K}$, define

$$V(\mu) = \text{largest subspace of } V \text{ which is invariant under } K, \quad (2.4)$$
$$\text{and isomorphic to a direct sum of copies of } \mu,$$

the μ-*primary component* or μ K-*type* of V. The multiplicity of μ in π is defined to be the dimension of $V(\mu)$ divided by the dimension of μ. If $V(\mu) \neq 0$, we say μ is a K-*type* of π.

Here is an example. Fix a complex number λ, and a sign $\varepsilon = \pm 1$. Define a representation $(\pi_{\varepsilon,\lambda}, \mathcal{K}_{\varepsilon,\lambda})$ of SL$(2n, \mathbf{R})$ by

$$\mathcal{K}_{\varepsilon,\lambda} = \{\text{functions } f \text{ on } \mathbf{R}^{2n} \text{ such that } f(x) = \varepsilon f(-x);$$
$$f(tx) = t^{\lambda} f(x) \text{ (all positive } t \in \mathbf{R}); \text{ and}$$
$$f \text{ restricted to the sphere } S^{2n-1} \text{ is square integrable}\}, \quad (2.5)$$
$$(\pi_{\varepsilon,\lambda}(g) \cdot f)(x) = f(g^{-1}x) \quad (g \in \text{SL}(2n, \mathbf{R}), x \in \mathbf{R}^{2n}, f \in \mathcal{K}_{\varepsilon,\lambda}).$$

Then $\mathcal{K}_{\varepsilon,\lambda}$ is a Hilbert space, with inner product

$$\langle f_1, f_2 \rangle = \int_{S^{2n-1}} f_1(x) \bar{f}_2(x)\, dx.$$

The representation $\pi_{\varepsilon,\lambda}$ is unitary (that is,

$$\langle \pi_{\varepsilon,\lambda}(g) f_1, f_2 \rangle = \langle f_1, \pi_{\varepsilon,\lambda}(g)^{-1} f_2 \rangle)$$

if and only if the real part of λ is exactly $-n$, as one can check by calculus. To understand how $\pi_{\varepsilon,\lambda}$ restricts to SO$(2n)$, we need to understand how to decompose the representation of SO$(2n)$ on functions on S^{2n-1}. This is substantially the theory of spherical harmonics; the conclusion is that

the SO$(2n)$-types of $\mathcal{K}_{+1,\lambda}$ have highest weights $(2m, 0, \ldots, 0)$; those of $\mathcal{K}_{-1,\lambda}$ have highest weights $(2m + 1, 0, \ldots, 0)$. All SO$(2n)$-types have multiplicity one.

The "picture" of $\mathcal{K}_{\varepsilon,\lambda}$ is therefore a string of points on a line, marching off to infinity inside $i(\mathfrak{b}_0)^*$. This picture does not depend on λ. Since $\mathcal{K}_{\varepsilon,\lambda}$ and $\mathcal{K}_{\varepsilon,\lambda'}$ are inequivalent unless $\lambda = \lambda'$ (at least if $n \geqslant 2$), we get a continuous family of distinct representations with the same restriction to SO$(2n)$. In contrast with the $K \supseteq T$ situation, we cannot hope to identify a representation by its restriction to K alone. Nevertheless, the K-types provide some useful preliminary information.

Having come so far with this example, we may as well state which of the $\pi_{\varepsilon,\lambda}$ are irreducible representations. For simplicity, assume $n \geqslant 2$. If r is a nonnegative integer, and $\varepsilon = (-1)^r$, then

$$V_r = \text{polynomial functions of degree } r \text{ on } \mathbf{R}^{2n}$$

is obviously a closed, G-invariant subspace of $\mathcal{H}_{\varepsilon,r}$. Its K-types have highest weights $(r - 2k, 0, \ldots, 0)$, with k a nonnegative integer. Dually,

$$\mathcal{H}^0_{\varepsilon, -r-2n} = \{ f \in \mathcal{H}_{\varepsilon, -r-2n} \mid \text{for all } p \in V_r, \int_{S^{2n-1}} f(x) p(x) \, dx = 0 \}$$

is a closed, G-invariant subspace of $\mathcal{H}_{\varepsilon, -r-2n}$. There are no other nontrivial closed G-invariant subspaces of any of the $\mathcal{H}_{\varepsilon, \lambda}$, as is not hard to prove. In particular, $\pi_{\varepsilon, \lambda}$ is irreducible unless (ε, λ) belongs to the set

$$\{ ((-1)^r, r), ((-1)^r, -r-2n) \mid r = 0, 1, 2, \ldots \}.$$

Here is Harish-Chandra's basic result.

THEOREM 2.6 (HARISH-CHANDRA). *Suppose* (π, V) *is a quasisimple irreducible representation of* G (*Definition 2.2*).

(a) *Each primary component* $V(\mu)$ (*for* $\mu \in \hat{K}$) *is finite dimensional; that is, every* μ *has finite multiplicity in* π.

(b) *Write* $V_K = \bigoplus_{\mu \in \hat{K}} V(\mu)$ (*algebraic direct sum—no closures or limits*). *If* X *belongs to the Lie algebra* $\mathfrak{g}_0$ *of* G, *and* $v \in V_K$, *then*

$$\pi(X)v = \lim_{t \to 0} \frac{1}{t} (\pi(\exp tX)v - v)$$

exists and belongs to V_K. *This defines a Lie algebra representation of* $\mathfrak{g}_0$ *on* V_K, *which is irreducible in the algebraic sense.*

DEFINITION 2.7. Suppose (π, V) is a quasisimple irreducible representation of G. The representation (π, V_K) of $\mathfrak{g}_0$ is called the *Harish-Chandra module* of (π, V). Two representations of G are called *infinitesimally equivalent* if their Harish-Chandra modules are isomorphic. Define $\hat{G}$ to be the set of infinitesimal equivalence classes of irreducible quasisimple representations of G.

The reason we need a fancy definition of equivalence is that the definition of representation is very loose. In (2.5), for example, we could define $\mathcal{C}_{\varepsilon, \lambda}$ (instead of $\mathcal{H}_{\varepsilon, \lambda}$) by replacing the words "square integrable" by "continuous". Then $\mathcal{C}_{\varepsilon, \lambda}$ is a Banach space, with norm

$$\|f\| = \max_{x \in S^{2n-1}} |f(x)|.$$

The representations $(\pi_{\varepsilon, \lambda}, \mathcal{H}_{\varepsilon, \lambda})$ and $(\pi_{\varepsilon, \lambda}, \mathcal{C}_{\varepsilon, \lambda})$ obviously ought to be equivalent, by the natural inclusion of $\mathcal{C}_{\varepsilon, \lambda}$ in $\mathcal{H}_{\varepsilon, \lambda}$. This inclusion is not a Banach space isomorphism, however; so we have to ask for less. The goal is now to prove something like Propositions 1.10, 1.12 and 1.15 for $\hat{G}$, with K playing the role of T.

3. Lowest K-types. Suppose $\pi \in \hat{G}$. We want to find a single K-type of π which is distinguished, like the highest weight of a representation of a compact group. That analogy suggests that we should order $\hat{K}$ somehow and look at some sort of extremal K-type. To order $\hat{K}$, it is only reasonable to use highest weights. So we fix a maximal torus $T \subseteq K$, and a linear ordering $\leq$ of $i(t_0)^*$. (In the example $SL(2n, \mathbf{R}) \supseteq SO(2n)$, we can use the maximal torus B and the ordering defined in §1.) The most obvious idea is to use the ordering of $i(t_0)^*$ to order highest weights, and therefore $\hat{K}$. This leads to a theory but not the best one. (One reason is that different orderings of $i(t_0)^*$ can define very different orderings of $\hat{K}$; one gets not one but several descriptions of $\hat{G}$, difficult to compare with each other or with anything else.) The next most obvious idea is to use the lengths of highest weights, in analogy with Proposition 1.15. This almost works; the reason for introducing the slight modification of it defined below is rather technical, and will not be discussed.

DEFINITION 3.1. Let $\Delta(\mathfrak{k}, t)$, the roots of t in $\mathfrak{k}$, denote the set of nonzero weights of T in the (adjoint) representation of K on

$$\mathfrak{k} = \mathrm{Lie}(K)_{\mathbf{C}},$$

the complexified Lie algebra of K. Thus $\Delta(\mathfrak{k}, t)$ is a subset of $\hat{T}$ or $i(t_0)^*$. Put

$$\Delta^+(\mathfrak{k}, t) = \{\alpha \in \Delta(\mathfrak{k}, t) \mid \alpha > 0\},$$

the *positive roots* (in the fixed ordering of $i(t_0)^*$); and

$$2\rho_c = \sum_{\alpha \in \Delta^+(\mathfrak{k}, t)} \alpha \in i(t_0)^*.$$

DEFINITION 3.2. Write $\langle \, , \, \rangle$ for the Killing form on $\mathfrak{g}_0$. By restriction to t_0 and dualization, it defines an inner product on $i(t_0)^*$, invariant under $W(K, T)$. If $\mu \in \hat{K}$ has highest weight χ, define the *norm* of μ

$$\|\mu\| = \langle \chi + 2\rho_c, \chi + 2\rho_c \rangle.$$

If $\pi \in \hat{G}$ and μ is a K-type of π, we say that μ is a *lowest K-type* if $\|\mu\|$ is as small as possible (among K-types of π).

In the running example, the complexified Lie algebra $\mathfrak{so}(2n, \mathbf{C})$ consists of $2n \times 2n$ skew symmetric matrices. The conjugation action of B on these matrices is not hard to compute; and one finds (in analogy with the example after (1.8)) that

$$\Delta(\mathfrak{k}, t) = \{\chi(0, \ldots, \pm 1, 0, \ldots, \pm 1, 0, \ldots, 0)\},$$
$$\Delta^+(\mathfrak{k}, t) = \{\chi(\ldots, 1, \ldots, \pm 1, \ldots)\},$$
$$2\rho_c = \chi(2n - 2, 2n - 4, \ldots, 2, 0).$$

In the example after (2.5), we see that $\mathcal{H}_{1, \lambda}$ has the trivial representation of $SO(2n)$ (highest weight $\chi(0, \ldots, 0)$) as its unique lowest K-type. If $n \geq 2$, $\mathcal{H}_{-1, \lambda}$ has the $\mathbf{C}^{2n}$ representation of $SO(2n)$ (highest weight $\chi(1, 0, \ldots, 0)$) as its unique lowest K-type. If $n = 1$, the two components

$\chi(\pm 1)$ of the $\mathbf{C}^2$ representation of SO(2) are the lowest K-types of $\mathcal{H}_{-1,\lambda}$. (Almost any definition will give the same lowest K-types in this example; difficulties arise only in more subtle situations.)

For any G, the norm of the trivial representation μ_0 is strictly less than that of any other representation of K; so μ_0 is the unique lowest K-type of any $\pi \in \hat{G}$ containing it. For these representations, the concept of lowest K-type is not very helpful. A typical representation μ of K, however, cannot easily be a lowest K-type of a representation π in which it occurs (because there is so much competition from K-types of smaller norm). To say that μ *is* a lowest K-type means that these other K-types must not occur in π at all. From such information, one can deduce a lot about the structure of π. The basic result is analogous to the second half of Proposition 1.10.

THEOREM 3.3. *Suppose $\pi \in \hat{G}$, and $\mu \in \hat{K}$ is a lowest K-type of π. Then μ has multiplicity one in π.*

If one takes a sufficiently sophisticated view of the proof Proposition 1.10, then one can say that this result is proved in the same way. For example, for Proposition 1.10 one has to show that a certain Lie algebra $\mathfrak{n}$ annihilates something. This amounts to constructing an element of $H^0(\mathfrak{n}, V)$. For Theorem 3.3, one constructs instead an element of some higher cohomology $H^R(\mathfrak{n}, V_K)$. I will say no more about proofs.

It is natural to ask how unique the lowest K-type is.

DEFINITION 3.4. Two representations μ and μ' of K are called *associate* if they are both lowest K-types of a single $\pi \in \hat{G}$. Write $\mathcal{C}(\mu)$ for the set of representations of K associate to μ.

In the running example, write $\mu(m_1,\ldots,m_n)$ for the SO($2n$) representation of highest weight $\chi(m_1,\ldots,m_n)$; this is defined only if

$$m_1 \geqslant \cdots \geqslant m_{n-1} \geqslant |m_n|.$$

Then

$$\mathcal{C}(\mu(m_1,\ldots,m_n)) = \begin{cases} \{\mu(m_1,\ldots,m_n)\} & \text{if } m_n \neq \pm 1, \\ \{\mu(m_1,\ldots,\pm m_n)\} & \text{if } m_n = \pm 1. \end{cases} \quad (3.5)$$

(This is not at all obvious.) As motivation, consider the (irreducible) representation π of SL($2n$, $\mathbf{R}$) on $\Lambda^n(\mathbf{C}^{2n})$. Using the calculation after (1.8), it is easy to check that the B weights of π are

$$\{\chi(m_1,\ldots,m_n) \mid \text{all } m_i \text{ are } 0 \text{ or } \pm 1\}.$$

Using Proposition 1.15 and (1.13), we conclude that $\mu(1,1,\ldots,1,\pm 1)$ are both K-types of π. In fact they are the only ones, as follows, for example, from Weyl's dimension formula. Since they obviously have the same norm, they must be associate (as asserted in (3.5)).

THEOREM 3.6. *If $\mu, \mu' \in \hat{K}$, then $\mathcal{Q}(\mu) = \mathcal{Q}(\mu')$ or $\mathcal{Q}(\mu) \cap \mathcal{Q}(\mu')$ is empty. Fix μ. The set $\mathcal{Q} = \mathcal{Q}(\mu)$ has order 2^l (for some l). There is a group $R = R(\mathcal{Q})$, isomorphic to $(\mathbf{Z}/2\mathbf{Z})^l$, which acts on $\mathcal{Q}$ in a simply transitive way. Suppose $\pi \in \hat{G}$ has μ as a lowest K-type. Then there is a subgroup $R_\pi \subseteq R(\mathcal{Q})$, such that the set of lowest K-types of π is exactly*

$$\{\rho \cdot \mu \,|\, \rho \in R_\pi\} \subseteq \mathcal{Q}.$$

In particular, this set has 2^k elements for some $k \leq l$. For "most" π having μ as lowest K-type, $R_\pi = R$; that is, the set of lowest K-types of π is $\mathcal{Q}$.

The word "most" of course requires explanation; we postpone that to §5. This result is roughly analogous to Proposition 1.15, with R replacing $W(K, T)$. By (3.5), the group R has order one or two for $\mathrm{SL}(2n, \mathbf{R})$. (Given l, one can find a simple group and an associate class $\mathcal{Q}$ with $R(\mathcal{Q}) \cong (\mathbf{Z}/2\mathbf{Z})^l$; but this is done by gluing together simpler cases. The largest "interesting" $R(\mathcal{Q})$ has order 4.)

We can now picture a representation of G as consisting of an array of K-types outside a certain sphere (centered at $-2\rho_c$) in $i(\mathfrak{t}_0)^*$. The array touches the sphere in 2^k points, which will be arranged in a rectangular form.

4. Standard representations. In this section, we will describe how to find some representations with a given lowest K-type. This material is quite technical, and the reader may prefer simply to look at Proposition 4.11 before passing to §5. We need some structure theory first. Define

$$\mathfrak{p}_0 = \text{orthogonal complement of } \mathfrak{k}_0 \text{ in } \mathfrak{g}_0; \qquad (4.1)$$

"orthogonal" means with respect to the Killing form. If $G = \mathrm{SL}(2n, \mathbf{R})$, then

$\mathfrak{g}_0 = 2n \times 2n$ real matrices of trace zero,

$\mathfrak{k}_0 = 2n \times 2n$ real skew-symmetric matrices,

$\mathfrak{p}_0 = 2n \times 2n$ real symmetric matrices of trace zero.

In general, $[\mathfrak{p}_0, \mathfrak{p}_0] \subseteq \mathfrak{k}_0$. Any Lie subalgebra of $\mathfrak{p}_0$ is therefore abelian. Let $\mathfrak{a}_0$ be such a subalgebra. Then $A = \exp(\mathfrak{a}_0)$ is a closed vector subgroup of G. Define

$$L = \text{centralizer of } A \text{ in } G,$$
$$\mathfrak{l}_0 = \mathrm{Lie}(L) = \text{centralizer of } \mathfrak{a}_0 \text{ in } \mathfrak{g}_0,$$
$$\tilde{\mathfrak{a}}_0 = (\text{center of } \mathfrak{l}_0) \cap \mathfrak{p}_0. \qquad (4.2)$$

Obviously $\tilde{\mathfrak{a}}_0 \supseteq \mathfrak{a}_0$. If $\tilde{\mathfrak{a}}_0 = \mathfrak{a}_0$, we call $\mathfrak{a}_0$ a *special subalgebra* of $\mathfrak{p}_0$, and A a *special vector subgroup* of G. Here is an example for $G = \mathrm{SL}(n, \mathbf{R})$.

Write $d(x_1,\ldots,x_n)$ for the diagonal matrix with entries $x_1,\ldots,x_n$. Define

$$\mathfrak{a}_0 = \{d(3t,-2t,-t,0,\ldots,0)\,|\,t \in \mathbf{R}\}.$$

Then $\mathfrak{l}_0$ (defined by (4.2)) is

$$\mathfrak{l}_0 = \left\{ \left. \begin{pmatrix} \begin{matrix} x & 0 & 0 \\ 0 & y & 0 \\ 0 & 0 & z \end{matrix} & \quad 0 \\ \quad 0 & U \end{pmatrix} \right| - \operatorname{tr} U = x + y + z \right\}.$$

Therefore

$$\tilde{\mathfrak{a}}_0 = (\text{center of } \mathfrak{l}_0) \cap \mathfrak{p}_0$$
$$= \{(t_1, t_2, t_3, -t_0,\ldots,-t_0)\,|\,(2n-3)t_0 = t_1 + t_2 + t_3\}.$$

Thus $\mathfrak{a}_0$ is not a special vector subalgebra, but $\tilde{\mathfrak{a}}_0$ is.

Suppose A is a special vector subgroup of G. Define L as in (4.2). Put

$$\mathfrak{m}_0 = \text{orthogonal complement of } \mathfrak{a}_0 \text{ in } \mathfrak{l}_0,$$

$$M = \text{group generated by } \exp(\mathfrak{m}_0) \text{ and } L \cap K. \qquad (4.3a)$$

Then the group L is a direct product M and A:

$$L = M \times A. \qquad (4.3b)$$

In the example above, M consists of matrices of the form

$$\begin{pmatrix} \begin{matrix} \pm 1 & 0 & 0 \\ 0 & \pm 1 & 0 \\ 0 & 0 & \pm 1 \end{matrix} & \quad 0 \\ \quad 0 & X \end{pmatrix}$$

having determinant one.

PROPOSITION 4.4. *Let $\mathcal{Q} \subseteq \hat{K}$ be an associate class of representations of K (Definition 3.4). Then there is attached to $\mathcal{Q}$ a special vector subgroup $A = A(\mathcal{Q})$ of G, defined up to conjugation by K. If $\mathcal{Q}$ consists of the trivial representation of K alone, then the Lie algebra $\mathfrak{a}_0$ of A is a maximal subalgebra of $\mathfrak{p}_0$. For "most" $\mathcal{Q}$, $\mathfrak{a}_0$ is the centralizer of $\mathfrak{t}_0$ (the Lie algebra of the maximal torus of K) in $\mathfrak{p}_0$.*

The group A is constructed in a very concrete way ([2, Proposition 5.3.3 and (5.3.23)]; it is what is called A^r there). Unfortunately, the construction is a little involved to recall. Instead, we will simply state what it gives for $\mathrm{SL}(2n, \mathbf{R})$. Recall the notation introduced before (3.5) for representations of $\mathrm{SO}(2n)$; and suppose

$$\mathcal{Q} = \mathcal{Q}(\mu(m_1,\ldots,m_n)).$$

Let r be the greatest integer such that

$$|m_{n-r+1}| \leqslant 1; \qquad (4.5a)$$

if $|m_n| > 1$, we set $r = 0$. (Thus r is the number of m_i's which are equal to zero or ± 1.) The special vector subgroup attached to $\mathcal{C}$ is

$$A = \{d(s_1, s_1, \ldots, s_{2n-2r}, s_{2n-2r},$$
$$t_1, \ldots, t_{2r}) \mid s_i, t_j > 0, \text{ and } (\Pi_i s_i^2)(\Pi_j t_j) = 1\}. \quad (4.5b)$$

Thus if $\mathcal{C}$ consists of the trivial representation, A is all diagonal matrices with positive entries having determinant one. For most $\mathcal{C}$, r is zero; so A consists of diagonal matrices with positive entries equal in pairs, having determinant one. (This agrees with the claim of Proposition 4.4.) The group M of (4.3) is

$$M = \left\{ g = \begin{bmatrix} \begin{bmatrix} X_1 & & & \\ & \ddots & & \\ & & X_{n-r} & \\ & & & \varepsilon_1 \\ & & & & \ddots \\ & & & & & \varepsilon_{2r} \end{bmatrix} \end{bmatrix} \middle| \begin{array}{l} X_i \text{ is a } 2 \times 2 \text{ matrix of} \\ \text{determinant } \pm 1, \varepsilon_i = \pm 1, \\ \text{and } \det g = 1 \end{array} \right\}.$$

$$(4.5c)$$

The identity component of M is a direct product of $n - r$ copies of $SL(2, \mathbf{R})$.

PROPOSITION 4.6. *In the setting of Proposition* 4.4, *write $M = M(\mathcal{C})$ for the group constructed from $A(\mathcal{C})$ by* (4.3). *Then there is a representation $\delta \in \hat{M}$ with the property that $\mathcal{C}$ is precisely the set of lowest K-types of*

$$\text{Ind}_{M \cap K}^K \left(\delta \mid_{M \cap K} \right).$$

This means that

(a) *If $\mu \in \mathcal{C}$, there is a $\gamma \in (M \cap K)\hat{}$ such that γ occurs in both $\delta \mid_{M \cap K}$ and $\mu \mid_{M \cap K}$.*

(b) *Suppose $\lambda \in \hat{K}$, and there is a $\gamma \in (M \cap K)\hat{}$ occurring in both δ and λ. Then if $\mu \in \mathcal{C}$, $\|\lambda\| \geqslant \|\mu\|$ (Definition* 3.2). *Equality holds if and only if $\lambda \in \mathcal{C}$.*

The representation δ is determined up to conjugation under the normalizer of A in K by these properties. It belongs to the discrete series of M.

In the example (4.5), δ is roughly the tensor product of the discrete series representations of $SL(2, \mathbf{R})$ with parameters $m_i - m_i/|m_i|$ ($i = 1, 2, \ldots n - r$). (That is the discrete series of lowest weight m_i.) Of course

δ has to be defined on the disconnected part of M as well. This is governed by how many of the integers $m_{n-r+1}, \ldots, m_n$ are 0 and how many are ± 1. To be precise, set

$$
M^{\#} = \left\{ g = \begin{bmatrix} \begin{bmatrix} X_1 & & \\ & \ddots & \\ & & X_{n-r} \end{bmatrix} & & \\ & \begin{matrix} \varepsilon_1 & & \\ & \ddots & \\ & & \varepsilon_{2r} \end{matrix} \end{bmatrix} \middle| \det X_i = 1, \det g = 1 \right\}.
$$

$$\tag{4.7a}$$

Then

$$
M^{\#} \cong [\mathrm{SL}(2,\,R)]^{n-r} \times (\mathbf{Z}/2\mathbf{Z})^{2r-1}. \tag{4.7b}
$$

Suppose exactly s of the m_i's are ± 1. Define a character of the $(\mathbf{Z}/2\mathbf{Z})^{2r-1}$ factor of $M^{\#}$ by

$$
\chi_s(d(1, \ldots, 1, \varepsilon_1, \ldots, \varepsilon_{2r})) = \prod_{j=1}^{s} \varepsilon_j. \tag{4.7c}
$$

Define δ_0 to be the discrete series representation of $[\mathrm{SL}(2,\mathbf{R})]^{n-r}$ mentioned above (having lowest weight m_i on the ith factor). Set

$$
\delta^{\#} = \delta_0 \otimes \chi_s \in (M^{\#})\hat{\,}, \qquad \delta = \mathrm{Ind}_{M^{\#}}^{M}(\delta^{\#}). \tag{4.7d}
$$

Verification of Proposition 4.6 in this case is now routine. The general case is similar—see [1, §7].

Suppose A is a special vector subgroup of G, and MA is the centralizer of A in G (cf. (4.3)). Then we can find a nilpotent subgroup N of G such that

$$
MA \text{ normalizes } N, \quad MA \cap N = \{1\}, \quad \dim N = \frac{1}{2} \dim \frac{G}{MA}. \tag{4.8}
$$

The group $P = MAN$ is then called a *parabolic subgroup* of G. In the example (4.5), we can take N to consist of the upper triangular matrices with 1's on the diagonal and 0's in the $(2i, 2i + 1)$ entries for $i = 1, \ldots, n - r$. Since A is a vector group, the set $\hat{A}$ (or homomrphisms from A to the nonzero complex numbers) is naturally isomorphic to the complex dual $\mathfrak{a}^*$ of $\mathrm{Lie}(A)$. Fix $\delta \in \hat{M}$ and $\nu \in \hat{A}$. Since M and A commute, $\delta \otimes \nu$ is a representation of MA (on the space of δ); and since MA normalizes N, it extends to MAN by making N act trivially. This extension is denoted $\delta \otimes \nu \otimes 1$. Define

$$
\pi(P, \delta \otimes \nu) = \pi(\delta \otimes \nu) = \mathrm{Ind}_{MAN}^{G} \delta \otimes \nu \otimes 1. \tag{4.9}
$$

Then it can be shown that

$$\pi(\delta \otimes \nu)\big|_K \cong \operatorname{Ind}_{M \cap K}^K \big(\delta\big|_{M \cap K}\big). \tag{4.10}$$

This construction, in conjunction with Proposition 4.6, proves the following result. (The multiplicity one assertion was omitted from Proposition 4.6 for brevity.)

PROPOSITION 4.11. *Let $\mathcal{Q} \subseteq \hat{K}$ be an associate class of representations of K (Definition 3.4), and $A = A(\mathcal{Q})$ a corresponding special vector subgroup (Proposition 4.4). Then there is a continuous family*

$$\big\{\pi(\nu) = \pi(\mathcal{Q}, \nu)\,\big|\,\nu \in \hat{A} \cong \mathfrak{a}^*\big\}$$

of (possibly reducible) representations of G, each having $\mathcal{Q}$ as its set of lowest K-types; and these have multiplicity one. For almost all ν, $\pi(\nu)$ is irreducible.

The representations of Proposition 4.11 are called *standard*. In $SL(2n, \mathbf{R})$, suppoose $\mu = \mu(m_1, \ldots, m_n)$, and r of the m_i's are 0 or ± 1. Then we have produced an $n + r - 1$ (complex) parameter family of representations with lowest $SO(2n)$-type μ.

Here are some auxiliary facts, less directly related to the classification problem.

PROPOSITION 4.12. *Write Ψ for the correspondence of Propositions 4.4 and 4.6, from associate classes of representations of K, to K-conjugacy classes of triples (A, M, δ). Here A is a special vector subgroup, M is given by (4.3), and δ is a discrete series representation of M. Then Ψ is bijective.*

These K-conjugacy classes of triples parametrize the various continuous series of representations in Harish-Chandra's Plancherel formula. These representations in turn are just those of Proposition 4.11, with ν a unitary character of A—that is, $\nu \in i\mathfrak{a}_0^*$. The Plancherel formula can therefore be written

$$L^2(G) = \sum_{\mathcal{Q}} \int_{[i\mathfrak{a}_0(\mathcal{Q})]^*} \pi(\mathcal{Q}, \nu) \otimes [\pi(\mathcal{Q}, \nu)]^* \, dm(\nu),$$

in its crudest form. Proposition 4.12 also justifies the omission of the construction of $A(\mathcal{Q})$: it is uniquely determined by the requirements of Proposition 4.6. Computationally, this is not as ridiculous an approach (to finding $A(\mathcal{Q})$) as it sounds.

5. Classification.

DEFINITION 5.1. Suppose $\mu \in \hat{K}$. Put $\mathcal{Q} = \mathcal{Q}(\mu)$, the associate class of μ (Definition 3.4), and let $A = A(\mathcal{Q})$ be special vector subgroup attached to $\mathcal{Q}$. Write $\{\pi(\nu)\,|\,\nu \in \hat{A}\}$ for a series of representations as in Proposition

4.11, constructed using $\delta \in \hat{M}$ (Proposition 4.6). Fix $\nu \in \hat{A}$, and write $\mathcal{H}$ for the space of $\pi(\nu)$. (It is a Hilbert space.) Let $\mathcal{H}_1$ be the intersection of all the closed, G-invariant subspaces of $\mathcal{H}$ containing the K-type μ; since μ has multiplicity one in π, it has multiplicity one in $\mathcal{H}_1$. Let $\mathcal{H}_0$ be the largest closed, G-invariant subspace of $\mathcal{H}_1$ *not* containing μ. Define $\bar{\pi}(\mu, \nu) =$ representation of G on

$$\bar{\mathcal{H}} = \bar{\mathcal{H}}_{\mu,\nu} = \mathcal{H}_1/\mathcal{H}_0;$$

$\bar{\pi}(\mu, \nu)$ is a quasisimple, irreducible representation of G, containing the K-type μ (with multiplicity one) as a lowest K-type.

THEOREM 5.2. *In the setting of Definition 5.1, suppose π is an irreducible quasisimple representation of G having μ as a lowest K-type. Then there is a $\nu \in \hat{A}$ such that π is infinitesimally equivalent to $\bar{\pi}(\mu, \nu)$ (Definition 5.1).*

This is a version of the *Langlands classification* of $\hat{G}$: it says that any irreducible representation of G can be realized in a special way inside a standard representation. To identify a representation this way in practice, one first computes a lowest K-type μ. The parameter ν determines the eigenvalues of all the Casimir operators; so if these are known, one can deduce something about ν. It is possible for two different representations to have the same lowest K-types and the same eigenvalues for all Casimir operators, so the method is not perfect; but it is often useful.

As an example, consider the representation π of $SL(2n, \mathbf{R})$ on the space $\mathcal{H}^0_{+1,-2n}$ defined before (2.5). Its $SO(2n)$-types are those of $\mathcal{H}_{+1,-2n}$ minus these of V_0:

$$\{\mu(2k, 0, 0, \ldots, 0) \mid k = 1, 2, 3 \ldots\}.$$

The lowest is evidently $\mu = \mu(2, 0, \ldots, 0)$. The integer r of (4.5) is $n - 1$, so the group A attached to μ is

$$A = \{d(s, s, t_1, t_2, \ldots, t_{2n-2}) \mid s, t_j > 0, \text{ and } s^2(\Pi t_i) = 1\}.$$

Therefore

$$M = \left\{ g = \begin{pmatrix} X & & & \\ & \varepsilon_1 & & \\ & & \ddots & \\ & & & \varepsilon_{2n-2} \end{pmatrix} \middle| \begin{array}{l} X \text{ is a } 2 \times 2 \text{ matrix} \\ \text{of determinant } \pm 1, \\ \varepsilon_i = \pm 1, \text{ and } \det g = 1 \end{array} \right\}.$$

The representation δ is described by (4.7). Its restriction to the identity component of $M \cap K$, which is isomorphic to $SO(2)$, is the sum of the representations $\mu(2k)$, with k a nonzero integer; and each of these occurs with multiplicity one. Also, δ is trivial on the diagonal matrices in $M \cap K$. These two properties characterize δ.

Suppose $\nu \in \hat{A}$; write

$$\nu\big(d(s, s, t_1, \ldots, t_{2n-2})\big) = (s^{2\nu_0}) \prod t_i^{\nu_i},$$

with all $\nu_i \in \mathbb{C}$, and $2\nu_0 + \Sigma \nu_i = 0$. Since $\mathcal{K}_{+1, -2n}/\mathcal{K}_{+1, -2n}^0$ is the trivial representation, all the Casimir operators of $SL(2n, \mathbb{R})$ act in π as in the trivial representation. If $\pi \cong \bar{\pi}(\mu, \nu)$, Harish-Chandra's theory of Casimir operators implies that

$$\left(\nu_0 + \tfrac{1}{2}, \nu_0 - \tfrac{1}{2}, \nu_1, \ldots, \nu_{2n-2}\right)$$

must be a permutation of

$$\left(n - \tfrac{1}{2}, n - \tfrac{3}{2}, \ldots, -\left(n - \tfrac{1}{2}\right)\right).$$

Up to permutation of $\nu_1, \ldots, \nu_{2n-2}$, there are $2n - 1$ possibilities for ν satisfying the conditions given so far: for $j = 1, 2, \ldots, 2n - 1$, we can take

$$\nu_0 = n - j,$$

$$\{\nu_1, \ldots, \nu_{2n-2}\} = \left\{-\tfrac{1}{2}, \ldots, -\left(n - \tfrac{1}{2}\right)\right\} - \left\{\left(n - j + \tfrac{1}{2}, n - j - \tfrac{1}{2}\right)\right\}.$$

We will see in a moment that permutation of $\{\nu_1, \ldots, \nu_{2n-2}\}$ does not affect $\bar{\pi}(\mu, \nu)$; so we have $2n - 1$ possibilities for π in the Langlands classification. Any of several more technical arguments show that, in fact, the correct ν is given by the formulas above with $j = 2n - 1$. Notice that $\pi(\nu)$ is a very large representation: it contains about half the representations of $SO(2n)$, usually with large multiplicity. The K-types of $\bar{\pi}(\mu, \nu) \cong \pi$ all have multiplicity one and lie on a single line in $\hat{K}$. Therefore if $n \geqslant 2$, $\bar{\pi}(\mu, \nu)$ is a very small piece of $\pi(\nu)$; a result like Theorem 5.2 contains almost no hint of the subtleties of structure of the representations $\bar{\pi}(\mu, \nu)$.

To make Theorem 5.2 into a classification of $\hat{G}$, we need to describe the equivalences among the various $\bar{\pi}(\mu, \nu)$. Define (with $\mathcal{C} = \mathcal{C}(\mu)$)

$$N(\mathcal{C}) = \text{normalizer of the triple } (A, M, \delta) \text{ in } K,$$

$$W(\mathcal{C}) = N(\mathcal{C})/M \cap K. \tag{5.3}$$

Obviously $M \cap K$ is a normal subgroup of $N(\mathcal{C})$, and the quotient $W(\mathcal{C})$ (which is finite) acts on A and $\hat{A}$.

PROPOSITION 5.4. *In the setting of Definition 5.1 and (5.3), the representations $\bar{\pi}(\mu, \nu)$ and $\bar{\pi}(\mu, \nu')$ are infinitesimally equivalent if and only if ν and ν' are conjugate by $W(\mathcal{C})$.*

The result might suggest that if $w \in W(\mathcal{C})$, then $\pi(\nu)$ and $\pi(w \cdot \nu)$ are infinitesimally equivalent. This is not quite true: the two representations are built out of the same irreducible pieces, but they may be put together differently.

We will describe $W(\mathcal{C})$ for $SL(2n, \mathbf{R})$. If A is as in (4.5), and $\nu \in \hat{A}$, write

$$\nu\big(d(s_1, s_1, \ldots, t_{2r})\big) = \left(\prod_i s_i^{2\phi_i}\right)\left(\prod_j t_j^{\psi_j}\right), \qquad (5.5a)$$

with $\phi_i, \psi_j \in \mathbf{C}$, and

$$2\sum \phi_i + \sum \psi_j = 0. \qquad (5.5b)$$

Then $W(\mathcal{C})$ acts on ν by permuting the ϕ's and ψ's:

$$w(\phi_1, \ldots, \phi_{n-r}, \psi_1, \ldots, \psi_{2r})$$
$$= \big(\phi_{\sigma(1)}, \ldots, \phi_{\sigma(n-r)}, \psi_{\tau(1)}, \ldots, \psi_{\tau(2r)}\big). \qquad (5.5c)$$

The permutations σ and τ are required only to satisfy the following conditions. Recall that $\mu = \mu(m_1, \ldots, m_n)$, with s of the m_i's equal to 1. Then we require (of σ and τ)

$$m_i = m_{\sigma(i)}, \qquad i = 1, \ldots, n - r,$$
$$\tau\{1, \ldots, s\} = \{1, \ldots, s\} \text{ or } \{s + 1, \ldots, 2r\}. \qquad (5.5d)$$

We now have an extremely explicit parametrization of the representations of $SL(2n, \mathbf{R})$ having a specified lowest $SO(2n)$-type, and a fairly reasonable answer to the same question in general (Theorem 5.2 and Proposition 5.4). To complete the classification of $\hat{G}$, we only need to know the complete set of lowest K-types of each representation $\bar{\pi}(\mu, \nu)$. Recall the group $R(\mathcal{C})$ of Theorem 3.6.

PROPOSITION 5.6. $R(\mathcal{C})$ *may be regarded as a group of homomorphisms from* $W(\mathcal{C})$ *to* $\{\pm 1\}$. *That is, if* $\rho \in R(\mathcal{C})$, *and* $w \in W(\mathcal{C})$, *then there is a natural definition of* $\rho(w) \in \{\pm 1\}$. *It satisfies*

$$\rho(w_1 w_2) = \rho(w_1)\rho(w_2), \qquad \rho_1\rho_2(w) = \rho_1(w)\rho_2(w)$$

for $\rho, \rho_1, \rho_2 \in R(\mathcal{C})$, *and* $w, w_1, w_2 \in W(\mathcal{C})$.

In $SL(2n, \mathbf{R})$ if $\mu = \mu(m_1, \ldots, m_n)$, then $R(\mu)$ is nontrivial only if $m_n = \pm 1$; in that case it has order 2 (cf. (3.5)). Suppose $m_n = \pm 1$; write $R(\mathcal{C}) = \{1, \rho\}$. If $w \in W(\mathcal{C})$ corresponds to (σ, τ) by (5.5), then the pairing of Proposition 5.6 is given by

$$\rho(w) = \begin{cases} 1 & \text{if } \tau\{1, \ldots, s\} = \{1, \ldots, s\}, \\ -1 & \text{if } \tau\{1, \ldots, s\} = \{s + 1, \ldots, 2r\}. \end{cases} \qquad (5.7)$$

(Our hypothesis on m_n forces $r = s$, so the second case can really arise.)

In the general case, define for $\nu \in \hat{A}$

$$W(\mathcal{C}, \nu) = \{w \in W(\mathcal{C}) \mid w\nu = \nu\},$$

$$R(\mathcal{C}, \nu) = \{\rho \in R(\mathcal{C}) \mid \text{for all } w \in W(\mathcal{C}, \nu) \, \rho(w) = 1\}. \quad (5.8)$$

For most ν, $W(\mathcal{C}, \nu) = \{1\}$, and $R(\mathcal{C}, \nu) = R(\mathcal{C})$.

PROPOSITION 5.9. *In the setting of Definition 5.1, the set of lowest K-types of $\bar{\pi}(\mu, \nu)$ is*

$$R(\mathcal{C}, \nu) \cdot \mu \subseteq \mathcal{C}(\mu)$$

(Theorem 3.6 and (5.8)); that is, the group $R_{\bar{\pi}(\mu,\nu)}$ of Theorem 3.6 is just $R(\mathcal{C}, \nu)$.

Suppose $G = \mathrm{SL}(2n, \mathbf{R})$, $\mu = \mu(m_1, \ldots, m_n)$, and $m_n = \pm 1$. Let $\nu \in \hat{A}$ correspond to (ϕ_i, ψ_j) as in (5.5). To compute $R(\mathcal{C}, \nu)$, we need to know when there is an element $w = (\sigma, \tau)$ of $W(\mathcal{C})$ which fixes ν, but satisfies

$$\tau\{1, \ldots, s\} = \{s + 1, \ldots, 2r\}.$$

Clearly this is the case if and only if

$$\{\psi_1, \ldots, \psi_s\} = \{\psi_{s+1}, \ldots, \psi_{2r}\}. \quad (5.10)$$

So the lowest K-types of $\bar{\pi}(\mu, \nu)$ are

$$\begin{aligned}
\{\mu(m_1, m_2, \ldots, \pm m_n)\} \quad &\text{if (5.10) fails,} \\
\{\mu(m_1, m_2, \ldots, m_n)\} \quad &\text{if (5.10) holds.}
\end{aligned} \quad (5.11)$$

The results of this section may be summarized in the following classification of $\hat{G}$ (compare Corollary 1.16).

THEOREM 5.12. *To each associate class $\mathcal{C}$ of representations of K (Definition 3.4), attach a special vector subgroup $A(\mathcal{C})$ and a series of representations $\{\pi(\nu) \mid \nu \in A(\mathcal{C})\hat{\ }\}$ (Propositions 4.4 and 4.11). Define $W(\mathcal{C})$ by (5.3), and write*

$$A(\mathcal{C})\hat{\ }/W(\mathcal{C}) = \text{orbits of } W(\mathcal{C}) \text{ on } A(\mathcal{C})\hat{\ }.$$

Then there is a one-to-finite correspondence from

$$\{\text{pairs } (\mathcal{C}, \bar{\nu}) \mid \mathcal{C} \text{ is an associate class in } \hat{K}, \text{ and } \bar{\nu} \in A(\mathcal{C})\hat{\ }/W(\mathcal{C})\}$$

onto $\hat{G}$; the pair $(\mathcal{C}, \nu)$ goes to the set

$$\{\bar{\pi}(\mu, \nu) \mid \mu \in \mathcal{C}\} = \Phi(\mathcal{C}, \nu) \subseteq \hat{G};$$

here $\nu \in \hat{A}$ belongs to the orbit $\bar{\nu}$. The set $\Phi(\mathcal{C}, \bar{\nu})$ has order

$$|R(\mathcal{C})/R(\mathcal{C}, \nu)|$$

(Theorem 3.6 and (5.8)). Each element of $\Phi(\mathcal{C}, \bar{\nu})$ has as its set of lowest K-types a single orbit of $R(\mathcal{C}, \nu)$ on $\mathcal{C}$.

References

1. D. Vogan, *Algebraic structure of the representations of semisimple Lie groups*. I, Ann. of Math. (2) **109** (1979), 1–60.

2. _____, *Representations of real reductive Lie groups*, Birkhäuser, Boston, Basel and Stuttgart, 1981.

3. G. Warner, *Harmonic analysis on semi-simple Lie groups*, I, Springer-Verlag, Berlin, Heidelberg and New York, 1972.

DEPARTMENT OF MATHEMATICS, MASSACHUSETTS INSTITUTE OF TECHNOLOGY, CAMBRIDGE, MASSACHUSETTS 02139

Lectures in Applied Mathematics
Volume **21**, 1985

Indefinite Harmonic Theory
and Unitary Representations

Joseph A. Wolf

Wilfried Schmid's lecture ended with an indication of our uniform construction of (possibly) singular representations of semisimple and reductive Lie groups. The positive energy representations and Gupta-Bleuler triples, described by Chris Fronsdal, typify our situation, though the correspondence is not yet exact. Here I will describe the construction in some detail and try to indicate the direction this work is now taking. Complete details of much of this can be found in the paper, *Singular unitary representations and indefinite harmonic theory*, November 1981, by Rawnsley, Schmid and myself.

Some notation is needed before I can begin. G will be a reductive Lie group, assumed connected to avoid technicalities, such as a unitary group $U(k, l)$ or the universal cover $SO(4, 2)\tilde{}$ of the conformal group. Lower case gothic letters denote complexified Lie algebras. Thus $\mathfrak{g}_0$ is the real Lie algebra of G and $\mathfrak{g} = \mathfrak{g}_0 \otimes \mathbf{C}$. So, if H is a Lie subgroup of G we also have subalgebras $\mathfrak{h}_0 \subset \mathfrak{g}_0$ and $\mathfrak{h} \subset \mathfrak{g}$. A "grammar" of this sort is useful because we deal with many subgroups of G.

Let me remind you of the "tempered" or "regular" or "Harish-Chandra" series of representations of G, in order to indicate the "location" of the representations I will be constructing. Let B be a Cartan subgroup of G. In other words, $\mathfrak{b}_0$ is a subalgebra of $\mathfrak{g}_0$ which is maximal for the property that

$\mathfrak{b}_0$ is abelian and $\mathrm{ad}(\mathfrak{b})$ is diagonalizable, and
$B = \{g \in G, \mathrm{Ad}(g)\xi = \xi$ for every $\xi \in \mathfrak{b}_0\}$.

1980 *Mathematics Subject Classification.* Primary 22E46; Secondary 22E70.

Then $B = T \times A$ where T is the compactly embedded (compact modulo the center of G) part and A is the split part (ad($\mathfrak{a}_0$) has all eigenvalues real). Further, there are subgroups $P = MAN$, called cuspidal parabolic subgroups, such that

    $MA = M \times A$ is a reductive subgroup of G,

    T is a compactly embedded Cartan subgroup of M,

    N is the nilpotent radical of P.

M is specified because $MA = \{g \in G: \mathrm{Ad}(g)a = a \text{ for all } a \in A\}$. N is not unique, but the result does not depend on the choice of N. Now one considers

    μ: (relative) discrete series representation of M,

    α: element of $\mathfrak{a}_0^*$, i.e. $e^{i\alpha}$ is a unitary character on A.

That gives a unitary representation of P on the space of μ,

$$\left(\mu \otimes e^{i\alpha}\right)(man) = e^{i\alpha}(a)\mu(m)$$

and thus defines the unitarily induced representation

$$\pi(\mu, \alpha) = \underset{P \uparrow G}{\mathrm{Ind}}\left(\mu \otimes e^{i\alpha}\right)$$

of G. There are about a half dozen "standard" names and notations for these representations. I call the set of all $\pi(\mu, \alpha)$, constructed from B, the "B-series". If B is as noncompact as possible, it usually is called the "principal series". When B is as compact as possible it usually is called the "fundamental series". If B is compactly embedded, then $A = 1$, $M = G = P$, and we have the (relative) discrete series.

One can try to get other representations by letting the parameters μ, α go "out of range". If $\alpha \in \mathfrak{a}^*$ but $\alpha \notin \mathfrak{a}_0^*$, one still has $\pi(\mu, \alpha)$, but it is not unitary. Sometimes it can be unitarized when α satisfies a technical condition. That gives a "complementary" series. I will avoid those. If μ is a continued discrete series representation of M of some sort, one still has a unitary representation $\pi(\mu, \alpha)$. The uniform geometric construction, which I will describe, gives some continued discrete series representations (as well as the usual discrete series), and thus gives the sort of singular $\pi(\mu, \alpha)$ obtained by letting μ go singular.

Here is the setting for the geometric construction. Let H be the centralizer of some torus subgroup of G, and let ψ be a unitary representation of H. The space G/H has a number of structures as complex manifolds; they come from embeddings of G/H as open G-orbits in a certain compact complex manifold $G_{\mathbf{C}}/H_{\mathbf{C}}Q_-$. On the Lie algebra level, $\mathfrak{h}$ is the reductive part of a parabolic subalgebra $\mathfrak{h} + \mathfrak{q}_-$ of $\mathfrak{g}$, $\mathfrak{q}_+ = \bar{\mathfrak{q}}_-$ represents the holomorphic tangent space, and necessarily $\mathfrak{h}_0$ contains a CSA $\mathfrak{b}_0$ of $\mathfrak{g}_0$ such that B is as compact as possible. Now ψ defines a holomorphic vector bundle $\mathbf{V} \to G/H$, fiber $V =$ representation space of ψ. We look at the representation of G on L_2 cohomologies of $\mathbf{V} \to G/H$.

All this is classical when H is compactly embedded in G. There, the statement is mostly contained in the Kostant-Langlands Conjecture of the 1960s, which was refined and proved by Schmid in the 1970s. The interest here will be when H is not compactly embedded in G.

Classically, with H compact modulo the center of G, there is a (positive definite) G-invariant hermitian metric on G/H, so we have a positive definite pointwise inner product $\langle \varphi(x), \varphi'(x) \rangle$ of $\mathbf{V}$-valued differential forms, thus a global linear product

$$\langle \varphi, \varphi' \rangle_{G/H} = \int_{G/H} \langle \varphi(x), \varphi'(x) \rangle \, d(xH),$$

and genuine Hilbert spaces

$$L_2^q(G/H, \mathbf{V}) = \{\mathbf{V}\text{-valued } (0,q)\text{-forms } \omega \text{ on } G/H: \langle \omega, \omega \rangle_{G/H} < \infty\}$$

on which G acts by unitary representations. Let $\bar{\partial}^*$ denote the formal adjoint of $\bar{\partial}$ on $\mathbf{V}$-valued forms. Then the Kodaira-Hodge-Laplace operator is

$\square$: closure of $\bar{\partial}\bar{\partial}^* + \bar{\partial}^*\bar{\partial}$ on $L_2^q(G/H, \mathbf{V})$ from the dense subspaces consisting of C_c^∞ forms.

The kernel

$$\mathcal{H}_2^q(G/H, \mathbf{V}) = \{\omega \in L_2^q(G/H, \mathbf{V}): \square\omega = 0\}$$

is the space of harmonic forms. It is closed in L_2^q, so G acts on it by a unitary representation. Note that compactness of H is crucial at the very start of this construction.

The "compactly embedded" condition on H forces $\dim V < \infty$, so, in particular, ψ has a highest weight, say λ. Let ρ be half the sum of the positive roots. The Kostant-Langlands Conjecture is as follows.

1. If $\lambda + \rho$ is orthogonal to some root of G then $\mathcal{H}_2^q(G/H, \mathbf{V}) = 0$ for all q.

2. Suppose that $\lambda + \rho$ is not orthogonal to any root of G. A root is called "compact" if it is a root of the maximal compact subgroup, "noncompact" otherwise. Let

$$q(\lambda + \rho) = (\text{number of compact positive roots } \alpha \text{ with } (\lambda + \rho, \alpha) < 0)$$
$$+ (\text{number of noncompact positive roots } \beta$$
$$\text{with } (\lambda + \rho, \beta) > 0).$$

Then $\mathcal{H}_2^q(G/H, \mathbf{V}) = 0$ for $q \neq q(\lambda + \rho)$, and G acts on $\mathcal{H}_2^{q(\lambda+\rho)}(G/H, \mathbf{V})$ irreducibly by the discrete series representation with Harish-Chandra parameter $\lambda + \rho$.

In general, there is no good relation between the harmonic L_2 spaces $\mathcal{H}_2^q(G/H, \mathbf{V})$ and ordinary Dolbeault cohomology $H^q(G/H, \mathbf{V})$, though there is of course a natural map

$$\mathcal{K}_2^q(G/H, \mathbf{V}) \to H^q(G/H, \mathbf{V}) \quad \text{by } \omega \mapsto [\omega]$$

which simply sends a harmonic form to its Dolbeault class. In the course of his work that resulted in proving the Kostant-Langlands Conjecture stated above, W. Schmid also showed that

3. If $(\lambda + \rho, \alpha) < 0$ for every noncompact positive root, i.e. if the bundle $\mathbf{V} \to G/H$ is negative, then $\mathcal{K}_2^{q(\lambda+\rho)}(G/H, \mathbf{V}) \to H^{q(\lambda+\rho)}(G/H, \mathbf{V})$ is an isomorphism on the subspaces of vectors that have finite expansion under the maximal compact subgroup K of G.

The hypotheses of (3) can always be arranged by suitable (depending on λ) choice of the complex structure, and then $q(\lambda + \rho)$ is equal to

$$s = \dim_{\mathbb{C}} K/H, \text{ dimension of the maximal compact}$$
$$\text{complex submanifold } K/H \text{ of } G/H.$$

The resulting isomorphism $\mathcal{K}_2^s(G/H, \mathbf{V})_K \to H^s(G/H, \mathbf{V})_K$ of Harish-Chandra modules is useful in a number of contexts. In our situation with H noncompact it leads to a K-type analysis of $\mathcal{K}_2^s(G/H, \mathbf{V})$.

We want to carry this work over to the case where H is not necessarily compact. The first major problem is to define the appropriate analogues of the spaces $\mathcal{K}_2^q(G/H, \mathbf{V})$. Neither "harmonic" or "square integrable" has a completely obvious definition here, where the only invariant hermitian metrics are indefinite. Some technical tricks, using the fibration I will describe below, allow us to define an auxiliary positive definite hermitian metric on G/H, which is not G-invariant except in the rather special case, described above, where H is compact. But this auxiliary metric only suffers bounded distortion under any element of G. So, if we use it to define the $L_2^q(G/H, \mathbf{V})$, then G acts on that Hilbert spaces by bounded linear transformations. Then we say that a form

$\omega \in L_2^q(G/H, \mathbf{V})$ is harmonic if $\bar{\partial}\omega = 0$ and $\bar{\partial}^*\omega = 0$ where $\bar{\partial}^*$ is the formal adjoint of $\bar{\partial}$ relative to the G-invariant indefinite-hermitian metric on G/H

and harmonic forms are understood as distribution solutions to the hyperbolic system $\bar{\partial}\omega = 0$, $\bar{\partial}^*\omega = 0$. Thus, we have a closed G-invariant subspace of $L_2^q(G/H, \mathbf{V})$,

$$\mathcal{K}_2^q(G/H, \mathbf{V}) = \begin{cases} \text{all } \mathbf{V}\text{-valued } (0, q)\text{-forms on } G/H \text{ that are } L_2 \\ \text{relative to the auxiliary positive definite metric and} \\ \text{harmonic relative to the invariant indefinite metric.} \end{cases}$$

It is a Hilbert space on which G acts continuously by bounded linear operators. Since Dolbeault cohomology can be done with distribution forms just as well as with smooth forms, we still have the canonical maps $\mathcal{K}_2^q(G/H, \mathbf{V}) \to H^q(G/H, \mathbf{V})$.

The invariant indefinite metric defines a hermitian inner product $\langle \, , \, \rangle_{G/H}$ just as in the classical setting. Let $s = \dim_{\mathbf{C}} K/K \cap H$, dimension of the maximal compact subvariety. Calculating in a special case, Rawnsley and I were amazed to see that, for negative $\mathbf{V}$,

(a) $(-1)^s \langle \, , \, \rangle_{G/H}$ is positive semidefinite on $\mathcal{K}_2^s(G/H, \mathbf{V})$,

(b) the canonical map $\mathcal{K}_2^s(G/H, \mathbf{V}) \to H^s(G/H, \mathbf{V})$ is surjective on the level of K-finite vectors,

(c) the kernel of $\mathcal{K}_2^s(G/H, \mathbf{V}) \to H^s(G/H, \mathbf{V})$ coincides with the kernel of $\langle \, , \, \rangle_{G/H}$ on $\mathcal{K}_2^s(G/H, \mathbf{V})$.

Thus, in the special case, passage to the quotient defined a unitary representation

π_V: action of G on $\mathcal{K}_2^s(G/H, \mathbf{V})/(\text{kernel of } \langle \, , \, \rangle_{G/H})$

which unitarized the Fréchet representation of G on $H^s(G/H, \mathbf{V})$. These representations π_V were irreducible and singular in the sense described earlier.

Schmid and I now discovered that (a), (b) and (c) hold in some generality, and for essentially geometrical reasons. There is a C^∞ fibration

π: $G/H \to K/L$ with structure group $L = K \cap H$.

When G/H is indefinite-hermitian symmetric, i.e. is a semisimple symmetric space with invariant complex structure, the base and fiber are complex manifolds. The projection need not be holomorphic, but still we find an analogue of the Leray spectral sequence and show that

$$H^q(G/H, \mathbf{V})_K \cong H_{\bar{d}}^q\big(K/L, \mathbf{H}^0(\text{fiber}, \mathbf{V})\big)_K$$

as a K-module. Here d is the $\bar{\partial}$ operator of K/L modified by a term that measures the failure of π: $G/H \to K/L$ to be holomorphic. This, some estimates and some tensoring arguments lead to a complete analysis of the global character, the K-character and the K-spectrum of the Dolbeault cohomologies $H^q(G/H, \mathbf{V})$.

When π: $G/H \to K/L$ is holomorphic, the K-decomposition $H^s(G/H, \mathbf{V})_K = H^s(K/L, \mathbf{H}^0(\text{fiber}, \mathbf{V}))_K$ can be done on the level of harmonic forms, even L_2-harmonic forms for negative $\mathbf{V}$. Thus, when $\mathbf{V} \to G/H$ is negative, we obtain (a), (b) and (c). Since we understand the Dolbeault space $H^s(G/H, \mathbf{V})$, we then have complete character and spectral information on the unitary representations π_V. That is the content of the Rawnsley-Schmid-Wolf paper which I mentioned at the beginning of this talk.

Schmid and I are now trying to get rid of various restrictions, e.g. that π: $G/H \to K/L$ be holomorphic, or even that rank K = rank G. So far, we have made some progress in obtaining the fundamental series representations of G in this way. To be precise we have that now on the Dolbeault level, modulo correctness of a result of Schmid that has not yet been written down in complete detail. This, incidently, completes the proof that Zuckerman's derived modules are the same as the corresponding Dolbeault cohomologies, and we are now investigating the L_2 properties of certain harmonic representatives.

The unitary representations π_V, for $G/H \to K/L$ holomorphic, are highest weight representations. So their duals π_V^* are lowest weight representations, that is, positive energy representations. The scheme for the indecomposable G-module $\mathcal{H}_2^s(G/H, \mathbf{V})$ given by

$$\mathcal{H}_2^s(G/H, \mathbf{V})/(\ker\langle\,,\,\rangle_{G/H}) \rightsquigarrow (\ker\langle\,,\,\rangle_{G/H}) \rightsquigarrow 0$$

has formal similarity to the scheme

$$(\text{transverse photons}) \rightsquigarrow (\text{longitudinal photons}) \rightsquigarrow 0$$

of Gupta-Bleuler quantization. This appears to be tied to our definition of harmonic. More generally, if one defines a generalized harmonic space

$$\tilde{\mathcal{H}}_2^s(G/H, \mathbf{V}) = \left\{\omega \in L_2^s(G/H, \mathbf{V}) : (\bar{\partial}\bar{\partial}^* + \bar{\partial}^*\bar{\partial})^N \omega = 0 \text{ for } N \gg 0\right\},$$

then the scheme of quotients of

$$\tilde{\mathcal{H}}_2^s(G/H, \mathbf{V}) \supset \mathcal{H}_2^s(G/H, \mathbf{V}) \supset (\ker\langle\,,\,\rangle_{G/H})$$

has strong formal similarity to the Gupta-Bleuler triples of C. Fronsdal's talk: singletons, de Sitter electrodynamics, conformal QED and possibly conformal gravity. The intriguing fact here is that, in Fronsdal's setting, an inspection of the table printed with his lecture shows that in each case he has the formal analogue of

$$\tilde{\mathcal{H}}_2^s(G/H, \mathbf{V})/\mathcal{H}_2^s(G/H, \mathbf{V}) \cong (\ker\langle\,,\,\rangle_{G/H})$$

and in each case his module is an irreducible positive energy module. It will be interesting to understand this from the viewpoint of our L_2 harmonic forms.

DEPARTMENT OF MATHEMATICS, UNIVERSITY OF CALIFORNIA, BERKELEY, CALIFORNIA 94720

Lectures in Applied Mathematics
Volume 21, 1985

Induced Representations and Quantum Fields

Gregg J. Zuckerman[1]

Induced representations of groups arise whenever we study coset spaces of groups by subgroups. Quantum fields usually transform via one particular physical group, the Poincaré group,

$$O(3,1) \ltimes \mathbf{R}^{3,1}.$$

The physical significant coset space for the Poincaré group is the Minkowski space

$$[O(3,1) \ltimes \mathbf{R}^{3,1}]/O(3,1),$$

which we will call $\mathfrak{M}$. Quantum field theory on $\mathfrak{M}$ implicitly involves induced representations of the Poincaré group, induced from finite-dimensional representations of $O(3,1)$. However, there is not much point to applying a very general mathematical theory of induced representations to *one* physical example.

It turns out that some physicists have been considering groups and coset spaces alternative to the Poincaré group and Minkowski space. Dirac [2], following earlier work of de Sitter on cosmology, studied field theory on $O(3,2)/O(3,1)$. Dirac again [3], following earlier work of Bateman and Cunningham on electrodynamics, studied field theory on conformal space,

$$O(4,2)/[\mathbf{R}^x \times O(3,1)] \ltimes \mathbf{R}^{3,1}.$$

Schrödinger, the creator of Newtonian wave mechanics, was implicitly studying field theory on a coset space which is the "speed of light goes to

1980 *Mathematics Subject Classification.* Primary 81E20, 22E70.
[1] Supported by NSF Grant MCS80-05151.

infinity" limit of Minkowski space. Schrödinger's space-time is best described as a quotient of a certain pair of subgroups of $O(5, 2)$ (see [6]).

We have found that the language of induced representations and induced modules is very useful for the study of classical fields on coset space models for space-time. In addition, we have found a setting for quantum fields in the context of induced modules. We will introduce the general notions of induced module theory in §1. We specialize in §2 to a modern treatment of Dirac's work on the example of conformal space. Finally, we make some definitions in §3 of free quantum fields and Fock spaces in a general group theoretical framework. We hope to expand the material in §3 to a more complete theory at some later time.

We thank Paul Sally and AMS-SIAM for the opportunity to lecture at this conference and contribute to the proceedings.

1. As we explained in the introduction, the mathematical theory of classical and quantized fields is usually presented in the context of Minkowski space. However, following the pioneering work of Dirac and others, we will establish the basic notion of field theory in a wider class of geometries. At first we will take for our "space-time" any finite-dimensional homogeneous manifold X, which carries a smooth transitive action of a finite-dimensional Lie group G. We will write this action abstractly: given $x \in X$ and $g \in G$, g sends x to the new point gx. If $g_1 \in G$, then $g(g_1 x) = (g g_1)x$. We choose arbitrarily a base point $x_0 \in X$, and let H be the subgroup of G consisting of elements h such that $hx_0 = x_0$. The map $g \mapsto gx_0$ is a smooth map of G onto X, and identifies the *coset space G/H* with X via the rule $gH \mapsto gx_0$.

For example, Minkowski $\mathfrak{M}$ space itself can be viewed as 2×2 Hermitian matrices. The group $\mathrm{SL}(2, \mathbf{C})$ operates via the rule $h \mapsto ghg^*$, where g^* is the adjoint to g. The semidirect product

$$\mathbf{P} = \mathrm{SL}(2, \mathbf{C}) \ltimes \mathfrak{M}$$

operates on $\mathfrak{M}$ via the given action of $\mathrm{SL}(2, \mathbf{C})$ and via the translation action of $\mathfrak{M}$ on itself. $\mathbf{P}$ is the (proper) Poincaré group. We can regard $\mathfrak{M}$ as a homogeneous manifold for $\mathbf{P}$, and we can identify $\mathfrak{M}$ with $\mathbf{P}/\mathrm{SL}(2, \mathbf{C})$. Note that the element $-I$ in $\mathrm{SL}(2, \mathbf{C})$ acts trivially on $\mathfrak{M}$.

We will discuss Dirac's examples in §2. In the meantime, an easier example to visualize is the two-sphere as a homogeneous space for $\mathrm{SU}(2)$: think of $\mathrm{SU}(2)$ operating on traceless 2×2 Hermitian matrices via $H \mapsto gHg^{-1}$. The set of H with $\det H = -1$ can be thought of as a two-sphere, carrying a transitive action of $\mathrm{SU}(2)$. Again, $-I$ acts trivially.

The key group-theoretic notion of physical field theory is that of an induced G-module. The concept of a G-module arises in the context of quantities that transform linearly under some action of G. Suppose V is a real vector space, possibly infinite dimensional. Suppose G operates linearly on V: thus, for each $g \in G$ and $v \in V$, we have a rule for defining a new vector gv in V, subject to the following conditions:

$$g(g_1 v) = (gg_1)v \quad \text{for } g_1 \in G, \tag{1.1a}$$

$$g(v_1 + v_2) = gv_1 + gv_2, \tag{1.1b}$$

$$g(\lambda v_1) = \lambda gv_1 \quad \text{for } \lambda \text{ a real number.} \tag{1.1c}$$

Thus, we can view the elements of V as "quantities that transform linearly via $v \mapsto gv, g \in G$."

Clearly, any G-module V is associated to a representation of G by linear transformations of V: each g is represented by $\pi(g)$, where $\pi(g)v = gv$ for any v in V. However, we prefer not to choose a particular basis for V, so that we do not have particular matrix representations of G associated to the G-module V. In this sense, a G-module is more abstract than the typical representations used in physics.

Let us, therefore, immediately introduce the notion of an induced G-module, which is a more concrete instance of the general notion of a module. Fix again a homogeneous manifold $X = G/H$. Let E be a finite-dimensional H-module. Thus, for each $h \in H$ and $e \in E$, we have a rule for defining a new vector he in E, subject to the conditions (1.1). We assume further that $h \mapsto he$ depends smoothly on h. We can always choose a basis for the finite-dimensional vector space E, and then construct a matrix representation of H, but for now we choose to view E abstractly, with no preferred basis.

Let $C^\infty(G, E)$ be the linear space of all E-valued infinitely differentiable functions on G. For $g \in G$ and ψ in $C^\infty(G, E)$, define a new function $g\psi$ by

$$(g\psi)(g_1) = \psi(g^{-1}g_1). \tag{1.2}$$

By the above rule, $C^\infty(G, E)$ becomes an infinite-dimensional G-module. Now, let $I_H^G(E) = I(E)$ be the linear subspace of $C^\infty(G, E)$ consisting of functions ψ such that

$$\psi(g_1 h) = h^{-1}\psi(g_1) \tag{1.3}$$

for $g_1 \in G$ and $h \in H$. One checks easily that if ψ is in $I(E)$, then so is $g\psi$ for any $g \in G$. Thus, $I(E)$ is a G-submodule of $C^\infty(G, E)$. We call $I(E)$ an induced G-module, and the representation of G by linear transformations of $I(E)$, an induced representation.

$I(E)$ is a very large space of functions. Intuitively, condition (1.3) constrains ψ only along directions tangent to the left cosets of H in G, i.e., the submanifolds $g_1 H$, g_1 in G. For any point x in X, there is some neighborhood $\mathfrak{U}$ and a smooth map $\gamma: \mathfrak{U} \to G$ such that if $y \in \mathfrak{U}$,

$$\gamma(y)x_0 = y.$$

(γ is a local cross-section of the fiber bundle $G \to X$, with projection map $g \mapsto gx_0$.) Let $C^\infty(\mathcal{U}, E)$ be the smooth functions from $\mathcal{U}$ to E. If ψ is in $I(E)$, we can form the composite function $\psi \circ \gamma$ in $C^\infty(\mathcal{U}, E)$. Then, every function in $C^\infty(\mathcal{U}, E)$ can be approximated as closely as one likes by a function of the form $\psi \circ \gamma$, ψ in $I(E)$.

Consider the example of the two-sphere, regarded as the coset space $SU(2)/U(1)$, where $U(1)$ is the diagonal subgroup of matrices

$$\begin{bmatrix} \alpha & 0 \\ 0 & \bar{\alpha} \end{bmatrix}, \quad \alpha \in \mathbf{C}, |\alpha| = 1.$$

Let E be the one-dimensional complex $U(1)$-module with the transformation rule $e \mapsto \alpha e$. The induced module $I(E)$ consists of functions ψ from $SU(2)$ to E such that

$$\psi\left(g_1 \begin{bmatrix} \alpha & 0 \\ 0 & \bar{\alpha} \end{bmatrix}\right) = \bar{\alpha}\psi(g_1). \tag{1.4}$$

If $g = \begin{bmatrix} -1 & 0 \\ 0 & -1 \end{bmatrix}$, then

$$(g\psi)(g_1) = \psi(g^{-1}g_1) = \psi(g_1 g^{-1}) = -\psi(g_1).$$

In other words, if we make a 2π radian rotation of the two-sphere, the quantity ψ changes sign. Thus we certainly cannot regard ψ as a function on the two-sphere itself. We can try to pull ψ down to the two-sphere by choosing a section ψ of the projection $SU(2) \to SU(2)/U(1)$. However, there are *no* global smooth cross-sections to this projection. If γ is a local cross-section, we can form $\psi \circ \gamma$. But then the transformation law of $\psi \circ \gamma$ will be a bit tricky: we must remember that *both* ψ and γ are transforming.

Physicists (e.g. Pauli and Dirac) introduced quantities similar to ψ in the theory of electron spin. We choose to regard these fields as constrained (via (1.3)) functions on an appropriate Lie group. Thus, a Pauli two-component spinor field will be thought of as living in

$$I_{SU(2)}^{SU(2) \ltimes \mathbf{R}^3}(\mathbf{C}^2)$$

where $\mathbf{C}^2$ is the defining $SU(2)$ module, and $SU(2) \ltimes \mathbf{R}^3$ is the universal covering group of the (proper) Euclidean motion group, $SO(3) \ltimes \mathbf{R}^3$.

Our group-theoretic approach to physical fields can now be summarized by the following definition:

1.5. A classical linear physical field ψ "on X" is a function in some induced G-module $I(E)$, for some given H-module (usually finite dimensional) E. The G-transformation law for ψ is given by (1.2). The type of ψ is determined by the H-module E.

Before considering some more examples, let us enumerate some mathematical operations on physical fields. Suppose first that S is a closed subgroup of G, and let $Y = Sx_0$ be the orbit of x_0 under S acting on X. The stability group of x_0 in S is the intersection $S \cap H$. We can regard $S/S \cap H$ as a model for Y. Now let E be an H-module. We can restrict E to an $S \cap H$-module E and form the induced S-module $I_{S \cap H}^S(\tilde{E})$.

Now let ψ be a physical field of type E on X. Thus, ψ is a function in $I_H^G(E)$. We can restrict ψ to a function $\tilde{\psi}$ in $C^\infty(S, E)$. One checks that $\tilde{\psi}$ lies in the S-submodule $I_{S \cap H}^S(\tilde{E})$. One sees also that the map $\psi \mapsto \tilde{\psi}$ is an S-intertwining or S-module map from $I_H^G(\tilde{E})$ to $I_{S \cap H}^S(\tilde{E})$. We say that $\tilde{\psi}$ is the restriction of ψ from X to Y. Note that $\tilde{E}$ can be reducible as an $S \cap H$ module even if E is irreducible as an H-module.

Our next operation involves changing the type of a field ψ on a field X. Let F be a fixed finite-dimensional G-module. We restrict F to an H-module, which we also denote by F. Suppose ψ has type E. Form the tensor product H-module $F \otimes E$. We will denote by ψ^F any function in $I_H^G(F \otimes E)$. The transformation properties of ψ^F are simply related to those of ψ:

LEMMA 1.6. *The G-module $F \otimes I_H^G(E)$ is naturally isomorphic to the G-module $I_H^G(F \otimes E)$ via the map T defined by*

$$[T(f \otimes \psi)](g) = (g^{-1}f) \otimes \psi(g).$$

EXAMPLE. Suppose E_0 is the trivial H-module: E_0 is one dimensional, and H acts trivially. Then, $I_H^G(E_0)$ is isomorphic to the G-module of functions $C^\infty(G/H)$, the scalar fields ϕ on X. $I_H^G(F)$ will be isomorphic to the G-module $C^\infty(G/H, F)$ where G acts by the formula

$$(g\phi^F)(x) = g\phi^F(g^{-1}x), \tag{1.7}$$

ϕ^F a function from G/H to F. In case $G = P$ the Poincaré group, and $H = SL(2, \mathbf{C})$ the Lorentz group, formula (1.7) above is (in basis free form) the usual transformation law for an F-valued field on Minkowski space, P/H. Since every finite-dimensional H-module E extends to a P module F, all the physical fields of the standard Minkowski theory are included in this example.

2. Let us now discuss Dirac's 1936 Annals of Mathematics, *Wave equations in conformal space*. Let $\mathbf{R}^{4,2}$ be a six-dimensional real vector space with a quadratic form of signature $(4, 2)$. Let H be the group of linear transformations of $\mathbf{R}^{4,2}$ that preserve the quadratic form up to a positive scalar. Let D be the subgroup of H consisting of scalar multiples of the identity, I_6. Then, D is the center of H, and $H = O(4, 2)D$, with $D \cap O(4, 2) = \{\pm I_6\}$. Let G be the semidirect product of H with $\mathbf{R}^{4,2}$, the translation group. We can identify $\mathbf{R}^{4,2}$ as G/H.

Let $\mathfrak{N}$ be the set of nonzero null vectors in $\mathbf{R}^{4,2}$. $\mathfrak{N}$ is a connected manifold, and H acts transitively on $\mathfrak{N}$. Choose a vector x_1 in $\mathfrak{N}$, and let P_1 be the stabilizer of x_1 in H. Then $\mathfrak{N}$ can be identified as H/P_1.

Next, let $\overline{\mathfrak{M}}$ be the set of null *lines* through the origin of $\mathbf{R}^{4,2}$. We can think of $\overline{\mathfrak{M}}$ as the quotient of $\mathfrak{N}$ by the action of D, and we can identify $\overline{\mathfrak{M}}$ with $H/P_1 D$. The four-dimensional manifold $\overline{\mathfrak{M}}$ is what Dirac calls "conformal space." $\overline{\mathfrak{M}}$ is related to Minkowski space $\mathfrak{M}$ as follows: let x_2 be a second null vector chosen so that the inner product of x_1 and x_2 is one. Let P_2 be the stabilizer of x_2 in H. The vector x_1 generates a line $\bar{x}_1$ in $\overline{\mathfrak{M}}$. The orbit $P_2 \bar{x}_1$ in $\overline{\mathfrak{M}}$ can be identified with $P_2/P_2 \cap (P_1 D)$. It is an elementary mathematical exercise to show that

$$P_2 \cong [\mathbf{R}^x \times O(3,1)] \ltimes R^{3,1} \quad \text{and} \quad P_2 \cap (P_1 D) \cong \mathbf{R}^x \times O(3,1),$$

so that $P_2 \bar{x}_1$ can be identified with

$$O(3,1) \ltimes \mathbf{R}^{3,1}/O(3,1)$$

which is a group-theoretic model for Minkowski space $\mathfrak{M}$. Thus, $P_2 \bar{x}_1$ is a copy of $\mathfrak{M}$ sitting inside $\overline{\mathfrak{M}}$. In fact, $P_2 \bar{x}_1$ is open and dense in $\overline{\mathfrak{M}}$, so that $\overline{\mathfrak{M}}$ is a compactification of $\mathfrak{M}$. (For a traditional coordinate dependent explanation of this compactification, see Dirac's *Wave equations in conformal space*.)

Dirac considers various types of fields on G/H: he then relates them to fields on $\mathfrak{N}$, $\overline{\mathfrak{M}}$, and finally $\mathfrak{M}$. Choose some finite-dimensional H-module E under the assumption that if λI_6 is in D, λ operates on E by the rule $e \mapsto \lambda^N e$, N a fixed real number. Form the induced G-module $I_H^G(E)$, and let $I(E)^0$ denote the D-invariant fields in $I(E)$: if we regard $I(E)$ as E-valued functions on $\mathbf{R}^{4,2}$, $I(E)^0$ will be the functions homogeneous of degree N (we will ignore the problem of smoothness at the origin). $I(E)^0$ will be an H-module (but no longer a G-module) in which D acts trivially.

Now let $\tilde{E}$ be the restriction of E to P_1. We can restrict fields in $I(E)^0$ to fields in $I_{P_1}^H(\tilde{E})$, by simply restricting the domain of a field ψ from G/H to $\mathfrak{N} = H/P_1$.

Let $\tilde{\psi}$ denote the restriction: then $\psi \mapsto \tilde{\psi}$ is an H-module map from $I(E)^0$ to $I_{P_1}^H(\tilde{E})$. The image of this restriction map will be the D-invariant fields in $I_{P_1}^H(\tilde{E})$. But now D is in the center of H, so the D-invariants in $I_{P_1}^H(\tilde{E})$ can be identified with the induced H-module $I_{P_1 D}^H \bar{E}$, where $\bar{E}$ is the extension of E to $P_1 D$ with D acting trivially. Thus, we have an H-module map of $I_H^G(E)^0$ onto $I_{P_1 D}^H \bar{E}$. In language closer to Dirac's paper, we have an association of fields on conformal space $\overline{\mathfrak{M}}$ to homogeneous fields on $\mathbf{R}^{4,2}$. These fields on $\overline{\mathfrak{M}}$ transform via $H = O(4,2)D$. If we further restrict to Minkowski space $\mathfrak{M}$ inside $\overline{\mathfrak{M}}$, we have the P_2-module map

$$I_{P_1 D}^H \bar{E} \mapsto I_{P_2 \cap P_1 D}^{P_2} \tilde{\bar{E}},$$

where $\tilde{\bar{E}}$ is the restriction of $\bar{E}$ to $P_2 \cap P_1 D$. (This map is not onto, since $\overline{\mathfrak{M}}$ is compact and $\mathfrak{M}$ is only open in $\overline{\mathfrak{M}}$.) However, this final restriction to $\mathfrak{M}$ is necessary only for understanding the Poincaré group transformation law of the fields which Dirac produces on conformal space $\overline{\mathfrak{M}}$. What Dirac really wants to capture is the transformation law under H of fields on conformal space. For him, the transition from $\mathbf{R}^{4,2}$ to $\overline{\mathfrak{M}}$ via $\mathfrak{M}$ is a geometric convenience, since $\mathbf{R}^{4,2}$ carries a linear action of H whereas $\overline{\mathfrak{M}}$ carries a nonlinear action of H.

From the modern, the admittedly abstract point of view of induced modules, $\overline{\mathfrak{M}}$ is no more difficult to handle than $\mathbf{R}^{4,2}$. One can now study intrinsically on $\overline{\mathfrak{M}}$ problems that Dirac cleverly reduced in some cases to problems on $\mathbf{R}^{4,2}$. There appears to be one striking advantage to the intrinsic approach: we can construct many *more* types of fields on $\overline{\mathfrak{M}}$ than can Dirac, by his Annals paper methods. Thus, Dirac begins with an H-module E which (up to a twist) restricts to the $P_1 D$ module $\bar{E}$ above. We can start with an arbitrary finite-dimensional $P_1 D$-module W and construct fields of type W in the induced H-module $I_{P_1 D}^H W$. In general, W will *not* extend to a module over H, so that we cannot think of W as an $\bar{E}$.

For example, the four-dimensional half-spin module S^+ of (the double cover of the identity component of) H is irreducible as an H-module, but reducible as a $P_1 D$-module. S^+ has a two-dimensional $P_1 D$-submodule W_1 and a two-dimensional quotient module W_2. Fields of type W_1 (with a suitable twist) are now known to describe neutrino-type fields on $\overline{\mathfrak{M}}$ (massless helicity one-half fields). Interestingly, Dirac works out fields of type S^+ and runs into problems interpreting his calculations. The module W does not extend to H: $O(4,2)$ has no two-dimensional (even multi-valued) representations. Hence, fields of type W_1 do not arise in Dirac's treatment.

The reader may ask at this point, just exactly what is our modern treatment of neutrino-type fields? We can view such fields as certain constrained functions on H with values in W_1. Now $H (= O(4,2)D)$ may be realized concretely as a sixteen-dimensional manifold inside 6×6 real matrices. Just as $\mathfrak{M}$ is an algebraic subvariety of six-dimensional space, H is an algebraic subvariety of thirty-six-dimensional space. The point is that as long as we care to use more than the customary four dimensions, our induced module approach is no more abstract than Dirac's "five-cone" approach, and we can construct more fields to boot.

3. We would like to conclude this paper with a brief introduction to *quantum* fields in the context of induced modules. We continue to take for our "space-time" a connected homogeneous manifold $X = G/H$. For our classical fields we fix an induced G-module $I(E)$. For simplicity, E will be a real finite-dimensional H-module.

Now, let $\mathcal{S}$ be an infinite-dimensional G-module carrying a complex vector space structure and a G-invariant positive definite Hermitian inner product $\langle\ |\ \rangle$. Let Q be the algebra of linear operators on $\mathcal{S}$, and Q_0 the real linear subspace of Hermitian operators. Thus, Q_0 is a real G-module, infinite dimensional, via the action $A \mapsto U(g)AU(g^{-1})$.

We next form the algebraic tensor product of real G-modules, $Q_0 \otimes_{\mathbf{R}} I(E)$. If life were simple, we would then define a G-invariant quantum field of type E to be a G-invariant element of $Q_0 \otimes_{\mathbf{R}} I(E)$. The trouble with this definition is that there may not exist *any* G-invariant elements. Let us illustrate the problem by sketching a theory of *free* G-invariant quantum fields.

Suppose V is a G submodule of $I(E)$. We suppose further that there exists a nondegenerate G-invariant antisymmetric real-valued bilinear form $\Omega(\ ,\ )$ on V. We say that (V, Ω) is a symplectic submodule. As a further condition on V, assume that the complexification $V \otimes_{\mathbf{R}} \mathbf{C}$ of V decomposes into a direct sum of two complex G-modules V^+ and V^- in such a way that Ω vanishes for pairs of functions in V^+ as well as for pairs of functions in V^-. Finally, assume that complex conjugation interchanges the subspace V^+ and V^-.

We can then define a G-invariant Hermitian form B on V^+ via

$$B(\psi_1, \psi_2) = \Omega(\psi_1, \bar{\psi}_2). \tag{3.1}$$

Let $\mathcal{S}$ be the symmetric tensor algebra of V^+ (alternatively, $\mathcal{S}$ is the algebra of polynomial functions on V^-). In a standard fashion, going back to Fock, we can build from B a G-invariant Hermitian inner product $\langle\ |\ \rangle$ on $\mathcal{S}$, and a G-module map $\rho\colon V \to Q_0$, the Hermitian linear operators on $\mathcal{S}$, such that ρ is a representation of the Heisenberg algebra $V \oplus \mathbf{R}$ built from Ω.

We now have the following situation: on the one hand, we have the embedding $\iota\colon V \hookrightarrow I(E)$ of V into $I(E)$. The R-linear map ι is a G-module map. On the other hand, we have Fock's G-module map $\rho\colon V \to Q_0$. Mathematically, we have a new G-module map,

$$\rho \otimes \iota\colon V \otimes_{\mathbf{R}} V \to Q_0 \otimes_{\mathbf{R}} I(E).$$

If V were finite dimensional, the existence of Ω would ensure the existence of a G-invariant vector ξ in $V \otimes_{\mathbf{R}} V$. We could then construct a G-invariant in $Q_0 \otimes_{\mathbf{R}} I(E)$ by the formula

$$(\rho \otimes \iota)(\xi).$$

Unfortunately, the physically interesting symplectic submodules V in $I(E)$ are generally infinite dimensional, and $V \otimes_{\mathbf{R}} V$ does *not* contain a G-invariant vector. One way out of this mess is to weaken our notion of a quantum field.

Let $\overline{Q}_0$ be the collection of all Hermitian inner products on $\mathbb{S}$. If A is in Q_0, A defines a Hermitian inner product α by the formula

$$\alpha(s, t) = \langle s|A|t\rangle = \langle As|t\rangle = \langle As|t\rangle = \langle s|At\rangle.$$

We obtain a G-module structure on $\overline{Q}_0$ and a G-module map

$$Q_0 \rightsquigarrow \overline{Q}_0, \qquad A \mapsto \alpha.$$

Because $\mathbb{S}$ is infinite dimensional, Q_0 is not all of $\overline{Q}_0$. At the very least, a quantum field should be some sort of $\overline{Q}_0 \otimes_\mathbf{R} E$ valued function on G.

Let E, V, Ω, $\mathbb{S}$ and ρ be as before. Let s and t be arbitrary states, i.e., vectors in $\mathbb{S}$. It is an elementary property of the Fock representation ρ that there exists a unique element $\eta_{s,t}$ in $V \otimes_\mathbf{R} C$ such that for any element ψ in V,

$$\langle s|\rho(\psi)|t\rangle = \omega(\psi, \eta_{s,t}). \tag{3.2}$$

Now, $\eta_{s,t}$ is in V_C and hence in $I(E_C)$. Thus, $\eta_{s,t}$ is a function from G to E_C. If e^* lies in the dual space of E, then $(\eta_{s,t}(g), e^*)$ is a complex number.

We now define a function

$$\hat{\psi}: G \rightarrow \overline{Q}_0 \otimes_\mathbf{R} E \tag{3.3}$$

by the formula

$$\langle s|(\hat{\psi}(g), e^*)|t\rangle = (\eta_{s,t}(g), e^*). \tag{3.4}$$

By $(\hat{\psi}(g), e^*)$ we mean the contraction of $\hat{\psi}(g)$ in $\overline{Q}_0 \otimes_\mathbf{R} E$ with e^* in E^*. Thus, $(\hat{\psi}(g), e^*)$ lives in $\overline{Q}_0$. One checks that because $\eta_{s,t}$ lives in $I(E_C)$, the function $\hat{\psi}$ satisfies the equation

$$\hat{\psi}(gh) = h^{-1}\hat{\psi}(g) \tag{3.5}$$

where $h^{-1}\hat{\psi}(g)$ is calculated via the H-module action on $\overline{Q}_0 \otimes_\mathbf{R} E_C$. One also checks the equation between "operators"

$$(\hat{\psi}(g_1^{-1}g), e^*) = U(g_1)^{-1}(\hat{\psi}(g), e^*)U(g_1), \tag{3.6}$$

where $g_1 \mapsto U(g_1)$ is the operator representation of G on the Fock space $\mathbb{S}$. On the other hand, since $\hat{\psi}(g)$ is not really an operator, but only an Hermitian form, we should write the above as

$$\langle s|(\hat{\psi}(g_1^{-1}g), e^*)|t\rangle = \langle g_1 s|(\hat{\psi}(g), e^*)|g_1 t\rangle. \tag{3.7}$$

If we ignore the difference between Q_0 and $\overline{Q}_0$, then $g \mapsto \hat{\psi}(g)$ is an operator-valued quantum field "on" G/H. (3.6) says that the "components" $(\hat{\psi}(g), e^*)$ of $\hat{\psi}(g)$ transform in a specific fashion under left translation by G, i.e., $\hat{\psi}$ is a quantum field of type E. The discrepancy between Hermitian operators in Q_0 and Hermitian inner products in $\overline{Q}_0$

shows up dramatically when we try to form products of field "operators" at a single point. $\overline{Q}_0 \otimes_{\mathbf{R}} \mathbf{C}$ is *not* an algebra, so that a product

$$\left(\hat{\psi}(g), e_1^* \right) \cdots \left(\hat{\psi}(g), e_m^* \right) \tag{3.8}$$

has no a priori mathematical meaning. On the other hand, physicists have rules, at least in perturbation theory, for making computational sense out of products such as (3.8). For now, we venture no further into the jungle of field products.

REFERENCES

1. A. S. Davydov, *Quantum mechanics*, Chapter XIV, Technical Translation Series, NEU Press, 1966.

2. P. A. M. Dirac, *The electron wave equation in deSitter space*, Ann. of Math. (2) **36** (1935), 657–669.

3. _____, *Wave equations in conformal space*, Ann. of Math. (2) **37** (1936), 429–442.

4. C. Fronsdal, *Semisimple gauge theories and conformal gravity*, these PROCEEDINGS.

5. R. Jost, *General theory of quantized fields*, Lectures in Appl. Math., vol. 4, Amer. Math. Soc., Providence, R.I., 1965.

6. G. Zuckerman, *Nonrelativistic quantum mechanics and five dimensional conformal geometry*, Preprint (in progress).

DEPARTMENT OF MATHEMATICS, YALE UNIVERSITY, NEW HAVEN, CONNECTICUT 06520

IV. KAC-MOODY ALGEBRAS AND NONLINEAR THEORIES

Lectures in Applied Mathematics
Volume **21**, 1985

Why Kac-Moody Subalgebras
Are Interesting in Physics

L. Dolan[1]

I. Introduction. The theory of elementary particle physics is the study of what we currently believe to be the four fundamental interactions of nature: the familiar forces of electromagnetism and gravity acting both at large and small distances; and the two more anti-intuitive forces of the weak and strong interactions, which only occur at very short distances.

Each of these interactions can be described by a field theory loaded with infinities that can sometimes be consistently removed in a weak coupling, i.e. perturbative expansion. The strong interactions, distinguished by their large coupling at distances on the order of the size of elementary particles, so far have no controlled approximation in which to study the nonperturbative phenomena such as quark confinement and spontaneous symmetry breakdown.

Symmetry in a theory has always been a tool which gives information about the solution independent of the approximation scheme. Recently a new symmetry group has been identified [1] in the nonlinear sigma model as the hidden symmetry which led to the calculation of its S-matrix [2]. This new symmetry is a Kac-Moody [3] subalgebra. Because of the extensive connection between two-dimensional chiral models and four-dimensional nonabelian gauge theories [4,5], a similar algebra appears on a restricted class of gauge fields [6]. If this extra symmetry can be found for the complete gauge theory, a conjecture supported by the Kac-Moody representations of the dual string model [7], it may provide a tool for

---

1980 *Mathematics Subject Classification.* Primary 81E99, 81G05, 81G20.

[1]Work supported in part by the U. S. Department of Energy under Contract Grant No. DE-Ac02-82ER40033.B000.

nonperturbative solvability in the spirit of the stunning successes of one-dimensional mathematical physics.

In these lectures, §II introduces Kac-Moody algebras. §III lists the present state of the occurrence of these algebras. §§IV and V provide a brief tour through what we know and expect from solvability of lower-dimensional models. §VI discusses loop space gauge theory. §§VII and VIII give the Kac-Moody subalgebras and their applications for chiral and gauge theories.

II. What is a Kac-Moody or associated affine algebra? A Lie algebra is completely determined by its structure constants C_{abc}. For example, SU(2) has three generators T^a, $a = 1, 2, 3$ and $[T^a, T^b] = \varepsilon_{abc} T^c$. The structure constants are $C_{abc} = \varepsilon_{abc}$. The defining or spinor representation is $T^a = \sigma^a/2i$, where σ^a are the Pauli matrices.

An infinite parameter, i.e. infinite-dimensional algebra, has an infinite number of generators. The Kac-Moody algebra [3] associated with a finite-parameter semisimple Lie group G is $G \otimes \mathbf{C}[t, t^{-1}] \times C_c$. The generator commutator relations are

$$\left[M_a^{(n)}, M_b^{(m)} \right] = C_{abc} M_c^{(n+m)} + n\delta_{n,-m} c (\mathrm{Tr}\, T^a T^b) P,$$
$$\left[P, M_a^{(n)} \right] = 0. \tag{1}$$

Here $n, m = -\infty, \ldots, -1, 0, 1, 2, \ldots, \infty$, T^a are matrix generators of G, C_{abc} are structure constants of G, and c is some constant number. The term $cn\delta_{n,-m} \mathrm{Tr}(T^a T^b) P$ is called the central extension and vanishes identically for $n, m = 0, 1, \ldots, \infty$. What occurs naturally in some of the physical models discussed in these lectures is "half" of a Kac-Moody algebra, $G \otimes C[t]$:

$$\left[M_a^{(n)}, M_b^{(m)} \right] = C_{abc} M_c^{(n+m)} \quad \text{for } n, m = 0, 1, 2, \ldots, \infty. \tag{2}$$

A defining representation for these generators ($n \geqslant 0$) is $M_a^{(n)} = T^a \otimes t^n$. T^a is a generator of G and t is a variable. For example, when $G = \mathrm{SU}(2)$ and $T^a = \sigma^a/2i$, then

$$M_3^{(n)} = \frac{\sigma^3}{2i} \otimes t^n = \frac{1}{2i} \begin{pmatrix} t^n & 0 \\ 0 & -t^n \end{pmatrix},$$

etc. Different theories give different realizations of $M_a^{(n)}$. For $t = e^{i\theta}$, the Kac-Moody subalgebra $G \otimes C[t]$ describes a mapping from S^1 to elements of G and the elements of $G \otimes C[t]$ are called loops in G.

III. Theories in which a Kac-Moody algebra or subalgebra have appeared so far. These are

2-dim spin systems,
2-dim integrable continuum theories,
2-dim chiral theories (including the nonlinear sigma model),

3-dim Yang-Mills in loop space,
4-dim self-dual Yang-Mills in local variables $A_\mu^a(x)$,
4-dim self-dual Yang-Mills in loop space,
Dual Resonance model,
Current algebra.

IV. How this infinite-parameter algebra is useful in solving the theory nonperturbatively. This occurs in several different ways, and the approach applicable for Yang-Mills will depend on how the algebra appears in the theory.

A. From the mathematics literature, since 1978, various relationships have been established between exactly integrable systems and generators of a Kac-Moody algebra.

The Lax pair $\dot{L} = [L, M]$ takes place in the algebra, and for the K-dV equation this can be used to prove exact integrability of the system. Also, the affine algebra has been used to construct explicitly a general class of solutions for K-dV and to linearize the periodic Toda lattice. In the case of a finite-parameter algebra, the method of orbits of Lie groups has led to the quantization of the integrable generalized Toda chains [8].

B. Exactly solvable systems are described most often by the *inverse scattering problem* method.

This method has unified many of the techniques of one-dimensional mathematical physics, namely,

(1) the transfer matrix formalism,
(2) the exact solution of the Ising and Baxter models,
(3) infinite sets of conserved charges,
(4) the Bethe Ansatz solution of quantum continuous field theories,
(5) the diagonalization of quantum Hamiltonians.

The method solves classical field theories, quantum field theories and lattice spin models. It can be thought of as a canonical transformation to action-angle variables; and it has as a hallmark an infinite set of commuting charges $[Q_n, Q_m] = 0$. These form an infinite-parameter commuting algebra, a Kac-Moody subalgebra where, for example,

$$Q_n = M_1^{(n)}, \qquad n \geq 0.$$

Then $[Q_n, Q_m] = [M_1^{(n)}, M_1^{(m)}] = \varepsilon_{11c} M_c^{(n+m)} = 0$ since the structure constants are antisymmetric. Therefore, theories which have charges in a Kac-Moody algebra may be solvable by the inverse method.

Depending on which inverse method is used, the questions being answered are slightly different and, therefore, the role of the infinite set of charges is different. (Some of the material presented in this section is reviewed with further references in [9].)

The Classical Inverse Method. This method solves the initial value problem of the classical nonlinear field equations: given $\varphi(x, 0)$, what is

$\varphi(x, t)$? The way it does it is to map the initial data $\varphi(x, 0)$ into the scattering data $a(k, 0)$ of an associated linear eigenvalue problem. Then $a(k, 0)$ is evolved to $a(k, t)$ which is mapped back to $\varphi(x, t)$.

EXAMPLE. The nonlinear Schrödinger equation is a $(1 + 1)$-dim classical field theory of one complex scalar nonrelativistic field:

$$i\partial_t\varphi = -\partial_x^2\varphi + 2c\varphi^*\varphi\varphi. \tag{3}$$

The linear scattering problem is

$$
\begin{matrix}
L_{op}\,\psi & = & k\psi \\
\| & & \| \\
\begin{pmatrix} i\partial_x & \sqrt{c}\,\varphi \\ \sqrt{c}\,\varphi^* & -i\partial_x \end{pmatrix} & & \begin{pmatrix} \psi_1 \\ \psi_2 \end{pmatrix}
\end{matrix}
$$

The Lax pair is $[L_{op}, M_{op}] = \dot{L}_{op}$ where $\partial_t\psi = M_{op}\psi$ and M_{op} is defined such that $\dot{k} = 0$. The integrability condition $\partial_x\partial_t\psi = \partial_t\partial_x\psi$ is a condition on $\varphi(x, t)$ given by the NLSE. The solution to the linear scattering problem has the form

$$\lim_{x\to\infty} \begin{pmatrix} \psi_1(x, 0; k) \\ \psi_2(x, 0; k) \end{pmatrix} = \begin{pmatrix} e^{ikx} & a(k, 0) \\ e^{-ikx} & b(k, 0) \end{pmatrix}. \tag{4}$$

Here $a(k, 0)$, $b(k, 0)$ are functions of $\varphi(x, 0)$, $\varphi^*(x, 0)$. Since $\dot{k} = 0$, $a(k, t)$ and $b(k, t)$ are determined from $\varphi(x, 0)$:

$$a(k, t) = a(k, 0), \qquad b(k, t) = e^{-ik^2 t}b(k, 0). \tag{5}$$

The action angle variables are $P(k, t) = (1/\sqrt{c})\ln|a(k, t)|$, $Q(k, t) = (1/\pi\sqrt{c})\arg b(k, t)$ since the NLSE Hamiltonian can be written as

$$H = \int_{-\infty}^{\infty} dx\big(|\partial_x\varphi|^2 + c|\varphi|^4\big) = \frac{1}{\pi\sqrt{c}}\int_{-\infty}^{\infty} dk\, k^2 P.$$

That is to say, in terms of P and Q, the Hamiltonian is only a function of P. Thus, it is always true that $\delta H/\delta Q(k, t) = -\dot{P}(k, t) = 0$, and an infinite set of conserved charges M_l arise from the expansion of $P(k, t)$ in a power series in k, or $\ln a(k, t) = -ic\sum_{l=0}^{\infty} M_l 1/k^{l+1}$. Then

$$M_l = \frac{1}{\pi\sqrt{c}}\int_{-\infty}^{\infty} dk\, k^l P(k, t), \quad \text{all } \dot{M}_l = 0.$$

To derive the NLSE as Hamilton's equations, define $\pi(x, t) = i\varphi^*(x, t)$ and the Poisson bracket $\{\pi(x, t), \varphi(y, t)\} = \delta(x - y)$. The $\dot{\varphi} = \{H, \varphi\}$, $\dot{\pi} = \{H, \pi\}$ are the NLSE. We see that the transformation from $\varphi(x, t)$, $\pi(x, t)$ to $P(k, t)$, $Q(k, t)$ is canonical, i.e. the two sets of variables obey the same commutation relations

$$\{P(k, t), Q(k', t)\} = \delta(k - k').$$ (6)

Since $\{P(k, t), P(k', t)\} = 0$, the conserved charges M_l satisfy $\{M_l, M_n\} = 0$. They form an infinite-parameter abelian algebra.

The Quantum Inverse Method. The fields $\varphi(x, t)$ are now quantum Schrödinger picture operators, and solving the quantum theory means to diagonalize the quantum Hamiltonian, i.e. to find its eigenvalues and eigenfunctions.

For the example above, the quantum Hamiltonian is made finite by normal ordering

$$H = \int_{-\infty}^{\infty} dx : |\partial_x \varphi|^2 + c |\varphi|^4 :.$$ (7)

The associated linear operator problem is $\partial_x \Psi = -i : \tilde{Q}\Psi :$ where

$$\tilde{Q} = \begin{pmatrix} -k & -\sqrt{c}\,\varphi \\ \sqrt{c}\,\varphi^* & k \end{pmatrix} \quad \text{and} \quad \Psi = \begin{pmatrix} \psi_1 & \psi_2^* \\ \psi_2 & \psi_1^* \end{pmatrix}.$$ (8)

The Poisson brackets are replaced by commutators.

The scattering data a, b are now operators which, although complicated functions of φ and φ^*, have simple commutation relations between themselves and the Hamiltonian: $[H, a] = 0$ and $[H, b] \sim b$. Therefore $a(k, t)$ generates an infinite set of conserved quantum charges, and $b(k, t)$ creates energy eigenstates which coincide with Bethe's Ansatz.

The NLSE is the second-quantized form of an N-body quantum mechanics problem with a δ-function potential:

$$H = \sum_{i=1}^{N} \frac{-\partial^2}{\partial x_i^2} + c \sum_{i<j} \delta(x_i - x_j).$$ (9)

Bethe's Ansatz for the N-particle wave function $\psi(x_1, \ldots, x_n)$, where $H\psi = E\psi$, was

$$\psi(x_1, \ldots, x_n) \sim \exp\left(-\sum_{i<j} |x_i - x_j|\right).$$ (10)

In the NLSE, the solution to the associated linear problem can be written as a path-ordered exponential

$$\Psi(x, t) = : P\exp\left(-i\int_y^x dz\, \tilde{Q}(z, t; k)\right)\Psi(y, t):,$$

$$\lim_{x \to \infty} \Psi = \begin{pmatrix} a & b^* \\ b & a^* \end{pmatrix}.$$ (11)

((11) is the scattering data.) To make contact with a finite lattice, observe that Ψ can be written as

$$\Psi = \lim_{\substack{N \to \infty \\ \varepsilon \to 0}} \prod_{n=1}^{n+1} L_n(k), \qquad N_\varepsilon = x - y = \text{length of lattice} = L,$$

where $L_n(k) = \exp(-i\varepsilon\tilde{Q}(y - (n - 1)\varepsilon, t; k))$. Then define

$$T_L(k) = \prod_{n=1}^{n+1} L_n(k) = \begin{pmatrix} A(k) & B(k) \\ C(k) & D(k) \end{pmatrix}.$$

On the lattice, the trace of the scattering data is the transfer matrix operator: for e.g., for the Baxter model [10]

$$\hat{T}(v) = \text{tr} \prod_{n=1}^{N} \sum_{j=1}^{4} W_j \hat{\sigma}_j(n)\sigma_j \sim A + D. \tag{12}$$

The Baxter model or symmetric eight-vertex model is a two-dimensional theory of classical statistical mechanics. A comprehensive treatment of an infinite-dimensional algebra of commuting charges, quantum lattice integrability, the transfer matrix method and Kramers-Wannier duality transformations can be made in this model.

The partition function is

$$Z = \sum_{\text{all } N_j} \exp\left(-\sum_{j=1}^{8} \varepsilon_j N_j\right). \tag{13}$$

Each site (n_0, n_1) of a rectangular lattice has an associated energy ε_j depending on the vertex configuration and N_j is the number of vertices of type j in an allowed set of configurations. The site configurations are

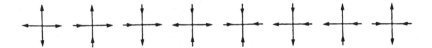

and the weights ε_j of the first four are set equal to those of the last four, respectively (hence the symmetric eight-vertex model).

If we define $\vec{\alpha}_{n_0} = (\alpha_{n_0 1}, \alpha_{n_0 2}, \ldots, \alpha_{n_0 N})$ where $\alpha_{n_0 n_1} = \pm 1$ represents a single vertical lattice link, the partition function Z can be expanded in terms of direct-product vector states $|\vec{\alpha}_{n_0}\rangle = |\alpha_{n_0 1}\rangle \otimes \cdots \otimes |\alpha_{n_0 N}\rangle$ where $|\alpha_{n_0 n_1}\rangle = \binom{0}{1}$ or $\binom{1}{0}$. For periodic boundary conditions,

$$Z = \sum_{\vec{\alpha}_1 \cdots \vec{\alpha}_m} \langle \vec{\alpha}_1 | \hat{T} | \vec{\alpha}_2 \rangle \langle \vec{\alpha}_2 | \hat{T} | \vec{\alpha}_3 \rangle \cdots \langle \vec{\alpha}_M | \hat{T} | \vec{\alpha}_1 \rangle = \text{Tr} \, \hat{T}^M \tag{14}$$

where $\hat{T}$, the transfer matrix, is given by (12) and w_j are functions of ε_j.

The transfer matrix $\hat{T}$ is the unifying concept between statistical mechanics in d dimensions and quantum field theory on the lattice in $d - 1$ dimensions. If there is some quantum Hamiltonian H which commutes with $\hat{T}$, then diagonalizing $\hat{T}$ is equivalent to solving the quantum theory. On the other hand, since $Z = \text{Tr } \hat{T}$, this solves Z.

Baxter found a parameterization of $w_j \rightarrow \zeta, V, l$, such that for fixed ζ and l,

$$[\hat{T}(V), \hat{T}(V')] = 0. \tag{15}$$

This one-parameter family of commuting transfer matrices led to the diagonalization of $\hat{T}$, a set of conserved commuting charges $\hat{Q}_n$, and the diagonalization of $\hat{H}_{XYZ}$ since

$$\hat{Q}_n = \frac{\partial^{n+1}}{\partial V^{n+1}} \ln \hat{T}(V) \big|_{V=\zeta}$$

and

$$\hat{Q}_0 = -\frac{1}{J_3 \text{sn}(2\zeta, l)} \hat{H}_{XYZ} + \text{constant}, \tag{16}$$

where

$$\hat{H}_{XYZ} = -\frac{1}{2} \sum_{n=1}^{N} \sum_{a=1}^{3} J_a \hat{\sigma}_a(n) \hat{\sigma}_a(n + 1)$$

and $J_1/J_3 = \text{cn}(2\zeta, l), J_y/J_z = \text{sn}(2\zeta, l)$. Then

$$[\hat{H}_{XYZ}, \hat{Q}_n] = 0, \qquad [\hat{Q}_n, \hat{Q}_m] = 0. \tag{17}$$

V. Construction of a set of commuting charges from Kramers-Wannier self-duality. Although the inverse scattering method has been associated with the transfer matrix formalism of statistical mechanics, we cannot apply either of these techniques to the gluon theory because we have no clue as to what is the relevant linear problem, such as the form of the commuting charges.

We now give a theorem [9] which uses another property of the spin models, i.e. Kramers-Wannier self-duality, to construct an infinite commuting algebra.

Consider any quantum Hamiltonian $\hat{H} = \kappa B + \Gamma \tilde{B}$, where $\tilde{B}$ is the dual of B such that the dual of $\tilde{B}$ is B, $(B + A) = \tilde{B} + \tilde{A}$ and the dual of $(BA) = \tilde{B}\tilde{A}$. Γ and κ are coupling constants.

If $[B, [B, [B, \tilde{B}]]] = 16[B, \tilde{B}]$, then for $Q_{2n} = \kappa(W_{2n} - \tilde{W}_{2n-2}) + \Gamma(W_{2n} - \tilde{W}_{2n-2})$ where $W_{2n+2} = -\frac{1}{8}[B, [\tilde{B}, W_{2n}]] - \tilde{W}_{2n}, W_0 \equiv B$, there exists a set of conserved commuting charges

$$[Q_{2n}, \hat{H}] = 0, \qquad [Q_{2n}, Q_{2m}] = 0. \tag{18}$$

Notice the power of this result. It does not refer to
(1) dimension of space-time,
(2) lattice or continuum, local or loop space.
To make it more familiar, some examples are:
(a) Ising model.

$$B = \sum_{n=1}^{N} \hat{\sigma}_3(n)\hat{\sigma}_3(n+1).$$

$n = 1\ 2 \qquad\qquad N$

The fundamental variable is defined on a lattice site; the eigenvalues take on 2 discrete values ± 1. The dual transformation is $\tilde{\sigma}_3(n) = \hat{\sigma}_1(1)\cdots\hat{\sigma}_1(n)$ and to make sure that the dual of $\tilde{B}$ is B then $\tilde{\sigma}_1(n) = \hat{\sigma}_3(n)\hat{\sigma}_3(n+1)$. Then

$$\tilde{B} = \sum_{n=1}^{N} \hat{\sigma}_1(n),$$

$$\frac{1}{\kappa}\hat{H} = \Sigma\sigma_3\sigma_3 + \frac{\Gamma}{\kappa}\Sigma\sigma_1 = \frac{\Gamma}{\kappa}\left(\frac{\kappa}{\Gamma}\Sigma\tilde{\sigma}_1 + \Sigma\tilde{\sigma}_3\tilde{\sigma}_3\right)$$

$$\equiv \hat{h}\left(\sigma; \frac{\Gamma}{\kappa}\right) = \frac{\Gamma}{\kappa}\hat{h}\left(\tilde{\sigma}; \frac{1}{\Gamma/\kappa}\right).$$

Under this dual transformation, the theory for weak coupling $\Gamma/\kappa \ll 1$ is related to the same theory for strong coupling $(\Gamma/\kappa)^{-1} \ll 1$. It is a powerful nonperturbative tool, since information about weak coupling can be used to calculate in the strong sector. In the underlying 2-dim Ising partition function, this operator transformation corresponds to the original Kramers-Wannier transformation which they used to calculate the critical temperature.

(b) $X - Z$ model. $B = \sum_{n=1}^{N}\hat{\sigma}_3(n)\hat{\sigma}_3(n+1)$.
The dual transformation is $\tilde{\sigma}_3(n) = \hat{\sigma}_1(n)$, $\tilde{\sigma}_1(n) = \hat{\sigma}_3(n)$. Then $\tilde{B} = \sum_{n=1}^{N}\hat{\sigma}_1(n)\hat{\sigma}_1(n+1)$. In both cases $[B, [B, [B, \tilde{B}]]] = 16[B, \tilde{B}]$ and the commuting charges constructed from the theorem are those generated by a specific choice of Baxter's transfer matrix.

VI. Loop space Yang-Mills theory. It is reasonable that a set of commuting charges should occur in 4-dim $SU(N)$ gauge theory since a Kramers-Wannier-like duality property exists for this case.
Instead of a local Lagrangian density $\mathcal{L}(x) = \frac{1}{4}F_{\mu\nu}^a(x)F_{\mu\nu}^a(x)$, the theory can also be described in a functional formulation whose fundamental variable is, for example, the element of the holonomy group $\psi[\xi] = P\exp(\oint A \cdot d\xi)$.

There are various applications of loop space.

A. G. 't Hooft's idea [11] was to concentrate on the formation of electric and magnetic flux tubes directly:

$A(C) = \frac{1}{2} \operatorname{tr} P \exp(i\oint \vec{A} \cdot d\vec{x}) = e^{i\phi_B}$ measures magnetic flux,

$B(C) = e^{i\phi_E}$ measures electric flux.

1. The operator $B(C)$ was defined by its commutation rules with A: $A(C)B(C') = B(C')A(C)e^{2\pi i n/N}$ where n is the number of times C encircles C', and N is the N of SU(N).

2. For Z_2 gauge theory, on the lattice, $B(C)$ is the K-W dual of $A(C)$ [12]. From the commutation relations, $[A, A, [A, \tilde{B}]] \sim A^2[A, \tilde{B}]$. For Z_2, $A^2 = 1$, so the theorem may be appliciable. For SU(2), B is not the exact dual of A and the condition $[A, [A, [A, [\tilde{A}]]]] \sim [A, \tilde{A}]$ may be a guide in finding $\tilde{A}$ [9].

B. Another use of loop space is Polyakov string theory [4,13]. Now the functional field $\psi[\xi]$ is thought of as a string whose dynamics are to be determined.

The simplest object, which plays the role of the free propagator in particle language is

$$G[\xi] = \langle 0 | \operatorname{tr} \psi[\xi] | 0 \rangle \Leftrightarrow G(x) = \langle 0 | T\varphi(x)\varphi(0) | 0 \rangle.$$

Just as the interacting n-point function is expressed in terms $G(x)$, so is $\langle 0 | \operatorname{tr} \psi[\xi_1] \operatorname{tr} \psi[\xi_2] \cdots | 0 \rangle$ given in terms of $G[\xi]$.

The program is to (1) describe free string theory, (2) find hidden symmetries of free strings, and (3) build interactions to preserve integrability.

In analogy with the free particle propagator, free string theory is given in terms of the two-dimensional Liouville model [4,13], which will be described in D. Friedan's lecture in these proceedings.

$$G(x - x') = \langle x | \hat{G} | x' \rangle = \langle x | (\hat{p}^2 + m^2)^{-1} | x' \rangle$$

$$= \langle x | \int_0^T dt\, e^{-t(\hat{p}^2 + m^2)} | x' \rangle$$

$$= \int_0^T dt\, e^{-m^2 t} \int Dx(t) \exp\left(-\int_0^t dt'\, \dot{x}_\mu^2\right)$$

$$\sim \sum_{\text{paths } P} e^{-L(P)},$$

$$G[\xi] \sim \sum_{\text{surfaces } S_c} e^{-A(S_c)} = \int d\mu(S_c)$$

$$= \int D\varphi(u, v) \exp(-(26 - D/48\pi)S[\varphi])$$

where $S[\varphi] = \int d^2u \{\frac{1}{2}(\partial_\mu\varphi)^2 + \mu^2 e^\varphi\}$ is the Liouville action.

In integrable models, the number of degrees of freedom is equal to the number of constants of the motion. In free theories, there is always an infinite number of conserved quantities, so to preserve integrability in an interacting system, the interactions must be such that the symmetry of the free system is not destroyed.

Even if the interacting theory is not exactly integrable, symmetries could be used to set up Ward identities which may restrict loop space Green's functions enough so they are completely determined. This is what happens in the 2-dim nonlinear sigma model, where the existence of half of a Kac-Moody charge algebra implies S-matrix factorizability and no particle production and a calculation of the exact S-matrix [2].

C. The classical loop space equations are similar to the chiral model 2-dim equations [4,14]:

$$\frac{\delta}{\delta\xi_\mu(s)}\left(\psi^{-1}\frac{\delta\psi}{\delta\xi_\mu(s)}\right) \sim \psi^{-1}_{x:\xi(s)}D_\mu F_{\mu\nu}(\xi)\dot{\xi}_\nu\psi_{\xi(s):x}. \tag{19}$$

Here $\psi = P\exp(\oint A \cdot d\xi)$. For A_μ satisfying the Yang-Mills equations $D_\mu F_{\mu\nu} = 0$,

$$\frac{\delta}{\delta\xi_\mu}\left(\psi^{-1}\frac{\delta\psi}{\delta\xi_\mu}\right) = 0 \Leftrightarrow \partial_\mu\left(g^{-1}\partial_\mu g\right) = 0. \tag{20}$$

In 3-dim, the loop equation leads to an explicit realization of hidden symmetry of interacting "strings".

We now discuss the concrete results for the models interesting to particle theorists.

VII. Evidence for a Kac-Moody algebra in Yang-Mills theory from the chiral models. The Lagrangian density in 2 Euclidean dimensions is

$$\mathcal{L}(x) = \tfrac{1}{16}\operatorname{tr}\partial_\mu g\partial_\mu g^{-1}. \tag{21}$$

The matrix field $g(x)$ is an element of some finite-dimensional group G which has generators T_e. $[T_e, T_f] = C_{efg}T_g$. The equations of motion are $\partial_\mu(g^{-1}\partial_\mu g) = 0$.

In a field theory, a symmetry is a transformation on the field $g(x) \rightarrow g'(x)$ which leaves the action I invariant. $I = \int d^2x\,\mathcal{L}(x)$. If the transformations form a group, it is called a symmetry group of the theory.

Symmetry is extremely valuable nonperturbative information, since it tells us something about the solution of the theory even though we may not be able to solve the theory exactly.

In this spirit, it is thought that some as yet unknown or hidden symmetry of SU(3) gauge theory may be responsible for or give some clue for calculating, in a controlled approximation, such nonperturbative phenomena as quark or gluon confinement.

The 2-dimensional chiral models, and in particular the $O(N)$ nonlinear sigma model NLσM, have many features in common with 4-dimensional SU(N) gauge theory. They are both asymptotically free and renormalizable as quantum field theories. On the lattice the $O(4)$ NLσM has the same Migdal recursion relations as the SU(2) gauge theory. The $O(3)$ NLσM has instanton solutions. And the loop space Yang-Mills equations look like $\partial_\mu(g^{-1}\partial_\mu g)$ for the classical theory.

Thus, when it was understood that the NLσM and the principal chiral models had an infinite set of conserved charges or currents, it was natural to ask if the 4-dim NAGT (nonabelian gauge theory) also possessed them.

Since there is no systematic procedure to translate properties from one theory to the other, it seemed reasonable to try to identify the abstract algebra responsible for the "hidden" symmetry currents of the 2-dim models, with the hunch that the *same* algebra would occur in the 4-dim case. That is to say, once the algebra itself was unhidden in the chiral theory, one could then ask what representation of it is carried by the nonabelian gauge field.

The surprise, of course, was that the hidden symmetry turned out to be "half" of a Kac-Moody algebra—a sophisticated structure already being used by mathematicians in connection with integrable systems *and* in a different approach, the dual string model. This model is an alternative description to NAGT of the strong interactions, and the fact that the same symmetry appears is intriguing. In some sense the symmetry is more fundamental to the physics than is the particular model we choose to describe it.

Also, it became clear that the chiral charges carried a representation of half of the Kac-Moody algebra which appears rather different from the representations familiar to mathematicians.

Noether's theorem states that for any continuous symmetry transformation, there is an associated constant of motion. For a symmetry, the infinitesimal transformation $(\lambda^a \ll 1)g' \sim g + \lambda^a \Delta^a g$ shifts $\mathcal{L}(x) \to \mathcal{L}(x) + \partial_\mu \Lambda_\mu$, a total divergence, *without* use of the equations of motion (off shell). On shell, $\Delta^a \mathcal{L} = \partial_\mu \mathrm{tr}(\delta \mathcal{L}/\delta \partial_\mu g)\Delta^a g$ for any $\Delta^a g$ [15].

Therefore the conserved current is $J_\mu^a = \mathrm{tr}(\delta\mathcal{L}/\delta\partial_\mu g)\Delta_g^a = -\Lambda_\mu^a$. In d-dimensions, the charge is

$$Q^a = \int_{-\infty}^{\infty} d^{d-1}x \, J_0(x_1,\ldots,x_{d-1},t). \tag{22}$$

It is conserved: $\dot{Q}^a = \int d_x^{d-1}\partial_0 J_0^a = -\int d_x^{d-1}\partial_i J_i^a = 0$, if

$$J_i^a(\pm\infty, t) = 0. \tag{23}$$

In quantum theory, g and $\pi = \delta\mathcal{L}/\delta\dot{g}$ are operators and Q^a are generators of the algebra

$$[Q^a, g(x,t)] = \Delta^a g(x,t), \qquad [Q^a, Q^b] = C_{abc}Q^c.$$

(The bracket is defined by the quantum commutator, for example, for no constraints $[\pi, g] \sim \delta$). That is to say, if the symmetry transformations $\Delta^a g$ close an algebra, the finite transformations $g \to g'$ will form a group.

In classical field theory, the charge is a c-number (not an operator). The generator of the infinitesimal transformation can be written as

$$M_a = -\int d^d x \, \Delta^a g(x) \frac{\delta}{\delta g(x)}. \tag{24}$$

Then classically, $[M_a, g(y)] = \Delta^a g(y)$ and $[M_a, M_b] = C_{abc}M_c$.

This classical realization of the generators will be used in the following.

If $\Delta^a g$ is a symmetry of I, then $g \to g + \lambda^a \Delta^a g$ will leave the equations of motion invariant, i.e. if g is a solution to $\partial_\mu(g^{-1}\partial_\mu g)$, so is $g + \lambda^a \Delta^a g$ to lowest order in the group parameters λ^a.

The chiral model has an infinite set of infinitesimal symmetry transformations

$$\Delta_a^{(n)}g = -g\Lambda_a^{(n)}. \tag{25}$$

Here $a = 1, 2, \ldots$, the number of generators of G; and $n = 0, 1, 2, \ldots, \infty$:

$$\Lambda_a^{(n+1)}(x,t) = \int_{-\infty}^x dy \, D_0 \Lambda_a^{(n)}(y,t)$$

$$= \int_{-\infty}^x dy \big(\partial_0 \Lambda_a^{(n)}(y,t) + [A_0(y,t), \Lambda_a^{(n)}(y,t)]\big),$$

$$A_\mu \equiv g^{-1}\partial_\mu g,$$

$$\Lambda_a^{(0)} = T^a \quad \text{the generators of } G,$$

$$\Lambda_a^{(1)}(x,t) = \int_{-\infty}^x dy [A_0(y,t), T^a]. \tag{26}$$

The $\Delta_a^{(n)}g$ shift $\mathcal{L}(x)$ by a total divergence, leave the equations of motion invariant, and generate "half" of a Kac-Moody algebra [1].

$$M_a^{(n)} = -\int d^2x \, \Delta_a^{(n)}g(x) \frac{\delta}{\delta g(x)},$$

$$[M_a^{(n)}, M_b^{(m)}] = C_{abc}M_c^{(n+m)} \quad \text{for } n, m = 0, 1, 2, \ldots, \infty. \tag{27}$$

For the Noether charges associated with $\Delta_a^{(n)}g$ to be conserved, a boundary condition on the fields is always required: $A_\mu(\pm\infty, t) = 0$, i.e. $g(\pm\infty, t) = \text{constant}$. (See (23).)

In the specific case of the $O(N)$ NLσM, $g_{ab}(x) = \delta_{ab} - 2\varphi_a(x)\varphi_b(x)$ with $\sum_{a=1}^N \varphi_a^2(x) = 1$, and therefore $g = g^{-1}$. In this case the Noether charges associated with $\Delta_a^{(n)}g$ are equal to the famous nonlocal charges originally found by Luscher and Pohlmeyer [16] from inverse scattering techniques. In the derivation of (25) (for $n = 1, 2$), we were naturally led to a second set of symmetry transformations corresponding to "left-multiplications" (see [15, (2.14a)]):

$$\tilde{\Delta}_a^{(n)}g = \tilde{\Lambda}_a^{(n)}g \quad \text{for } n = 0, 1, \ldots, \infty. \tag{28}$$

Here

$$\tilde{\Lambda}_a^{(n+1)}(x, t) = \int_{-\infty}^x dy\, \tilde{D}_0 \tilde{\Lambda}_a^{(n)}(y, t) \equiv \int_{-\infty}^x dy \big(\partial_0 \hat{\Lambda}_a^{(n)} + \big[\tilde{A}_0, \tilde{\Lambda}_a^{(n)}\big]\big)$$

and

$$\tilde{A}_0 \equiv g\partial_0 g^{-1}, \qquad \tilde{\Lambda}_a^{(0)} = \tilde{T}^a \quad \text{where } [\tilde{T}^a, \tilde{T}^b] = C_{abc}\tilde{T}^c. \tag{29}$$

In the NLσM, $\tilde{A}_0 = A_0$, and the Noether charges associated with $\tilde{\Delta}_a^{(n)}g$ are proportional to those associated with $\Delta_a^{(n)}g$, thus leading to no new information about constants of the motion.

Furthermore, in the general chiral models and for $\tilde{T}^a = T^a$, (27) can be rederived for $\tilde{\Delta}_a^{(n)}g$, so that even when the second set of charges is different, they just form another half Kac-Moody algebra among themselves. The commutation relations between the two sets are, in general, complicated, giving rise to symmetry transformations which on shell $(\partial_\mu(g^{-1}\partial_\mu g) = 0)$ differ from $\Delta_a^{(n)}g$ and $\tilde{\Delta}_a^{(n)}g$ by T^a replaced with other constant but g-dependent generators of G and no new conserved charges result. Clearly what is fundamental to the physics of chiral charges is the "half" Kac-Moody structure.

For the mathematical purists, we note that by a clever choice $\tilde{T}_a = T'_a \equiv g(-\infty, t)T_a g^{-1}(-\infty, t)$ for constant $g(\pm\infty, t)$, and a redefinition $\overline{\Delta}_a^{(n)}g = \Delta_a^{(n)}g$ for $n = 1, 2, \ldots, \infty$, $\overline{\Delta}_a^{(n)}g = \tilde{\Delta}_a^{(-n)}g$ for $n = -1, -2, \ldots, -\infty$, $\overline{\Delta}_a^{(0)}g = \Delta_a^{(0)}g + \tilde{\Delta}_a^{(0)}g$; then *on shell* the two sets can be combined to give, for integer n and m,

$$\big[\overline{M}_a^{(n)}, \overline{M}_b^{(m)}\big] = C_{abc}\overline{M}_c^{(n+m)}, \qquad \big[\tilde{M}_a^{(0)}, \overline{M}_b^{(n)}\big] = 0. \tag{30}$$

In light cone coordinates, a different off-shell set of Kac-Moody-like transformations was defined subsequent to [1]. Off shell, these transformations are not at fixed time. But they provide a mathematical mechanism to give the algebra of (30) off shell [17].

From the symmetry transformations, an associated linear problem can be constructed. For example, for $\psi = \sum_{n=0}^\infty l^{-n}\chi^{(n)}$, $\chi^{(0)} \equiv 1$, and on shell $\partial_\mu \chi^{(n+1)} = -\varepsilon_{\mu\nu}(\partial_\nu + A_\nu)\chi^{(n)}$, then $\Lambda^{(1)} = [\chi^{(1)}, T]$ and

$$\partial_\mu \psi = -\frac{1}{l}\varepsilon_{\mu\nu}(\partial_\nu + A_\nu)\psi.$$

The linear problem can be written as

$$\partial_\mu \psi = - (1 + l^2)^{-1} (A_\mu + l\varepsilon_{\mu\nu} A_\nu)\psi. \tag{31}$$

In Lax form,

$$\partial_0 \psi = -M\psi, \qquad \partial_1 \psi + \frac{1}{1 + l^2}(A_1 - lA_0)\psi = L\psi,$$

so the consistency condition for (31) is

$$[L, M] = \dot{L}. \tag{32}$$

Here

$$M = -\frac{1}{1 + l^2}(A_0 + lA_1)$$

and (32) is valid for $\partial_\mu A_\mu = \partial_\mu(g^{-1}\partial_\mu g) = 0$. When one attempts to solve the initial value problem (for example for the NLσM) with the inverse scattering method, the scattering data remain constant for all time. The model is not yet exactly integrated classically or quantum mechnically, i.e. the correct linear problem has not yet been found [18].

Nevertheless, in the quantum theory, some progress has been made using a quantum version of these nonlocal charges.

The exact on-shell matrix can be calculated using the nonlocal charges. The n-particle S-matrix factorizes into a product of a 2-particle S-matrices. There is no particle production, but the S-matrix is nontrivial due to time delay (phase shift) and interchange of interrnal symmetry quantum numbers.

EXAMPLE. The NLσM fields φ_a (from $g_{ab} = \delta_{ab} - 2\varphi_a\varphi_b$ with $\Sigma_{a=1}^N \varphi_a^2 = 1$) carry isospin $c = a$ and have rapidity $\theta = \ln((P^0 + P^1)/m)$.

For elastic scattering, the nonlocal conservation laws Q_n^a are used to show that

$$\begin{aligned}
\langle \theta_1' C_1', \theta_2' C_2' \text{ out} | Q_1^a | \theta_1 C_1, \theta_2 C_2 \text{ in} \rangle \\
= (Q_1^a \text{ out}) c_1' c_2' c_1 c_2 \langle \theta_1' c_1', Q_2' c_2' \text{ out} | \theta_1 c_1, \theta_2 c_2 \text{ in} \rangle \\
= \langle \theta_1' c_1', \theta_2' c_2' \text{ out} | \theta_1 c_1, \theta_2 c_2 \text{ in} \rangle (Q_1^a \text{ in}) c_1' c_2' c_1 c_2,
\end{aligned} \tag{33}$$

which are equivalent to the factorization equations after some algebra [2].

VIII. Half of a Kac-Moody algebra in real self-dual Yang-Mills. In 4-Euclidean dimensions, the self-dual equations are

$$F_{\mu\nu}(x) = \tfrac{1}{2}\varepsilon_{\mu\nu\alpha\beta} F_{\alpha\beta}(x) \equiv \tilde{F}_{\mu\nu}(x),$$

$$F_{\mu\nu} \equiv \partial_\mu A_\nu - \partial_\nu A_\mu + [A_\mu, A_\nu],$$

$$A_\mu \equiv A_\mu^a(x) T^a,$$

and T^a are the antihermitian generators of SU(N), with structure constants $[T^a, T^b] = C_{abc}T^c$. Real solutions of SU(N) correspond to real $A_\mu^a(x)$. The symbol $\varepsilon_{\mu\nu\alpha\beta}$ is totally antisymmetric, $\varepsilon_{1234} = 1$. Solutions of $F = \tilde{F}$ imply $D_\mu F_{\mu\nu} = 0$. Change variables to

$$\sqrt{2}\, y = x_4 - ix_3, \qquad \sqrt{2}\, z = x_2 - ix_1,$$
$$\sqrt{2}\, \bar{y} = x_4 + ix_3, \qquad \sqrt{2}\, \bar{z} = x_2 + ix_1.$$

The self-dual equations become

$$F_{y\bar{y}} + F_{z\bar{z}} = 0; \quad F_{\bar{y}z} = 0; \quad F_{yz} = 0. \tag{34}$$

It will be helpful to define some new functions before we give the Kac-Moody transformations.

What can we say when the curl is zero?

Consider our old friend $\psi[\xi] = P\exp(\int_x^y d\xi \cdot A)$,

$$\psi[\xi + \delta\xi] - \psi[\xi] = \delta y_\mu \psi A_\mu(y) - \delta x_\mu A_\mu(x)\psi$$
$$+ \int_0^1 ds\, \psi_{x:\xi(s)} F_{\mu\nu}(\xi(s))\dot{\xi}\nu(s)\psi_{\xi(s):x}.$$

Then, for $F_{\mu\nu} = 0$, ψ depends only on endpoints. The nonintegrable phase factor becomes integrable [19]. The Yang construction for the new functions is as follows: $F_{z\bar{y}} = 0$ allows the definition of *local* objects in terms of $Pe^{\int A \cdot d\xi}$, since if we fix z, $\bar{y}$, then $\dot{\xi}_z = \dot{\xi}_{\bar{y}} = 0$ and $\delta\xi_z = \delta\xi_{\bar{y}} = 0$ so

$$D[\xi] = P\exp\left(\int_Q^P \{d\bar{z}'A_{\bar{z}} + dy'A_y\}\right),$$

$$D[\xi + \delta\xi] - D[\xi] = \delta P_\mu DA_\mu(P) - \delta Q_\mu A_\mu(Q)D$$
$$+ \int ds\, \psi_{x:\xi(s)}\{F_{\mu\nu}\dot{\xi}_\nu\delta\xi_\mu\}\psi_{\xi(s):x}$$

$$\downarrow$$

$$= F_{\mu z}\dot{\xi}_{\bar{z}}\delta\xi_\mu + F_{\mu\bar{y}}\dot{\xi}_y\delta\xi_\mu$$
$$= F_{yz}\dot{\xi}_{\bar{z}}\delta\xi_y + F_{z\bar{y}}\dot{\xi}_y\delta\xi_{\bar{z}}$$
$$= 0. \tag{35}$$

So $D[\xi]$ depends only on the endpoints. Fix $Q = -\infty$ and let $P = (\bar{z}, y, z, \bar{y})$, then

$$D(\bar{z}, y, z, \bar{y}) = P\exp\left(\int_\infty^{(\bar{z},y,z,\bar{y})} (dy'A_y + d\bar{z}'A_{\bar{z}})\right),$$

$$\uparrow$$

path has z, $\bar{y}$ fixed

and similarly from $F_{\bar{z}y} = 0$,

$$\bar{D}(\bar{z}, y, z, \bar{y}) = P \exp\left(\int_{-\infty}^{(\bar{z},y,z,\bar{y})} (d\bar{y}'A\bar{y} + dz'Az)\right).$$

$$\uparrow$$

$$\text{path has } \bar{z}, y \text{ fixed} \qquad (36)$$

The Kac-Moody-like transformations [6] are most conveniently expressed in terms of D and $\bar{D}$:

$$\Delta^{(n)}A_y = D_y\Omega^{(n)},$$

$$\Delta^{(n)}A_{\bar{y}} = -D_{\bar{y}}\Omega^{(n)},$$

$$\Delta^{(n)}A_z = -D_z\Omega^{(n)},$$

$$\Delta^{(n)}A_{\bar{z}} = D_{\bar{z}}\Omega^{(n)}, \qquad (37)$$

$$\Omega^{(n)} = -\tfrac{1}{2}(\bar{D}^{-1}\Lambda^{(n)}\bar{D} + D^{-1}\Lambda^{(n)t}D),$$

$\Lambda^0 = T^e\rho^e \equiv T$, $\qquad \rho^e$ are constant infinitesimal parameters,

$$\Lambda^1(\bar{z}, y, z, \bar{y}) = \int_{-\infty}^z dz' \, \mathcal{D}_y T = \left[\int_{-a}^z dz' J^{-1}\partial_y J, T^e\right]\rho^e,$$

$$J \equiv D\bar{D}^{-1}, \qquad (38)$$

$$\partial_z\Lambda^{(n+1)} = \mathcal{D}_y\Lambda^{(n)}, \qquad \mathcal{D}_y \equiv \partial_y + [J^{-1}\partial_y J,$$

$$\partial_{\bar{y}}\Lambda^{(n+1)} = -\mathcal{D}_{\bar{z}}\Lambda^{(n)}, \qquad \tilde{\mathcal{D}}_{\bar{y}} \equiv \partial_{\bar{y}} + [J\partial_{\bar{y}}J^{-1}, \qquad (39)$$

$$\partial_{\bar{z}}\Lambda^{(n+1)\dagger} = \tilde{\mathcal{D}}_{\bar{y}}\Lambda^{(n)\dagger}, \qquad \partial_y\Lambda^{(n+1)t} = -\tilde{\mathcal{D}}_z\Lambda^{(n)\dagger}.$$

These transformations leave $A_\mu^a(x)$ real and if $A_\mu(x)$ is a solution to $F = \tilde{F}$, so is $A_\mu + \Delta^{(n)}A_\mu$.

A gauge transformation of the second kind can be written as

$$\delta A_\mu = D_\mu\lambda(x), \qquad \lambda^\dagger = -\lambda.$$

For $n = 0$ in the above set,

$$\Delta^{(n)}A_\mu = D_\mu\bar{\Omega}^{(0)} \quad \text{where } \bar{\Omega}^{(0)} = -\tfrac{1}{2}(\bar{D}^{-1}T\bar{D} + D^{-1}TD)$$

and therefore $\Delta^{(0)}A_\mu$ is recognized as a local gauge transformation. The other $\Delta^{(n)}A_\mu$ are of course not gauge transformations.

The algebra is $Q_c^{(m)} = -\int d^4x \, \bar{\Delta}_c^{(m)}A_\mu(x) \, \delta/\delta A_\mu(x)$,

$$[Q_c^{(m)}, Q_b^{(n)}] = C_{cba}Q_a^{(n+m)} + \int d^4x \, D_\mu[\Omega_c^{(m)}, \Omega_b^{(n)}]\frac{\delta}{\delta A_\mu}$$

where $n, m = 0, 1, \ldots, \infty$, where $\bar{\Delta}_c^{(m)}A_\mu$ are linear combinations of $\Delta_c^{(m)}$.

The second term spoils the integrability condition when it is not zero. The second term is a real local infinitesimal gauge transformation. Therefore, enlarge the set of transformations to

$$\delta_{p,\lambda}A_\mu = \rho_b^n\overline{\Delta}_b^{(n)}A_\mu(x) - D_\mu\lambda(x), \qquad \lambda = -\lambda^\dagger.$$

Then

$$\delta_{p,\lambda}\left(A_\mu + \delta_{\sigma_1\kappa}A_\mu\right) - \delta_{p,\lambda}A_\mu = \delta_{\tau,s}A_\mu \equiv \tau_a^l\Delta_a^{(l)}A_\mu - D_\mu S, \quad p \Leftrightarrow \sigma, \lambda \Leftrightarrow \kappa$$

where $\tau_a^l = \delta_{l_1 m + n}\varepsilon_{abc}\rho_b^n\sigma_c^m$, and

$$S(x) = [\lambda, \kappa] + \rho_b^n\sigma_c^m\left[\overline{\Omega}_b^n, \overline{\Omega}_c^m\right]$$
$$+\lambda[A + \delta_{\sigma,\kappa}A] - \lambda - \kappa[A + \delta_{p,\lambda}A] + \kappa.$$

$\overline{\Omega}_b^n$ are linear combinations of Ω_b^n. Therefore the new set $\delta_{p,\lambda}A_\mu$, including Kac-Moody-like transformations and local gauge transformations, generate finite values which form a group. Since $\Delta_b^{(n)}A_\mu$ and $\delta_{p,\lambda}A_\mu$ are symmetry transformations, they transform real self-dual solutions to real self-dual solutions with the same action infinitesimally. Presumably the instanton solutions with their $8n - 3$ parameters reflect the infinite-parameter symmetry algebra. Also, these transformations already include both "left and right multiplication" symmetries for real fields (see (13) in [6]).

Clearly, to be interesting for the strong interactions, it is necessary to find the symmetry transformations for the full Yang-Mills theory [20]. Given the occurrence of the whole Kac-Moody algebra in the dual string [7], perhaps the full quantum SU(N) gauge theory carries a representation of SU(N) $\otimes$ **C**$[t, t^{-1}] + C_c$.

REFERENCES

1. L. Dolan, Phys. Rev. Lett. **47** (1981), 1371.
2. M. Luscher, Nuclear Phys. B **135** (1978), 1.
3. V. G. Kac, Math USSR-Izv. **2** (1968), 1271; R. Moody, J. Algebra **10** (1968), 211.
4. A. M. Polyakov, Phys. Lett. B **82** (1979), 247; Nuclear Phys. B **164** (1980), 171.
5. D. B. Fairlie, J. Nuyts and R. G. Yates, J. Math. Phys. **19** (1978), 528.
6. L. Dolan, Phys. Lett. B **113** (1982), 387.
7. I. B. Frenkel and V. G. Kac, Invent. Math. **62** (1980), 23; J. Lepowsky and R. L. Wilson, Comm. Math. Phys. **62** (1978), 43. See also I. Frenkel's and J. Lepowsky's lectures, these PROCEEDINGS.
8. For early references, see B. Kostant, Invent. Math **48** (1978), 101; M. Alder and P. V. Moerbeke, Adv. in Math. **38** (1980), 267; and RIMS preprint no. 362, 1981, Kyoto.
9. L. Dolan and M. Grady, Phys. Rev. D **25** (1982), 1587.
10. R. Baxter, Ann. Physics **70** (1972), 193, 323 and **76** (1973), 1, 25, 48.
11. G. 't Hooft, Nuclear Phys. B **138** (1978), 1 and **153** (1979), 141.
12. A. Ukawa et al., Phys. Rev. D **21** (1980), 1013.
13. A. M. Polyakov, Phys. Lett. B **103** (1981), 207, 211.
14. L. Dolan, Phys. Rev. D **22** (1980), 3104.

15. Nonlocal chiral charges were first shown to be Noether charges in L. Dolan and A. Roos, Phys. Rev. D **22** (1980), 2018.

16. M. Luscher and K. Pohlmeyer, Nuclear Phys. B **137** (1978), 46.

17. Y.-S. Wu, Nucl. Phys. B **211** (1983), 160.

18. In a private communication, S. Michailou has stated that the classical initial value problem for the nonlinear sigma model can be solved using a Riemann-Hilbert analysis.

19. C. N. Yang, Phys. Rev. Lett. **38** (1979), 1377.

20. L. Dolan, Proceedings of XI Internat. Colloq. Group Theoretical Methods in Physics, Istanbul. Turkey, August 1982.

DEPARTMENT OF PHYSICS, ROCKEFELLER UNIVERSITY, NEW YORK, NEW YORK 10021

Lectures in Applied Mathematics
Volume **21**, 1985

Representations of Kac-Moody Algebras and Dual Resonance Models

I. B. Frenkel[1]

CONTENTS

Introduction. The theories of Kac-Moody algebras and dual resonance models were born at approximately the same time (1968). The second theory underwent enormous development until 1974 (see reviews [**25, 26**]) followed by years of decline, while the first theory moved slowly until the work of Kac [**14**] in 1974 followed by accelerated progress. Now both theories have gained considerable interest in their respective fields, mathematics and physics. Despite the fact that these theories have no common motivations, goals or problems, their formal similarity goes remarkably far. In this paper we discuss primarily the mathematical

1980 *Mathematics Subject Classification.* Primary 17B65, 81E99.
[1]Supported in part by NSF Grants MCS-8108814(A01) and MCS 81-02534.

theory. For a review of the physical theory see the paper of J. Schwarz in this volume [27].

At an early stage of the history of dual resonance models, it was understood that many results could be naturally interpreted in terms of the representation space V of the commutation relations

$$[\mathbf{x}^\mu, \mathbf{p}^\nu] = ig^{\mu\nu}, \qquad [\alpha^\mu(m), \alpha^\nu(n)] = mg^{\mu\nu}\delta_{m,-n} \qquad (1)$$

where $\mu, \nu \in \{0, 1, \ldots, d - 1\}$, $m, n \in \mathbf{Z}$, $g^{\mu\nu} = \text{diag}(-1, 1, \ldots, 1)$. Therefore, V is the linear span of the elements of type[2]

$$\prod_{i=1}^N \alpha^{\mu_i}(m_i)|p\rangle \qquad (2)$$

where $m_i < 0$, $\mathbf{p}^\mu|p\rangle = p^\mu|p\rangle$, $\mathbf{x}^\mu|p\rangle = ig^{\mu\nu}(\partial/\partial p^\mu)|p\rangle$, $p = p^\mu e_\mu \in \mathbf{R}^{1,d-1}$ and $\{e_\mu\}_{\mu=0}^{d-1}$ is a basis. Later it was observed that (1) is just the canonical quantization of a string moving in d-dimensional space-time; hence, the name of the theory was changed to dual string models. In order to obtain the dual amplitudes physicists defined a "vertex operator"

$$X(p, z) = \exp\left(p \cdot \sum_{n=1}^\infty \frac{\alpha(-n)}{n} z^n \right)$$

$$\cdot \exp(p \cdot (\mathbf{p} \log z - i\mathbf{x})) \exp\left(-p \cdot \sum_{n=1}^\infty \frac{\alpha(n)}{n} z^{-n} \right), \qquad (3)$$

where $p \cdot \alpha(-n)$ denotes

$$p_\mu \alpha^\mu(-n) = g_{\mu\nu} p^\mu \alpha^\nu(-n), \text{ etc.}$$

Ten years later this operator was reborn in the representation theory of Kac-Moody algebras. For some time mathematicians tried to obtain as simple a representation of the infinite-dimensional Lie algebras as they have for classical finite-dimensional Lie algebras. The solution was first given in [21] for one algebra, and Garland observed some similarity with dual resonance models. Then in [9, 28] the "vertex construction" was found for the whole class of affine Lie algebras and the similarity became a precise correspondence.

It happens that if one restricts p in (2) to be in the even integer lattice called the root lattice, then the space V is exactly the space of an irreducible representation of an affine Lie algebra $\hat{\mathfrak{g}}$. Furthermore, the operators $\alpha^\mu(m)$, $\mathbf{p}^\mu$, together with the modified vertex operators $X(p, z) \cdot \varepsilon_p$, $\|p\|^2 = 2$, provide the "basic" representation of the affine Lie

---

[2] $|p\rangle$ is a standard bracket notation used in physics. In mathematical literature this element is usually denoted by e^p.

algebra $\hat{g}$. The only modification is the operator ε_p of multiplication by ± 1 on vectors $|q\rangle$, which nevertheless is quite important. In this paper we will consider two further applications of these vertex representations. The first application is the study of all the representations of $\hat{g}$ with a dominant highest weight (standard representations) inside the basic representation V. This is possible thanks to the existence of subalgebras $\hat{g}_{[n]} \subset \hat{g}$, $n \in \mathbf{N}$, isomorphic to $\hat{g}$. We prove the conjecture about multiplicities of irreducible representations of $\hat{g}_{[n]}$ in V which was formulated in [8].

As a second application of vertex representations, we give a construction of an arbitrary Kac-Moody algebra with real roots of equal length. This representation is not irreducible, but it yields some new information about hyperbolic algebras. In particular, we obtain the explicit form of operators corresponding to imaginary root vectors of zero norm. Again we should acknowledge the priority of physicists, who first introduced these operators [4]

$$A^i(n) = \int_C \alpha^i(z) X(nk, z) \frac{dz}{z} \tag{4}$$

where $n \in \mathbf{Z}$, k, $\alpha^i \in \mathbf{R}^{1,d-1}$, $\langle k, k \rangle = \langle k, \alpha^i \rangle = 0$, $\langle \alpha^i, \alpha^j \rangle = \delta_{ij}$. They noticed, in particular, that for fixed k these operators form a Heisenberg subalgebra

$$\left[A^i(m), A^j(n)\right] = m\delta_{ij}\delta_{m,-n}k(0). \tag{5}$$

Another important object in the dual resonance models is the Virasoro algebra spanned by the operators

$$L(n) = \frac{1}{2} \sum_{m=-\infty}^{\infty} \; : \alpha(-m) \cdot \alpha(n+m): , \tag{6}$$

where : : denotes the normal ordering, i.e., $: \alpha(m) \cdot \alpha(n): \; = \alpha(m) \cdot \alpha(n)$ if $n \geqslant m$ and $= \alpha(n) \cdot \alpha(m)$ if $n < m$. These operators satisfy

$$[L(m), L(n)] = (m - n)L(m + n) + \frac{d}{12}(m^3 - m)\delta_{m,-n}. \tag{7}$$

This algebra is known to mathematicians as a central extension of the algebra of vector fields on a circle. In dual resonance models the Virasoro algebra is used for the definition of "physical" subspaces of V:

$$V_c = \left\{ v \in V : (L(n) - c\delta_{n,0})v = 0, n \geqslant 0 \right\} \tag{8}$$

where c is a constant. In the representation theory the Virasoro algebra commutes with the Kac-Moody algebra g. The subspace V_c becomes a representation of g. It turns out that as such, V_c is irreducible whenever $V_c \neq 0$, if $g = \hat{s}l(2)$, [16, 28]. This is no longer true in general. In this paper we introduce a maximal commuting algebra $\hat{S}^W$ for arbitrary simple finite-dimensional $\mathfrak{g}$, which contains the Virasoro algebra. We obtain the decomposition

$$V = \sum_{\mu} V(\mu) \otimes \Omega(\mu) \tag{9}$$

where $V(\mu)$ is an irreducible finite-dimensional representation of $\mathfrak{g}$, and $\Omega(\mu)$ is the corresponding irreducible representation of $\hat{S}^W$. We also conjecture a resolution of $\Omega(\mu)$.

One of the highest achievements of the dual resonance models is the no-ghost theorem which asserts that for the critical dimension $d = 26$, the physical space V_1 contains no vectors of negative norm. We use this theorem in order to obtain an upper bound for the multiplicities of hyperbolic algebras of rank 26

$$\dim \mathfrak{g}_\alpha \leqslant \mu(-\langle \alpha, \alpha \rangle / 2), \tag{10}$$

where

$$\sum_{n=-1}^{\infty} \mu(n)q^n = \frac{1}{q\Pi_{n=1}^{\infty}(1 - q^n)^{24}}. \tag{11}$$

In fact, (10) is true for an algebra containing all the hyperbolic algebras of rank 26. This algebra, called the Monster Lie algebra, has been introduced in [2] with the hope that it will "explain" the biggest sporadic group F_1 of Fisher and Griess, known as Monster. One of the proofs of the no-ghost theorem uses the space $V_{1,k,p}$, spanned by the elements of type

$$\prod_{i=1}^{N} A^{m_i}(-n_i)|p\rangle \tag{12}$$

where $A^{m_i}(-n_i)$ is defined in (4), $\langle k, k \rangle = 0$ and satisfies the properties

$$\langle p, k \rangle = 1, \qquad \langle p, p \rangle = 2. \tag{13}$$

In the case when p is in the unique even unimodular lattice in $\mathbf{R}^{1,25}$, the space $V_{1,k} = \oplus_p V_{1,k,p}$, where p runs through all the elements satisfying (13), has the character

$$\mathrm{Ch}\, V_{1,k} = q^{-1} + \mathrm{Const} + 196884q + \cdots. \tag{14}$$

Here, Const depends on k and its minimal value is 24. The number 196884 exceeds only by 1 the dimension of the minimal representation of F_1. It was conjectured in [3] that there is a "natural" representation of F_1 in a space with the character given by the right side of (14). Recently [10], considerable progress was achieved in the understanding of the connection of dual resonance models and the natural representation of the Fisher-Griess Monster.

Predicting this connection, F. Dyson, in his talk at the von Humboldt Foundation Colloquium, said:

"Stranger things have happened in the history of physics than the unexpected appearance of sporadic groups." And he continued: "We have strong evidence that the creator of the universe loves symmetry, what lovelier symmetry could he find than the symmetry of the Monster?"

1. Kac-Moody algebras and standard representations.

1.1. We begin by recalling first the definition of Kac-Moody algebras. Let I be a finite subset in $\mathbf{Z}$ and $|I| = l$. An $l \times l$ integer matrix $A = (a_{ij})_{i,j \in I}$ is called a Cartan matrix if it satisfies the following properties:

(i) $a_{ii} = 2$, a_{ij} are nonpositive integers for $i \neq j$, and $a_{ij} = 0$ implies $a_{ji} = 0$, $i, j \in I$.

(ii) There exists a nondegenerate diagonal $l \times l$ matrix D such that the matrix DA is symmetric.

We denote by $\mathfrak{g}(A)$ a Lie algebra over $\mathbf{C}$ with $3l$ generators e_i, f_i, h_i, $i \in I$, and the following defining relations:

$$\left[h_i h_j\right] = 0, \quad \left[e_i, f_j\right] = -\delta_{ij} h_i,$$

$$\left[h_i, e_j\right] = a_{ij} e_j, \quad \left[h_i f_j\right] = -a_{ij} f_j,$$

$$(\operatorname{ad} e_i)^{1-a_{ij}} e_j = (\operatorname{ad} f_i)^{1-a_{ij}} f_j = 0, \qquad i \neq j, \tag{1.1}$$

where $i, j \in I$. The algebra $\mathfrak{g}(A)$ is called a *Kac-Moody algebra* [13, 24]. In this paper we will suppose that A is symmetric.

We define the following notions: the Cartan subalgebra $\mathfrak{h} = \mathfrak{h}_{\mathbf{R}} \otimes \mathbf{C}$, where $\mathfrak{h}_{\mathbf{R}} = \Sigma_{i \in I} \mathbf{R} h_i$; the bilinear symmetric form $\langle , \rangle$ on $\mathfrak{h}$, such that $\langle h_i, h_j \rangle = a_{ij}$, $i, j \in I$; the Weyl group W generated by reflections with respect to h_i in $\mathfrak{h}$; the root lattice $Q = \Sigma_{i \in I} \mathbf{Z} \alpha_i \subset \mathfrak{h}_{\mathbf{R}}$ (for symmetric A we identify $\alpha_i = h_i$), the weight lattice $P = \Sigma_{i \in I} \mathbf{Z} \omega_i \subset \mathfrak{h}_{\mathbf{R}}^*$, $\langle \alpha_i, \omega_j \rangle = \delta_{ij}$; the system of real roots $\Delta_R = \cup_{i \in I} W \alpha_i \subset Q$. We will define several subalgebras of $\mathfrak{g}(A)$ as follows: $\mathfrak{n}^+$ generated by e_i, $i \in I$, $\mathfrak{n}$ generated by f_i, $i \in I$, $\mathfrak{b}^+ = \mathfrak{h} \oplus \mathfrak{n}^+$, $\mathfrak{b}^- = \mathfrak{h} \oplus \mathfrak{n}^-$. There is a unique bilinear invariant symmetric form on $\mathfrak{g}(A)$ [13, 24]. We choose such a form, which extends the bilinear symmetric form $\langle , \rangle$ on $\mathfrak{h}$ and denote it by the same symbol.

Let $\varepsilon: Q \times Q \to \{\pm 1\}$ be a bilinear function satisfying the following conditions:

$$\varepsilon(\alpha, \beta) \varepsilon(\beta, \alpha) = (-1)^{\langle \alpha, \beta \rangle}, \qquad \alpha, \beta \in Q,$$

$$\varepsilon(\alpha, \alpha) = -1 \quad \text{for } \langle \alpha, \alpha \rangle = 2, \alpha \in Q. \tag{1.2}$$

The existence of such a cocycle follows by extending its arbitrary definition on the basis elements $\varepsilon(\alpha_i, \alpha_j)$, $i < j$, $i, j \in I$.

Kac-Moody algebras admit a rather explicit description for a special choice of Cartan matrices. The first type of Kac-Moody algebras we are considering is distinguished by the condition that the Cartan matrix A is positive definite. The corresponding Lie algebras $\mathfrak{g} = \mathfrak{g}(A)$ are finite-dimensional simple Lie algebras with roots of equal length, i.e., of one of the types $A_l, l \geqslant 1, D_l, l \geqslant 4, E_l, l = 6, 7, 8$. These algebras have a simple realization in terms of a Chevalley basis:

$$\mathfrak{g} = \mathfrak{h} + \sum_{\alpha \in \Delta} \mathbf{C} x_\alpha,$$

$$[h, x_\alpha] = \langle h, \alpha \rangle x_\alpha, \qquad h \in \mathfrak{h},$$

$$[x_\alpha, x_\beta] = 0 \quad \text{if } \alpha + \beta \notin \Delta \cup 0,$$

$$[x_\alpha, x_\beta] = \varepsilon(\alpha, \beta) x_{\alpha+\beta} \quad \text{if } \alpha + \beta \in \Delta,$$

$$[x_\alpha, x_{-\alpha}] = -\alpha. \tag{1.3}$$

The identification with the first definition is determined by setting $x_{\alpha_i} = e_i, x_{-\alpha_i} = f_i, i \in I$.

The second type of Kac-Moody algebras we consider is distinguished by the condition that the Cartan matrix is positive semidefinite. It can be defined as the unique extension of a positive-definite symmetric Cartan matrix A and will be denoted by $\hat{A}$. The corresponding Lie algebra $\hat{\mathfrak{g}} = \mathfrak{g}(\hat{A})$ is the affine Lie algebra (or loop algebra) with real roots of equal length, i.e., of one of the types $A_l^{(1)}, l \geqslant 1, D_l^{(1)}, l \geqslant 4, E_l^{(1)}, l = 6, 7, 8$. The Lie algebra $\hat{\mathfrak{g}}$ admits the following realization:

$$\hat{\mathfrak{g}} = \mathfrak{g} \otimes \mathbf{C}[t, t^{-1}] \oplus \mathbf{C}c,$$

$$[x \otimes t^m, y \otimes t^n] = [x, y] \otimes t^{m+n} + m \langle x, y \rangle \delta_{m,-n} c \tag{1.4}$$

where $x, y \in \mathfrak{g}, m, n \in \mathbf{Z}, c$ is a central element in $\hat{\mathfrak{g}}$. We will identify $\mathfrak{g} = \mathfrak{g} \otimes 1$ and set $e_0 = x_{-\alpha} \otimes t, f_0 = x_\alpha \otimes t^{-1}, h_0 = c - \alpha$, where α is the maximal root in Δ. This provides one with the equivalence of the two definitions of $\hat{\mathfrak{g}}$. We denote by $\tilde{\mathfrak{g}}$ a semidirect product of $\hat{\mathfrak{g}}$ with $\mathbf{C}d$, where $d = t(d/dt)$ is a derivation of $\hat{\mathfrak{g}}$.

We will also consider a class of hyperbolic Kac-Moody algebras, which are determined by the condition that the signature of $\langle , \rangle$ on $\mathfrak{h}_\mathbf{R}$ has type $(1, l - 1)$. In particular, if we define the overextended Cartan matrix $\hat{A} = (a_{ij})_{i,j=-1}^l$ by the conditions $a_{0,-1} = -1, a_{0,i} = 0, i = 1, \ldots, l$, and $A = (a_{ij})_{i,j=0}^l$ is the extended Cartan matrix. Then the hyperbolic algebra $\widehat{\mathfrak{g}} = \mathfrak{g}(\hat{A})$ admits the following construction [5]. Let V be the basic representation of $\tilde{\mathfrak{g}}$ with the grading starting from 1, i.e., the irreducible representation with the vector $v_0 \in V$ satisfying the conditions

$$\mathfrak{g} \otimes \mathbf{C}[t]v_0 = 0, \qquad cv_0 = dv_0 = v_0, \tag{1.5}$$

and let V^* be the contragradient representation. We define a map ϕ: $V^* \times V \to \tilde{\mathfrak{g}}$ by

$$\phi(v^*, v) = -\sum_{j \in J} \langle v^*, x_j v \rangle x_j \tag{1.6}$$

where $v^* \in V^*$, $v \in V$, $\{x_j: j \in J\}$ is an orthonormal basis for $\tilde{\mathfrak{g}}$ with respect to the form. We define a $\mathbf{Z}$-graded Lie algebra $\mathfrak{a} = \Sigma_{n \in \mathbf{Z}} \mathfrak{a}$ where $\mathfrak{a}_0 = \tilde{\mathfrak{g}}$, $\mathfrak{a}_{-1} = V$, $\mathfrak{a}_1 = V^*$, and where $\mathfrak{a}^+ = \Sigma_{n \geqslant 1} \mathfrak{a}_n$ and $\mathfrak{a}^- = \Sigma_{n \geqslant 1} \mathfrak{a}_{-n}$ are the free Lie algebras generated by $\mathfrak{a}_1$ and $\mathfrak{a}_{-1}$, respectively. The Lie brackets between $\mathfrak{a}$, and $\mathfrak{a}_{-1}$ are defined by $[v^*, v] = \phi(v^*, v)$ and between $\mathfrak{a}_k$ and $\mathfrak{a}_{-l}$, $k, l \geqslant 1$, are defined inductively. Let $\mathcal{I} = \mathcal{I}^+ + \mathcal{I}^-$, $\mathcal{I}^\pm = \Sigma_{k \geqslant 1} \mathcal{I}_k^\pm$, $\mathcal{I}_k^\pm = \{x \in \mathfrak{a}_{\pm k}: (\text{ad } \mathfrak{a}_{\pm 1})^{k-1} x = 0\}$. Then we have

THEOREM 1.1 [5]. *$\mathcal{I}$ is an ideal in $\mathfrak{a}$ and $\mathfrak{a}/\mathcal{I} \approx \mathfrak{g}(\widehat{A})$.*
In particular one has for $\alpha \in \Delta$, such that $\langle \alpha, \omega_{-1} \rangle = 1$, that

$$\dim \widehat{\mathfrak{g}}_\alpha = p^{(l)}(-\langle \alpha, \alpha \rangle / 2 + 1) \tag{1.7}$$

where

$$\sum_{n=0}^\infty p^{(l)}(n) q^n = \varphi(q)^{-l} \tag{1.8}$$

and we denote $\varphi(q) = \Pi_{m=1}^\infty (1 - q^m)$.

1.2. We recall Kac's results about standard representations [14]. Let us denote $Q^+ = \{\alpha = \Sigma_{i \in I} n_i \alpha_i, \ n_i \in \mathbf{Z}_+, \ i \in I\}$, $P^{++} = \{\lambda = \Sigma_{i \in I} h_i \omega_i, \ n_i \in \mathbf{Z}_+, \ i \in I\}$. An element $\lambda \in P^{++}$ is called a dominant highest weight. The standard representation on irreducible representation with dominant highest weight λ of Kac-Moody algebra $\mathfrak{g}$ is by definition the unique irreducible representation $V(\lambda)$ satisfying the following property: there exists a vector $v_0 \in V(\lambda)$ such that

$$\mathfrak{n}^+ v_0 = 0, \qquad h v_0 = \langle \lambda, h \rangle v_0, \quad h \in \mathfrak{h}. \tag{1.9}$$

Let $\mathbf{C}(P)$ denote the algebra of all formal sums of elements of the group algebra $\mathbf{C}[P]$ with support in a finite union of the sets $\nu + P^+$. Let V be a representation of $\mathfrak{g}$. We call the character of V the formal sum

$$\text{ch } V = \sum_{\mu \in P} \dim V_\mu \cdot e^\mu \tag{1.10}$$

where $V_\mu = \{v \in V: hv = \langle \mu, h \rangle v \text{ for all } h \in \mathfrak{h}\}$. We will consider the category of representations V with $\text{ch } V \in \mathbf{C}(P)$.

PROPOSITION 1.2 (KAC). (i) *For $\lambda \in P^{++}$ one has*

$$\text{ch } V(\lambda) = \frac{\Sigma_{w \in W} \det(w) e^{w(\lambda + \rho)}}{e^\rho \Pi_{\alpha \in \Delta_+} (1 - e^{-\alpha})}, \tag{1.11}$$

or equivalently for $\mu \in P^{++}$, $\mu \neq \lambda$,

$$\sum_{w \in W} \det(w) \dim V(\lambda)_{\mu + \rho - w\rho} = 0. \tag{1.12}$$

(ii) *In particular, for* $\lambda = 0$

$$\sum_{w \in W} \det(w) e^{w\rho} = e^{\rho} \prod_{\alpha \in \Delta_+} (1 - e^{-\alpha}), \tag{1.13}$$

where $\rho = \sum_{i \in I} \omega_i$.

When $\mathfrak{g}$ is a finite-dimensional simple Lie algebra, Proposition 1.2 becomes the classical result of H. Weyl. Later on we will use the following corollary of (ii), in this case taking the value at $t(\lambda + \rho)$ and setting $q = e^{-t}$,

$$q^{\|\lambda\|^2/2} \prod_{\alpha \in \Delta_+} \left(1 - q^{\langle \alpha, \lambda + \rho \rangle}\right) = \sum_{w \in W} \det(w) q^{\|w(\lambda+\rho) - \rho\|^2/2}. \tag{1.14}$$

When the Kac-Moody algebra is an affine Lie algebra $\hat{\mathfrak{g}}$ (ii) is known as the Macdonald identity. Again taking the value at $t(\lambda + \hat{\rho})$ and setting $q = e^{-t}$, we get

$$q^{\|\lambda\|^2/2} \prod_{\alpha \in \hat{\Delta}_+} \left(1 - q^{\langle \alpha, \lambda + \hat{\rho} \rangle}\right) = \sum_{w \in \hat{W}} \det(w) q^{\|w(\lambda+\hat{\rho}) - \hat{\rho}\|^2/2}. \tag{1.15}$$

In subsection 3.1 we will show how to "divide" the numerator of (1.11) by the denominator in the affine case. For the representations of level 1 the character of $V(\lambda)$ has an especially simple and beautiful form [6, 15]:

PROPOSITION 1.3. *Let* $\hat{\omega} = d + \omega$ *be a level 1 weight of* $\hat{\mathfrak{g}}$. *Then*

$$\operatorname{ch} V(\hat{\omega}) = \varphi(q)^{-l} \sum_{\gamma \in \omega + Q} e^{\gamma} \cdot q^{\langle \gamma, \gamma \rangle/2}, \tag{1.16}$$

where $q = e^{-c}$.

2. Vertex representations of Kac-Moody algebras.

2.1. Kac-Moody algebras admit a particular representation, which plays an important role in the theory. We will call it the vertex representation and we will use it for several applications later on. Now we will give a construction of the vertex representation.

Let $\mathfrak{g}(A)$ be a Kac-Moody algebra with the nondegenerate invariant bilinear symmetric form $\langle , \rangle$. We call the Lie algebra $\hat{\mathfrak{g}}(A) = \mathfrak{g}(A) \otimes C[T, T^{-1}] \oplus CC$, with the Lie bracket given by (1.4), the affinization of Kac-Moody algebra $\mathfrak{g}(A)$. In particular, the subalgebra $\hat{\mathfrak{h}} = \mathfrak{h} \otimes C[T, T^{-1}] \oplus CC$ is called the Heisenberg subalgebra of $\hat{\mathfrak{g}}(A)$. We define the action of Q on $\hat{\mathfrak{h}}$ by the formula

$$\alpha \cdot h \otimes T^m = h \otimes T^m - \langle \alpha, h \rangle \delta_{0,m} C, \qquad \alpha \cdot C = C \tag{2.1}$$

where $\alpha \in Q$, $h \in \mathfrak{h}$. The pair $(\hat{\mathfrak{h}}, Q)$ is called the Heisenberg system of $\hat{\mathfrak{g}}(A)$. In this paper we will always use the notation $h(n)$ for $h \otimes T^n$.

Let Γ be a lattice in $\mathfrak{h}_\mathbf{R}$ such that $Q \subset \Gamma \subset P$, and let $\varepsilon: Q \times \Gamma \to \mathbf{C}_1^* = \{ z \in \mathbf{C}: |z| = 1 \}$ be a bilinear cocycle, whose restriction to $Q \times Q$ satisfies (1.2). We will fix a polarization of $\hat{\mathfrak{h}} = \mathfrak{h}^+ + (\mathfrak{h} + \mathbf{C}C) + \mathfrak{h}^-$, where $\mathfrak{h}^{\pm} = \mathfrak{h} \otimes T^{\pm}\mathbf{C}[T^{\pm 1}]$. Let $S(\mathfrak{h}^-)$ denote a symmetric algebra of $\mathfrak{h}^-$ and $\mathbf{C}[\Gamma]$ be a group algebra of Γ. We construct an irreducible projective representation with the cocycle ε of the Heisenberg system $(\hat{\mathfrak{h}}, Q)$ in the space $V = S(\mathfrak{h}^-) \otimes \mathbf{C}[\Gamma]$ in a standard way [9]:

$$h(n) \cdot v \otimes e^\alpha = n(\partial_{h(-n)}v) \otimes e^\alpha, \qquad n > 0,$$

$$h(-n) \cdot v \otimes e^\alpha = (h(-n)v) \otimes e^\alpha, \qquad n > 0,$$

$$h(0) \cdot v \otimes e^\alpha = v \otimes (\partial_h e^\alpha) = \langle h, \alpha \rangle v \otimes e^\alpha,$$

$$\beta \cdot v \otimes e^\alpha = v \otimes e^{\alpha + \beta}, \qquad C \cdot v \otimes e^\alpha = v \otimes e^\alpha \qquad (2.2)$$

where $\alpha \in \Gamma$, $\beta \in Q$, $v \in S(\mathfrak{h}^-)$.

We define a class of *vertex operators* which was first introduced in dual resonance models:

$$X(\alpha, z) = \exp\left(\sum_{n=1}^\infty \frac{z^n}{n} \alpha(-n) \right)$$

$$\cdot \exp(\log z\alpha(0) + \alpha)\exp\left(-\sum_{n=1}^\infty \frac{z^{-n}}{n} \alpha(n) \right) \qquad (2.3)$$

where $\alpha \in Q$, $z \in \mathbf{C} \backslash 0$. Generally speaking $X(\alpha, z)$ maps V into V', the space of formal series of elements from V. However, the homogeneous components $X_n(\alpha)$, $n \in \mathbf{Z}$, defined by the decomposition $X(\alpha, z) = \sum_{n \in \mathbf{Z}} X_n(\alpha)z^{-n}$ are well-defined operators on V. We will often denote $X_n(\alpha)$ by $\int_C X(\alpha, z)z^n \, dz/z$, where C is a circle containing the origin. This notation has a precise meaning in the following sense. Let $v \in V$, then

$$X_n(\alpha)v = \int_C X(\alpha, z)v \cdot z^n \frac{dz}{z}.$$

We will always operate with vertex operators having in mind this fact.

Let $\varepsilon_\alpha: \mathbf{C}[\Gamma] \to \mathbf{C}[\Gamma]$, $\alpha \in Q$, be given by $\varepsilon_\alpha \cdot e^\beta = \varepsilon(\alpha, \beta)e^\beta$. We introduce the operators $X^\varepsilon(\alpha, z) = X(\alpha, z)\varepsilon_\alpha$. One can find directly the commutation relations between the homogeneous components of vertex operators; in particular, one has [9]

$$\left[h(m), X_n^\varepsilon(\alpha) \right] = \langle h, \alpha \rangle X_{n+m}^\varepsilon(\alpha),$$

$$\left[X_n^\varepsilon(\alpha), X_m^\varepsilon(\beta) \right] = 0, \qquad \langle \alpha, \beta \rangle \geq 0,$$

$$\left[X_n^\varepsilon(\alpha), X_m^\varepsilon(\beta) \right] = \varepsilon(\alpha, \beta) X_{n+m}^\varepsilon(\alpha + \beta), \qquad \langle \alpha, \beta \rangle = -1,$$

$$\left[X_n^\varepsilon(\alpha), X_m^\varepsilon(-\alpha) \right] = -(\alpha(n + m) + n\delta_{m, -m}) \qquad (2.4)$$

where α, $\beta \in \Delta$, $n, m \in \mathbf{Z}$. The above calculations are based on the following formula for $|z| > |z_0|$:

$$X^\varepsilon(\alpha, z) X^\varepsilon(\beta, z_0) = \varepsilon(\alpha, \beta)(z - z_0)^{\langle \alpha, \beta \rangle}$$
$$\cdot (zz_0)^{-\langle \alpha, \beta \rangle / 2} : X^\varepsilon(\alpha, z) X^\varepsilon(\beta, z_0): \quad (2.5)$$

where $: :$ denotes the normal ordering, and the operator analogue of the Cauchy residue formula. Setting $h_i = h_i(0)$, $e_i = X_0^\varepsilon(\alpha_i)$, $f_i = X_0^\varepsilon(-\alpha_i)$, one can deduce from (2.4) the commutation relations (1.1).

THEOREM 2.1. *Let $\mathfrak{g}$ be a nonaffine Kac-Moody algebra. Then its representation π in V is defined by*

$$\pi(e_i) = X_0^\varepsilon(\alpha_i),$$
$$\pi(f_i) = X_0^\varepsilon(-\alpha_i), \quad i \in I,$$
$$\pi(h_i) = h_i(0).$$

In particular, the operators $X_0^\varepsilon(\alpha)$, $\alpha \in \Delta_R$, define the representation of all real root elements.

PROOF. The relations (1.1) follow immediately from the commutation relations of vertex operators (2.4). Then the simplicity of $\mathfrak{g}$ [11] implies the result.

It is interesting to note that although the vertex representation has appeared first in the affine case, affine Lie algebras are excluded from the statement of Theorem 2.1. We will formulate this result separately.

THEOREM 2.2 [9]. *Let $\mathfrak{g}$ be a simple finite-dimensional Lie algebra. Then the representation π of the corresponding affine Lie algebra $\hat{\mathfrak{g}}$ is defined by*

$$\pi(x_\alpha(n)) = X_n^\varepsilon(\alpha), \quad \alpha \in \Delta, n \in \mathbf{Z},$$
$$\pi(h(n)) = h(n), \quad \pi(c) = \mathrm{Id}.$$

The representation V_Γ decomposes into the sum of fundamental representations of $\hat{\mathfrak{g}}$ corresponding to the orbits of Q in Γ.

We define a Hermitian form $(\, ,\,)$ in V by the conditions

$$\left(1 \otimes e^\alpha, 1 \otimes e^\beta\right) = \delta_{\alpha, \beta}, \quad h(n)^* = h(-n), \quad h \in \mathfrak{h}_\mathbf{R}. \quad (2.6)$$

The form $(\, ,\,)$ is positive definite on V if and only if the form $\langle \, , \rangle$ is positive definite on $\mathfrak{h}_\mathbf{R}$. One can check the following property:

$$X_n^\varepsilon(\alpha)^* = X_{-n}^\varepsilon(-\alpha). \quad (2.7)$$

We introduce now an important algebra acting on V which is called the Virasoro algebra. This algebra is, in fact, a central extension of the algebra of vector fields on the unit circle and can be defined as follows:

$$\mathfrak{v} = \sum_{h \in \mathbf{Z}} \mathbf{C}L(n) \oplus \mathbf{C}c, \quad (2.8)$$

$$[L(m), L(n)] = (m - n)L(m + n) + \frac{m^3 - m}{12}\delta_{m,-n}c \quad (2.9)$$

where c is a central element of $\mathfrak{v}$. Let h_i, $i = 1, \ldots, l$, be an orthonormal basis of $\mathfrak{h}$. One obtains a representation of the Virasoro algebra $\mathfrak{v}$ on V by setting

$$L(n) = \frac{1}{2} \sum_{i=1}^{l} \sum_{m\in\mathbf{Z}} : h_i(h - m)h_i(m):. \quad (2.10)$$

In this case $c = l \cdot \mathrm{Id}$. One can verify the following commutation relations [9]:

$$[L(m), h(n)] = -nh(m + n), \quad h \in \mathfrak{h}, \quad (2.11)$$

$$[L(m), X_n^\varepsilon(\alpha)] = -nX_{m+n}^\varepsilon(\alpha), \quad \alpha \in \Delta_R. \quad (2.12)$$

The Virasoro algebra $\mathfrak{v}$ commutes with $\mathfrak{g}$ and therefore allows one to significantly "decrease" the reducible representation V. We define for $k \in \mathbf{C}$

$$V_k = \{v \in V: (L(n) - k\delta_{0,n})v = 0, n \in \mathbf{Z}\}. \quad (2.13)$$

Then for each $k \in \mathbf{C}$ such that $V_k \neq \phi$ we obtain a representation of $\mathfrak{g}$. There representations are irreducible only in the simplest case of $\mathfrak{g} = \mathfrak{sl}(2)$ [16, 28].

Let $\hat{\mathfrak{g}}$ be an affinization of a Kac-Moody algebra $\mathfrak{g}$. We can extend it to a semidirect product with the Virasoro algebra $\mathfrak{v}$ by using the following commutation relations:

$$[L(m), x \otimes t^n] = -t^{m+1}\frac{d}{dt}x \otimes t^n = -nx \otimes t^{n+m}. \quad (2.14)$$

Then (2.11), (2.12) imply that the vertex representation of $\hat{\mathfrak{g}}$ can be extended to the semidirect product if one defines $L(n)$ by (2.10).

2.2. Real root vectors exhaust the root system of Kac-Moody algebra only in the finite-dimensional case. In general, we have in addition infinitely many imaginary root vectors, and it is not known yet which operators represent a general vector of this type. We will find now the form of operators corresponding to the light-cone (isotropic) imaginary root vectors for a hyperbolic Lie algebra. Physicists can recognize "photon" vertex operators introduced in dual resonance models [4].

PROPOSITION 2.3. *Let $\mathfrak{g}$ be a hyperbolic algebra of rank l, and let $\gamma \in \Delta$ be an isotropic imaginary root vector. Then any $x \in \mathfrak{g}_\gamma$ has the following representation on V:*

$$\pi(x) = \int_C \alpha(z)X^\varepsilon(\gamma, z)\frac{dz}{z} \quad (2.14)$$

where $\alpha \in \mathfrak{h}$ is orthogonal to γ.

PROOF. First we will check the identity

$$\left[\int_C X^\varepsilon(\alpha, z)\,\frac{dz}{z}, \int_C X^\varepsilon(\beta, z)\,\frac{dz}{z}\right] = \varepsilon(\alpha, \beta)\int_C \alpha(z) X^\varepsilon(\alpha + \beta, z)\,\frac{dz}{z}$$

(2.15)

where $\alpha, \beta \in \Delta_R$, $\alpha + \beta = \gamma$. One has

$$\left[\int_C X^\varepsilon(\alpha, z)\,\frac{dz}{z}, \int_{C_0} X^\varepsilon(\beta, z_0)\,\frac{dz_0}{z_0}\right]$$

$$= \int_{C_0}\left(\int_{C_R} X^\varepsilon(\alpha, z) X^\varepsilon(\beta, z_0)\,\frac{dz}{z} - \int_{C_r} X^\varepsilon(\beta, z_0) X^\varepsilon(\alpha, z)\,\frac{dz}{z}\right)\frac{dz_0}{z_0}$$

$$= \int_{C_0}\left(\int_{C_R\backslash C_r} \varepsilon(\alpha, \beta)\frac{zz_0}{(z - z_0)^2} : X^\varepsilon(\alpha, z) X^\varepsilon(\beta, z_0): \frac{dz}{z}\right)\frac{dz_0}{z_0}$$

$$= \varepsilon(\alpha, \beta)\int_{C_0} \alpha(z_0) X^\varepsilon(\alpha + \beta, z_0)\,\frac{dz_0}{z_0}.$$

This implies that $\mathfrak{g}_\gamma$ consists of elements of the form (2.14), where $\alpha \in \mathfrak{h}$ is orthogonal to γ. Theorem 1.1 implies that $\dim \mathfrak{g}_\gamma = l - 2$, and we note that $\int_C \gamma(z) X^\varepsilon(\gamma, z)\,dz/z = 0$. Therefore, every operator of this form belongs to $\mathfrak{g}_\gamma$.

Let us consider an affine Lie algebra as a subalgebra of hyperbolic algebra acting on the vertex representation. Then one has

PROPOSITION 2.4. *The operators*

$$\int_C X^\varepsilon(\alpha + nc, z)\,\frac{dz}{z}, \qquad \int_C h(z) X^\varepsilon(nc, z)\,\frac{dz}{z},$$

$$\alpha \in \Delta, n \in \mathbf{Z}, h \in \mathfrak{h}, \text{ and } h(0),$$

where $h \in \mathfrak{h}$ is orthogonal to c ($\langle c, c \rangle = 0$), define a representation of the affine Lie algebra $\hat{\mathfrak{g}}$ on V. In particular,

(i)

$$\left[\int_C h(z) X^\varepsilon(mc, z)\,\frac{dz}{z}, \int_C g(z) X^\varepsilon(nc, z)\,\frac{dz}{z}\right] = m\langle h, g \rangle \delta_{m,-n} c(0),$$

(ii)

$$\left[\int_C h(z) X^\varepsilon(mc, z)\,\frac{dz}{z}, \int_C X^\varepsilon(\alpha + nc, z)\,\frac{dz}{z}\right]$$

$$= \varepsilon(mc, \alpha + nc)\langle h, \alpha \rangle \int_C X^\varepsilon(\alpha + (m + n)c, z)\,\frac{dz}{z},$$

(iii)

$$\left[\int_C X^\varepsilon(\alpha + mc, z)\frac{dz}{z}, \int_C X(\beta + nc, z)\frac{dz}{z} \right]$$

$$= \varepsilon(\alpha + mc, \beta + mc)$$

$$\cdot \begin{cases} 0, & \langle \alpha, \beta \rangle \geq 0, \\ \int_C X^\varepsilon(\alpha + \beta + (m + n)c)\frac{dz}{z}, & \langle \alpha, \beta \rangle = -1, \\ \int_C \alpha(z)X^\varepsilon((m + n)c, z)\frac{dz}{z} + m\delta_{m,-n}c(0), & \alpha = -\beta. \end{cases}$$

PROOF. The proof follows from Proposition 2.3 and Theorem 2.1; also, it is not difficult to check the commutation relations directly, e.g. (i)

$$h(z)g(z_0) = \, :h(z)g(z_0): \, + \frac{zz_0}{(z - z_0)^2}\langle h, g \rangle,$$

$$\left[\int_C h(z)X^\varepsilon(mc, z)\frac{dz}{z}, \int_C g(z_0)X^\varepsilon(nc, z_0)\frac{dz_0}{z_0} \right]$$

$$= \int_{C_0}\left(\int_{C_R} h(z)g(z_0)X^\varepsilon(mc, z)X(nc, z_0)\frac{dz}{z} \right.$$

$$\left. - \int_{C_r} g(z_0)h(z)X^\varepsilon(mc, z)X^\varepsilon(nc, z_0)\frac{dz_0}{z} \right) \frac{dz_0}{z_0}$$

$$= \int_{C_0}\left(\int_{C_R \backslash C_r} X^\varepsilon(mc, z)X^\varepsilon(nc, z_0)\langle h, g \rangle \frac{zz_0}{(z - z_0)^2} \right) \frac{dz_0}{z_0}$$

$$= \langle h, g \rangle \int_{C_0}\left(z\frac{d}{dz}X^\varepsilon(mc, z) \right)_{z=z_0} X^\varepsilon(nc, z_0)\frac{dz_0}{z_0}$$

$$= m\langle h, g \rangle \int_{C_0} c(z_0)X^\varepsilon((m + n)c, z_0)\frac{dz_0}{z_0}$$

$$= m\langle h, g \rangle \delta_{m,-n}c(0)$$

because

$$\int_C c(z)X^\varepsilon(kc, z)\frac{dz}{z} = \delta_{k,0}c(0). \quad \text{Q.E.D.}$$

If we consider the dimension of the space of operators of the form $\int_C h(z)X^\varepsilon(nc, z)\, dz/z$, then we can notice it is one more than dim $\mathfrak{g}_c$.

One can ask a question: What is the meaning of this additional one-dimensional space? It turns out that this is a derivation of the affine Lie algebra $\hat{g}$.

PROPOSITION 2.5. *Under the conditions of Proposition 2.4, the operators* $\int_C d(z)X^\varepsilon(nc, z)\, dz/z$ *where* $\langle d, c\rangle = -1$, $\langle d, \alpha\rangle = 0$, $\alpha \in \Delta$, *define the extension of* $\hat{g}$ *by the Virasoro algebra. In particular,*

(i)

$$\left[\int_C d(z)X^\varepsilon(mc, z)\,\frac{dz}{z}, \int_C X^\varepsilon(\alpha + nc, z)\,\frac{dz}{z}\right]$$
$$= -n\varepsilon(mc, \alpha + nc)\int_C X^\varepsilon(\alpha + (m+n)c, z)\frac{dz}{z},$$

(ii)

$$\left[\int_C d(z)X^\varepsilon(mc, z)\,\frac{dz}{z}, \int h(z)X^\varepsilon(nc, z)\,\frac{dz}{z}\right]$$
$$= -n\int_C X^\varepsilon((m+n)c, z)\,\frac{dz}{z},$$

(iii)

$$\left[\int_C d(z)X^\varepsilon(mc, z)\,\frac{dz}{z}, \int_C d(z)X^\varepsilon(nc, z)\,\frac{dz}{z}\right]$$
$$= (m-n)\int_C (z)X^\varepsilon((m+n)c, z)\,\frac{dz}{z} + 2m^3\delta_{m,-n}.$$

3. Standard representations of finite-dimensional and affine Lie algebras in the vertex.

3.1. In this section we will consider the vertex representation $V = V_P$ of finite-dimensional Lie algeba g. One can see that any irreducible finite-dimensional representation $V(\lambda)$ with the highest weight λ occurs in V, e.g. such subrepresentation of g is generated by the highest weight vector $1 \otimes e^\lambda \in V$. Let $\Omega(\lambda)$ be the vector space of all the highest weight vectors of irreducible representations isomorphic to $V(\lambda)$ in V. One has a natural isomorphism

$$V \approx \sum_{\lambda \in P^{++}} V(\lambda) \otimes \Omega(\lambda). \tag{3.1}$$

There is a natural duality between vector spaces $V(\lambda)$ and $\Omega(\lambda)$ which we will make more apparent in the next subsection. In particular, we will define a Lie algebra acting irreducibly on $\Omega(\lambda)$. Now we will determine the multiplicities of $V(\lambda)$ in V, i.e., we will find the character of $\Omega(\lambda)$.

THEOREM 3.1 [17]. *Let $\mathfrak{g}$ be a simple Lie algebra. For any dominant λ the multiplicities of $V(\lambda)$ in the vertex representation V are given by the generating function*

$$\text{ch}\,\Omega(\lambda) = q^{\langle\lambda,\lambda\rangle/2}\frac{\prod_{\alpha\in\Delta_+}\left(1 - q^{\langle\alpha,\lambda+\rho\rangle}\right)}{\varphi(q)^l}. \tag{3.2}$$

PROOF. We give here a proof different from the one given in [17], which has a simple generalization to the case when $\mathfrak{g}$ is affine. Let us denote $S(\mu) = S(\hat{\mathfrak{h}}^-) \otimes e^\mu$ and consider any irreducible representation U of $\mathfrak{g}$ and its intersections with the spaces $S(\lambda + \rho - w\rho)$. We let

$$n_{\lambda+\rho-w\rho} = \dim(U \cap S(\lambda + \rho - w\rho)). \tag{3.3}$$

The character formula (1.12) for $\mathfrak{g}$ implies that, except in the case when $U \approx V(\lambda)$,

$$\sum_{w\in W} \det w \cdot n_{\lambda+\rho-w\rho} = 0. \tag{3.4}$$

In the case when $U \approx V(\lambda)$, $n_\lambda = 1$, and $n_{\lambda+\rho-w\rho} = 0$, for $w \neq 1$. We multiply the equality (3.4) by q^k, where k is the level of the highest weight vector in $U \subset V$. Then summing up these equalities for all irreducible $U \neq V(\lambda)$ we obtain

$$\text{ch}\,\Omega(\lambda) = \sum_{w\in W} \det w \cdot \text{ch}\,S(\lambda + \rho - w\rho). \tag{3.5}$$

Now, using the fact that

$$\text{ch}\,S(\mu) = \varphi(q)^{-l}q^{\langle\mu,\mu\rangle/2} \tag{3.6}$$

and the specialized form of Weyl's identity (1.14) we obtain the result.

It is a remarkable fact that in the same space $V = V_p$ we can study not only all the standard representations of $\mathfrak{g}$ but also all the standard representations of $\hat{\mathfrak{g}}$. Let us define a subalgebra $\hat{\mathfrak{g}}_{[n]} \subset \hat{\mathfrak{g}}$:

$$\hat{\mathfrak{g}}_{[n]} = \mathfrak{g} \otimes \mathbf{C}[t^n, t^{-n}] + \mathbf{C}c. \tag{3.7}$$

Then clearly $\hat{\mathfrak{g}}_{[n]} \approx \hat{\mathfrak{g}}$ where the isomorphism is given by

$$i_n(x \otimes t^{kn}) = x \otimes t^k, \qquad i_n(nc) = c \tag{3.8}$$

where $k \in \mathbf{Z}$, $x \in \mathfrak{g}$. Therefore, if we restrict the representation V to $\hat{\mathfrak{g}}_{[n]}$, then taking into account isomorphism (3.8) we obtain a representation of level n. Moreover, this representation is reducible (the only exception is when $n = 1$, and $\mathfrak{g}$ is of type E_8), and any irreducible level n standard representation $V(\lambda)$ with the highest weight λ occurs in V, e.g. such subrepresentation is generated by the highest weight vector $1 \otimes e^{\lambda_0} \in V$, where $\lambda = nd + \lambda_0 - \langle \lambda_0 \lambda_0 \rangle c/2$. We define $\Omega(\lambda)$ as in the finite-dimensional case and we get the decomposition

$$V \approx \sum_{\substack{\lambda \in \hat{P}^{++} \\ \text{level } \lambda = n}} V(\lambda) \otimes \Omega(\lambda). \tag{3.9}$$

In [8] we formulated a conjecture about the multiplicites of $V(\lambda)$ in V which we will now prove.

THEOREM 3.2. *Let $\hat{\mathfrak{g}}$ be an affine Lie algebra acting as the algebra $\mathfrak{g}_{[n]}$ in the vertex representation V. For any dominant λ the multiplicities of $V(\lambda)$ in V are given by the generating function*

$$\text{ch}\,\Omega(\lambda) = q^{\langle \lambda, \lambda \rangle /2} \frac{\prod_{\alpha \in \hat{\Delta}_+} \left(1 - q^{\langle \alpha, \lambda + \hat{\rho} \rangle} \right)}{\varphi(q)^l}. \tag{3.10}$$

PROOF. The proof is a literal repetition of the proof of Theorem 3.1 where we only change ρ to $\hat{\rho}$, W to $\hat{W}$ and we use (1.15) instead of (1.14).

Note that Theorem 3.2 implies immediately Theorem 1.6 of [8].

We remarked in [8] about the miraculous coincidence

$$\text{ch}\,\Omega(\lambda) = \text{ch}_q \Omega(\lambda)' \tag{3.11}$$

where $\Omega(\lambda)'$ and ch_q are defined as follows (see [22]). Let $\hat{\mathfrak{h}}'$ be a principal Heisenberg subalgebra [19]; then the restriction of any irreducible level 1 representation of $\hat{\mathfrak{g}}$ to $\hat{\mathfrak{h}}'$ is still irreducible and isomorphic to the canonical Fock space V_0. The general standard representation $V(\lambda)$ of $\hat{\mathfrak{g}}$ has the decomposition

$$V(\lambda) = V_0 \otimes \Omega(\lambda)' \tag{3.12}$$

where $\Omega(\lambda)'$ is the vector space of vacuum vectors in $V(\lambda)$ for $\hat{\mathfrak{h}}'$. One can define a character ch_q of the spaces in (3.12) with respect to a differentiation d' so that

$$\text{ch}_q V(\lambda) = \text{ch}_q V_0 \cdot \text{ch}_q \Omega(\lambda)' \tag{3.13}$$

where

$$\text{ch}_q V(\lambda) = \text{ch}\,V(\lambda)\big|_{e^{-\alpha_i} = q, i = 1, \ldots, l}$$

is the so-called principally specialized character. Then the character $\text{ch}_q \Omega(\lambda)'$ is given by the right side of (3.10). Theorem 3.2, (3.11) and (3.13) imply that

$$\text{Ch} = \sum_{\substack{\lambda \in \hat{P}_{++} \\ \text{level } \lambda = n}} \text{ch}\,V(\lambda)\big|_{e^{-c} = q^n} \cdot \text{ch}_q V(\lambda) \tag{3.14}$$

does not depend on level n, and thanks to (1.16) has a very simple form. An explanation of this fact from the represenation theory point of view would be very important.

3.2. We know from (2.11) and (2.12) that the Virasoro algebra commutes with the action of $\mathfrak{g}$. Therefore, starting from one element $1 \otimes e^\lambda \in \Omega(\lambda)$, we can generate an infinite-dimensional subspace of $\Omega(\lambda)$ by the action of the Virasoro algebra. However, only in one special case, when $\mathfrak{g} = \mathfrak{sl}(2)$, do we obtain the whole space $\Omega(\lambda)$, i.e., the Virasoro algebra acts irreducibly on $\Omega(\lambda)$. In order to generate $\Omega(\lambda)$ for an arbitrary $\mathfrak{g}$, we need an extension of the Virasoro algebra. Before we give the definition of this algebra we recall Wick's theorem well known in physics literature.

The product of two linear operators $a_1, a_2 \in \hat{\mathfrak{h}}$ differ from the normally ordered product by some scalar operator. We call this scalar operator contraction and we will denote it by a lower bracket

$$a_1 a_2 = \; :a_1 a_2: \; + \underbracket{a_1 a_2}. \tag{3.15}$$

We define a normally ordered product with pairing by

$$:a_1 \cdots \underbracket{a_j \cdots a}_k \cdots a_n: \; = \underbracket{a_j a_k}: a_1 \cdots a_{j-1} a_{j+1} \cdots a_{k-1} a_{k+1} \cdots a_n:.$$

$$\tag{3.16}$$

Now we can give a simple formulation of Wick's theorem.

THEOREM 3.3. *The product of normally ordered products of linear operators*

$$\left(: a_1 \cdots a_{k_1} :\right)\left(: a_{k_1+1} \cdots a_{k_2} :\right) \cdots \left(: a_{k_{n-1}+1} \cdots a_{k_n} :\right)$$

is equal to the sum of all normally ordered products of these operators

$$: a_1 \cdots a_{k_1} a_{k_1+1} \cdots a_{k_n} :$$

with all possible pairings between elements of the n sets

$$\left(a_1, \ldots, a_{k_1}\right), \left(a_{k_1+1}, \ldots, a_{k_2}\right), \ldots, \left(a_{k_{n-1}+1}, \ldots, a_{k_n}\right)$$

including the normal product without pairings.

PROOF. The proof is a straightforward induction.

We will apply the Wick theorem to the case when $n = 2$ and $A_k = a_k(z) = \Sigma_{h \in \mathbf{Z}} a_k(n) z^{-n}$, $a_k \in \mathfrak{h}$. Simple calculations show that

$$\underbracket{a_1(z_1) a_2(z_2)} = \frac{z_1 z_2}{\left(z_1 - z_2\right)^2} \langle a_1, a_2 \rangle, \qquad |z_1| > |z_2|. \tag{3.17}$$

For example we can prove, using Wick's theorem, the commutation relations (2.9) with $c = l \cdot \mathrm{Id}$. One has

$$\sum_{i=1}^{l} \sum_{j=1}^{l} : h_i(z)h_i(z): : h_j(z_0)h_j(z_0):$$

$$= \sum_{i=1}^{l} \sum_{j=1}^{l} \delta_{ij} \frac{zz_0}{(z-z_0)^2} \delta_{ij} \frac{zz_0}{(z-z_0)^2}$$

$$= l \cdot \frac{z^2 z_0^2}{(z-z_0)^4}.$$

We get four terms with one contraction and two terms with two contractions; therefore,

$$\sum_{i=1}^{l} \sum_{j=1}^{l} : h_i^2(z): : h_j^2(z_0):$$

$$= \sum_{i=1}^{l} \sum_{j=1}^{l} : h_i^2(z)h_j^2(z_0):$$

$$+ \frac{4zz_0}{(z-z_0)^2} \sum_{i=1}^{l} : h_i(z)h_i(z_0): + 2l \frac{z^2 z_0^2}{(z-z_0)^4}. \qquad (3.18)$$

Now (2.9) follows from the Cauchy residue formula

$$[L(m), L(z_0)] = \int_{C_R \backslash C_r} \left(\frac{zz_0}{(z-z_0)^2} \sum_{i=1}^{l} : h_i(z)h_i(z_0): \right.$$

$$\left. + \frac{l}{2} \frac{z^2 z_0^2}{(z-z_0)^4} \right) z^m \frac{dz}{z}$$

$$= \left(z_0 \frac{d}{dz} z^m \sum_{i=1}^{l} : h_i(z)h_i(z_0): + \frac{l}{12} z_0^2 \frac{d^3}{dz^3} z^{m+1} \right)_{z=z_0}$$

$$= 2m z_0^m L(z_0) + z_0^{m+1} \frac{d}{dz_0} L(z_0) + \frac{l}{12}(m+1)m(m-1)z_0^m.$$

$$(3.19)$$

We define now a new class of algebras generalizing the Virasoro algebra. Let $h \in \mathfrak{h}$, we denote

$$h^{(n)}(z) = D_z^n h(z) = \left(z \frac{d}{dz} \right)^n h(z). \qquad (3.20)$$

Let $P(z)$ be a W-invariant polynomial in $h_i^{(n)}(z)$, $i = 1, \ldots, l$, $n \in \mathbf{Z}_+$, and let $P(n)$, $n \in \mathbf{Z}$, be defined by

$$: P(z) := \sum_{n \in \mathbf{Z}} P(n) z^{-n}. \tag{3.21}$$

The algebra of operators $\hat{S}^W$ is by definition the linear span of all $P(n)$, $n \in \mathbf{Z}$, and Id, such that $: P(z) :$ commutes with $\mathfrak{g}$. The operators $P(n)$ are well defined, and $\hat{S}^W$ has a natural grading.

PROPOSITION 3.4. *The algebra of operators $\hat{S}^W$ is closed under the Lie bracket.*

PROOF. We have to show that for any $P(m), Q(n) \in \hat{S}^W$, $m, n \in \mathbf{Z}$,

$$[P(m), Q(n)] = \sum_{j \in J} f_j(m, n) R_j(m + n) + f(m, n) \delta_{m, -n} \tag{3.22}$$

where $R_j(m + n) \in \hat{S}^W$, f_j, f are some functions on $\mathbf{Z} \times \mathbf{Z}$, and J is a finite set. This form follows from the Wick theorem and (3.21). In fact, the typical term of

$$\left[\int_C : P(z) : z^m \frac{dz}{z}, \int_{C_0} : Q(z_0) : z_0^n \frac{dz_0}{z_0} \right]$$

will be

$$\int_{C_0} \left(\int_{C_R \setminus C_r} \prod_i \left(D_z^{m_i} D_{z_0}^{n_i} \frac{z z_0}{(z - z_0)^2} \right) : P_1(z) Q_1(z_0) : z^m \frac{dz}{z} \right) z_0^n \frac{dz_0}{z_0} \tag{3.23}$$

where i runs through a finite number of contractions, and P_1, Q_1 are contracted P and Q. The typical term in (3.22) will be

$$\int_{C_0} \left(\int_{C_R \setminus C_r} \frac{z^{1+M} z_0^{1+N}}{(z - z_0)^{2+M+N}} : P_1(z) Q_1(z_0) : z^m \frac{dz}{z} \right) z_0^n \frac{dz_0}{z_0}$$

$$= \int_{C_0} \frac{1}{(M + N + 1)!} \frac{d^{M+N+1}}{dz^{M+N+1}} \left(z^{M+m} : P_1(z) Q_1(z_0) : \right)_{z=z_0} z_0^{n+N} dz_0,$$

which is clearly of the form in (3.22).

The Virasoro algebra certainly is a subalgebra of $\hat{S}^W$, and spanned by $P(n)$, $n \in \mathbf{Z}$, and Id, where P is the Casimir element in $S(\mathfrak{h})^W$. In order to illustrate what type of elements occur in $\hat{S}^W$ we will introduce a generalization of Segal's operators (see [7]).

Let $\{x_j\}$ be an orthonormal basis of $\mathfrak{g}$, and let $[x_i, x_j] = \Sigma_k C_{ij}^k x_k$. For any $\mathfrak{g}$-module V from the category $\mathfrak{M}$ [15, p. 102] we introduce the operator

$$L(z_1, z_2) = \Sigma x_j(z_1)x_j(z_2) = \Sigma_j :x_j(z_1)x_j(z_2): + \dim \mathfrak{g}\, \frac{z_1 z_2}{(z_1 - z_2)^2},$$

(3.24)

where $|z_1| > |z_2|$. The second expression in (3.24) allows us to extend the definition of the operator $L(z_1, z_2)$ for every $z_1 \neq z_2$. Clearly the homogeneous components of $L(z_1, z_2)$ are well-defined operators in V.

PROPOSITION 3.5. The operator $L(z_1, z_2)$ commutes with $\mathfrak{g}$. If, in addition, $z_1^n = z_2^n$, $n = 2, 3, \ldots$, then $L(z_1, z_2)$ commutes with $\hat{\mathfrak{g}}_{[n]}$.

PROOF.

$$\left[x_i(m), \Sigma_j x_j(z_1)x_j(z_2) \right]$$

$$= \Sigma_j \left([x_i(m), x_j(z_1)]x_j(z_2) + x_j(z_1)[x_i(m), x_j(z_2)] \right)$$

$$= \Sigma_{jk} \left(C_{ij}^k x_k(z_1)x_j(z_2)z_1^m + C_{ij}^k x_k(z_1)x_k(z_2)z_2^m \right)$$

$$= \Sigma_{ji} C_{ij}^k x_k(z_1)x_j(z_2)(z_1^m - z_2^m),$$

which obviously implies the result.

In the vertex representation we have

$$L(z_1, z_2) = \sum_{i=1}^{l} h_i(z_1)h_i(z_2) - \sum_{\alpha \in \Delta} X^\varepsilon(\alpha_1 z_1)X^\varepsilon(-\alpha_1 z_2)$$

$$= \sum_{i=1}^{l} :h_i(z_1)h_i(z_2): + l\frac{z_1 z_2}{(z_1 - z_2)^2}$$

$$- \frac{z_1 z_2}{(z_1 - z_2)^2} \sum_{\alpha \in \Delta} :X^\varepsilon(\alpha_0 z_1)X^\varepsilon(-\alpha_1 z_2):.$$ (3.25)

Therefore the operators

$$-D_{z_1}^{2n}\left(\frac{(z_1 - z_2)^2}{z_1 z_2} L(z_1, z_2) \right)\Bigg|_{z_1 = z_2 = z} = \sum_{\alpha \in \Delta} :\alpha^{2n}(z): + Q$$ (3.26)

commute with $\mathfrak{g}$, belong to $\hat{S}^W$, and $\deg Q < 2n$. We formulate without proof a generalization of this result.

THEOREM 3.6. (i) *For every $P \in S(\mathfrak{h})^W$ there exist Q, $\deg Q < \deg P$, so that* $: P(z): + : Q(z):$ *commutes with* $\mathfrak{g}$.

(ii) *For any dominant highest weight λ, $\Omega(\lambda)$ is an irreducible representation of $\hat{S}^W$ satisfying the following properties:*

$$P(n)v_0 = 0, \quad n > 0, \quad P(0)v_0 = \langle P, \lambda \rangle v_0, \qquad (3.27)$$

where $v_0 = 1 \otimes e^\lambda \in V$, and $P(n) \in \hat{S}^W$, $n \in \mathbf{Z}$.

Here we will note only that the proof of Theorem 3.6 for $\mathfrak{g} = \mathfrak{sl}(n)$ follows easily from Theorem 1.6 in [8].

We remarked already that the decomposition (3.1) displays some kind of duality between the representations of the algebras $\mathfrak{g}$ and $\hat{S}^W$. In fact, the irreducible finite-dimensional representations $V(\lambda)$ of $\mathfrak{g}$ naturally correspond to the irreducible representations $\Omega(\lambda)$ of $\hat{S}^W$. The duality can be extended further if one considers another natural class of representations, namely the induced representations called Verma modules. By definition a Verma module of $\mathfrak{g}$ is defined by any $\lambda \in \mathfrak{h}^*$ as the induced module

$$M(\lambda) = U(\mathfrak{g}) \otimes_{U(\mathfrak{b}^+)} \mathbf{C}_\lambda \qquad (3.28)$$

where U denotes the universal enveloping algebra and $\mathbf{C}_\lambda$ is a $\mathfrak{b}^+$-module via the action

$$(h + x)z = \langle \lambda, h \rangle z, \quad h \in \mathfrak{h}, x \in \mathfrak{n}^+, z \in \mathbf{C}. \qquad (3.29)$$

One can define an analogue of Verma modules for $\hat{S}^W$ by fixing a polarization

$$\hat{S}^W = \hat{S}^W_+ + \left(\hat{S}^W_0 + \mathbf{C}\,\mathrm{Id} \right) + \hat{S}^W_- \qquad (3.30)$$

where S^W_+ is spanned by $P(n)$, $n > 0$, and similarly for $\hat{S}^W_0$, $\hat{S}^W_-$. Then for any $\lambda \in \mathfrak{h}^*$,

$$\mathfrak{M}(\lambda + \rho) = U(\hat{S}^W) \otimes_{U(\hat{S}^W_+ + \hat{S}^W_0)} \mathbf{C}^\lambda \qquad (3.31)$$

where $\mathbf{C}^\lambda$ is a $(\hat{S}^W_+ + \hat{S}^W_0)$-module via the action

$$(h + x)z = \langle \lambda, h \rangle z, \quad h \in \hat{S}^W_0, x \in \hat{S}^W_+, z \in \mathbf{C}, \qquad (3.32)$$

and $\langle \lambda, h \rangle$ is the value of the polynomial h on λ. Using the Chevalley theorem that $S(\mathfrak{h})^W \approx \mathbf{C}[P_1, \ldots, P_l]$ and Theorem 3.6, one has

$$\mathrm{ch}\,\mathfrak{M}(\lambda + \rho) = q^{\langle \lambda, \lambda \rangle / 2} / \varphi(q)^l. \qquad (3.33)$$

Therefore Theorem 3.2 implies that

$$\mathrm{ch}\,\Omega(\lambda) = \sum_{w \in W} \det w \, \mathrm{ch}\,\mathfrak{M}(w(\lambda + \rho)) \qquad (3.34)$$

which is similar to the Weyl formula (1.11)

$$\operatorname{ch} V(\lambda) = \sum_{w \in W} \det w \operatorname{ch} M(w(\lambda + \rho)). \qquad (3.35)$$

We know from the results of Bernstein, Gelfand and Gelfand [1], that (3.35) reflects a much stronger result about the existence of the exact sequence of $\mathfrak{g}$-modules

$$0 \leftarrow V(\lambda) \leftarrow C_0 \leftarrow C_1 \leftarrow \cdots \leftarrow C_s \leftarrow 0 \qquad (3.36)$$

where $s = \dim \mathfrak{n}_-$, $C_k = \oplus_{w \in W^{(k)}} M(w(\lambda + \rho))$. Therefore, one is led to conjecture that the formula (3.34) also is not just a coincidence.

Conjecture 3.7. There exists an exact sequence of $\hat{S}^W$-modules

$$0 \leftarrow \Omega(\lambda) \leftarrow C_0 \leftarrow C_1 \leftarrow \cdots \leftarrow C_s \leftarrow 0 \qquad (3.37)$$

where s is as in (3.35), $C_k = \oplus_{w \in W^{(k)}} \mathfrak{M}(w(\lambda + \rho))$.

In particular, for $\mathfrak{g} = \mathfrak{sl}(2)$ one should have, according to Conjecture 3.7, the exact sequence

$$0 \leftarrow \Omega(\lambda) \leftarrow \mathfrak{M}(\lambda + \rho) \leftarrow \mathfrak{M}(-\lambda - \rho) \leftarrow 0 \qquad (3.38)$$

where $\Omega(\lambda)$, $\mathfrak{M}(\,)$ are modules of $\hat{S}^W$ of rank 1. We noted above that in this case the Virasoro subalgebra contains the main information about $\hat{S}^W$. Rocha-Caridi has proved that (3.38) is in fact an exact sequence of Virasoro modules.

Finally, we will note that the strange duality between Lie algebras $\mathfrak{g}$ and $\hat{S}^W$ becomes more transparent in the affine case. Let us consider the decomposition (3.9). The vector space $V(\lambda)$ is by definition the representation space of the affine Lie algebra $\hat{\mathfrak{g}}$. The vector space $\Omega(\lambda)$, thanks to the miracle in (3.11), can be considered as the representation space of the Z-algebras studied in [23]. It is shown in [23] that some categories of representations (which include standard representations) of Z-algebras and the corresponding affine Lie algebras are equivalent. This implies immediately the equivalence of the affine analogues of (3.35) and (3.36).

4. The dual model in the critical dimension and hyperbolic algebras of rank 26. One of the main achievements of dual models is the no-ghost theorem, which reveals the critical number 26. This beautiful result implies the existence of one special representation of hyperbolic Lie algebras of rank 26 and allows us to obtain an upper bound for root multiplicities. For the sake of completeness we recall here the main steps of the proofs of the no-ghost theorem.

We define V_n, $n \in \mathbf{Z}$, to be the space of highest weight vectors of the Virasoro algebra, i.e. $v \in V_n$ iff

$$L(m)v = 0, \quad m \geqslant 1, \quad L(0)v = nv. \qquad (4.1)$$

The space V_1 plays the most important role in the dual theories. Let $V_i^\flat = \{v \in V_1 : v = h(-1) \otimes 1, h \in \mathfrak{h}\}$. Clearly $V_1^\flat$ contains elements of negative norm if and only if $\mathfrak{h}_\mathbf{R}$ contains such elements. Let us denote $V_1^\circ = V_1 \ominus V_1^\flat$. Then there is no reason to expect that there are no other elements with negative norm in V_1°. However, physicists obtained the following remarkable result:

THEOREM 4.1 (NO-GHOST THEOREM). *Let V be a vertex representation of a hyperbolic Lie algebra* $\mathfrak{g}$, *then V_1° is a positive semidefinite space if and only if* rank $\mathfrak{g} \leqslant 26$.

PROOF. We will indicate the main steps of the proof. (See details in [25].)

Step 1. We have to prove that for every $\alpha \in Q \backslash 0$, the space $V_1^\circ \cap S(\alpha)$ is positive semidefinite. We consider $S(\alpha)$ naturally graded of $S(\alpha)$ with respect to the degree operator

$$S(\alpha) = \sum_{M=0}^{\infty} S(\alpha)_M. \qquad (4.2)$$

Then $v \in S(\alpha)_M$ belongs to V_1° only if $\langle \alpha, \alpha \rangle = 2(1 - M)$. Now let us fix any isotropic vector $c \in \mathfrak{h}$, normalized by the condition $\langle c, \alpha \rangle = -1$, and let us define the transverse space $T(\alpha, c)$ by the condition that $v \in T(\alpha, c)$ if and only if

$$L(n)v = c(n)v = 0, \qquad n \geqslant 1. \qquad (4.3)$$

Let us define the subspace $G(\alpha, c)$ of $S(\alpha)$ as the linear span of the elements of type

$$v_{\{\lambda_j, \mu_j\}} = L(-1)^{\lambda_1} \cdots L(-n)^{\lambda_n} c(-1)^{\mu_1} \cdots c(-m)^{\mu_m} v \qquad (4.4)$$

where $v \in T(\alpha, c)$, $\lambda_j, \mu_j \in \mathbf{Z}_+$, and $\Sigma \lambda_j + \Sigma \mu_j > 0$.

LEMMA 4.2. (i) *The space $T(\alpha, c)$ is positive definite.*
(ii) *The subspaces $T(\alpha, c)$ and $G(\alpha, c)$ have zero intersection.*
(iii) $S(\alpha) = T(\alpha, c) \oplus G(\alpha, c)$.

Step 2. Now we can construct an operator which projects the space $S(\alpha)_N$ to $T(\alpha, c)_N$. Let $c(z) = \Sigma_{n \in \mathbf{Z}} c(n) z^{-n}$ be the generating function; then the operator

$$D(n) = \int_C c(z)^{-1} z^n \frac{dz}{z} \qquad (4.5)$$

is well defined in $S(\alpha)$ thanks to the condition $\langle c, \alpha \rangle = -1$, namely, $c(z)^{-1} = (1 + c_0(z))^{-1} = 1 - c_0(z) + c_0^2(z) - \cdots$. Moreover,

$$(D(n) - \delta_{0,n})v = 0, \qquad n \geqslant 0, v \in T(\alpha, c), \qquad (4.6)$$

because $D(n) - \delta_{0,n}$, $n \geq 0$, is given by a series of terms each of which contains at least one $c(m)$, $m > 0$. We define now an important operator

$$E = (D(0) - 1)(L(0) - 1) + \sum_{n=1}^{\infty} (D(-n)L(n) + L(-n)D(n)). \quad (4.7)$$

It is clear that $Ev = 0$ for $v \in T(\alpha, c)$. However, only when rank $\mathfrak{g}$ is 26 do all solutions come from $T(\alpha, c)$.

LEMMA 4.3. *Let* rank $\mathfrak{g}$ *be* 26, *then*
  (i) *the eigenvectors of* E *span all of* $S(\alpha)$,
  (ii) *the eigenvalues of* E *are nonpositive integers*,
  (iii) *the eigenvalue zero corresponds to, precisely, the transverse subspace* $T(\alpha, c)$.

The critical dimension appears in the formula

$$[L(n), E] = -nL(n) + \frac{\dim \mathfrak{h} - 26}{12}(n^3 - n)D(n),$$

which provides the key to the proof of the lemma.

Let us define the projection operator onto $T(\alpha, c)_N$

$$P = \int_C z^E \frac{dz}{z}.$$

Then the definition (4.7) of E implies that $\langle v, v \rangle = \langle P_v, P_v \rangle \geq 0$. This ends the proof of the theorem for rank $\mathfrak{g} = 26$.

We also note how to construct an explicit basis of the space $T(\alpha - Nc, c)_N$. Let us denote

$$A_i(n) = \int_C \alpha_i(z) X^\varepsilon(nc, z) \frac{dz}{z} \quad (4.8)$$

where $i = 1, \ldots, l$, α_i orthogonal to c (and not proportional to c). Then the elements of type

$$\prod_i A_{m_i}(-n_i) 1 \otimes e^\alpha, \qquad \sum_i n_i = N, \quad n_i > 0, \quad (4.9)$$

provides a basis of $T(\alpha - Nc, c)_N$. We denote by $\overline{V}_{1,c,\alpha}$ the space spanned by the elements (4.9) with arbitrary N.

Before we use the no-ghost theorem for hyperbolic Lie algebras of rank 26, we will prove one general fact about the subspace V_1. We note that when $\mathfrak{h}_\mathbf{R}$ is positive definite, the subspace V_1 is isomorphic to the adjoint representation of $\mathfrak{g}$. It is no longer true in general; however, one has

PROPOSITION 4.4. *Let A be a subspace of V_1 generated by the action of Kac-Moody algebra $\mathfrak{g}$ acting on $1 \otimes e^{\alpha}$, $\alpha \in \Delta_R$. Then there is an epimorphism of representations of $\mathfrak{g}$,*

$$p: A \to \mathfrak{g}, \tag{4.10}$$

preserving the weight subspaces.

PROOF. Let $\{\alpha_i\}_{i=1}^{l}$ be a root basis of Δ_R. Any root vector of $\mathfrak{g} \setminus \mathfrak{h}$ is of the form

$$\text{ad } X_0^{\varepsilon}(\alpha_{i_1}) \cdots \text{ad } X_0^{\varepsilon}(\alpha_{i_{n-1}}) \cdot X_0^{\varepsilon}(\alpha_{i_n}) \quad \text{or}$$

$$\text{ad } X_0^{\varepsilon}(-\alpha_{i_1}) \cdots \text{ad } X_0^{\varepsilon}(-\alpha_{i_{n-1}}) \cdot X_0^{\varepsilon}(-\alpha_{i_n}). \tag{4.11}$$

We will consider the following elements of A:

$$X_0^{\varepsilon}(\alpha_{i_1}) \cdots X_0^{\varepsilon}(\alpha_{i_{n-1}}) \cdot 1 \otimes e^{\alpha_{i_n}} \quad \text{or}$$

$$X_0^{\varepsilon}(-\alpha_{i_1}) \cdots X_0^{\varepsilon}(-\alpha_{i_{n-1}}) \cdot 1 \otimes e^{-\alpha_{i_n}}. \tag{4.12}$$

By definition, the map p sends the elements of the form (4.12) to the elements of the form (4.11), respectively. Also we set

$$p: \alpha(-1) \otimes 1 \to \alpha(0), \quad \alpha \in \mathfrak{h}. \tag{4.13}$$

We will prove by induction on n that p is a linear map. This is true for $n = 1$; suppose it is true for $n - 1$. Suppose that for n we have a linear dependence of elements (4.12) (denote it by Y) but the corresponding linear combination of elements (4.11) (denote it by Z) does not vanish. There exists $i \in \{1, \ldots, l\}$ so that $[X_0^{\varepsilon}(-\alpha_i), Z] \neq 0$, because otherwise $\mathfrak{g} \cdot Z$ would be a nontrivial ideal in the simple Lie algebra $\mathfrak{g}$. But $X_0^{\varepsilon}(-\alpha_i) \cdot Y = 0$, which contradicts the induction assumption. Finally the homomorphism property of p follows from simple calculations

$$p: X_0^{\varepsilon}(-\alpha) \cdot 1 \otimes e^{\alpha} = \alpha(-1) \otimes 1 \to \text{ad } X_0^{\varepsilon}(-\alpha) \cdot X_0^{\varepsilon}(\alpha) = \alpha(0),$$

$$p: \left(X_0^{\varepsilon}(-\alpha) \right)^2 \cdot 1 \otimes e^{\alpha} = 2 \cdot 1 \otimes e^{-\alpha}$$

$$\to \left(\text{ad } X_0^{\varepsilon}(-\alpha) \right)^2 \cdot X_0^{\varepsilon}(\alpha) = 2 X_0^{\varepsilon}(-\alpha).$$

The no-ghost theorem and Proposition 4.4 allow us to get an upper bound for root multiplicities of hyperbolic algebras of rank 26.

PROPOSITION 4.5. *Let $\mathfrak{g}$ be a hyperbolic algebra of rank 26 and $\alpha \in \Delta$. Then*

$$\dim \mathfrak{g}_{\alpha} \leqslant \mu(-\langle \alpha, \alpha \rangle / 2) \tag{4.14}$$

where

$$\sum_{n=-1}^{\infty} \mu(n) q^n = \frac{1}{q \prod_{n=1}^{\infty} (1 - q^n)^{24}}. \tag{4.15}$$

PROOF. Let us consider the null radical $\{0\}$ of the form $\langle , \rangle$ in V_1. Clearly it is a subrepresentation of $\mathfrak{g}$. We denote by $\overline{V}_1$ the factor representation $V_1/\{0\}$ of $\mathfrak{g}$.

We will prove that for $\alpha \neq 0$,

$$\dim(\overline{V}_1)_\alpha = \mu(-\langle \alpha, \alpha \rangle/2). \tag{4.16}$$

Lemma 4.2 and the no-ghost theorem imply that $(V_1)_\alpha \approx S(\alpha)_M$ and $(\overline{V}_1)_\alpha \approx T(\alpha, c)_M$, where $\langle \alpha, \alpha \rangle = 2(1 - M)$. Let us define $p^{(l)}(n)$ by

$$\sum_{n=0}^{\infty} p^{(l)}(n)q^n = \varphi(q)^{-l}. \tag{4.17}$$

We will show by induction that

$$\dim T(\alpha, C)_N = p^{(24)}(N). \tag{4.18}$$

It follows from the definition of $S(\alpha)$ that $\dim S(\alpha)_N = p^{(26)}(N)$. Also we deduce from Lemma 4.2 that $\dim G(\alpha, c)_N = \sum_{M<N} \dim T(\alpha, c)_M$. $p^{(2)}(N - M) = p^{(26)}(N) - p^{(24)}(N)$ by induction (obviously $T(\alpha, c)_0 = 1$). This implies that $\dim T(\alpha, c)_N = p^{(24)}(N)$.

Now we consider the epimorphism $p: A \to \mathfrak{g}$ from Proposition 4.4. The subrepresentation $A \cap \{0\}$ is mapped to zero because $\mathfrak{g}$ does not contain nontrivial ideals. Thus we have an epimorphism $\overline{p}: \overline{A} = A/A \cap \{0\} \to \mathfrak{g}$. Thus $\dim \mathfrak{g}_\alpha \leq \dim \overline{A}_\alpha \leq \dim(\overline{V}_1)_\alpha = \mu(-\langle \alpha, \alpha \rangle/2)$. Q.E.D.

Proposition 4.5 and the results of [5] allow us to conjecture that for every hyperbolic algebra $\mathfrak{g}$ of rank l one has

$$\dim \mathfrak{g}_\alpha \leq p^{(l-2)}(-\langle \alpha, \alpha \rangle/2 + 1). \tag{4.19}$$

Moreover, the upper bound (4.16) is the best possible (when the function depends only on $\langle \alpha, \alpha \rangle$). In fact, Theorem 1.1 and Proposition 1.3 imply that for $\alpha \in \Delta_R$, isotropic $c \in \Delta$, $\langle c, \alpha \rangle = 1$,

$$\dim \mathfrak{g}_{\alpha-nc} = p^{(l-2)}(n) = p^{(l-2)}\left(-\frac{\|\alpha - nc\|^2}{2} + 1\right). \tag{4.20}$$

The no-ghost theorem distinctly displays the critical dimension 26 for dual resonance models. It is amazing that the same critical number appears in a theory which seemed to be absolutely remote from any physics, namely, the theory of finite simple groups. Recently Conway and collaborators [2] defined one infinite-dimensional Lie algebra, which they called the Monster Lie algebra, with the hope that it would "explain" the biggest sporadic group F_1 of Fisher and Griess. We will give another definition of this algebra based on the vertex representation. Let $\mathfrak{h}_\mathbf{R} \approx \mathbf{R}^{25,1}$; then there is a unique even unimodular lattice Q in $\mathfrak{h}_\mathbf{R}$. We set $\Delta_R = \{\alpha \in Q: \langle \alpha, \alpha \rangle = 2\}$. Then we can construct the space V as in subsection 2.1. The Monster Lie algebra $\mathfrak{g}$ is by definition the Lie algebra generated by $\{X_0^\varepsilon(\alpha), \alpha \in \Delta_R\}$. Clearly $\mathfrak{g}$ contains all the hyperbolic algebras of rank 26.

Let us fix a light-cone element $c \in \Delta$ such that there are no real roots orthogonal to it. Such a vector exists and the set $L = \{\alpha \in \Delta_R: \langle \alpha, c \rangle = 1\}$ is isomorphic to the unique even unimodular lattice of rank 24, which does not contain elements of length $\sqrt{2}$ [2]. We denote by $\overline{V}_{1,c}$ the space $\Sigma_{\alpha \in L} \overline{V}_{1,c,\alpha}$. Then the character of $\overline{V}_{1,c}$ is

$$j(q) = \theta_L(q)/\eta(q)^{24} = q^{-1} + 24 + 196884q + \cdots . \quad (4.21)$$

It was noticed by McKay that the number 196884 exceeds by only one the dimension of the minimal representation of F_1. Conway and Norton [3] conjectured that there is a natural graded representation of F_1 with the character (4.21) minus 24. First Garland [12] and Kac [17] independently tried to construct F_1 in a space isomorphic to $\overline{V}_{1,c}$. The first problem was to obtain a representation of one important subgroup $C = 2_+^{24+1} \cdot (\cdot 0)/\pm 1$, where $\cdot 0$ is the automorphism group of the Leech lattice. It is easy to construct another group $C' = 2^{24} \cdot (\cdot 0)$ $(= (2_+^{24+1}/\pm 1) \cdot (\cdot 0))$. Using one observation of Griess, Kac [18] succeeded in passing from C' to C. The last question is: Where is the whole group F_1? Recently, important progress has been made in answer to this question [10]. Turning again to the dual resonance models gives a hint as to the answer. Physicists know that in the continuous version of $\overline{V}_{1,c}$ the obvious action of the group $O(24)$ can be extended to the bigger group $O(25)$. This extension becomes apparent only if we return to the bigger space $\overline{V}_1$. Whether this unusual phenomenon corresponds to the extension of C to F_1 will become clear in the future.

ACKNOWLEDGMENTS. I am grateful to A. Feingold, J. Lepowsky and A. Meurman for stimulating discussions. I thank A. Feingold for reading the manuscript and A. Rocha-Caridi for proving the exactness of (3.38).

ADDED IN PROOF. I have learned from P. Goddard and D. Olive that they independently discovered the vertex representation of the Lie algebra associated with an even unimodular lattice (see Section 2). I am grateful to P. Goddard for correction of Proposition 4.4.

The representation of F_1 in the graded space with the character $j(q)-24$ has been recently constructed by J. Lepowsky, A. Meurman and myself. However, the relation of this construction to the dual resonance model in the critical dimension 26 is still waiting for its explanation.

REFERENCES

1. I. N. Bernstein, I. M. Gelfand and S. I. Gelfand, *Differential operators on the base affine space and a study of g-modules, Lie groups and their representations*, Proc. Summer School on Group Representations (I. M. Gelfand, Ed.), Bolyai János Math. Soc., Wiley, New York, 1975, pp. 39–64.

2. J. H. Conway, L. Queen and N. J. A. Sloane, *A Monster Lie algebra?* (to appear).

3. J. H. Conway and S. P. Norton, *Monstrous moonshine*, Bull. London Math. Soc. **11** (1979), 308–339.

4. E. Del Guidice, P. DiVecchia and S. Fubini, *General properties of the dual resonance model*, Ann. Physics **70** (1972), 378.

5. A. J. Feingold and I. B. Frenkel, *A hyperbolic Kac-Moody algebra and the theory of Siegel modular forms of genus 2*, Math. Ann. **263** (1983), 82–144.

6. A. J. Feingold, and J. Lepowsky, *The Weyl-Kac character formula and power series identities*, Adv. in Math. **29** (1978), 271–309.

7. I. B. Frenkel, *Two constructions of affine Lie algebra representations and boson-fermion correspondence in quantum field theory*, J. Funct. Anal. **44** (1981), 259–327.

8. _____, *Representations of affine Lie algebras, Hecke modular forms and Korteweg-de Vries type equations* (Proc. 1981 Rutgers Conf. Lie Algebras and Related Topics), Lecture Notes in Math., vol. 933, Springer-Verlag, Berlin and New York, 1982, pp. 71–110.

9. I. B. Frenkel and V. G. Kac, *Basic representations of affine Lie algebras and dual resonance models*, Invent. Math. **62** (1980), 23–66.

10. I. B. Frenkel, J. Lepowsky and A. Meurman, *An E_8-approach to F_1*, Proc. 1982 Montreal Conf. on Finite Group Theory (to appear).

11. O. Gabber and V. G. Kac, *On defining relations of certain infinite-dimensional Lie algebras*, Bull. Amer. Math. Soc. (N.S.) **5** (1981), 185–189.

12. H. Garland, *Lectures on loop algebras and Leech lattice*, Yale University, Spring 1980.

13. V. G. Kac, *Simple irreducible graded Lie algebras of finite growth*, Math. USSR-Izv. **2** (1968), 1271–1311.

14. _____, *Infinite-dimensional Lie algebras and Dedekind's η-function*, J. Funct. Anal. Appl. **8** (1974), 68–70.

15. _____, *Infinite-dimensional algebras, Dedekind's η-function, classical Möbius function and the very strange formula*, Adv. in Math. **30** (1978), 85–136.

16. _____, *Contravariant form for infinite dimensional Lie algebras and superalgebras*, Lecture Notes in Physics, vol. 94, Springer-Verlag, Berlin and New York, 1979, pp. 441–445.

17. _____, *An elucidation of "Infinite-dimensional algebras...and the very strange formula." $E_8^{(1)}$ and the cube root of the modular invariant j*, Adv. in Math. **35** (1980), 264–273.

18. _____, *A remark on the Conway-Norton conjecture about the "Monster" simple group*, Proc. Nat. Acad. Sci. U.S.A. **77** (1980), 5048–5049.

19. V. G. Kac, D. A. Kazhdan, J. Lepowsky and R. L. Wilson, *Realization of the basic representations of the Euclidean Lie algebras*, Adv. in Math. **42** (1981), 83–112.

20. J. Lepowsky, *Application of the numerator formula to k-rowed plane partitions*, Adv. in Math. **35** (1980), 179–194.

21. J. Lepowsky and R. L. Wilson, *Construction of the affine Lie algebra $A_1^{(1)}$*, Comm. Math. Phys. **62** (1978), 43–63.

22. _____, *A Lie theoretic interpretation and proof of the Rogers-Ramanujan identities*, Adv. in Math. **43** (1982).

23. _____, *The structure of standard modules. I. Universal algebras and the Rogers-Ramanujan identities* (to appear).

24. R. V. Moody, *A new class of Lie algebras*, J. Algebra **10** (1968), 211–230.

25. J. Scherk, *An introduction to the theory of dual models and strings*, Rev. Modern Physics **47** (1975), 123–164.

26. J. H. Schwarz, *Dual-resonance theory*, Phys. Rep. **8** (1973), 269–335.

27. _____, *Mathematical issues in superstring theory*, these PROCEEDINGS.

28. G. Segal, *Unitary representations of some infinite dimensional groups*, Comm. Math. Phys. **80** (1981), 301–342.

DEPARTMENT OF MATHEMATICS, YALE UNIVERSITY, NEW HAVEN, CONNECTICUT 06520

SCHOOL OF MATHEMATICS, INSTITUTE FOR ADVANCED STUDY, PRINCETON, NEW JERSEY 08540

Current address: Department of Mathematics, Rutgers University, New Brunswick, New Jersey 08903

Lectures in Applied Mathematics
Volume **21**, 1985

Kac-Moody Symmetry of Gravitation and Supergravity Theories

Bernard Julia

ABSTRACT. The equations of motion for ($D = 4$) gravitational plane waves $\{g_{\mu\nu}(t, x^1)\}$, $\mu, \nu = 0, 1, 2, 3$, and more generally the sets of classical solutions of supergravities $N' = 1, 2, \ldots, 7$, that depend only on 2 of the 4 coordinates, are invariant under transformations of Kac-Moody (affine type 1) algebras. These invariances are merely verified by explicit computation; however, they form a regular pattern, and the table of "internal" symmetries of these supergravities in various dimensions suggests a deeper and wider role for Kac-Moody algebras. We shall provide as many definitions as possible and review the "dimensional reduction" from the dual string model of Ramond, Neveu and Schwarz ($d = 10$) to plane waves with effective dimension $d = 2$. In this process, the algebra of symmetries that preserve the effective coordinates (internal symmetries) grows regularly by increasing its real rank. The same property can be observed in some real forms of the magic square of Freudenthal and Tits. The connection with completely integrable systems and, in particular, nonlocal charges will be briefly mentioned. Finally we shall describe the converse of dimensional reduction called "group disintegration" (increasing space-time dimension) and make a few comments about relevant superalgebras.

I. Dual string models and local field theories. It is not only of pedagogical interest to recall the connections between the dual resonance models and local finite component field theories. In fact, the two maximal supersymmetric theories: $N = 4$ supersymmetric Yang-Mills theory [1] and $N = 8$ ($N' = 7$) supergravity [1, 2] are low energy limits of the open (resp. closed) fermionic string model of Ramond and Neveu-Schwarz [3] (R.N.S. model). One describes the dynamics of such a single free string by a generalization of the massive 1-particle action $S = m \int ds$. The new

---

1980 *Mathematics Subject Classification*. Primary 81G20, 22E65; Secondary 20F05.

action reduces to the area spanned by an open (resp. closed) string when the spinorial coordinates $\chi^{M\varepsilon}$ vanish (hence the name "dual string model") and it can be written as

$$
S = \frac{1}{4\pi\alpha'} \int \int d\sigma \, d\tau \sqrt{\det g_{ij}} \, \eta_{MN} \left[\partial_i Y^M \partial_j Y^N g^{ij} + i\bar{\chi}^M \gamma^i \partial_i \chi^N \right.
$$

$$
\left. + \bar{\psi}_i \gamma^j \gamma^i \chi^M \partial_j Y^N - \frac{1}{2} \bar{\chi}^M \chi^N \bar{\psi}_i \gamma^j \gamma^i \psi_j \right]
$$

$$
(i = \sigma, \tau). \quad (1)
$$

$g^{ij}(\sigma, \tau)$ and $\psi_i^\varepsilon(\sigma, \tau)$ ($\varepsilon = 1, 2$ is a spinorial index on the string) appear only algebraically and are auxiliary fields. The important fields of the classical theory are the coordinates of the string $Y^M(\sigma, \tau)$, $\sigma \in [0, \pi]$ (resp. periodic), and the spinorial coordinates $\chi^{M\varepsilon}(\sigma, \tau)$, $M = 0, 1, \ldots, 9$ ($\varepsilon = 1, 2$). One may also associate to each open string a matrix degree of freedom λ in the Lie algebra of a compact Lie group G [3] which is conserved in the absence of interactions.

The interaction of strings can be described (after first quantization of the one string theory) by addition of nonlinear terms to the Schrödinger-like equation satisfied by the string wave-functional. The most complete results have been obtained in the "light cone gauge": the Lagrangian (1) leads to a constrained Hamiltonian system because of the (σ, τ) reparametrization (gauge) invariance of the surface $Y^M(\sigma, \tau)$. The choice of isothermal coordinates (σ, τ) with $\tau = Y^+ = y^+$ ($Y^\pm = 2^{-1/2}(Y^0 \pm Y^1)$) leads to a wave-functional $\Psi[y^+, y^-, Y^\perp(\sigma)]$ where $Y^\perp(\sigma)$ are the 8 residual coordinates, $Y^-(\sigma)$ can be expressed in terms of $y^- = C'$ and the other dynamical variables; we have ignored the fermionic coordinates for simplicity. The dual vertex operators describe the splitting and joining of strings; the nonlinear terms of the first quantized open string action are schematically $g\Psi^3$ and $g^2\Psi^4$ [4].

The vertex V_0 for the emission of a string in its lowest (spinless) excited state has been used recently in the so-called "homogeneous" construction of the basic representation of some affine Kac-Moody algebras $\mathcal{G}^{(1)}$ [5]. For completeness we must mention that for $\mathcal{A}_8^{(1)}$, $\mathcal{D}_8^{(1)}$ and $\mathcal{E}_8^{(1)}$ the space of this (simplest) representation is very closely related to a subset of the bosonic open string states of the R.N.S. model: one should ignore the fermionic coordinates, fix k_+ and k_- (the 2 longitudinal components of total momentum), and consider only the Fock space of transverse oscillators as well as discrete values of the transverse total momentum corresponding to the root lattice of $\mathcal{G}$. Let us note also that the "Spectrum Generating Algebra" [6] (constructed with the next vertex operator V_1^M [3c]) can be seen as a subalgebra of the semidirect product of the diffeomorphism group of the circle (Virasoro algebra) by the algebra $\mathcal{G}^{(1)}$

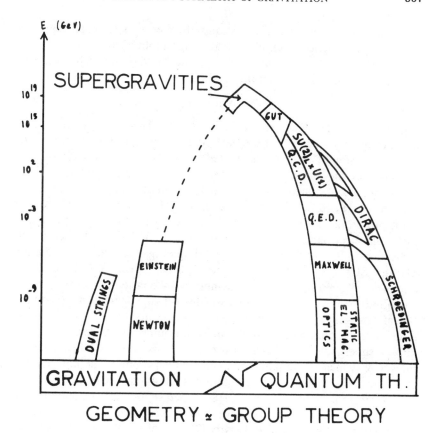

FIGURE 1. A theoretical cathedral.

which was introduced in [7]. This paragraph was meant to stimulate interactions between physicists familiar with [3 and 6] and mathematicians interested in [5 and 7], but it might also encourage the reader to follow our next arguments because he will find $\mathcal{E}_8^{(1)}$ realized completely differently in another sector of the closed string R.N.S. model [32].

Let us now present, rather schematically, the status of theoretical physics and motivate our work on the missing blocks of Figure 1. In 50 years nobody has been able to reconcile Einstein's General Relativity and quantum mechanics. This problem led to a separate evolution of quantum field theory to describe electrons and photons (quantum electrodynamics), later including weak interactions (Glashow-Salam-Weinberg: $SU(2)_L \times U(1)$ model). Most people believe that Yang-Mills theory with "color" gauge group SU(3) coupled to quarks (Q.C.D.) describes not only high energy phenomena (parton model, asymptotic freedom) but also low energy strong (nuclear) interactions; the latter still defy analytic computations but the former also require further work. In particular, the high energy behavior (Regge) of scattering amplitudes and the experimental observation that bound states of quarks lie on linear trajectories, i.e. that their angular momentum satisfies

$$J = \alpha' m^2 + \alpha_0 \tag{2}$$

suggest that a stringlike structure is present. This string is the color electric flux tube between the quark and the antiquark of a meson. A singular version of this tube is created by the Wilson loop operator and the study of the dynamics of this loop reveals the analogy with a dual string model in 4 dimensions (not 10). The numerical value of α' (it is the same α' in (1) and (2)) is approximately 1 $(GeV)^{-2}$, a number of order 1 if the proton mass is set equal to 1. The next idea is to unify Q.C.D. and weak and electromagnetic interactions in a "Grand Unified Theory" of the Yang-Mills type again with a (simple) gauge group containing $SU(3) \times SU(2)_L \times U(1)$. At each step Q.E.D., $SU(2)_L \times U(1)$, G.U.T., one may or may not enlarge the symmetry by considering supersymmetries [8]. It is, however, mandatory to consider supersymmetries (actually "local" or gauge supersymmetries) if one wants to *truly unify* all 4 interactions (including also gravitation) in a universal theory. Today we know 7 supergravities in 4 space-time dimensions; they are labelled by an integer N' which is the naïve number of supersymmetry spinor generators. For $N' = 7$ there is an eighth set of 4 odd generators in the superalgebra of global (or "rigid" [8]) supersymmetries; the spinors are sO(3, 1) spinors. These supergravities have been conceived as generalizations of Einstein's General Relativity [9]; one of the motivations was to bridge the 50-year-old gap. Both the unification of forces and the consistency of quantum supergravity (especially $N' = 7$) look promising but lack solid confirmation.

However, in the process of constructing the $N' = 7$ supergravity Lagrangian, the authors of [2] stumbled across a mysterious E_7 $(+7)$ symmetry group; it is a group acting on the set of classical solutions of the Euler-Lagrange equations by pointwise action on the fields $g_{\mu\nu}(x^\mu)$, etc. and some dual fields (pseudo-potentials). A systematic study of these groups was undertaken in [10 and 11] (see also [12]). The most exciting feature is the appearance of rules relating symmetries for various N' and various numbers of effective dimensions (to be defined below). In particular, this allowed me to recognize $A_1^{(1)}$ as a symmetry of gravitational plane waves [11]: supergravity theory can be useful to its mother, Einstein's theory, which happens to be a realistic and immediately relevant physical theory. It can also be useful in the study of completely integrable systems [11, 13] (and below).

We shall now describe the long march from the intrinsically 10-dimensional R.N.S. model to a local field theory in 4 dimensions namely the seventh supergravity ($N = 8$). The first step is a limiting process: the small slope limit [14]. We have at our disposal 2 parameters α' which will be taken very small in the above units and g the string coupling strength. If α' tends to zero and α_0 stays fixed in (2), J is half integer only for $m^2 = 0$ or $m^2 \to +\infty$; in [1] the lowest excitation ($J = 0$, $m^2 < 0$) is excluded by a clever truncation of the Hilbert space. If we consider only "tree graphs", i.e. if we forbid the appearance of holes in the surface spanned by the collection of strings in interaction (it corresponds to the classical approximation to the S matrix), then the S matrix for the zero mass string states can be obtained by an alternative action. This action is that of a classical field theory in 10 dimensions which is local (involves a finite number of derivatives of the fields) to each order in α'. Neglecting the high mass string states is a low energy approximation so we may as well keep only the terms with no more than 2 derivatives in this Lagrangian. The fields generated in this limit by closed string states are: a metric $g_{MN}(x^M)$ [14], an antisymmetric generalized gauge field $A_{[MN]}(x)$, and a scalar $\varphi(x)$; the generalized gauge invariance is naturally $A'_{MN} = A_{MN} + \partial_M \Lambda_N - \partial_N \Lambda_M$. Here we ignored the fermionic $\chi^{M\varepsilon}(\sigma, \tau)$; the dimension of space-time in this case must be 26. We are mostly interested in the 10-dimensional R.N.S. model; its "bosonic" closed strings are described at low energy by the above 3 fields [1] plus a vector gauge field $A_M(x)$ and a generalized gauge field $A_{MNP}(x)$ [15]; the fermionic field states (in the sense of 10-dimensional field theory) are $\psi_M^A(x)$, $\chi^A(x)$ a (gauge) vector spinor field and a spinor field; the spinors of sO(1, 9) have 32 components ($A = 1, \ldots, 32$). General arguments show that the reparametrization invariance of the string implies "gauge invariance of the S matrix": for a graviton of momentum k_M this means that a polarization tensor h_{MN} is equivalent to a polarization $h_{MN} + k_M \varepsilon_N + k_N \varepsilon_M$ ($g_{MN} = \eta_{MN} + G_{10}^{1/2} h_{MN} +$ higher order). By further arguments originally due to R. P. Feynman, one can derive from this property of the S-matrix the coordinate reparametrization invariance of the 10-dimensional local Lagrangian field theory. It is an urgent but open question to deduce directly the last invariance from the first. The simplest example might be the Yang-Mills gauge invariance of low energy open string states; the most complicated application would be to derive from the 2-*dimensional* "local" supersymmetry of (1) and the careful choice of Hilbert space of [1] the 10-*dimensional* "local" supersymmetry of the Lagrangian describing the interaction of the zero mass closed string states listed above. We shall conclude this section by giving the relation between the Newton constant of gravitation G_{10} appearing in the new effective action and our original parameters α' and g:

$$S^{\text{eff}} = \int - \frac{R}{16\pi G_{10}} \sqrt{\det g_{MN}} \, dx^{10} + \text{other terms} \qquad (3)$$

with $G_{10} \sim (g\alpha')^4$ up to a numerical constant. The appearance of the Riemann scalar curvature R follows from covariance and from the exclusion of higher derivative terms [14]. The full action S^{eff} was also constructed directly by assuming "supercovariance" [16]. If one takes for g some value of order 1 one sees that $\sqrt{\alpha'} \sim G_{10}^{1/8}$; the typical length scale of this new type of string can be related to the known strength of gravitation G_4 in 4 dimensions when G_{10} is related to G_4 as in the next section. The problem of quantum corrections (loops) is discussed in [3].

II. Dimensional reduction. We have succeeded in reducing our infinite component nonlocal field theory to a local theory with a finite number of fields: 128 bosonic oscillators and 128 fermionic oscillators for each value of the 10-dimensional momentum. Unfortunately, we are presently aware of only 1 time- and 3 space-dimensions. The second step in our long march is to reduce the dimension of a local field theory from 10 to 4.

(a) *The principle of symmetric criticality.* We shall discuss primarily the simplest version of dimensional reduction, namely the appearance of an abelian algebra of r Killing vector fields corresponding to ignorable coordinates. But we shall begin with a warning: the unwritten rule that all symmetric solutions of some Euler-Lagrange equations can be obtained by extremizing the symmetric action is not always valid [17]. This procedure amounts to taking a general action (invariant under a group G) to replace the "field" variables by the most general symmetrical Ansatz and to finally extremize the resulting functional. It has been justified [17] for a smooth action: $G \times \mathcal{F} \to \mathcal{F}$ of a compact Lie group on the smooth space $\mathcal{F}$ of field variables. We can place ourselves in such a situation by assuming the topology of the space-time manifold M_{10} to be $M_{10-r} \times T_r$ where T_r is the r-dimensional torus. If one considers axially symmetric solutions one does not even need to compactify some directions, the group $U(1)$ is already compact. Returning to the first case, we can identify G and T_r and the image of a field $f(x^\mu, x^i)$, $\mu = 0, 1, \ldots,$ $(9 - r)$, $i = 10 - r, \ldots, 9$, is the translated field $f(x^\mu, x^i - a^i)$. In fact the Euler-Lagrange equations are local in M_{10} and they are not affected by global topological properties; boundary conditions or asymptotic behavior of the fields on the other hand are modified.

(b) 11-*dimensional supergravity.* The conjecture of [1, 15 and 16] was that for $r = 6$ the symmetric action obtained from S^{eff} in (3) by dimensional reduction would be the action of the seventh supergravity (also called $N = 8$ supergravity) in dimension 4 [18]. It was motivated by two facts: firstly, the coincidence of the sets of oscillators of the linearized theories including their spins, and secondly, the remarkable identity $8 \times 4 = 1 \times 32$. More precisely, one notes the coincidence of the global supersymmetry algebra ($N = 8$ in 4 dimensions):

$$\{Q_\alpha^k, \overline{Q}_l^\beta\}_+ = (\gamma^\mu)_\alpha^\beta \delta_l^k P_\mu, \qquad (4)$$

$k, l = 1,\ldots,8$; $\alpha, \beta = 1,\ldots,4$, Majorana (real) spinor indices; $\overline{Q} \equiv Q^+ \gamma^0$ and γ^μ the 4 Dirac matrices with the superalgebra in dimension 10 whose 32 odd generators form a real spinor of sO(9, 1) and satisfy

$$\{Q_A, \overline{Q}^B\} = (\Gamma^M)_A^B P_M, \qquad A, B = 1,\ldots,32, M = 0,\ldots,9. \qquad (5)$$

This coincidence holds for a special choice of the first 4 Γ^M matrices. These matrices act on the tensor product space of the vector representation of SO(8) by the spinor representation of sO(3, 1) with indices $A = (k, \alpha)$. One must also assume cyclicity of $x^4, x^5,\ldots,x^9$ or equivalently $P_4 = P_5 = \cdots = P_9 = 0$: the vanishing of the translation generators. In fact, previous experience in supergravity theory suggested the existence of a global (called rigid in [8]) SU(8) invariance of the equations of motion of the 4-dimensional theory. The real spinors Q_α^k can be projected onto the irreducible (chiral or Weyl) representations of sO(3, 1): $Q_\alpha^k = Q_{\text{Right}}^k + Q_{k\,\text{Left}}$; the components Q_R, Q_L transform as 8 and $\overline{8}$ of SU(8). The whole game will be to destroy the space-time diffeomorphism group of symmetries to build up an "internal" symmetry group, i.e. a group acting at each space-time point on the various components of the fields. The internal symmetry group for $r = 6$ (in dimension 4) must have grown up to include SU(8).

It turned out that the 10-dimensional theory was much easier to construct as a dimensional reduction of supergravity in 11 dimensions than directly in 10 dimensions. This suggested the existence of a new closed string dual model in 11 dimensions; in fact, it is very likely that a dual model with manifest supersymmetry in the embedding space ($\mathbf{R}^{10}$ here) will have different (or maybe no) critical dimension below which it has no ghosts (i.e. particles with negative probability). Ignoring these interesting questions we went ahead and constructed the unique supergravity with 256 degrees of freedom in 11 dimensions [16]. Its Bose field content is a gauge 3-form A_{MNP} plus a metric field g_{MN}, $M, N = 0,\ldots,10$, that contains 3 10-dimensional tensors $g_{M'N'}, g_{M'10}, g_{10\,10}$. The generalized gauge field splits into 2 gauge fields in 10 dimensions: $A_{M'N'P'}$ and $A_{M'N'10}$. We recognize the fields predicted by the string model experts, the fermionic fields, behave analogously. The construction of the Lagrangian was not much more difficult than in [9] except for the proliferation of spinor indices. It led to the remarkable bosonic terms:

$$L_{11}^{\text{Bose}} = -\frac{1}{4}\sqrt{g}\,R - \frac{1}{48}\sqrt{g}\,F_{MNPQ}F^{MNPQ}$$

$$+ \frac{2\sqrt{g}}{(12)^4}\varepsilon^{PQ\cdots Z}F_{PQRS}F_{TUVW}A_{XYZ}. \tag{6}$$

The last term was imposed by local supersymmetry invariance, it transforms by a total derivative under the transformation $A_{MNP} \to A_{MNP} + \partial_{[M}\Lambda_{NP]}(F_{MNPQ} = 4\partial_{[M}A_{NPQ]})$; it suggests a 12-dimensional formulation of the theory and a special role for $(1 + 2)$-dimensional bubbles. Let us, however, start our descent: as it already appears from the correspondence between 11 and 10 dimensions, the gauge fields generate scalar fields and gauge fields with fewer tensor indices upon dimensional reduction. In the sequel we shall ignore the fermionic fields and the supersymmetries except for occasional remarks.

The 11-dimensional diffeomorphism group gets broken down to the d-dimensional diffeomorphism group. The internal (bosonic) symmetries I_d after reduction of 11-d spatial dimensions are given in column $N' = 7$ of Table 1. They are still conjectural for $d = 2$ and 1. In dimensions 8, 6, 4, 2 they are only symmetries of the equations of motion. In the odd dimensions 7, 5 (resp. 3) the equations of motion can be formulated using "dual" potentials $A_{MNP} \to B_{MN}$, $A_{MNP} \to B$ and $A_{MN} \to B_M$ (resp. $A_M \to B$); the dual equations are the Euler-Lagrange equations for an action that is invariant under the groups listed in Table 1. Let us try to understand these groups and the discrete "dualities" one uses to exhibit them. The latter are discussed in greater generality in [12]; here we shall only treat the case $d = 3$.

(c) *The growth of the internal symmetry group.* In space-time dimension d we called the internal symmetry group I_d. We may distinguish two mechanisms for the growth of I_d. Let us first consider linear (constant coefficients) changes of coordinates in the internal space which do not introduce in the fields any dependence on these internal coordinates: they act locally as the isotropy group $GL(r, \mathbf{R})$ of the tangent space to T_r. Upon dimensional reduction an action such as (3) becomes

$$S^{\text{eff}} = \int_{T_r} \Pi\, dx^i\, S_{\text{sym}}^{\text{eff}} = \int \Pi\, dx^i \int \Pi\, dx^\mu \sqrt{\det g_{\mu\nu}}\,\mathcal{L}_{(\varphi)}^{\text{sym}}. \tag{7}$$

The factor coming from the integral of the internal space volume element can be assumed to be constant as a function of x^μ by a suitable choice of internal coordinates. Clearly S^{eff} is invariant under $GL(r)$ but $S_{\text{sym}}^{\text{eff}}$ will only be invariant under the subgroup $SL(r)$ that preserves the volume element; other invariances must be explained otherwise. (Clearly the

TABLE 1. The disintegration triangle.

d \ N'	7	6	5	4	3	2	1	0
						l_G		x_G
								r_G
						$G/K \begin{vmatrix}\dim G\\\dim K\end{vmatrix}$		
						$(G = I_d)$		
11	$g_{MN},\, A_{MNP}\ {}^0_0$							
10	$\mathbf{R}\begin{vmatrix}1\\0\end{vmatrix}{}^1_1$							
9	$A_1 \times \mathbf{R}/U(1)\begin{vmatrix}4\\1\end{vmatrix}{}^2_2$							
8	$(A_2 \times A_1)/(A_1 \times U(1))\begin{vmatrix}11\\4\end{vmatrix}{}^3_3$							
7	$A_4/B_2\begin{vmatrix}24\\10\end{vmatrix}{}^4_4$							
6	$D_5/B_2\begin{vmatrix}45\\20\end{vmatrix}{}^5_5$	$(D_3 \times A_1)/(B_2 \times A_1)\begin{vmatrix}18\\13\end{vmatrix}{}^{-8}_{\ 1}$						
5	$E_6/C_4\begin{vmatrix}78\\26\end{vmatrix}{}^6_6$	$A_5^*/C_3\begin{vmatrix}35\\21\end{vmatrix}{}^{-7}_{\ 2}$						
4	$E_7/A_7\begin{vmatrix}133\\63\end{vmatrix}{}^7_7$	$D_6^*/U(6)\begin{vmatrix}66\\36\end{vmatrix}{}^{-6}_{\ 3}$	$A_5/U(5)\begin{vmatrix}35\\25\end{vmatrix}{}^{-15}_{\ \ 1}$	$(A_3 \times A_1)/(A_3 \times U(1))\begin{vmatrix}18\\16\end{vmatrix}{}^{-14}_{\ \ 1}$	$U(3)/U(3)\begin{vmatrix}9\\9\end{vmatrix}{}^{-9}_{\ 0}$	$U(2)/U(2)\begin{vmatrix}4\\4\end{vmatrix}{}^{-4}_{\ 0}$	$U(1)/U(1)\begin{vmatrix}1\\1\end{vmatrix}{}^{-1}_{\ 0}$	$g_{\mu\nu}\begin{vmatrix}0\\0\end{vmatrix}{}^0_0$
3	$E_8/D_8\begin{vmatrix}248\\120\end{vmatrix}{}^8_8$	$E_7/(D_6 \times A_1)\begin{vmatrix}133\\69\end{vmatrix}{}^{-5}_{\ 4}$	$E_6/(D_5 \times U(1))\begin{vmatrix}78\\46\end{vmatrix}{}^{-14}_{\ \ 2}$	$D_5/(D_4 \times U(1))\begin{vmatrix}45\\29\end{vmatrix}{}^{-13}_{\ \ 2}$	$A_4/(D_3 \times U(1))\begin{vmatrix}29\\16\end{vmatrix}{}^{-8}_{\ 1}$	$(A_2 \times A_1)/(S(U(2) \times U(1)))\begin{vmatrix}11\\7\end{vmatrix}{}^{-3}_{\ 1}$	$(A_1 \times U(1))/U(1)^2\begin{vmatrix}4\\2\end{vmatrix}{}^0_1$	$A_1/U(1)\begin{vmatrix}3\\1\end{vmatrix}{}^1_1$
2	$E_8^{(1)}$	$E_7^{(1)}$	$E_6^{(1)}$	$D_5^{(1)}$	$A_4^{(1)}$	$A_2^{(1)} \times A_1 \times \ldots$	$A_1^{(1)} \times U(1) \times \ldots$	$A_1^{(1)} \times \mathbf{R} \ldots$

whole discussion of dimensional reduction ought to be formulated in the language of foliations; a first step in this direction was [19].)

The second mechanism is more subtle and still under active investigation [20]. Let us describe it in a simple case which is completely understood. A naïve observer does not see any invariance of the free Maxwell equations beyond Lorentz covariance. After dimensional reduction from 4 to 3 dimensions the situation worsens, the initial Lagrangian has become

$$L_{(3)} = -\frac{1}{4}F_{\mu\nu}^2 + \frac{1}{2}\partial_\mu A_3^2 \quad \text{(signature } + - - -\text{)}, \tag{8}$$

$F_{\mu\nu} = \partial_\mu A_\nu - \partial_\nu A_\mu$ (we chose the Killing vector spacelike; as long as it is not a null vector very little depends on this). Suppose that a 3-dimensional observer is brighter and that he notices that the classical Maxwell equations are obtained equivalently from

$$L'_{(3)} = \frac{1}{2}\left(\partial_\mu B^2 + \partial_\mu A_3^2\right). \tag{9}$$

B is defined by $\partial_\mu B = \frac{1}{2}\varepsilon_\mu^{\;\;\nu\rho} F_{\nu\rho}$ (a compatible system for solutions): it is a Bäcklund transformation. $L'_{(3)}$ is now obviously symmetric under a 3-parameter group

$$B \to B + C', \quad A_3 \to A_3 + C' \quad \text{and} \quad (A_3 + iB) \to e^{i\theta}(A_3 + iB). \tag{10}$$

This group is an invariance of $L'_{(3)}$ and a fortiori of its equations of motion, hence it is a symmetry of the equations of motion of $L_{(3)}$ (they are the same). There is a canonical way to go from (8) to (9) and back [2, 12]. The main difference between (9) and (8) is that the SO(2) symmetry exchanges 2 fields in the first case whereas it exchanges (essentially) a field F_{ij}, $i, j = 1, 2$, for a (nonconjugate) momentum $\partial_0 A_3$ or conversely in the second case.

In this toy model, the transformations of (10) can be traced back to canonical transformations in 4 *dimensions*. The last two can even be redefined as symmetries of Maxwell's action [21] and they generate Noether currents. If we now replace Maxwell's action by the Hilbert-Einstein action (first term of (3) in 4 space-time dimensions) we are exactly in the same situation. We can reduce the action to 3 effective dimensions, (8) is known, and Papapetrou [22] defined the analogue of B.

The analogue of (9) is quite surprising. It reveals a "hidden" SL(2**R**) symmetry and reads ([**10**] and references therein)

$$L'_3 = -\frac{1}{4}R_3\sqrt{g_3} + \frac{\sqrt{g_3}}{8} g_3^{\mu\nu} \frac{\partial_\mu Z\, \partial_\nu \bar{Z}}{(\mathrm{Im}\,(Z))^2} \tag{11}$$

with $Z = B - ig_{33}$. The SO(2) subgroup can be interpreted as the helicity subgroup in the linear approximation [**10**], and its nonlinear action is used to transform, for example, the Schwarzschild solution into a Taub-NUT metric. It is traditionally called the Ehlers symmetry. One conclusion of [**10**] was that the subset of symmetries we did not understand, for $N' = 7$ and say $d = 4$ supergravity, was connected with dualities. The manipulations involved are exactly of the same type as the above duality (for $N' = 0$), and it is important to clarify (11) before we try to "explain" E_7. Let us note some reason of optimism: the same duality group SL(2**R**) would appear for $N' = 0$ in Table 1 if we chose a timelike Killing vector to reduce from 4 to 3 dimensions; the same thing would be true for $d = 3$ and $N' = 1$ to 7. There is a minor change, however, in the symmetric space that appears in (11) (SL(2**R**)/SO(2)) or its generalization. (I_3/K) was of the noncompact type (K = maximal compact subgroup of I_3) for a spacelike Killing vector $\partial/\partial x^3$. This reflected the absence of ghosts; if we suppress the time direction this last constraint disappears and typically SO($2n$) gets replaced by SO*($2n$) inside a new K^*. We think this persistence of I_3 suggests its existence in 4 dimensions already contrary to previous beliefs.

Another observation is that these internal symmetries do not act on the graviton field $g_3^{\mu\nu}$ and consequently preserve the energy momentum tensor [**12**]; one can expect them to be special canonical transformations [**20, 23**] that preserve the Hamiltonian.

(d) *Some remarks about 4, 2 and 1 dimensions.* We have just discussed the odd-dimensional theories. In dimensions 8, 6, 4 and 2 the above discrete dualities preserve the tensor character of one kind of gauge fields: respectively A_{MNP}, A_{MN}, A_μ and φ. Consequently, there is no action that is invariant under the exchange of $F_{\mu\nu}^{ij}$ for their duals $(G_{\mu\nu[ij]} = \frac{1}{2}\varepsilon_{\mu\nu\rho\sigma}F^{\rho\sigma ij} +$ nonlinear terms) in the 4-dimensional example. Yet the equations of motion and the energy momentum tensor are preserved by continuous dualities that generalize the SO(2) invariance of the Maxwell equations in 4 dimensions. We expect [**20**] that it is possible to redefine these dualities in various non-Lorentz covariant ways as in [**21**] in order to exhibit them as invariances of the action and not only of the equations of motion. To be precise let us now write the equations of motion of $N' = 7$ supergravity in manifestly $E_7(+7)$ invariant form [2, 12]:

$$R_{\mu\nu} - \frac{1}{2} g_{\mu\nu} R = \frac{1}{4} \left(\mathcal{V} \mathcal{F}_{\mu\rho} \right)' g^{\rho\sigma} \left(\mathcal{V} \mathcal{F}_{\sigma\nu} \right) + \frac{1}{24} \mathrm{Tr} \left(D_\mu \mathcal{V} \cdot \mathcal{V}^{-1} D_\nu \mathcal{V} \cdot \mathcal{V}^{-1} \right)$$

$$- \frac{g_{\mu\nu}}{48} \mathrm{Tr} \left(D_\rho \mathcal{V} \cdot \mathcal{V}^{-1} \right)^2,$$

$$\varepsilon^{\mu\nu\rho\sigma} \partial_\nu \mathcal{F}_{\rho\sigma} = 0, \qquad \mathcal{V} \mathcal{F}_{\mu\nu} = \frac{1}{2} \varepsilon_{\mu\nu\rho\sigma} \begin{bmatrix} 0 & 1 \\ -1 & 0 \end{bmatrix} \mathcal{F}^{\rho\sigma},$$

$$D^\mu \left(D_\mu \mathcal{V} \cdot \mathcal{V}^{-1} \right) = -3 \left[\mathcal{V} \mathcal{F}_{\mu\nu} \otimes \left(\mathcal{V} \mathcal{F}^{\mu\nu} \right)^t \right]^\perp, \qquad (12)$$

where

$$F_{\mu\nu}(x) = \begin{pmatrix} \partial_\mu A_\nu^{ij} - \partial_\nu A_\mu^{ij} \\ G_{\mu\nu ij} \end{pmatrix}$$

transforms as the 56-dimensional fundamental representation of E_7; $\mathcal{V}(x)$ is a generic group element of E_7 written in the 56-representation and parametrized (at each space-time point x^μ) by 70 physical fields plus 63 gauge degrees of freedom. $E_7(+7)$ was, in fact, discovered by assuming and then checking the existence of an SU(8) gauge invariance that gauges away 63 components to leave us with the 70 scalar degrees of freedom of $N' = 7$ supergravity. There is a Lagrangian for the scalar fields. It is invariant under the product of a right global action of E_7 and the left local (gauge) action of SU(8)

$$L_4^{\mathrm{Scalar}} = \frac{\sqrt{g}}{48} \mathrm{Tr} \left(D_\mu \mathcal{V} \cdot \mathcal{V}^{-1} \right)^2 \qquad (13)$$

with $D_\mu \mathcal{V} \equiv \partial_\mu \mathcal{V} - h_\mu \mathcal{V}$ where h_μ is an SU(8) valued 1-form to be varied independently of $\mathcal{V}$. (13) uses the trivial fiber bundle

$$I_4 = E_7(+7) \overset{\pi}{\to} E_7/\mathrm{SU}(8) = I_4/K_4$$

without choosing a section that would directly give the 70 physical fields on I_4/K_4. After extremizing L_4^{Scalar} over h_μ one obtains

$$L_4^s = \frac{\sqrt{g}}{48} \mathrm{Tr} \left[\left(\partial_\mu \mathcal{V} \cdot \mathcal{V}^{-1} \right)^\perp \right]^2$$

where $\perp$ means the component of the Lie algebra element orthogonal to SU(8) in E_7. This extremization has been performed in (12) and seems crucial for the elimination of ghosts. It has been speculated that SU(8) or a subgroup of it could be used for a G.U.T. theory [2]; this is still on very shaky grounds but time will tell. Let us just conclude these comments with a precise counting of states. The boson fields g_{MN} and A_{MNP} in 11 dimensions reduce to 1 graviton $g_{\mu\nu}$, 28 vectors and 70 scalars all massless in this form of the theory; the spinor field ψ_M^A reduces to $(\psi_\mu)_\alpha^k$ and to $(\psi_{i'}^k)_\alpha \sim \psi_\alpha^{[jkl]}$ ($i' = 4, 5, \ldots, 10$) which are, respectively, gauge fields for 8 local supersymmetries (the gauge theory of (4)) and 56 real spinor fields (spin $\frac{1}{2}$). To build up $E_7(+7)$ from SL(7**R**) we had to use the famous Cartan triality of SO(8) [2]. It is not clear at present what its connection with quark triality is.

We are now in 4 dimensions, but why could we not continue our descent to 3 or even 2 and 1 dimensions? It is now a purely mathematical dimensional reduction; previously we could have argued with O. Klein that T_r was very, very small compared to the size of reasonable elementary particles; by the uncertainty principle we would then have deduced that fields depending periodically and nontrivially on the internal coordinates were effectively so massive that we could not see them at our comparatively low energies. But it is extremely useful to assume the existence of more Killing vectors and to reduce the number of effective space-time dimensions below 4: it leads to simpler equations and solutions. The symmetry group I_3 or I_2 (or I_1) acts on the set of solutions and I_2 has been exploited by many people (for $N' = 1$ or 2); one can ignore the fermions which do not mix with the bosons under I_d anyhow.

Let us briefly recall how it was recognized that supergravities could in fact be useful for better established theories like Einstein's or coupled Einstein-Maxwell equations. In [2] we guessed from very simple counting arguments the line $d = 4$ of Table 1 and also the appearance of $E_8(+8)$ for $d = 3$ and $E_6(+6)$ for $d = 5$. The similarity with the magic square of Freudenthal and Tits [24] was striking. J. Schwarz puzzled us by mentioning the existence of a group E_9; I learned later what it is from W. Nahm and J. Tits. E_8 is already heavy to handle [12] and E_9 has not yet been confirmed. However, in the meantime I learned about [25–27]. They suggested a deep connection between the coset structures encountered for $N' = 0$ and $N' \neq 0$. A rule of extrapolation was then discovered for $N' = 7, 6, 5, 4$ [10]. It said that in all four cases the group I_2 should be $I_3^{(1)}$, namely the affine Kac-Moody group of first type associated to the internal symmetry in 3 dimensions. It took a few months [11] to take the idea seriously for $N' = 3$ and especially 2 and 1 and to check that $A_1^{(1)}$ ($A_2^{(1)}$) was indeed a symmetry of the set (electro-) gravitational plane waves and to recognize its loop subalgebra $A_1 \times$ **R**$[t, 1/t]$ ($A_2 \times$ **R**$[t, 1/t]$) under the name of Geroch [28] (Kinnersley-Chitre [29]).

We leave a discussion of the central charge of I_2 for [13]; let us just mention here that it was conjectured in [11] that the full loop algebra of G should be an invariance of symmetric space G/H "chiral models" and not only $G \times$ **R**$[t]$ [30]. The previous paragraph illustrates how progress can be delayed by insufficient interaction between "specialists".

Finally, we can go to 1 (time) dimension: we are now considering the so-called homogeneous space-times. Could it be that E_{10} is a symmetry of homogeneous $N' = 7$ supergravity? We expect SL(10) to be a symmetry because we assumed $r = 10$ commuting Killing vectors. We have conjectured in [2] that the full $d = 4$, $N' = 7$ supergravity should be a superchiral model in which all 256 fields would parametrize a super coset space; in particular, the bosons play a symmetrical role, the metric $g_{\mu\nu}$ parametrizes $GL(d)/SO(1, d - 1)$, the scalars parametrize I_d/K, and the vectors are in-between transforming under $GL(d)$ and I_d. This idea received a precise realization in the first rule of group disintegration (see next section). Pushing it a little further we find that the bosonic part of the supergroup I_0 is of "rank" 11 [10] and I_1 of "rank" 10. I_1 is now being studied in view of its interest for mathematicians [31, 32]; it is sometimes called hyperbolic and does not have a simple interpretation like some extension of a loop group. We are in a situation where physics could provide concrete and simple realizations of highly abstract mathematical objects [33].

Finally, we shall see in the next section that the extrapolation to E_{10} suggests an extrapolation from $D_8^{(1)}$ to "overextended D_8" noted $D^{\hat{\hat{}}}$ for the type I superstrings [1, 3, 34]. These have also 1 supersymmetry spinor generator on the closed string but possess half as many states and half as many supersymmetry generators in the embedding space as we had before. In 10 dimensions one has only 1 real Weyl (irreducible) 16-component spinor charge. It is a nontrivial coincidence that the Dynkin diagrams of $E_{10} = E_8^{\hat{\hat{}}}$ and $D_8^{\hat{\hat{}}}$ are two of the three strictly hyperbolic graphs of "highest" rank (equal to 10) [35].

III. Real disintegration and the magic square.

(a) *The "real" disintegration of groups.* We should recall some basic facts from Lie algebra theory: the Chevalley-Harish Chandra-Serre presentation. One can define a Lie algebra by generators and relations like finite groups. We choose an $(l \times l)$ matrix A_{ij} satisfying the following conditions:

$$A_{ii} = 2 \quad (\text{no sum over } i),$$

$$A_{ij} = \text{a nonpositive integer} \quad \text{for } i \neq j,$$

$$\text{if } A_{ij} = 0 \text{ then } A_{ji} = 0. \tag{14}$$

We shall need only symmetric matrices; the general theory requires them only to be "symmetrizable" [36, 37]. And we shall use irreducible matrices of this type. Let us now define a Lie algebra by choosing l commuting (or "Cartan") generators h_i and $2l$ simple roots: l positive ones e_i and l negative ones f_i. And let us consider a Lie algebra that they generate which is restricted only by relations that one can deduce from the following finite set:

$$[h_i, h_j] = 0, \qquad [h_i, e_j] = A_{ij} e_j,$$

$$[h_i, f_j] = -A_{ij} f_j, \qquad [e_i, f_j] = \delta_{ij} h_i,$$

and for $i \neq j$

$$\underbrace{[e_i, [e_i \cdots [e_i, e_j]] \cdots]}_{(-A_{ij}+1) \text{ times}} \quad \text{same } e_i \leftrightarrow f_i. \tag{15}$$

(1) The simple Lie algebras of finite dimensions over an algebraically closed field are in one-to-one correspondence with positive definite "Cartan matrices" A_{ij} of the above type.

(2) If A_{ij} is only positive semidefinite, it corresponds to an *affine* Kac-Moody algebra; if it is symmetric it can be obtained from a positive definite submatrix of rank $l - 1$. It is then called "affine of type 1" and "associated" to this submatrix.

(3) If A_{ij} has $(l - 1)$ positive eigenvalues and one negative one, it is called strictly hyperbolic.

(4) Dynkin diagrams are an equivalent way of describing A_{ij}.

(5) Affine Lie algebras are graded Lie algebras of finite Gel'fand-Kirillov dimension.

(6) An affine type 1 algebra $\mathcal{G}^{(1)}$ is the (universal) central extension of the loop algebra of $\mathcal{G}$ where $\mathcal{G}$ is defined by the principal positive definite submatrix. The center is one dimensional and corresponds to Weyl transformations in supergravity theories [11].

If the field of coefficients is the field of the real numbers, the above construction defines the "normal real form" (maximally noncompact) of the Lie algebra. The compact real form is obtained by replacing h_j by ih_j, and $(e_j + f_j)$ by $i(e_j + f_j)$ without changing $(e_j - f_j)$; more generally, multiplying by i $(l - r)h_j$'s and consistent combinations of the other generators leads to a real form of *real rank* equal to r. The real rank has a geometric interpretation [36]; it is the dimension of the maximal flat and totally geodesic submanifolds of the symmetric space G/K where K is the maximal compact subgroup of the real group G. The number of noncompact generators minus the number of compact generators is the *index* of the Killing form. For the theory of real forms and real Dynkin diagrams (Tits-Satake diagrams) we refer to Tits [24 and 36].

We called (real) group disintegration the converse of dimensional reduction. It satisfies a number of simple rules.

Rule 1. If a supergravity is described by the chiral model I_3/K_3 in 3 dimensions, it can come only from a d-dimensional theory with symmetry group I_d where I_d is determined by the following operations:

(a) Take the extended Dynkin diagram of I_3.

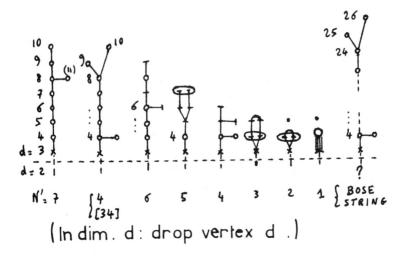

$\left(\text{In dim. } d \text{: drop vertex } d \text{ .}\right)$

FIGURE 2.

(b) Remove the $(d - 2)$th dot of the diagram starting from the dot corresponding to the smallest root of I_3 (the "new" dot of $I_2 = I_3^{(1)}$). It has to be a "real" root (circled in Figure 2), i.e. it is a combination of the h_j's with *real* coefficients.

(c) The resulting group can be read off the resulting diagram; it is the product of $SL(d - 2, \mathbf{R})$ by I_d. One should think of $SL(d - 2)$ as a subgroup of $GL(d)$, the group of linear coordinate transformations in d dimensions [10]. The group I_3 disintegrates into $I_d \times SL(d - 2)$ and representations of this product that correspond to *twice* [20] the number of degrees of freedom of the (generalized) gauge fields A_M, A_{MN}, A_{MNP}.

Rule 2. The real rank r and the index decrease by 1 in the process of going from dimension d to $(d + 1)$. The second property can be shown to be a consequence of the first (J. Tits has an unpublished formula for the index χ: $\chi = $ (relative real rank − "defect") − dim(anisotropic kernel); it is either self-explanatory or too long to explain).

Rule 3. $l + d - N' = 4$. l is the total (complex) rank and N' is the naive number of spinor generators evaluated in $d = 4$.

Rule 4. There is an approximate symmetry $(7 - N') \leftrightarrow (d - 3)$ as puzzling as the symmetry of the magic square (or of Cvitanovic's triangle [38]).

In Figure 2 one finds the Dynkin-Tits-Satake diagrams of the symmetries of pure supergravities, of $N = 4$ supergravity + matter [34] including $d = 3$ and 2, and a guess for the pure string (no spinor χ_M^ε) (critical dimension 26). The theory of *closed* strings with $n > 1$ supersymmetries on the string [39] has not been fully worked out yet. We can mention that

TYPE

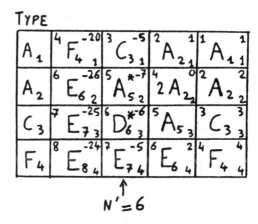

FIGURE 3.

for $n > 2$ its critical dimension is negative, but one discovers there a Kac-Moody symmetry on the string (with a central charge in the quantum theory).

(b) *Some interesting facts.* The Kac-Moody algebras $\mathcal{G}^{(1)}$ have finite codimension ideals and subalgebras. For example, one can require that $g(t_i)$ belongs to $\mathcal{K}_i$ for some values t_i of the loop parameter and some subalgebras $\mathcal{K}_i$ of $\mathcal{G}$ [11] and [Julia and Kac, unpublished].

There is a natural complex structure on the coset spaces of Table 1 including dimension 2 [28, 29].

The magic square may have some connection with supergravities [38] or at least dimensional reduction. (I am grateful to I. Frenkel for telling me about [40].) The geometrical magic square (Figure 3) has a remarkable (3×3) part satisfying Rules 1–4 above; we recall that it is formed with the real geometries of type F_4, C_3, A_2 and A_1 (Tits [24]). There are several constructions of real magic squares; one is due to Tits [40], others are due to Rosenfeld and Vinberg. Several forms satisfy Rules 1–4; one Tits form reproduces the line $d = 3$ of Table 1 for $N' = 7, 6$ and 5, and one notices the property of the corresponding symmetric spaces to have dimension $2^{N'}$. Can one define hypersymplectic geometries on homogeneous spaces of Kac-Moody (affine type 1) groups?

The most mysterious properties, like Rules 3 and 4, involve various values of N'. Until now only B. Morel and J. Thierry-Mieg have made some progress on the understanding of possible superalgebras generalizing I_d (see their report in the same book as [10]). It seems clear [2] that a magic supergroup is present; it is probably infinite dimensional. One can learn about the classification of simple superalgebras from [41]. We can conclude with two remarks. One is that dualities are paired by supersymmetry with chiralities (linear transformations) on the Fermi fields; the latter are manifest. The second remark is that there are close connections between infinite-dimensional Lie algebras and Lie superalgebras [42 and Kyoto school]....

REFERENCES

1. F. Gliozzi, D. Olive and J. Scherk, Nucl. Phys. B **122** (1977), 253.

2. E. Cremmer and B. Julia, Nucl. Phys. B **159** (1979), 141.

3. J. H. Schwarz, (a) these PROCEEDINGS, (b) CALT 68-911 and (c) Phys. Rep. **8** (1973), 269.

4. See R. Giles and C. B. Thorn, Phys. Rev. D **16** (1977), 366 (and references therein).

5. I. B. Frenkel and V. G. Kac, Invent. Math. **62** (1980), 23; see also I. B. Frenkel and J. Lepowsky, these PROCEEDINGS.

6. R. C. Brower, Phys. Rev. D **6** (1972), 1655; see also E. Del Giudice, P. Di Vecchia and S. Fubini, Ann. Physics **70** (1972), 378 and references therein.

7. G. Segal, Comm. Math. Phys. **80** (1981), 301.

8. B. Zumino, Lecture, AMS-SIAM Summer Seminar on Applications of Group Theory in Physics and Mathematical Physics, 1982.

9. D. Z. Freedman, P. Van Nieuwenhuizen and S. Ferrara, Phys. Rev. D **13** (1976), 3214; S. Deser and B. Zumino, Phys. Lett. B **62** (1976), 335.

10. B. Julia, Superspace and Supergravity (S. Hawking and M. Rocek, eds.), Cambridge Univ. Press, London and New York, 1981.

11. _____, Infinite Lie algebras in physics, Proc. Johns Hopkins Workshop on Particle Theory, Baltimore, May 1981.

12. _____, Erice Lectures, LPTENS preprint 82-18, 1982.

13. _____, Gravity, supergravity and integrable systems (Istanbul Conf. on Group Theoretical Methods in Physics), LPTENS preprint 82-23, 1982.

14. T. Yoneya, Nuovo Cimento **8** (1973), 951; J. Scherk and J. H. Schwarz, Nucl. Phys. B **81** (1974), 118.

15. W. Nahm, Nucl. Phys. B **135** (1978), 149.

16. E. Cremmer, B. Julia and J. Scherk, Phys. Lett. B **76** (1978), 409.

17. R. S. Palais, Comm. Math. Phys. **69** (1979), 19.

18. B. de Wit and D. Z. Freedman, Nucl. Phys. B **130** (1977), 105.

19. R. Geroch, J. Math. Phys. **12** (1971), 918.

20. B. Julia, Phys. Rep. (to appear).

21. S. Deser and C. Teitelboim, Phys. Rev. D **13** (1976), 1592.

22. A. Papapetrou, C. R. Acad. Sci. Paris Sér. A-B **257** (1963), 2797.

23. I am grateful to P. van Nieuwenhuizen and I. Bars for stressing the importance of a canonical action of these transformations. It is also suggested by the discussion of duality groups Sp($2n$, **R**) in M. K. Gaillard and B. Zumino, Nucl. Phys. B **193** (1981), 221.

24. See, for example, H. Freudenthal, ed., Algebraical and topological foundations of geometry, Pergamon Press, Oxford, 1962.

25. D. Maison, J. Math. Phys. **20** (1978), 871.

26. G. W. Gibbons and S. W. Hawking, Comm. Math. Phys. **66** (1979), 291.

27. M. Adler, Invent. Math. **50** (1979), 219.

28. R. Geroch, J. Math. Phys. **13** (1972), 394.

29. W. Kinnersley and D. M. Chitre, J. Math. Phys. **18** (1977), 1538.

30. M. Lüscher and K. Pohlmeyer, Nucl. Phys. B **137** (1978), 46.

31. A. Feingold and I. B. Frenkel, Math. Ann. **263** (1983), 87–144.

32. See also I. B. Frenkel, J. Lepowski and A. Meurman, Proc. Montreal Conf., 1983 (to appear).

33. B. Julia, in preparation.

34. A. H. Chamseddine, Nucl. Phys. B **185** (1981), 403.

35. N. Bourbaki, *Groupes et algèbre de Lie*, Chapters 4, 5, 6, Hermann, Paris, 1968.

36. S. Helgason, *Differential geometry, Lie groups and symmetric spaces*, Academic Press, New York, 1978.

37. H. Garland, *Arithmetic theory of loop groups*, Inst. Haute Étude Sci. Publ. Math., no. 52, Bures sur Yvette, 1980.

38. P. Cvitanovic, Nucl. Phys. B **188** (1981), 373.

39. P. Di Vecchia, Many Degrees of Freedom in Particle Theory (H. Satz, ed.), Plenum Press, New York, 1978.

40. D. Drucker, Mem. Amer. Math. Soc., no. 208, Amer. Matn. Soc., Providence, R. I., 1978 and J. Tits quoted therein.

41. V. G. Kac, Comm. Math. Phys. **53** (1977), 31.

42. G. Segal, unpublished.

LABORATOIRE DE PHYSIQUE THÉORIQUE DE L'ECOLE NORMALE SUPÉRIEURE, 24 RUE LHOMOND, 75231 PARIS CEDEX 05, FRANCE

Lectures in Applied Mathematics
Volume 21, 1985

Some Constructions of
the Affine Lie Algebra $A_1^{(1)}$

J. Lepowsky[1]

1. Introduction. This is an expository account of certain constructions of the affine Lie algebra $A_1^{(1)}$ using "vertex operators" and operators resembling them. A certain operator was invented by R. Wilson and the author in [11(a)], in the course of solving the problem of constructing $A_1^{(1)}$ as a Lie algebra of "concrete" operators. The surprising fact that operators similar to ours (namely, vertex operators) had been considered in the theory of dual resonance models (see e.g. [12, 15]) was pointed out to us by H. Garland.

The main result of [11(a)] is presented as Theorem 1 (§3) below. A simplification of the original proof [11(a)] of this theorem was achieved in [6], where the construction in [11(a)] was generalized to "most" affine Lie algebras. The best proof of the result of [11(a)] was found by I. Bars and Garland [unpublished] and by E. Date, M. Kashiwara and T. Miwa [2] (who discovered the connection between [11(a)] and the infinitesimal Bäcklund transformation for the Korteweg-de Vries equation; see also [1]); similar techniques were introduced in [4] and [16]. It is this "best" proof, in its algebraic form, that we give in §3.

An analogous construction of $A_1^{(1)}$ was discovered by I. Frenkel and V. Kac [4] and by G. Segal [16], using the vertex operators actually occurring in dual resonance theory. We treat this construction in §5.

The construction problem posed and solved in [11(a)] was originally motivated by a desire to develop some kind of algebraic structure

1980 *Mathematics Subject Classification.* Primary 17B65; Secondary 05A19, 15A66, 35Q20, 81E99.
[1]Partially supported by NSF grant MCS 80-03000.

underlying the classical Rogers-Ramanujan identities; a link between the product sides of these identities and certain "standard" representations of $A_1^{(1)}$ had been noticed in [9]. Such an algebraic structure was developed (using [11(a)]) in [11(b), (c)], and was considerably improved by the introduction of new algebras ("$\mathfrak{Z}$-algebras") [11(d), (e)]. The theory of these algebras is very general, and it provides a framework for constructing affine Lie algebras in a great variety of ways. When the theory is applied in the "principal picture" to the "basic" (level 1) $A_1^{(1)}$-representations, it yields the theorem of [11(a)] (see §3). For the "level 2 standard representations" in the principal picture, the theory gives constructions of $A_1^{(1)}$ discovered in [11(d)] and presented in §4. In the "homogeneous picture" for the level 1 standard representations, the theory leads [10] to the construction (mentioned above) of $A_1^{(1)}$ via the basic representation [4, 16], and also via the other level 1 standard representation [3, 16] (see §5). For the level 2 standard representations in the homogeneous picture, the theory gives still more constructions of $A_1^{(1)}$ [10] (see §6). These last constructions can be interpreted via the "boson-fermion correspondence" (cf. [3]).

It is interesting that for the level 2 representations in the principal picture (§4), $A_1^{(1)}$ appears naturally as a Lie algebra of operators acting irreducibly on a space of mixed boson-fermion states (the tensor product of a polynomial algebra with an exterior algebra). A similar but more complicated situation occurs for the level 2 representations in the homogeneous picture (§6).

For the standard $A_1^{(1)}$-representations of levels higher than 2, the theory of the new algebras becomes much deeper (see [11(d), (e)]). The Rogers-Ramanujan identities follow from the level 3 case [11(d), (e)].

The Lie algebra $A_1^{(1)}$ is the simplest of the affine Lie algebras, which form an important subclass of the class of Kac-Moody Lie algebras [5(a), 14]. (The papers [7(a), (b)], which are not well known, are also very interesting.) Discoveries in the case $A_1^{(1)}$ have frequently led to unexpected and fruitful new directions for the general theory. The standard representations ([5(b)]; see also [13]) are the analogues for Kac-Moody algebras of the finite-dimensional irreducible representations of the complex semisimple Lie algebras.

The reader is referred to the exposition [8] for more detailed motivation of the results discussed in this paper.

§§2–6 below are written in a self-contained way, with virtually no prerequisites.

We shall use the notations $\mathbf{N}$ for the set $\{0, 1, 2, \ldots\}$ of natural numbers, $\mathbf{Z}$ for the set of integers, and $\mathbf{C}$ for the set of complex numbers.

2. The Lie algebra $A_1^{(1)}$. Let $\mathfrak{g}$ be the 3-dimensional simple Lie algebra $\mathfrak{sl}(2, \mathbf{C})$, which consists of the 2×2 complex matrices of trace 0. The Lie bracket operation is the commutator of matrices. We shall use the following basis of $\mathfrak{g}$:

$$h = \begin{pmatrix} 1 & 0 \\ 0 & -1 \end{pmatrix}, \quad e = \begin{pmatrix} 0 & 1 \\ 0 & 0 \end{pmatrix}, \quad f = \begin{pmatrix} 0 & 0 \\ 1 & 0 \end{pmatrix}.$$

Then

$$[h, e] = 2e, \quad [h, f] = -2f, \quad [e, f] = h.$$

Consider the infinite-dimensional commutative associative algebra $\mathbf{C}[t, t^{-1}]$ of Laurent polynomials (terminating Laurent series) in the indeterminate t. This algebra consists of the polynomials

$$\sum_{j \in \mathbf{Z}} a_j t^j \qquad (a_j \in \mathbf{C}; a_j = 0 \text{ for all but finitely many } j)$$

in t and t^{-1} with complex coefficients.

Let $\hat{\mathfrak{g}}$ denote the infinite-dimensional complex Lie algebra

$$\hat{\mathfrak{g}} = \mathfrak{g} \otimes \mathbf{C}[t, t^{-1}] \oplus \mathbf{C}c,$$

with basis

$$\{h \otimes t^j, e \otimes t^j, f \otimes t^j, c \,|\, j \in \mathbf{Z}\},$$

and with bracket operation determined by the conditions

$$[c, \hat{\mathfrak{g}}] = 0, \quad [x \otimes t^i, y \otimes t^j] = [x, y] \otimes t^{i+j} + i\delta_{i+j,0}(\text{tr}(xy))c$$

for all $x, y \in \mathfrak{g}$ and $i, j \in \mathbf{Z}$. Here the symbol $\delta_{i+j,0}$ is the Kronecker delta, and $\text{tr}(xy)$ is the trace of the product of the 2×2-matrices x, y. It is easy to verify that the given bracket operation is skew-symmetric and satisfies the Jacobi identity, so that $\hat{\mathfrak{g}}$ forms a Lie algebra. This Lie algebra $\hat{\mathfrak{g}}$ ($= \mathfrak{sl}(2, \mathbf{C})\hat{\ }$) is also denoted $A_1^{(1)}$. It is the simplest affine Kac-Moody Lie algebra.

We observe that the structure of $\hat{\mathfrak{g}}$ can be described using the given basis as follows:

$$[c, h \otimes t^j] = [c, e \otimes t^j] = [c, f \otimes t^j] = 0,$$
$$[h \otimes t^i, h \otimes t^j] = 2i\delta_{i+j,0}c,$$
$$[h \otimes t^i, e \otimes t^j] = 2e \otimes t^{i+j},$$
$$[h \otimes t^i, f \otimes t^j] = -2f \otimes t^{i+j},$$
$$[e \otimes t^i, e \otimes t^j] = [f \otimes t^i, f \otimes t^j] = 0,$$
$$[e \otimes t^i, f \otimes t^j] = h \otimes t^{i+j} + i\delta_{i+j,0}c \qquad (1)$$

for all $i, j \in \mathbf{Z}$.

This information can in turn be reformulated using commuting formal variables ζ, ζ_1, ζ_2. Define

$$h(\zeta) = \sum_{j \in \mathbf{Z}} (h \otimes t^j)\zeta^j,$$

$$e(\zeta) = \sum_{j \in \mathbf{Z}} (e \otimes t^j)\zeta^j,$$

$$f(\zeta) = \sum_{j \in \mathbf{Z}} (f \otimes t^j)\zeta^j,$$

formal Laurent series in ζ with coefficients in $\hat{\mathfrak{g}}$. Also define the formal Laurent series

$$\delta(\zeta) = \sum_{j \in \mathbf{Z}} \zeta^j,$$

and let D be the formal differential operator

$$D = \zeta(d/d\zeta),$$

so that

$$(D\delta)(\zeta) = \sum_{j \in \mathbf{Z}} j\zeta^j. \tag{2}$$

(The notation δ is justified by the fact that $\delta(\zeta)$ "behaves like the δ-function at $\zeta = 1$"; see Lemma 1 below.) The bracket structure of $\hat{\mathfrak{g}}$ is completely described by the formulas

$$[c, h(\zeta_1)] = [c, e(\zeta_1)] = [c, f(\zeta_1)] = 0,$$
$$[h(\zeta_1), h(\zeta_2)] = 2(D\delta)(\zeta_1/\zeta_2)c,$$
$$[h(\zeta_1), e(\zeta_2)] = 2e(\zeta_2)\delta(\zeta_1/\zeta_2),$$
$$[h(\zeta_1), f(\zeta_2)] = -2f(\zeta_2)\delta(\zeta_1/\zeta_2),$$
$$[e(\zeta_1), e(\zeta_2)] = [f(\zeta_1), f(\zeta_1)] = 0,$$
$$[e(\zeta_1), f(\zeta_2)] = h(\zeta_2)\delta(\zeta_1/\zeta_2) + (D\delta)(\zeta_1/\zeta_2)c. \tag{3}$$

To verify this assertion, one computes and then compares the coefficients of $\zeta_1^i\zeta_2^j$ ($i, j \in \mathbf{Z}$) on the two sides of each equation.

The Lie algebra $\hat{\mathfrak{g}}$ contains interesting proper subalgebras isomorphic to $\hat{\mathfrak{g}}$ itself, in particular, the subalgebra $\mathfrak{k}$ with basis

$$\{h \otimes t^{2j}, e \otimes t^{2j+1}, f \otimes t^{2j+1}, c \mid j \in \mathbf{Z}\}.$$

By computing brackets, it is in fact easy to check that $\mathfrak{k}$ is a Lie subalgebra of $\hat{\mathfrak{g}}$, and that the following correspondence defines a Lie algebra isomorphism from $\hat{\mathfrak{g}}$ to $\mathfrak{k}$:

$$h \otimes t^j \mapsto h \otimes t^{2j} + \delta_{j,0}c,$$

$$e \otimes t^j \mapsto e \otimes t^{2j+1},$$

$$f \otimes t^j \mapsto f \otimes t^{2j-1}, \qquad c \mapsto 2c$$

for all $j \in \mathbf{Z}$.

The reason for passing to this subalgebra is to exhibit the $\mathbf{Z}$-grading of $\hat{\mathfrak{g}}$ associated with the following basis of $\mathfrak{k}$:

$$\{B_{2j+1}, X_j, c \mid j \in \mathbf{Z}\},$$

where

$$B_{2j+1} = (e + f) \otimes t^{2j+1},$$

$$X_{2j+1} = (-e + f) \otimes t^{2j+1},$$

$$X_{2j} = h \otimes t^{2j}$$

for all $j \in \mathbf{Z}$. The corresponding bracket structure of $\mathfrak{k}$ is

$$[c, B_i] = [c, X_k] = 0, \tag{4}$$

$$[B_i, B_j] = 2i\delta_{i+j,0}c, \tag{5}$$

$$[B_i, X_k] = 2X_{i+k}, \tag{6}$$

$$[X_k, X_l] = \begin{cases} 2(-1)^{k+1}B_{k+l} & \text{if } k + l \in 2\mathbf{Z} + 1, \\ 2(-1)^k k\delta_{k+l,0}c & \text{if } k + l \in 2\mathbf{Z} \end{cases} \tag{7}$$

for all $i, j \in 2\mathbf{Z} + 1$, $k, l \in \mathbf{Z}$. Note that $\mathfrak{k}$ is a $\mathbf{Z}$-graded Lie algebra, that is,

$$\mathfrak{k} = \oplus_{j \in \mathbf{Z}}\mathfrak{k}_j, \quad \text{where } [\mathfrak{k}_i, \mathfrak{k}_j] \subset \mathfrak{k}_{i+j};$$

the subspaces $\mathfrak{k}_j$ are as follows:

$$\mathfrak{k}_0 = \text{span}\{c, X_0\},$$

$$\mathfrak{k}_j = \text{span}\{B_j, X_j\}, \quad j \in 2\mathbf{Z} + 1,$$

$$\mathfrak{k}_j = \text{span}\{X_j\}, \qquad j \in 2\mathbf{Z}.$$

Observe that each basis element B_j, X_j has degree equal to its subscript.

In terms of formal variables, the bracket structure of $\mathfrak{k}$ may be described as follows: Let

$$B(\zeta) = \sum_{j \in 2\mathbf{Z}+1} B_j \zeta^j, \qquad X(\zeta) = \sum_{k \in \mathbf{Z}} X_k \zeta^k.$$

Then

$$[c, B(\zeta_1)] = [c, X(\zeta_1)] = 0,$$
$$[B(\zeta_1), B(\zeta_2)] = ((D\delta)(\zeta_1/\zeta_2) - (D\delta)(-\zeta_1/\zeta_2))c,$$
$$[B(\zeta_1), X(\zeta_2)] = X(\zeta_2)(\delta(\zeta_1/\zeta_2) - \delta(-\zeta_1/\zeta_2)),$$
$$[X(\zeta_1), X(\zeta_2)] = -2B(\zeta_2)\delta(-\zeta_1/\zeta_2) + 2(D\delta)(-\zeta_1/\zeta_2)c. \tag{8}$$

3. Construction by differential operators on $S(\hat{s}_-)$. We are now ready to construct $A_1^{(1)}$ by differential operators. Let $\hat{s}$ and $\hat{s}_-$ be the (**Z**-graded) Lie subalgebras of $\mathfrak{k}$ with bases

$$\{B_j, c | j \in 2\mathbf{Z} + 1\} \quad \text{and} \quad \{B_j | j \in -(2\mathbf{N} + 1)\},$$

respectively. The brackets among these elements show that $\hat{s}$ is an infinite-dimensional Heisenberg Lie algebra, and that $\hat{s}_-$ is an abelian Lie subalgebra of $\hat{s}$. Hence $\hat{s}$ has an irreducible representation on the symmetric algebra $S(\hat{s}_-)$, which is also the polynomial algebra

$$S(\hat{s}_-) = \mathbf{C}[x_1, x_3, x_5, \dots] \tag{9}$$

on variables $x_j, j \in 2\mathbf{N} + 1$ (identified with the elements B_{-j} of $\hat{s}_-$), by means of the following operators:

$$c \mapsto \tfrac{1}{2} \qquad \text{(scalar multiplication operator),}$$
$$B_j \mapsto j(\partial/\partial x_j) \quad \text{(annihilation operator),} \tag{10}$$
$$B_{-j} \mapsto x_j \qquad \text{(creation operator)}$$

for $j \in 2\mathbf{N} + 1$. Here x_j denotes the multiplication operator. The algebra $S(\hat{s}_-)$ is **Z**-graded (in fact, nonpositively graded) by the requirement that $\deg x_j = -j$ for $j \in 2\mathbf{N} + 1$. That is,

$$S(\hat{s}_-) = \bigoplus_{j \in \mathbf{Z}} S(\hat{s}_-)_j,$$

where $S(\hat{s}_-)_j$ is the span of the monomials $x_1^{i_1} x_3^{i_3} \cdots$ for which

$$\sum_{k \in 2\mathbf{N}+1} k i_k = -j.$$

In particular,

$$S(\hat{s}_-)_j = (0) \quad \text{for } j > 0, \qquad S(\hat{s}_-)_0 = \mathbf{C} \cdot 1.$$

For all $j \in \mathbf{Z}$, the operators c and B_{2j+1} on $S(\hat{s}_-)$ have degrees equal to their Lie algebra degrees 0 and $2j + 1$, respectively. (an operator of degree k on $S(\hat{s}_-)$ is an operator A which takes $S(\hat{s}_-)_j$ to $S(\hat{s}_-)_{j+k}$ for all $j \in \mathbf{Z}$.) Thus $S(\hat{s}_-)$ is a graded representation of the Lie algebra $\hat{s}$.

To describe the action of X_j ($j \in \mathbf{Z}$) as a differential operator of degree j on $S(\mathring{\mathfrak{g}}_-)$, set

$$E_1^+(\zeta) = \exp\left(\sum_{j \in 2\mathbf{N}+1} 2\zeta^j B_j/j \right), \tag{11}$$

$$E_1^-(\zeta) = \exp\left(-\sum_{j \in 2\mathbf{N}+1} 2\zeta^{-j} B_{-j}/j \right), \tag{12}$$

$$X'(\zeta) = -\frac{1}{2} E_1^-(-\zeta) E_1^+(-\zeta). \tag{13}$$

Here exp denotes the formal exponential power series. When $E_1^+(\zeta)$, $E_1^-(\zeta)$ or $X'(\zeta)$ is expanded as a formal Laurent series in ζ, the coefficient of ζ^j ($j \in \mathbf{Z}$) is a formal differential operator of operator degree j taking $S(\mathring{\mathfrak{g}}_-)$ to itself. For $j \in \mathbf{Z}$, denote by X_j' this coefficient of ζ^j for $X'(\zeta)$, so that

$$X'(\zeta) = \sum_{j \in \mathbf{Z}} X_j' \zeta^j. \tag{14}$$

It is certainly possible to write down an explicit formula for the operator X_j', as a formal infinite linear combination of monomials in the x_i's times monomials in the $\partial/\partial x_i$'s, but we shall not need this (complicated) formula. It is important to note that the infinite sum defining the operator X_j' is in fact a well-defined operator of degree j on $S(\mathring{\mathfrak{g}}_-)$, since only finitely many summands of this operator fail to vanish on any given element of $S(\mathring{\mathfrak{g}}_-)$.

THEOREM 1. *The operators c, B_j ($j \in 2\mathbf{Z}+1$) and X_j' ($j \in \mathbf{Z}$) (see (10), (13), (14)) satisfy the bracket relations (4)–(7), with X_j' in place of X_j. In particular, these operators span a Lie algebra isomorphic to $A_1^{(1)}$.*

PROOF. We already know that

$$[c, B_i] = [c, X_k'] = 0,$$
$$[B_i, B_j] = 2i\delta_{i+j,0} c$$

for $i, j \in 2\mathbf{Z}+1$, $k \in \mathbf{Z}$. To show that

$$[B_i, X_k'] = 2X_{i+k}' \tag{15}$$

for $i \in 2\mathbf{Z}+1$, $k \in \mathbf{Z}$, we compute as follows: For $i \in 2\mathbf{N}+1$,

$$\begin{aligned}
[B_i, E_1^-(-\zeta)] &= \left[B_i, \exp \sum_{j \in 2\mathbf{N}+1} 2\zeta^{-j} B_{-j}/j \right] \\
&= [B_i, \sum 2\zeta^{-j} B_{-j}/j] E_1^-(-\zeta) \\
&= 2\zeta^{-i}(2c) E_1^-(-\zeta) \\
&= 2\zeta^{-i} E_1^-(-\zeta).
\end{aligned}$$

The meaning of this computation is that when the expressions are applied to an arbitrary element of $S(\hat{\mathfrak{g}}_-)$, we get equal formal Laurent series in ζ with coefficients in $S(\hat{\mathfrak{g}}_-)$. We can use the formal rule

$$[B, \exp Y] = [B, Y] \exp Y$$

because $[B, Y]$ commutes with Y. Similarly, we have

$$[B_{-i}, E_1^+(-\zeta)] = 2\zeta^i E_1^+(-\zeta),$$
$$[B_i, E_1^+(-\zeta)] = [B_{-i}, E_1^-(-\zeta)] = 0.$$

Hence

$$[B_i, X'(\zeta)] = -\tfrac{1}{2}[B_i, E_1^-(-\zeta)]E_1^+(-\zeta) - \tfrac{1}{2}E_1^-(-\zeta)[B_i, E_1^+(-\zeta)]$$
$$= 2\zeta^{-i}X'(\zeta)$$

and similarly,

$$[B_{-i}, X'(\zeta)] = 2\zeta^i X'(\zeta).$$

Equating coefficients, we obtain (6).

The most interesting bracket relation is (7), or equivalently, (8), with X' in place of X. We have

$$4X'(\zeta_1)X'(\zeta_2) = E_1^-(-\zeta_1)E_1^+(-\zeta_1)E_1^-(-\zeta_2)E_1^+(-\zeta_2)$$
$$= E_1^-(-\zeta_1)E_1^-(-\zeta_2)E_1^+(-\zeta_1)E_1^+(-\zeta_2)$$
$$\cdot \exp\left[-\sum_{j\in 2\mathbf{N}+1} 2\zeta_1^j B_j/j, \sum_{k\in 2\mathbf{N}+1} 2\zeta_2^{-k}B_{-k}/k\right]$$
$$= E_1^-(-\zeta_1)E_1^-(-\zeta_2)E_1^+(-\zeta_1)E_1^+(-\zeta_2)$$
$$\cdot \exp\left(-\sum_{j\in 2\mathbf{N}+1} 4(\zeta_1/\zeta_2)^j/j\right)$$
$$= E_1^-(-\zeta_1)E_1^-(-\zeta_2)E_1^+(-\zeta_1)E_1^+(-\zeta_2)$$
$$\cdot \exp\left(\log(1 - \zeta_1/\zeta_2)^2(1 + \zeta_1/\zeta_2)^{-2}\right)$$
$$= E_1^-(-\zeta_1)E_1^-(-\zeta_2)E_1^+(-\zeta_1)E_1^+(-\zeta_2)$$
$$\cdot (1 - \zeta_1/\zeta_2)^2(1 + \zeta_1/\zeta_2)^{-2},$$

where we have used basic properties of exponential and logarithmic formal power series; the expression $(1 + \zeta_1/\zeta_2)^{-2}$ designates the binomial series in ζ_1/ζ_2. As above, all these expressions are to be applied to an arbitrary element of $S(\hat{\mathfrak{g}}_-)$, whereupon we obtain equal formal Laurent series in ζ_1 and ζ_2 with coefficients in $S(\hat{\mathfrak{g}}_-)$, at each step.

Reversing the roles of ζ_1 and ζ_2, we obtain

$$4X'(\zeta_2)X'(\zeta_1) = E_1^-(-\zeta_2)E_1^-(-\zeta_1)E_1^+(-\zeta_2)E_1^+(-\zeta_1)$$
$$\cdot (1 - \zeta_2/\zeta_1)^2(1 + \zeta_2/\zeta_1)^{-2}.$$

Since
$$E_1^\pm(-\zeta_1)E_1^\pm(-\zeta_2) = E_1^\pm(-\zeta_2)E_1^\pm(-\zeta_1),$$
we find that
$$4[X'(\zeta_1), X'(\zeta_2)]$$
$$= E_1^-(-\zeta_1)E_1^-(-\zeta_2)E_1^+(-\zeta_1)E_1^+(-\zeta_2)(4(D\delta)(-\zeta_1/\zeta_2)) \quad (16)$$
(see (2)). We shall use

LEMMA 1. *Let*
$$f(\zeta_1, \zeta_2) = \sum_{i,j \in \mathbf{Z}} v_{ij}\zeta_1^i\zeta_2^j$$

be a formal Laurent series in ζ_1, ζ_2 *with coefficients* v_{ij} *in a vector space, and suppose that for some* $n \in \mathbf{Z}$, *we have* $v_{ij} = 0$ *whenever* i *or* $j > n$. *Set*
$$(D_l f)(\zeta_1, \zeta_2) = \zeta_l(\partial f/\partial \zeta_l)(\zeta_1, \zeta_2)$$
for $l = 1, 2$. *Let* $a \in \mathbf{C}$, $a \neq 0$. *Then*
$$\delta(a\zeta_1/\zeta_2)f(\zeta_1, \zeta_2) = \delta(a\zeta_1/\zeta_2)f(a^{-1}\zeta_2, \zeta_2),$$
$$(D\delta)(a\zeta_1/\zeta_2)f(\zeta_1, \zeta_2) = (D\delta)(a\zeta_1/\zeta_2)f(a^{-1}\zeta_2, \zeta_2)$$
$$-\delta(a\zeta_1/\zeta_2)(D_1 f)(\zeta_1, \zeta_2).$$

PROOF. The first formula is easily checked directly, and the second follows upon applying the operator $\zeta_1(d/d\zeta_1)$. ∎

Letting both sides of (16) act on an arbitrary element of $S(\hat{\mathfrak{s}}_-)$, we find that Lemma 1 is applicable, and we obtain
$$[X'(\zeta_1), X'(\zeta_2)] = E_1^-(\zeta_2)E_1^-(-\zeta_2)E_1^+(\zeta_2)E_1^+(-\zeta_2)(D\delta)(-\zeta_1/\zeta_2)$$
$$-(((\zeta_1 d/d\zeta_1)E_1^-(-\zeta_1))E_1^-(-\zeta_2)E_1^+(-\zeta_1)E_1^+(-\zeta_2)$$
$$+E_1^-(-\zeta_1)E_1^-(-\zeta_2)(\zeta_1 d/d\zeta_1)$$
$$\cdot E_1^+(-\zeta_1)E_1^+(-\zeta_2))\delta(-\zeta_1/\zeta_2)$$
$$= (D\delta)(-\zeta_1/\zeta_2) + \left(\left(\sum_{j \in 2\mathbf{N}+1} 2\zeta_1^{-j}B_{-j}\right)\right.$$
$$\cdot E_1^-(-\zeta_1)E_1^-(-\zeta_2)E_1^+(-\zeta_1)E_1^+(-\zeta_2)$$
$$+E_1^-(-\zeta_1)E_1^-(-\zeta_2)E_1^+(-\zeta_1)E_1^+(-\zeta_2)$$
$$\left.\cdot \left(\sum_{j \in 2\mathbf{N}+1} 2\zeta_1^j B_j\right)\right)\delta(-\zeta_1/\zeta_2)$$
$$= (D\delta)(-\zeta_1/\zeta_2) - \left(\sum_{j \in 2\mathbf{Z}+1} 2\zeta_2^j B_j\right)\delta(-\zeta_1/\zeta_2),$$

which agrees with the right-hand side of (8). This completes the proof of Theorem 1. ∎

REMARK. By redefining $X'(\zeta)$ to be $+\frac{1}{2}E^-(-\zeta)E^+(-\zeta)$ (cf. (13)), we of course obtain another construction of $A_1^{(1)}$.

4. Construction by operators on $S(\hat{\mathfrak{s}}_-) \otimes \Lambda(\hat{\mathfrak{z}}_-^s)$. In order to describe our next realizations of the commutation relations (4)–(7), let $s = 0$ or 1, and denote by $\mathcal{C}^s$ the Clifford algebra on generators Z_j ($j \in 2\mathbf{Z} + s$) subject to the relations

$$\{Z_i, Z_j\} = 2(-1)^s \delta_{i+j,0} \tag{17}$$

for all $i, j \in 2\mathbf{Z} + s$, where $\{A, B\}$ denotes the anticommutator $AB + BA$. Let

$$\hat{\mathfrak{z}}_-^s = \mathrm{span}\{Z_j | j < 0\}.$$

The elements Z_j, $j < 0$, are linearly independent, and $\mathcal{C}^s$ has a standard irreducible representation on the exterior algebra $\Lambda(\hat{\mathfrak{z}}_-^s)$ on the generators Z_j, $j < 0$, determined by the anticommutation relations (17) and the conditions

$$Z_j \cdot 1 = 0, \qquad Z_{-j} \cdot w = Z_{-j} \wedge w$$

for $j \in 2\mathbf{Z} + s, j > 0$ and $w \in \Lambda(\hat{\mathfrak{z}}_-^s)$, together with the condition

$$Z_0 \cdot 1 = -1 \tag{18}$$

in the case $s = 0$. Set

$$Z(\zeta) = \sum_{j \in 2\mathbf{Z}+s} Z_j \zeta^j = \sum_{j \in \mathbf{Z}} Z_j \zeta^j,$$

where we put $Z_j = 0$ if $j \in 2\mathbf{Z} + 1 + s$.

The algebra $\Lambda(\hat{\mathfrak{z}}_-^s)$ is $\mathbf{Z}$-graded (and in fact nonpositively graded) by the requirement that

$$\deg Z_j = j$$

for $j < 0$, and for $j \in \mathbf{Z}$, the operator Z_j on $\Lambda(\hat{\mathfrak{z}}_-^s)$ has degree j.

We shall construct $A_1^{(1)}$ as a Lie algebra of operators on the (nonpositively $\mathbf{Z}$-graded) space

$$S(\hat{\mathfrak{s}}_-) \otimes \Lambda(\hat{\mathfrak{z}}_-^s)$$

(see (9)). Provide this space with the $\hat{\mathfrak{s}}$-module structure determined as follows:

$$c \mapsto 1, \quad B_j \mapsto 2j(\partial/\partial x_j) \otimes 1, \quad B_{-j} \mapsto x_j \otimes 1$$

for $j \in 2\mathbf{N} + 1$ (cf. (10)).

REMARK. The subspace

$$\Lambda(\hat{\mathfrak{z}}_-^s) = 1 \otimes \Lambda(\hat{\mathfrak{z}}_-^s)$$

of $S(\hat{\mathfrak{g}}_-) \otimes \Lambda(\hat{\mathfrak{g}}_-^s)$ is the vacuum space

$$\{v \in S(\hat{\mathfrak{g}}_-) \otimes \Lambda(\hat{\mathfrak{g}}_-^s) \mid B_j \cdot v = 0 \text{ for all } j \in 2\mathbf{N} + 1\}$$

for the action of $\hat{\mathfrak{g}}$ on $S(\hat{\mathfrak{g}}_-) \otimes \Lambda(\hat{\mathfrak{g}}_-^s)$.

Set

$$E_2^+(\zeta) = \exp\left(\sum_{j \in 2\mathbf{N}+1} \zeta^j B_j/j\right),$$

$$E_2^-(\zeta) = \exp\left(-\sum_{j \in 2\mathbf{N}+1} \zeta^{-j} B_{-j}/j\right),$$

$$X''(\zeta) = E_2^-(-\zeta)Z(\zeta)E_2^+(-\zeta),$$

where $Z(\zeta)$ is understood as $1 \otimes Z(\zeta)$. We may also write

$$X''(\zeta) = E_2^-(-\zeta)E_2^+(-\zeta)Z(\zeta) = Z(\zeta)E_2^-(-\zeta)E_2^+(-\zeta).$$

Define operators X_j'' ($j \in \mathbf{Z}$) on $S(\hat{\mathfrak{g}}_-) \otimes \Lambda(\hat{\mathfrak{g}}_-^s)$ by

$$X''(\zeta) = \sum_{j \in \mathbf{Z}} X_j'' \zeta^j.$$

Then for all $j \in \mathbf{Z}$, the operators B_j, $Z_j = 1 \otimes Z_j$ and X_j'' are of degree j.

THEOREM 2. *Let $s = 0$ or 1. The operators c, B_j ($j \in 2\mathbf{Z} + 1$) and X_j'' ($j \in \mathbf{Z}$) on the space $S(s_-) \otimes \Lambda(\hat{\mathfrak{g}}_-^s)$ satisfy the bracket relations (4)–(7), with X_j'' in place of X_j, so that these operators span a Lie algebra isomorphic to $A_1^{(1)}$. The space $S(\hat{\mathfrak{g}}_-) \otimes \Lambda(\hat{\mathfrak{g}}_-^s)$ is irreducible under these operators.*

PROOF. The relations corresponding to (4)–(6) are easily checked as in the proof of Theorem 1. To prove (7), or equivalently, (8), with X'' in place of X, we compute as follows:

$$X''(\zeta_1)X''(\zeta_2) = E_2^-(-\zeta_1)E_2^+(-\zeta_1)E_2^-(-\zeta_2)E_2^+(-\zeta_2)Z(\zeta_1)Z(\zeta_2)$$

$$= E_2^-(-\zeta_1)E_2^-(-\zeta_2)E_2^+(-\zeta_1)E_2^+(-\zeta_2)Z(\zeta_1)Z(\zeta_2)$$

$$\cdot \exp\left[-\sum_{j \in 2\mathbf{N}+1} \zeta_1^j B_j/j, \sum_{k \in 2\mathbf{N}+1} \zeta_2^{-k} B_{-k}/k\right]$$

$$= E_2^-(-\zeta_1)E_2^-(-\zeta_2)E_2^+(-\zeta_1)E_2^+(-\zeta_2)Z(\zeta_1)Z(\zeta_2)$$

$$\cdot \exp\left(-\sum_{j \in 2\mathbf{N}+1} 2(\zeta_1/\zeta_2)^j/j\right)$$

$$= E_2^-(-\zeta_1)E_2^-(-\zeta_2)E_2^+(-\zeta_1)E_2^+(-\zeta_2)Z(\zeta_1)Z(\zeta_2)$$

$$\cdot (1 - \zeta_1/\zeta_2)(1 + \zeta_1/\zeta_2)^{-1},$$

so that

$$[X''(\zeta_1), X''(\zeta_2)] = E_2^-(-\zeta_1)E_2^-(-\zeta_2)E_2^+(-\zeta_1)E_2^+(-\zeta_2)A,$$

where

$$A = Z(\zeta_1)Z(\zeta_2)(1 - \zeta_1/\zeta_2)(1 + \zeta_1/\zeta_2)^{-1}$$

$$-Z(\zeta_2)Z(\zeta_1)(1 - \zeta_2/\zeta_1)(1 + \zeta_2/\zeta_1)^{-1}$$

$$= \left(\sum_{i,j \in \mathbf{Z}} Z_i Z_j \zeta_1^i \zeta_2^j \right)\left(1 + 2 \sum_{k>0} (-1)^k \zeta_1^k \zeta_2^{-k} \right)$$

$$-\left(\sum_{i,j \in \mathbf{Z}} Z_j Z_i \zeta_1^i \zeta_2^j \right)\left(1 + 2 \sum_{k>0} (-1)^k \zeta_1^{-k} \zeta_2^k \right)$$

$$= \sum_{m,n \in \mathbf{Z}} a_{mn} \zeta_1^m \zeta_2^n,$$

where

$$a_{mn} = Z_m Z_n - Z_n Z_m + 2 \sum_{k>0} (-1)^k (Z_{m-k}Z_{n+k} - Z_{n-k}Z_{m+k}).$$

If $m + n \in 2\mathbf{Z} + 1$, then $a_{mn} = 0$. Suppose that $m + n \in 2\mathbf{Z}$. If $s + m \in 2\mathbf{Z}$,

$$a_{mn} = \begin{cases} Z_m Z_n + 2(Z_{m-2}Z_{n+2} + \cdots + Z_{n+2}Z_{m-2}) + Z_n Z_m & \text{for } m > n, \\ 0 & \text{for } m = n, \\ -Z_n Z_m - 2(Z_{n-2}Z_{m+2} + \cdots + Z_{m+2}Z_{n-2}) - Z_n Z_m & \text{for } m < n \end{cases}$$

and if $s + m \in 2\mathbf{Z} + 1$,

$$a_{mn} = \begin{cases} -2(Z_{m-1}Z_{n+1} + \cdots + Z_{n+1}Z_{m-1}) & \text{for } m > n, \\ 0 & \text{for } m = n, \\ 2(Z_{n-1}Z_{m+1} + \cdots + Z_{m+1}Z_{n-1}) & \text{for } m < n. \end{cases}$$

From the anticommutation relation (17), it follows that in all cases,

$$a_{mn} = 2(-1)^m m \delta_{m+n,0}.$$

Hence

$$A = 2(D\delta)(-\zeta_1/\zeta_2).$$

Using Lemma 1 as in the proof of Theorem 1, we obtain

$$[X''(\zeta_1), X''(\zeta_2)] = 2(D\delta)(-\zeta_1/\zeta_2) - 2\delta(-\zeta_1/\zeta_2)\left(\sum_{j \in 2\mathbf{Z}+1} \zeta_2^j B_j \right),$$

which agrees with (8).

The irreducibility follows from the irreducibility of $S(\hat{s}_-)$ under $\hat{s}$ and of $\Lambda(\hat{s}^s_-)$ under $\mathcal{C}^s$, together with the formula

$$Z(\zeta) = E_2^-(\zeta)X''(\zeta)E_2^+(\zeta),$$

which is a consequence of the definition of $X''(\zeta)$. ∎

REMARK. By redefining $Z_0 \cdot 1$ to be $+1$ in the case $s = 0$ (cf. (18)), we obtain another construction of $A_1^{(1)}$.

5. Construction by operators on $S(\hat{\mathfrak{h}}_-) \otimes \mathbf{C}[P]$. For the remaining constructions of $A_1^{(1)}$ that we present, we use the Lie algebra $\hat{\mathfrak{g}}$, with its bracket structure given by (1) and (3). The algebra $\hat{\mathfrak{g}}$ has a $\mathbf{Z}$-grading

$$\hat{\mathfrak{g}} = \bigoplus_{j \in \mathbf{Z}} \hat{\mathfrak{g}}_j$$

given by

$$\hat{\mathfrak{g}}_0 = \text{span}\{c, h, e, f\},$$
$$\hat{\mathfrak{g}}_j = \text{span}\{h \otimes t^j, e \otimes t^j, f \otimes t^j\} \quad \text{for } j \neq 0.$$

(We are identifying elements $x \in \mathfrak{g} = \mathfrak{sl}(2, \mathbf{C})$ with the corresponding elements $x \otimes t^0 = x \otimes 1$ of $\hat{\mathfrak{g}}$.) This $\mathbf{Z}$-grading is different from the $\mathbf{Z}$-grading of $\hat{\mathfrak{g}} \simeq \hat{\mathfrak{k}}$ considered earlier.

Let $\hat{\mathfrak{h}}$, $\hat{\mathfrak{h}}'$ and $\hat{\mathfrak{h}}_-$ be the Lie subalgebras of $\mathfrak{g}$ with bases

$$\{h \otimes t^j, c \mid j \in \mathbf{Z}\}, \quad \{h \otimes t^j, c \mid j \neq 0\} \quad \text{and} \quad \{h \otimes t^j \mid j < 0\},$$

respectively. Then $\hat{\mathfrak{h}}'$ is a Heisenberg Lie algebra, and $\hat{\mathfrak{h}}_-$ is an abelian Lie subalgebra. The Lie algebra $\hat{\mathfrak{h}}'$ has an irreducible representation on the symmetric algebra $S(\hat{\mathfrak{h}}_-)$, which may be identified with the polynomial algebra

$$S(\hat{\mathfrak{h}}_-) = \mathbf{C}[y_1, y_2, \ldots] \tag{19}$$

on variables y_j, $j > 0$ (identified with the elements $h \otimes t^{-j}$ of $\hat{\mathfrak{h}}_-$), determined as follows:

$$c \mapsto 1, \quad h \otimes t^j \mapsto 2j(\partial/\partial y_j), \quad h \otimes t^{-j} \mapsto y_j$$

for $j > 0$ (cf. (10)). The algebra $S(\hat{\mathfrak{h}}_-)$ is (nonpositively) $\mathbf{Z}$-graded by the condition that

$$\deg y_j = -j$$

for $j > 0$, and the operators c and $h \otimes t^j$ ($j \neq 0$) have degrees equal to their Lie algebra degrees.

Let P denote the abelian group $P = \frac{1}{2}\mathbf{Z}$ of half-integers, and let $\mathbf{C}[P]$ be the group algebra of P, with the element p of P written as a formal exponential e^p in $\mathbf{C}[P]$. That is, $\mathbf{C}[P]$ is an associative algebra with basis

$\{e^p | p \in P\}$ and with multiplication determined by the condition

$$e^p e^q = e^{p+q}$$

for $p, q \in P$.

Writing

$$P = \mathbf{Z} \cup (\mathbf{Z} + \tfrac{1}{2}),$$

we obtain a direct sum decomposition

$$\mathbf{C}[P] = \mathbf{C}[\mathbf{Z}] \oplus e^{1/2}\mathbf{C}[\mathbf{Z}], \qquad (20)$$

where we write $\mathbf{C}[\mathbf{Z}]$ for the subalgebra of $\mathbf{C}[P]$ spanned by the elements e^p for $p \in \mathbf{Z}$.

We make $\mathbf{C}[P]$ a graded vector space by the requirement

$$\deg(e^p) = -p^2. \qquad (21)$$

For $p \in \mathbf{Z}$ (respectively, $\mathbf{Z} + \tfrac{1}{2}$), $\deg(e^p) \in -\mathbf{N}$ (respectively, $-(\mathbf{N} + \tfrac{1}{4})$). We have

$$\mathbf{C}[P] = \bigoplus_{j \in \mathbf{Z}/4} \mathbf{C}[P]_j,$$

that is, $\mathbf{C}[P]$ is $\mathbf{Z}/4$-graded. Note that the pairs $(p, \deg(e^p))$ for $p \in \tfrac{1}{2}\mathbf{Z}$ all lie on a parabola.

We shall construct $\hat{\mathfrak{g}}$ as a Lie algebra of operators on the (nonpositively $\mathbf{Z}/4$-graded) space

$$S(\hat{\mathfrak{h}}_-) \otimes \mathbf{C}[P].$$

First, we make $\hat{\mathfrak{h}}'$ act according to its actions on $S(\hat{\mathfrak{h}}_-)$, that is,

$$c \mapsto 1, \quad h \otimes t^j \mapsto 2j(\partial/\partial y_j) \otimes 1, \quad h \otimes t^{-j} \mapsto \cdot_j \otimes 1 \qquad (22)$$

for $j > 0$. (Note that $\mathbf{C}[P]$ is the vacuum space for the action of $\hat{\mathfrak{h}}'$.) Next, for $y \in S(\hat{\mathfrak{h}}_-)$ and $p \in P$, we set

$$h \cdot (y \otimes e^p) = 2p(y \otimes e^p), \qquad (23)$$

where of course $h = h \otimes t^0 \in \hat{\mathfrak{g}}$. Then $S(\hat{\mathfrak{h}}_-) \otimes \mathbf{C}[P]$ becomes an $\hat{\mathfrak{h}}$-module.

It remains to describe the action of $e \otimes t^j$ and $f \otimes t^j$ for $j \in \mathbf{Z}$. Define operators e^+ and e^- on $S(\hat{\mathfrak{h}}_-) \otimes \mathbf{C}[P]$ as follows: For $\in S(\hat{\mathfrak{h}}_-)$ and $p \in P$,

$$e^+ \cdot (y \otimes e^p) = y \otimes e^{p+1}, \quad e^- \cdot (y \otimes e^p) = y \otimes e^{p-1}. \qquad (24)$$

Also define

$$\zeta^+ \cdot (y \otimes e^p) = \zeta^{2p}(y \otimes e^p), \quad \zeta^- \cdot (y \otimes e^p) = \zeta^{-2p}(y \otimes e^p). \qquad (25)$$

Then

$$e^+ \zeta^+ = \zeta^{-2}\zeta^+ e^+, \quad e^+ \zeta^- = \zeta^2 \zeta^- e^+,$$
$$e^- \zeta^+ = \zeta^2 \zeta^+ e^-, \quad e^- \zeta^- = \zeta^{-2}\zeta^- e^- \qquad (26)$$

In addition, define

$$E_1^+(+,\zeta) = \exp\left(\sum_{j>0} (h \otimes t^j)\zeta^j/j\right),$$

$$E_1^-(+,\zeta) = \exp\left(-\sum_{j>0} (h \otimes t^{-j})\zeta^{-j}/j\right),$$

$$E_1^+(-,\zeta) = \exp\left(-\sum_{j>0} (h \otimes t^j)\zeta^j/j\right),$$

$$E_1^-(-,\zeta) = \exp\left(\sum_{j>0} (h \otimes t^{-j})\zeta^{-j}/j\right). \qquad (27)$$

Then

$$E_1^+(+,\zeta)E_1^+(-,\zeta) = E_1^+(-,\zeta)E_1^+(+,\zeta) = 1,$$
$$E_1^-(+,\zeta)E_1^-(-,\zeta) = E_1^-(-,\zeta)E_1^-(+,\zeta) = 1. \qquad (28)$$

Set

$$X'(+,\zeta) = E_1^-(-,\zeta)\zeta\zeta^- e^+ E_1^+(-,\zeta),$$

$$X'(-,\zeta) = E_1^-(+,\zeta)\zeta\zeta^+ e^- E_1^+(+,\zeta). \qquad (29)$$

When $E_1^+(\pm,\zeta)$, $E_1^-(\pm,\zeta)$, $\zeta\zeta^\mp e^\pm$ or $X'(\pm,\zeta)$ is expanded as a formal Laurent series in ζ, the coefficient of ζ^j ($j \in \mathbf{Z}$) is an operator of degree j on $S(\hat{\mathfrak{h}}_-) \otimes \mathbf{C}[P]$. To verify this for $\zeta\zeta^- e^+$, let $y \in S(\hat{\mathfrak{h}}_-)$ be of degree $i \leq 0$, and let $p \in P$. Then

$$\zeta\zeta^- e^+ \cdot (y \otimes e^p) = \zeta^{-2p-1}(y \otimes e^{p+1}),$$

and by (21),

$$\deg(y \otimes e^p) = i - p^2, \qquad \deg(y \otimes e^{p+1}) = i - p^2 - 2p - 1.$$

Thus if A_k ($k \in \mathbf{Z}$) denotes the coefficient of ζ^k in the expansion of $\zeta\zeta^- e^+$, then we have

$$A_k \cdot (y \otimes e^p) = 0$$

unless $k = -2p - 1$, in which case

$$\deg(A_k \cdot (y \otimes e^p)) - \deg(y \otimes e^p) = k.$$

The verification for $\zeta\zeta^+ e^-$ is similar.

For $j \in \mathbf{Z}$, denote by $X_j'(\pm)$ the coefficient of ζ^j in the expansion of $X'(\pm,\zeta)$, so that

$$X'(+,\zeta) = \sum_{j\in\mathbf{Z}} X_j'(+)\zeta^j, \qquad X'(-,\zeta) = \sum_{j\in\mathbf{Z}} X_j'(-)\zeta^j, \qquad (30)$$

and $X_j'(\pm)$ are both operators of degree j.

THEOREM 3. *The operators* c, $h \otimes t^j$ *and* $X_j'(\pm)$ *($j \in \mathbf{Z}$) on* $S(\hat{\mathfrak{h}}_-) \otimes$ $\mathbf{C}[P]$ *satisfy the bracket relations* (1), *with* $X_j'(+)$ *in place of* $e \otimes t^j$ *and* $X_j'(-)$ *in place of* $f \otimes t^j$, *so that these operators span a Lie algebra isomorphic to* $\hat{\mathfrak{g}}$. *The subspaces*

$$S(\hat{\mathfrak{h}}_-) \otimes \mathbf{C}[\mathbf{Z}] \quad \text{and} \quad S(\hat{\mathfrak{h}}_-) \otimes e^{1/2}\mathbf{C}[\mathbf{Z}],$$

whose direct sum is $S(\hat{\mathfrak{h}}_-) \otimes \mathbf{C}[P]$, *are invariant and irreducible under* $\hat{\mathfrak{g}}$.

PROOF. The bracket relations corresponding to those in the first two lines of (1) follow from (22) and (23).

For $i \in \mathbf{Z}\backslash\{0\}$,

$$\left[h \otimes t^i, e^{\pm}\right] = 0, \qquad \left[h \otimes t^i, \zeta^{\pm}\right] = 0 \tag{31}$$

by (22), (24) and (25), and

$$\left[h, e^{\pm}\right] = \pm 2e^{\pm}, \qquad \left[h, \zeta^{\pm}\right] = 0 \tag{32}$$

by (23)–(25). Using (27) and (29) as in the proof of Theorem 1, we obtain

$$\left[h \otimes t^i, X'(\pm, \zeta)\right] = \pm 2\zeta^{-i}X'(\pm, \zeta)$$

for all $i \in \mathbf{Z}$, establishing the bracket relations corresponding to those in the third and fourth lines of (1) (see (30)).

To establish the relations corresponding to those in the last two lines of (3), we observe that

$$E_1^+(\pm, \zeta_1)E_1^-(\pm, \zeta_2) = E_1^-(\pm, \zeta_2)E_1^+(\pm, \zeta_1)$$

$$\cdot \exp\left(-\sum_{j>0} 2(\zeta_1/\zeta_2)^j/j\right)$$

$$= E_1^-(\pm, \zeta_2)E_1^+(\pm, \zeta_1) \exp\left(\log(1 - \zeta_1/\zeta_2)^2\right)$$

$$= E_1^-(\pm, \zeta_2)E_1^+(\pm, \zeta_1)(1 - \zeta_1/\zeta_2)^2,$$

$$E_1^+(\pm, \zeta_1)E_1^-(\mp, \zeta_2) = E_1^-(\mp, \zeta_2)E_1^+(\pm, \zeta_1)(1 - \zeta_1/\zeta_2)^{-2},$$

as in the proof of Theorem 1. Using (26) and (31), we thus obtain

$$X'(\pm, \zeta_1)X'(\pm, \zeta_2) = E_1^-(\mp, \zeta_1)E_1^-(\mp, \zeta_2)\zeta_1\zeta_1^{\mp}e^{\pm}\zeta_2\zeta^{\mp}e^{\pm}$$

$$\cdot E_1^+(\mp, \zeta_1)E_1^+(\mp, \zeta_2)(1 - \zeta_1/\zeta_2)^2$$

$$= E_1^-(\mp, \zeta_1)E_1^-(\mp, \zeta_2)\zeta_1^{\mp}\zeta_2^{\mp}e^{\pm}e^{\pm}$$

$$\cdot E_1^+(\mp, \zeta_1)E_1^+(\mp, \zeta_2)\zeta_1\zeta_2^3(1 - \zeta_1/\zeta_2)^2$$

$$= E_1^-(\mp, \zeta_1)E_1^-(\mp, \zeta_2)\zeta_1^{\mp}\zeta_2^{\mp}e^{\pm}e^{\pm}$$

$$\cdot E_1^+(\mp, \zeta_1)E_1^+(\mp, \zeta_2)\zeta_1\zeta_2(\zeta_2 - \zeta_1)^2$$

and similarly,

$$\begin{aligned}
X'(\pm, \zeta_1)X'(\mp, \zeta_2) &= E_1^-(\mp, \zeta_1)E_1^-(\pm, \zeta_2)\zeta_1^{\mp}\zeta_2^{\pm}e^{\pm}e^{\mp}E_1^+(\mp, \zeta_1) \\
&\quad \cdot E_1^+(\pm, \zeta_2)(\zeta_1/\zeta_2)(1 - \zeta_1/\zeta_2)^{-2} \\
&= E_1^-(\mp, \zeta_1)E_1^-(\pm, \zeta_2)E_1^+(\mp, \zeta_1)E_1^+(\pm, \zeta_2) \\
&\quad \cdot (\zeta_1/\zeta_2)^{\mp}(\zeta_1/\zeta_2)(1 - \zeta_1/\zeta_2)^{-2}.
\end{aligned}$$

Hence

$$[X'(\pm, \zeta_1), X'(\pm, \zeta_2)] = 0$$

and

$$\begin{aligned}
&[X'(+, \zeta_1), X'(-, \zeta_2)] \\
&= E_1^-(-, \zeta_1)E_1^-(+, \zeta_2)E_1^+(-, \zeta_1)E_1^+(+, \zeta_2) \\
&\quad \cdot \left((\zeta_1/\zeta_2)^-(\zeta_1/\zeta_2)(1 - \zeta_1/\zeta_2)^{-2} \right. \\
&\quad \left. - (\zeta_2/\zeta_1)^+(\zeta_2/\zeta_1)(1 - \zeta_2/\zeta_1)^{-2} \right) \\
&= E_1^-(-, \zeta_1)E_1^-(+, \zeta_2)E_1^+(-, \zeta_1)E_1^+(+, \zeta_2)(\zeta_1/\zeta_2)^- \\
&\quad \cdot \left((\zeta_1/\zeta_2)(1 - \zeta_1/\zeta_2)^{-2} - (\zeta_2/\zeta_1)(1 - \zeta_2/\zeta_1)^{-2} \right) \\
&= E_1^-(-, \zeta_1)E_1^-(+, \zeta_2)E_1^+(-, \zeta_1) \\
&\quad \cdot E_1^+(+, \zeta_2)(\zeta_1/\zeta_2)^-(D\delta)(\zeta_1/\zeta_2) \\
&= (D\delta)(\zeta_1/\zeta_2) - \delta(\zeta_1/\zeta_2) \\
&\quad \cdot \left(-\sum_{j>0} (h \otimes t^{-j})\zeta_1^{-j} - \sum_{j>0} (h \otimes t^j)\zeta_1^j - h \right) \\
&= (D\delta)(\zeta_1/\zeta_2) + \delta(\zeta_1/\zeta_2)\left(\sum_{j\in\mathbf{Z}} (h \otimes t^j)\zeta_1^j \right) \\
&= (D\delta)(\zeta_1/\zeta_2) + \delta(\zeta_1/\zeta_2)\left(\sum_{j\in\mathbf{Z}} (h \otimes t^j)\zeta_2^j \right),
\end{aligned}$$

by Lemma 1 and (28), together with the fact that

$$D_1(\zeta_1/\zeta_2)^- = -(\zeta_1/\zeta_2)^- h,$$

which we establish as usual by applying both sides to a vector; see (23) and (25). Since these brackets agree with those in the last two lines of (3), we have shown that the operators c, $h \otimes t^j$ and $X_j'(\pm)$ satisfy the bracket relations (1).

By (20), the subspaces

$$S(\hat{\mathfrak{h}}_-) \otimes \mathbf{C}[\mathbf{Z}] \quad \text{and} \quad S(\hat{\mathfrak{h}}_-) \otimes e^{1/2}\mathbf{C}[\mathbf{Z}]$$

have direct sum $S(\mathfrak{h}_-) \otimes \mathbf{C}[P]$, and these two spaces are clearly $\hat{\mathfrak{g}}$-invariant. For the $\hat{\mathfrak{g}}$-irreducibility of $S(\hat{\mathfrak{h}}_-) \otimes \mathbf{C}[\mathbf{Z}]$ (the irreducibility for the other direct summand being similar), let W be a nonzero invariant subspace. Since W is h-invariant, W contains an element of the form $y \otimes e^p$ for some nonzero $y \in S(\hat{\mathfrak{h}}_-)$ and $p \in \mathbf{Z}$. Applying a suitable succession of operators from $\hat{\mathfrak{h}}'$ (recalling the irreducibility of $\hat{\mathfrak{h}}'$ on $S(\hat{\mathfrak{h}}_-)$), we see that $e^p \in W$. Now the coefficients in the formal Laurent series

$$E_1^-(\pm, \zeta) X'(\pm, \zeta) E_1^+(\pm, \zeta) = \zeta \zeta^{\mp} e^{\pm}$$

map W to W, and so W is stable under $e^{\pm}$. But this shows that $\mathbf{C}[\mathbf{Z}] \subset W$, and thus that $W = S(\hat{\mathfrak{h}}_-) \otimes \mathbf{C}[\mathbf{Z}]$. ∎

REMARK. The vector

$$v_0 = 1 \otimes 1 \in S(\hat{\mathfrak{h}}_-) \otimes \mathbf{C}[\mathbf{Z}]$$

is nonzero and satisfies the conditions

$$\left(h \otimes t^j\right) \cdot v_0 = \left(e \otimes t^{j-1}\right) \cdot v_0$$
$$= \left(f \otimes t^j\right) \cdot v_0 = 0 \quad \text{for } j > 0, \tag{33}$$

$$c \cdot v_0 = v_0, \tag{34}$$

$$h \cdot v_0 = 0, \tag{35}$$

and it is not hard to see that any $\mathbf{Z}$-graded irreducible representation of $\hat{\mathfrak{g}}$ containing a nonzero homogeneous vector v_0 satisfying (33)–(35) must be equivalent to $S(\hat{\mathfrak{h}}_-) \otimes \mathbf{C}[\mathbf{Z}]$. But the representation of $\mathfrak{k}$ on $S(\hat{\mathfrak{s}}_-)$ given in Theorem 1, viewed as a representation of $\hat{\mathfrak{g}}$, also has this property, using the vector $1 \in S(\hat{\mathfrak{s}}_-)$; thus the two representations are equivalent. (Condition (34) holds because the isomorphism from $\hat{\mathfrak{g}}$ to $\mathfrak{k}$ takes c to $2c$. To establish that $S(\hat{\mathfrak{s}}_-)$ is $\mathbf{Z}$-graded as a $\hat{\mathfrak{g}}$-representation, let d be the derivation of $\hat{\mathfrak{g}}$ which acts as $1 \otimes t(d/dt)$ on $\mathfrak{g} \otimes \mathbf{C}[t, t^{-1}]$ and as 0 on $\mathbf{C}c$, and let d' be the restriction of d to the subalgebra $\mathfrak{k}$ of $\hat{\mathfrak{g}}$. Form the larger Lie algebras $\tilde{\mathfrak{g}} = \hat{\mathfrak{g}} \oplus \mathbf{C}d$ and $\mathfrak{k}' = \mathfrak{k} \oplus \mathbf{C}d'$. The isomorphism from $\hat{\mathfrak{g}}$ to $\mathfrak{k}$ extends to an isomorphism from $\tilde{\mathfrak{g}}$ to $\mathfrak{k}'$ taking d to $-\frac{1}{4}h + \frac{1}{2}d'$. The fact that $S(\hat{\mathfrak{s}}_-)$ is a $\mathbf{Z}$-graded $\mathfrak{k}$-representation implies that $S(\hat{\mathfrak{s}}_-)$ can be extended to a $\mathfrak{k}'$-representation for which d' acts as multiplication by the scalar j on the space $S(\hat{\mathfrak{s}}_-)_j$ for $j \in -\mathbf{N}$. But then $S(\hat{\mathfrak{s}}_-)$ is a $\tilde{\mathfrak{g}}$-representation generated by the d-annihilated vector 1, and so $S(\hat{\mathfrak{s}}_-)$ is a $\mathbf{Z}$-graded $\hat{\mathfrak{g}}$-representation.) Analogously, the irreducible representation $S(\hat{\mathfrak{h}}_-) \otimes e^{1/2}\mathbf{C}[\mathbf{Z}]$ of $\hat{\mathfrak{g}}$ in Theorem 3 is equivalent to the representation described in the remark after Theorem 1. In this case, conditions (33)–(35) are replaced by (33), (34) and $h \cdot v_0 = v_0$; for the vectors v_0, we take

$$1 \otimes e^{1/2} \in S(\hat{\mathfrak{h}}_-) \otimes e^{1/2}\mathbf{C}[\mathbf{Z}], \qquad 1 \in S(\hat{\mathfrak{s}}_-).$$

6. Construction by operators on $S(\hat{\mathfrak{h}}_-) \otimes \mathbf{C}[Z] \otimes \Lambda(\mathfrak{w}_-^{1/2})$ **and on** $S(\hat{\mathfrak{h}}_-) \otimes e^{1/2}\mathbf{C}[Z] \otimes \Lambda(\mathfrak{w}_-^0) \otimes \mathbf{C}^2$. Our final constructions of $A_1^{(1)}$ are related to those in §5 roughly as those in §4 are related to those in §3.

We use the basis and grading of $\hat{\mathfrak{g}}$ considered at the beginning of §5, and we also use the subalgebras $\hat{\mathfrak{h}}$, $\hat{\mathfrak{h}}'$ and $\hat{\mathfrak{h}}_-$, the $\mathbf{Z}$-graded polynomial algebra $S(\hat{\mathfrak{h}}_-)$ (see (19)), the group algebra $\mathbf{C}[P]$ and the operators $e^{\pm}$ and $\zeta^{\pm}$ on the space $S(\hat{\mathfrak{h}}_-) \otimes \mathbf{C}[P]$ (see (24) and (25)). Instead of using (21), we make $\mathbf{C}[P]$ a $\mathbf{Z}/8$-graded vector space by the condition

$$\deg(e^p) = -\tfrac{1}{2}p^2$$

for all $p \in \tfrac{1}{2}\mathbf{Z}$. We give $S(\hat{\mathfrak{h}}_-) \otimes \mathbf{C}[P]$ the $\hat{\mathfrak{h}}$-module structure determined by

$$c \mapsto 2, \quad h \otimes t^j \mapsto 4j(\partial/\partial y_j) \otimes 1, \quad h \otimes t^{-j} \mapsto y_j \otimes 1 \qquad (36)$$

for $j > 0$ (cf. (22)), and

$$h \cdot (y \otimes e^p) = 2p(y \otimes e^p) \qquad (37)$$

for $p \in P$, as in (23).

Let $t = 0$ or $\tfrac{1}{2}$, and denote by $\mathcal{D}^t$ (respectively, $\mathcal{D}_*^t$) the Clifford algebra on generators W_j, $j \in \mathbf{Z} + t$ (respectively, $j \in \mathbf{Z} + t, j \neq 0$) subject to the relations

$$\{W_i, W_j\} = 2\delta_{i+j,0} \qquad (38)$$

for all $i, j \in \mathbf{Z} + t$ (respectively, $i, j \in \mathbf{Z} + t$, $i, j \neq 0$). For $t = \tfrac{1}{2}$, $\mathcal{D}^t = \mathcal{D}_*^y$. Let $\mathfrak{w}_-^t = \mathrm{span}\{W_j | j < 0\}$. The elements $\{W_j, j < 0$, are linearly independent, and $\mathcal{D}_*^t$ has a standard irreducible representation on the exterior algebra $\Lambda(\mathfrak{w}_-^t)$ on the generators $W_j, j < 0$, determined by the anticommutation relations (38) and the conditions

$$W_j \cdot 1 = 0, \qquad W_{-j} \cdot w = W_{-j} \wedge w$$

for $j \in \mathbf{Z} + t, j > 0$ and $w \in \Lambda(\mathfrak{w}_-^t)$.

Denote by $\Lambda(\mathfrak{w}_-^t)_{\mathrm{even}}$ and $\Lambda(\mathfrak{w}_-^t)_{\mathrm{odd}}$ the subspaces of $\Lambda(\mathfrak{w}_-^t)$ spanned by the exterior products of even and odd numbers of W_j's, respectively, Then

$$\Lambda(\mathfrak{w}_-^t) = \Lambda(\mathfrak{w}_-^t)_{\mathrm{even}} \oplus \Lambda(\mathfrak{w}_-^t)_{\mathrm{odd}}.$$

For $t = 0$, define a representation of $\mathcal{D}^0$ on the space $\Lambda(\mathfrak{w}_-^0) \otimes \mathbf{C}^2$, where $\mathbf{C}^2$ is viewed as the space of two-rowed column vectors, by the relations (38) and the conditions

$$W_j \cdot (w \otimes u) = (W_j \cdot w) \otimes u \quad \text{for } j \neq 0, w \in \Lambda(\mathfrak{w}_-^0), u \in \mathbf{C}^2,$$

$$W_0 \cdot (1 \otimes u) = 1 \otimes \begin{pmatrix} 0 & 1 \\ 1 & 0 \end{pmatrix} u \quad \text{for } u \in \mathbf{C}^2.$$

The only proper nonzero $\mathcal{D}^0$-invariant subspaces of $\Lambda(\mathfrak{w}^0) \otimes \mathbf{C}^2$ are the two spaces

$$\Lambda(\mathfrak{w}^0_-) \otimes (v_1 + v_2) \quad \text{and} \quad \Lambda(\mathfrak{w}^0_-) \otimes (v_1 - v_2),$$

where v_1 and v_2 are the standard basis elements of $\mathbf{C}^2$.

For $t = 0$ or $\frac{1}{2}$, set

$$W(\zeta) = \sum_{j \in \mathbf{Z}+t} W_j \zeta^j = \sum_{j \in (1/2)\mathbf{Z}} W_j \zeta^j, \qquad (39)$$

where we put $W_j = 0$ if $j \in \mathbf{Z} + \frac{1}{2} + t$.

Equip the algebra $\Lambda(\mathfrak{w}^t_-)$ with the (nonpositive) $\frac{1}{2}\mathbf{Z}$-grading determined by the conditions $\deg W_j = j$ for $j < 0$. Then for all $k \in \mathbf{Z} + t$, $k \neq 0$, the operator W_k on $\Lambda(\mathfrak{w}_-)$ has degree k. For $t = 0$, give the space $\Lambda(\mathfrak{w}^0_-) \otimes \mathbf{C}^2$ the (nonpositive) $\mathbf{Z}$-grading such that

$$\deg(w \otimes u) = \deg w$$

for every homogeneous element $w \in \Lambda(\mathfrak{w}^0_-)$ and every $u \in \mathbf{C}^2$. Then for $k \in \mathbf{Z}$, the operator W_k on $\Lambda(\mathfrak{w}^0_-) \otimes \mathbf{C}^2$ has degree k.

We shall construct $\hat{\mathfrak{g}}$ as Lie algebras of operators on two subspaces of the (nonpositively) $\frac{1}{2}\mathbf{Z}$-graded vector space

$$S(\hat{\mathfrak{h}}_-) \otimes \mathbf{C}[\mathbf{Z}] \otimes \Lambda(\mathfrak{w}^{1/2}_-), \qquad (40)$$

and also on a subspace of the (negatively) $\mathbf{Z}/8$-graded vector space

$$S(\hat{\mathfrak{h}}_-) \otimes e^{1/2}\mathbf{C}[\mathbf{Z}] \otimes \Lambda(\mathfrak{w}^0_-) \otimes \mathbf{C}^2. \qquad (41)$$

We shall use the notation $\mathbf{C}[2\mathbf{Z}]$ for the subspace of $\mathbf{C}[\mathbf{Z}]$ spanned by the elements e^p for $p \in 2\mathbf{Z}$. Then

$$\mathbf{C}[\mathbf{Z}] = \mathbf{C}[2\mathbf{Z}] \oplus e^1\mathbf{C}[2\mathbf{Z}].$$

Define subspaces Λ_0 and Λ_1 of $\Lambda(\mathfrak{w}^0_-) \otimes \mathbf{C}^2$ by

$$\Lambda_0 = \Lambda(\mathfrak{w}^0_-)_{\text{even}} \otimes v_1 \oplus \Lambda(\mathfrak{w}^0_-)_{\text{odd}} \otimes v_2,$$

$$\Lambda_1 = \Lambda(\mathfrak{w}^0_-)_{\text{odd}} \otimes v_1 \oplus \Lambda(\mathfrak{w}^0_-)_{\text{even}} \otimes v_2.$$

Set

$$U^{(0)} = \mathbf{C}[2\mathbf{Z}] \otimes \Lambda(\mathfrak{w}^{1/2}_-)_{\text{even}} \oplus e^1\mathbf{C}[2\mathbf{Z}] \otimes \Lambda(\mathfrak{w}^{1/2}_-)_{\text{odd}},$$

$$U^{(1)} = \mathbf{C}[2\mathbf{Z}] \otimes \Lambda(\mathfrak{w}^{1/2}_-)_{\text{odd}} \oplus e^1\mathbf{C}[2\mathbf{Z}] \otimes \Lambda(\mathfrak{w}^{1/2}_-)_{\text{even}},$$

$$U^{(2)} = e^{1/2}\mathbf{C}[2\mathbf{Z}] \otimes \Lambda_0 \oplus e^{3/2}\mathbf{C}[2\mathbf{Z}] \otimes \Lambda_1,$$

and define

$$V^{(0)} = S(\hat{\mathfrak{h}}_-) \otimes U^{(0)}, \quad V^{(1)} = S(\hat{\mathfrak{h}}_-) \otimes U^{(1)}, \quad V^{(2)} = S(\hat{\mathfrak{h}}_-) \otimes U^{(2)}.$$

The spaces $V^{(0)}$ and $V^{(1)}$ are subspaces of the space (40), and $V^{(2)}$ is contained in the space (41). Using subscripts to designate homogeneous components, we have

$$V^{(0)} = \bigoplus_{j \in -\mathbf{N}} V_j^{(0)}, \quad V^{(1)} = \bigoplus_{j \in -1/2 - \mathbf{N}} V_j^{(1)}, \quad V^{(2)} = \bigoplus_{j \in -1/8 - \mathbf{N}} V_j^{(2)}.$$

Using the formulas (36) and (37), we provide the spaces (40) and (41) with the action of $\hat{\mathfrak{h}}$ in the obvious way. Then $\hat{\mathfrak{h}}$ also acts on the subspaces $V^{(0)}$, $V^{(1)}$ and $V^{(2)}$.

The operators $e^{\pm}$, $\zeta^{\pm}$ and $W(\zeta)$ (see (39)) also act in natural ways on the spaces (40) and (41).

Now define

$$E_2^{\pm}(+, \zeta) = \exp\left(\pm \sum_{j>0} \left(h \otimes t^{\pm j} \right) \zeta^{\pm j} / 2j \right),$$

$$E_2^{\pm}(-, \zeta) = \exp\left(\mp \sum_{j>0} \left(h \otimes t^{\pm j} \right) \zeta^{\pm j} / 2j \right)$$

(cf. (27)), and

$$X''(+, \zeta) = E_2^{-}(-, \zeta) \zeta^{1/2} \left(\zeta^{1/2} \right)^{-} e^{+} W(\zeta) E_2^{+}(-, \zeta),$$

$$X''(-, \zeta) = E_2^{-}(+, \zeta) \zeta^{1/2} \left(\zeta^{1/2} \right)^{+} e^{-} W(\zeta) E_2^{+}(+, \zeta)$$

(cf. (29)). Arguing as in §5, we see that when $E_2^{+}(\pm, \zeta)$, $E_2^{-}(\pm, \zeta)$, $\zeta^{1/2}(\zeta^{1/2})^{\mp} e^{\pm} W(\zeta)$ or $X''(\pm, \zeta)$ is expanded as a formal Laurent series in $\zeta^{1/2}$ on either the space (40) or the space (41), then the coefficient of ζ^{j} is zero if $j \in \mathbf{Z} + \frac{1}{2}$, and is an operator of degree j if $j \in \mathbf{Z}$.

For $j \in \mathbf{Z}$, define the operators $X_j''(\pm)$ of degree j on the space (40) or (41) by the expansions

$$X''(+, \zeta) = \sum_{j \in \mathbf{Z}} X_j''(+) \zeta^{j},$$

$$X''(-, \zeta) = \sum_{j \in \mathbf{Z}} X_j''(-) \zeta^{j}$$

(cf. (30)). It is easy to see that the operators $X_j''(\pm)$ preserve the subspaces $V^{(0)}$, $V^{(1)}$ and $V^{(2)}$.

THEOREM 4. *The operators c, $h \otimes t^j$ and $X_j''(\pm)$ ($j \in \mathbf{Z}$) on the space $V^{(0)}$, $V^{(1)}$ or $V^{(2)}$ satisfy the bracket relations (1), with $X_j''(+)$ in place of $e \otimes t^j$ and $X_j''(-)$ in place of $f \otimes t^j$, so that these operators span a Lie algebra isomorphic to $\hat{\mathfrak{g}}$. Each of the spaces $V^{(0)}$, $V^{(1)}$ and $V^{(2)}$ is irreducible under $\hat{\mathfrak{g}}$.*

The proof, using the techniques in the proofs of Theorems 1–3, is left as an exercise for the reader. ∎

REMARK. As in the remark at the end of §5, we have an equivalence between the representations constructed in §4 and those constructed in this section. Specifically, let $T^{(0)}$ be the representation $S(\hat{\mathfrak{s}}_-) \otimes \Lambda(\hat{\mathfrak{z}}_-^0)$ in Theorem 2, $T^{(1)}$ the representation with the same space but modified according to the remark after Theorem 2, and $T^{(2)}$ the representation $S(\hat{\mathfrak{s}}_-) \otimes \Lambda(\hat{\mathfrak{z}}_-^1)$ in Theorem 2. Then for $i = 0$, 1 or 2, the representations $T^{(i)}$ and $V^{(i)}$ of $\hat{\mathfrak{g}}$ are equivalent. In fact, condition (34) is replaced by the condition

$$c \cdot v_0 = 2v_0$$

in all three cases, and condition (35) is replaced by

$$h \cdot v_0 = 0, \quad h \cdot v_0 = 2v_0, \quad h \cdot v_0 = v_0,$$

respectively. We leave it as an exercise to exhibit the vectors v_0 in the six spaces.

REFERENCES

1. E. Date, M. Jimbo, M. Kashiwara and T. Miwa, *Transformation groups for soliton equations—Euclidean Lie algebras and reduction of the KP hierarchy*, Publ. Res. Inst. Math. Sci. **18** (1982), 1077–1110.

2. E. Date, M. Kashiwara and T. Miwa, *Vertex operators and τ functions—transformation groups for soliton equations*. I, Proc. Japan Acad. Ser. A Math. Sci. **57** (1981), 387–392.

3. I. B. Frenkel, *Two constructions of affine Lie algebra representations and boson-fermion correspondence in quantum field theory*, J. Funct. Anal. **44** (1981), 259–327.

4. I. B. Frenkel and V. G. Kac, *Basic representations of affine Lie algebras and dual resonance models*, Invent. Math. **62** (1980), 23–66.

5. V. G. Kac, (a) *Simple irreducible graded Lie algebras of finite growth*, Izv. Akad. Nauk SSSR Ser. Mat. **32** (1968), 1323–1367; English transl. in Math. USSR-Izv. **2** (1968), 1271–1311.

(b) *Infinite-dimensional Lie algebras and Dedekind's η-function*, Funkcional. Anal. i Prilozen. **8** (1974), 77–78; English transl. in Funkcional. Anal. Appl. **8** (1974), 68–70.

6. V. G. Kac, D. A. Kazhdan, J. Lepowsky and R. L. Wilson, *Realization of the basic representations of the Euclidean Lie algebras*, Adv. in Math. **42** (1981), 83–112.

7. I. L. Kantor, (a) *Simple graded infinite-dimensional Lie algebras*, Dokl. Akad. Nauk SSSR **179** (1968), 534–537; English transl. in Soviet Math. Dokl. **9** (1968), 409–412.

(b) *Graded Lie algebras*, Trudy Sem. Vektor. Tenzor. Anal. **15** (1970), 227–266. (Russian)

8. J. Lepowsky, *Affine Lie algebras and combinatorial identities* (Proc. 1981 Rutgers Conf. Lie Algebras and Related Topics), Lecture Notes in Math., vol. 933, Springer-Verlag, Berlin and New York, 1982, pp. 130–156.

9. J. Lepowsky and S. Milne, *Lie algebraic approaches to classical partition identities*, Adv. in Math. **29** (1978), 15–59.

10. J. Lepowsky and M. Primc, *Standard modules for type one affine Lie algebras* (Proc. 1982 New York Number Theory Seminar), Lecture Notes in Math., Springer-Verlag, Berlin and New York (to appear).

11. J. Lepowsky and R. L. Wilson, (a) *Construction of the affine Lie algebra $A_1^{(1)}$*, Comm. Math. Phys. **62** (1978), 43–53.

(b) *The Rogers-Ramanujan identities: Lie theoretic interpretation and proof*, Proc. Nat. Acad. Sci. U.S.A. **78** (1981), 699–701.

(c) *A Lie theoretic interpretation and proof of the Rogers-Ramanujan identities*, Adv. in Math. **45** (1982), 21–72.

(d) *A new family of algebras underlying the Rogers-Ramanujan identities and generalizations*, Proc. Nat. Acad. Sci. U.S.A. **78** (1981), 7254–7258.

(e) *The structure of standard modules*, I. *Universal algebras and the Rogers-Ramanujan identities*, Invent. Math. (to appear).

12. S. Mandelstam, *Dual resonance models*, Phys. Rep. C **13** (1974), 259–353.

13. R. Marcuson, *Tits' systems in generalized nonadjoint Chevalley groups*, J. Algebra **34** (1975), 84–96.

14. R. V. Moody, *A new class of Lie algebras*, J. Algebra **10** (1968), 211–230.

15. J. H. Schwarz, *Dual resonance theory*, Phys. Rep. C **8** (1973), 269–335.

16. G. Segal, *Unitary representations of some infinite-dimensional groups*, Comm. Math. Phys. **80** (1981), 301–342.

DEPARTMENT OF MATHEMATICS, RUTGERS UNIVERSITY, NEW BRUNSWICK, NEW JERSEY 08903

Lectures in Applied Mathematics
Volume **21**, 1985

Nonlinear Representations and the Affine Group of the Complex Plane

Jacques C. H. Simon

I. Introduction. Actions of Lie groups in topological vector spaces appear naturally in the theory of nonlinear autonomous partial differential equations. The one-parameter group of time evolution of the solutions is such an example. However, one-parameter groups do not contain many algebraic properties when considered as Lie groups of transformations. Local transformation groups of higher dimension appear when we deal with certain covariant equations of evolution.

Famous examples are given by relativistic wave equations. These equations are covariant under the Poincaré group which is a semidirect product $P = \mathbf{R}^4 \cdot \mathrm{SL}(2, \mathbf{C})$. Relativistic wave equations are of the form $d\psi_t/dt = A(\psi_t)$, where ψ_t belongs to a Banach space $\mathcal{H}$ of functions from $\mathbf{R}^3$ to a vector space E and $A = L + N$, where L is a linear (unbounded) operator in $\mathcal{H}$, and N is analytic around 0 in $\mathcal{H}$ with $N(0) = N'(0) = 0$. From a heuristic point of view, covariance means that given $g = (a, \Lambda) \in P$ and a local solution $\psi(t, x) = \psi_t(x)$ of the equation then $\psi'_t(x) = \psi'(t, x) = S(\Lambda)\psi(\tilde{\Lambda}^{-1}((t, x) - a))$ (where (S, E) is a linear representation of $\mathrm{SL}(2, \mathbf{C})$ in E and $\tilde{\Lambda}$ is the projection of Λ in $\mathrm{SO}_0(3, 1)$) is a local solution of $d\psi'_t/dt = A(\psi'_t)$. The mapping $(g, \psi_0) \to \psi'_0$ is a local action of P on $\mathcal{H}$ which leaves 0 invariant. This is the notion of local action of a Lie group on a Banach space that leaves 0 invariant which is our main subject of study. We shall make this notion more precise in the following:

Given two locally convex topological vector spaces $\mathcal{E}_1$ and $\mathcal{E}_2$, we denote by $\mathcal{F}(\mathcal{E}_1, \mathcal{E}_2)$ the space of formal power series from $\mathcal{E}_1$ to $\mathcal{E}_2$ of

1980 *Mathematics Subject Classification.* Primary 57S20, 22E99.
Key words and phrases. Affine group, nonlinear representation, wave equation.

the form $f = \Sigma_{n \geqslant 1} f^n$ where $f^n \in \mathcal{L}_n(\mathcal{E}_1, \mathcal{E}_2)$ (the space of continuous n-linear symmetric mappings from $\mathcal{E}_1$ to $\mathcal{E}_2$). We shall write $\mathcal{L}(\mathcal{E}_1, \mathcal{E}_2) = \mathcal{L}_1(\mathcal{E}_1, \mathcal{E}_2)$ and, when $\mathcal{E} = \mathcal{E}_1 = \mathcal{E}_2$, $\mathcal{F}(\mathcal{E}) = \mathcal{F}(\mathcal{E}, \mathcal{E})$ and $\mathcal{L}_n(\mathcal{E}) = \mathcal{L}_n(\mathcal{E}, \mathcal{E})$.

1.1. DEFINITION. An analytic representation $(S, \mathcal{E})$ of a real connected Lie group G in a Banach space $\mathcal{E}$ is a homomorphism $g \to S_g$ from G to the group of invertible elements in $\mathcal{F}(\mathcal{E})$ such that, if $S_g = \Sigma_{n \geqslant 1} S_g^n$, $S_g^n \in \mathcal{L}_n(\mathcal{E})$.

(1) The mapping $(g, \varphi_1, \ldots, \varphi_n) \to S_g^n(\varphi_1, \ldots, \varphi_n)$ is continuous from $G \times \mathcal{E}^n$ to $\mathcal{E}$.

(2) There exists an open neighbourhood V of the identity in G such that, for any $g \in V$, $S_g = \Sigma_{n \geqslant 1} S_g^n$ is an analytic mapping in an open neighbourhood θ_g of the origin in $\mathcal{E}$. The mapping $g \to S_g^1$ is then a continuous linear representation of G in $\mathcal{E}$ which we call the linear part of $(S, \mathcal{E})$.

Let us focus back to the Poincaré group. The Lie algebra $\mathfrak{t}_4$ of the group of space time translations is endowed with a natural quadratic form q of signature $(+ - - -)$; we choose an orthogonal basis (P_0, P_1, P_2, P_3) of $\mathfrak{t}_4$ with $q(P_0, P_0) = +1$, $q(P_k, P_k) = -1$ $(k = 1, 2, 3)$, we denote by $(U, \mathcal{H})$ a unitary representation of P such that $(dU_{P_0})^2 - \Sigma_{k=1}^3 (dU_{P_k})^2 = m^2 I$, $m^2 > 0$, and such that the spectrum of $id\, U_{P_0}$ has a definite sign. Given $a > 0$, we denote by $\mathcal{H}(a)$ the space of $\varphi \in \mathcal{H}$ such that, for any $n \geqslant 0$, $\|(dU_{P_k})^n \varphi\| \leqslant a^n \|\varphi\|$ $(k = 1, 2, 3)$; this is a closed subspace of $\mathcal{H}$. The strict inductive limit $\mathcal{H}_c$ of the family of Hilbert spaces $\mathcal{H}(a)$ $(a > 0)$ is dense in $\mathcal{H}$ and is invariant under U.

The following proposition was proved in [4].

1.2. PROPOSITION. *Given an analytic representation $(S, \mathcal{H})$ of P, the linear part of which is $(U, \mathcal{H})$, there exists a unique formal power series $A = \Sigma_{n \geqslant 1} A^n \in \mathcal{F}(\mathcal{H}_c, \mathcal{H})$ such that:*

(1) *A^1 is the canonical imbedding from $\mathcal{H}_c$ to $\mathcal{H}$.*

(2) *Given any $a > 0$, A is analytic in a neighbourhood of 0 in $\mathcal{H}(a)$.*

(3) *$S_g A = A S_g^1$ on $\mathcal{H}_c$.*

A variant of this proposition can be applied to the wave equation $(\partial/\partial t^2 - \Delta)\varphi + m^2 \varphi = J(\varphi)$, where $m^2 > 0$, J is analytic around 0 in $\mathbf{C}$ and valued in $\mathbf{C}$, $J(0) = J'(0)$, and φ is a function from $\mathbf{R}^3$ to $\mathbf{C}$. We take as the space of initial data the set of $(\varphi, \dot{\varphi})$ such that the Fourier transform $\hat{\varphi} \in L^1(\mathbf{R}^3)$ and the mapping $p \to (m^2 + |p|^2)^{-1/2}\hat{\dot{\varphi}}(p)$ is in $L^1(\mathbf{R}^3)$. It was proved [4] that there exists a subset (not reduced to 0) in the space of initial data which is a set of initial conditions of global solutions of the wave equation.

The proofs of these results, which depend on properties of the mass operator, fail when $m^2 \leqslant 0$. In fact, they do not really use the Lorentz part of P except in the fact that it commutes with the mass operator. We may, therefore, hope to get more information by studying analytic representations of $SO_0(3, 1)$. This is not yet performed but we can get results for some subgroups of $SO_0(3, 1)$. In the following we consider the affine group W of the complex plane which can be realized as the analytic subgroup of $SO_0(3, 1)$ associated to a Borel subalgebra of $\mathfrak{s}\mathfrak{o}(3, 1)$.

W is the group of transformations $z \to az + b$, $a \in \mathbf{C} - \{0\}$, $b \in \mathbf{C}$, in the complex plane. The (connected) Euclidean group of the plane is the subgroup of elements in W for which $|a| = 1$. The Lie algebra $\mathfrak{e}_2$ of E_2 has a basis $\{M, X_1, X_2\}$ on $\mathbf{R}$ with the following commutation relations:

$$[M, X_1] = -X_2, \qquad [M, X_2] = X_1, \qquad [X_1, X_2] = 0.$$

M generates a compact subgroup $K = \exp \mathbf{R} M$ of E_2 and X_1, X_2 generate the subgroup $R = \exp(\mathbf{R} X_1 + \mathbf{R} X_2)$ of the translations of the plane. A real basis of the Lie algebra $\mathfrak{w}$ of W is obtained by adding to $\{M, X_1, X_2\}$ the generator D of the one-parameter group of the homotheties $z \to az$, $a > 0$. We have

$$[D, M] = 0, \quad [D, X_i] = X_i \quad (i = 1, 2).$$

We suppose a unitary representation $(U, \mathcal{H})$ of W in a (countable) Hilbert space $\mathcal{H}$ is given, which does not contain one-dimensional representations of W. It can be seen from the knowledge of the unitary dual of $SO_0(3, 1)$ (see [3]) that any unitary representation of $SO_0(3, 1)$ which does not contain the trivial representation, has a restriction to W which satisfies the hypotheses of $(U, \mathcal{H})$.

Given $0 < a < b$, $\mathcal{D}_b(\mathcal{H})$ is the closed subspace of vectors $\varphi \in \mathcal{H}$ which are in the domain of $dU_{X_1 + iX_2}^n$ and $\|dU_{X_1 + iX_2}^n \varphi\| \leq b^n \|\varphi\|$ for any $n \geq 0$; $\mathcal{D}_{a,b}(\mathcal{H})$ is the closed subspace of $\varphi \in \mathcal{D}_b(\mathcal{H})$ which are in the domain of $dU_{X_1 + iX_2}^{-n}$ and $\|dU_{X_1 + iX_2}^{-n} \varphi\| \leq a^{-n} \|\varphi\|$ for any $n \geq 0$; we then define $\mathcal{D}(\mathcal{H}) = \bigcup_{0 < a < b} \mathcal{D}_{a,b}(\mathcal{H})$ which we endow with its natural LB space topology.

We denote by P_λ ($\lambda \in \mathbf{Z}$) the orthogonal projector on the eigenspace of dU_M associated to the eigenvalue $i\lambda$. Given $\alpha \geq 0$, we denote by $\mathcal{H}(K, \alpha)$ the Hilbert space of $\varphi \in \mathcal{H}$ such that

$$\|\varphi\|_{\alpha, K}^2 = \sum_{\lambda \in \mathbf{Z}} e^{2\alpha |\lambda|} \|P_\lambda \varphi\|^2 < +\infty.$$

We now introduce the Fréchet space $\mathcal{H}(K) = \bigcap_{\alpha \geq 0} \mathcal{H}(K, \alpha)$, and the LF space $\mathcal{H}_0 = \mathcal{H}(K) \cap \mathcal{D}(\mathcal{H})$. The topological vector space $\mathcal{H}_0$ is independent of the choice of the one-dimensional torus K in E_2 (see [5]). $\mathcal{H}_0$ is dense in $\mathcal{H}$ and is invariant under U_g ($g \in W$).

The main result of this article is the following:

1.3. PROPOSITION. *Given an analytic representation $(S, \mathcal{H})$ of W in a countable Hilbert space $\mathcal{H}$, the linear part of which $(S^1, \mathcal{H})$ is unitary and* *contains no one-dimensional representation of* W, *the following two properties are equivalent*:

(1) *There exists* $f \in \mathcal{F}(\mathcal{K}_0, \mathcal{K})$ *such that* f^1 *is the canonical injection from* $\mathcal{K}_0$ *to* $\mathcal{K}$ *and* $S_g f = f S_g^1$ *for any* $g \in W$.

(2) *There exists* $X \in \mathcal{L}_2(\mathcal{K}_0, \mathcal{K})$ *such that* $S_g^2 = S_g^1 X - X(S_g^1 \otimes S_g^1)$ *on* $\mathcal{K}_0$.

2. Preliminary lemmas.

2.1. LEMMA. *Suppose given an analytic representation* $(S, \mathcal{K})$ *of* W *in a Banach space* $\mathcal{K}$, *the linear part of which* $(S^1, \mathcal{K})$ *is isometric. We denote by* $\mathcal{K}^\infty$ *the Fréchet space of* C^∞ *vectors of* $(S^1, \mathcal{K})$. *There exists* $f \in \mathcal{F}(\mathcal{K})$ *analytic around the origin such that* $f^1 = I$ *and such that, if* $T_g = f S_g f^{-1}$ *and* $R_g = T_{g^{-1}}^1 T_g$,

(1) $\tilde{T}_k = T_k^1$ *when* $k \in K$,

(2) *for every* $n \geqslant 1$, $g \to R_g^n$ *is* C^∞ *from* W *to* $\mathcal{L}_n(\mathcal{K})$,

(3) *if* $x \in \mathfrak{w}$ *and* $dR_x^n = (d/dt)(R_{\exp tx}^n)_{t=0}$, *the mapping* $x \to dR_x^n$ *is linear from* $\mathfrak{w}$ *to* $\mathcal{L}_n(\mathcal{K}) \cap \mathcal{L}_n(\mathcal{K}^\infty)$,

(4) *with the notation* $(\text{ad } dS_x^1)Y = dS_x^1 Y - Y d(\otimes^n S^1)_x$ *if* $Y \in \mathcal{L}_n(\mathcal{K}^\infty)$, *we have*

$$\left\| \left(\text{ad } dS_{X_1}^1 \right)^{l_1} \left(\text{ad } dS_{X_2}^1 \right)^{l_2} \left(\text{ad } dS_D^1 \right)^{l_3} dR_x^n \right\| \leqslant K(n, l_3)^{l_1 + l_2} |x|,$$

with $l_1, l_2, l_3 \in \mathbf{N}$, $K(n, l_3) \geqslant 0$, $x \in \mathfrak{j} = \mathfrak{r} + \mathbf{R}D$ ($\mathfrak{r}$ *the Lie algebra of* R), *and* $|\ |$ *is a chosen norm on* $\mathfrak{j}$.

It results from [1, Proposition 5] that there exists $B \in \mathcal{F}(\mathcal{K})$ analytic around the origin in $\mathcal{K}$ and invertible such that if $\tilde{S}_g = B S_g B^{-1}$, the mapping $g \to \tilde{S}_{g^{-1}}^1 \tilde{S}_g$ is C^∞ from an open neighbourhood of the identity in W to the Banach space

$$H_r(\mathcal{K}) = \left\{ \varphi = \sum_{n \geqslant 1} \varphi^n \mid \varphi^n \in \mathcal{L}_n(\mathcal{K}), \ \sum_{n \geqslant 1} r^n \|\varphi^n\| < +\infty \right\}$$

for some $r > 0$;

and $\tilde{S}_k = \tilde{S}_k^1$ for all $k \in K$.

We can, therefore, suppose that $(S, \mathcal{K})$ has the properties of $(\tilde{S}, \mathcal{K})$. We then build a sequence $A^p \in \mathcal{L}_p(\mathcal{K}) \cap \mathcal{L}_p(\mathcal{K}^\infty)$ ($p \geqslant 1$), $A^1 = I$ such that

(a) $S_k^1 A^o = A^p S_k^1$ if $k \in K$,

(b) $\|A^p\| \leqslant (2r)^{1-p} (\sum_{q=0}^{p-2} 2^{-q})^{-p}$, $p \geqslant 2$ for some $r > 0$,

(c) $g \to S_{g^{-1}}^1 A^p S_g^1$ is C^∞ from W to $\mathcal{L}_p(\mathcal{K})$,

(d) defining $f_p = (I + A^p)(I + A^{p-1}) \cdots (I + A^2)$ for $p \geqslant 2$, $S(p)_g = f_p S_g f_p^{-1}$, and $Q^p(g) = S_{g^{-1}}^1 S(p)_g^p$, there exists $K(p, l_3) \geqslant 0$ such that

$$\left\| \partial_1^{l_1} \partial_2^{l_2} \partial_3^{l_3} Q^p(\exp(x_1 X_1 + x_2 X_2) \exp x_3 D)_{x_i = 0} \right\| \leqslant K(p, l_3)^{l_1 + l_2},$$

$$l_i \in \mathbf{N}.$$

It results from (b) that f_n converges to a limit $f \in H_r(\mathcal{H})$. We then define $T_g = fS_g f^{-1}$. The construction of the sequence (A^n) is inductive. Suppose we have built $A^2,\dots,A^p$. We then define

$$\theta^{p+1}(g) = S_g^1 S(p)_g^{p+1}, \qquad g \in W;$$

$j_\lambda \in \mathcal{D}(R), \lambda > 0$, by

$$j_\lambda(x) = \begin{cases} (2\pi)^{-r} \exp\left((\lambda - x_1^2 - x_2^2)^{-1}\right) & (x = x_1 X_1 + x_2 X_2) \\ & \text{if } x_1^2 + x_2^2 < \lambda, \\ 0 \quad \text{if } x_1^2 + x_2^2 \geqslant \lambda; \end{cases}$$

$$\hat{j}_\lambda(y) = \int_R \exp(i(x_1 y_1 + x_2 y_2)) j_\lambda(x)\, dx, \qquad y = y_1 X_1 + y_2 X_2;$$

and

$$A^{p+1} = \int_{R \times K} \hat{j}_{\lambda_p}(y) \theta^{p+1}(yk)\, dy\, dk.$$

It was proved in [5, Proposition 2.8] that the sequence (A^n) satisfies (a) and (b) for a good choice of the sequence (λ_p). Suppose now that (c) and (d) are true up to the order p. Writing at the order $p + 1$ the fact that $(S(p), \mathcal{H})$ is an analytic representation of W, we have $S_{x^{-1}}^1 A^{p+1} S_x^1 = F_1(x) - \theta^{p+1}(x) - F_2(x), x \in W$, with

$$F_1(x) = \int_{R \times K} \hat{j}_\lambda(y) \theta^{p+1}(ykx)\, dy\, dk$$

and

$$F_2(x) = \sum_{1 < q < p+1} \sum_{i_1 + \cdots + i_q = p+1} \int_{R \times K} \hat{j}_\lambda(y) \left(S_{x^{-1}}^1 R_{yk}^q S_x^1\right)$$
$$\times \left(R_x^{i_1} \otimes \cdots \otimes R_x^{i_q}\right) \sigma_{p+1}\, dy\, dk$$

where σ_{p+1} is the symmetrization operator in $\mathcal{H} \otimes \cdots \otimes \mathcal{H}$ ($p + 1$ times). It results from the induction hypothesis that θ^{p+1} is C^∞ from W to $\mathcal{L}_{p+1}(\mathcal{H})$. Any $x \in W$ writes $x = amc$ with $a \in R$, $m \in K$, and $c = \exp tD$. Therefore

$$F_1(x) = e^t \int_{R \times K} \hat{j}_\lambda(e^t y - k(m^{-1}(a))) \theta^{p+1}(cyk)\, dy\, dk. \quad (2.1.1)$$

Using again, at order $p + 1$, the fact that $(S(p), \mathcal{H})$ is an analytic representation of W, we have

$$\theta^{p+1}(cyk) = \theta^{p+1}(yk) + S_{(yk)^{-1}}^1 \theta^{p+1}(c) S_{yk}^1$$
$$+ \sum_{1 < q < p+1} \sum_{i_1 + \cdots + i_q = p+1} \left(S_{(yk)^{-1}}^1 R_c^q S_{yk}^1\right)\left(R_c^{i_1} \otimes \cdots \otimes R_c^{i_q}\right) \sigma_{p+1}.$$

$$(2.1.2)$$

Substituting (2.1.2) in (2.1.1) and since $\sup_{k \in K} \|\theta^{p+1}(yk)\|$ is at most of polynomial increase in $y \in R$ [5, Lemma 2.3], we see that F_1 is C^∞ from W to $\mathcal{L}_{p+1}(\mathcal{H})$.

In a similar way, an inductive use of the relation

$$S^1_{x^{-1}} R^n_{yk} S^1_x = R^n_{ykx} - R^n_x$$
$$- \sum_{1 < l < n} \sum_{i_1 + \cdots i_l = n} \left(S^1_{x^{-1}} R^l_{yk} S^1_x \right) \left(R^{i_1}_x \otimes \cdots \otimes R^{i_q}_x \right) \sigma_n$$

$$(2.1.3)$$

for $n < p + 1$ proves that F_2 is C^∞ from W to $\mathcal{L}_{p+1}(\mathcal{H})$.

Therefore, $x \to S^1_{x^{-1}} A^{p+1} S^1_x$ is C^∞ from W to $\mathcal{L}_{p+1}(\mathcal{H})$. As a consequence $A^{p+1}(\mathcal{H}^\infty) \subset \mathcal{H}^\infty$ and by the closed graph theorem $A^{p+1} \in \mathcal{L}_{p+1}(\mathcal{H}^\infty)$. Since, in addition,

$$dR^{p+1}_x \doteq \frac{d}{dt} \left(\theta^{p+1}(\exp tx) \right)_{t=0} - dS^1_x A^{p+1} + A^{p+1} d \left(\otimes^{n+1} S^1 \right)_x,$$

$x \in \mathfrak{w}$, we have $dR^{n+1}_x \in \mathcal{L}_{p+1}(\mathcal{H}^\infty)$.

It results from (2.1.1), (2.1.2) and the fact that $j_\lambda \in \mathcal{D}(R)$ that there exists $C_1(l_3) \geqslant 0$ such that

$$\left\| \partial^{l_1}_1 \partial^{l_2}_2 \partial^{l_3}_3 F_1 \left(\exp(x_1 X_1 + x_2 X_2) \exp x_3 D \right)_{x_i = 0} \right\| \leqslant (C_1(l_3))^{l_1 + l_2}.$$

Using (2.1.3) inductively together with the equalities

$$\int_{R \times K} \hat{j}_\lambda(y) R^n_{ykx} \, dy \, dk = e^t \int_{R \times K} \hat{j}_\lambda(e^t y - k(m^{-1}(a))) R^n_{cyk} \, dy \, dk$$

and

$$R^n_{cyk} = R^n_{yk} + \sum_{1 \leqslant q < n} \sum_{i_1 + \cdots + i_q = n} \left(S^1_{(yk)^{-1}} R^q_c S^1_{yk} \right) \left(R^{i_1}_c \otimes \cdots \otimes R^{i_q}_c \right) \sigma_n,$$

we see that there exists $C_2(l_3) \geqslant 0$ such that

$$\left\| \partial^{l_1}_1 \partial^{l_2}_2 \partial^{l_3}_3 F_2 \left(\exp(x_1 X_1 + x_2 X_2) \exp x_3 D \right)_{x_i = 0} \right\| \leqslant (C_2(l_3))^{l_1 + l_2}.$$

$$(2.1.4)$$

Since $Q^{p+1}(x) = F_1(x) - F_2(x) - A^{p+1}$, $x \in W$, there exists $K_{p+1} \geqslant 0$ such that property (d) is satisfied by Q^{p+1}. We denote by J the analytic subgroup of W associated to the Lie algebra $j = r + \mathbf{R}D$. In J, we have the following relations:

$$\exp t(a_1 X_1 + a_2 X_2 + D) = \exp((e^t - 1)(a_1 X_1 + a_2 X_2)) \exp tD$$

$$(2.1.5)$$

if $y(t) = \exp(t(a_1 X_1 + a_2 X_2 + D))$ and $x = \exp(\alpha_1 X_1 + \alpha_2 X_2)\exp uD$,

$$y(t)x = \exp\{((e^t - 1)a_1 + e^t\alpha_1)X_1$$
$$+ ((e^t - 1)a_1 + e^t\alpha_2)X_2\}\exp(t + u)D.$$

Writing $R^n_{y(t)x} = R^n_{\exp(x_1 X_1 + x_2 X_2)\exp x_3 D}$, we have

$$\frac{d}{dt}(R^n_{y(t)x})_{t=0} = (a_1 + \alpha_1)\partial_1 R^n_x + (a_2 + \alpha_2)\partial_2 R^n_x + \partial_3 R^n_x.$$

$$(2.1.6)$$

We then see inductively from inequalities of (d), relations (2.1.3), (2.1.6) and the relation

$$\mathrm{ad}(dS^1_{x_1})\cdots\mathrm{ad}(dS^1_{x_p})\,dR^n_x$$

$$= \frac{d}{du_1}\cdots\frac{d}{du_p}\frac{d}{du}\Big(S^1_{(\exp u_1 x_1)\cdots(\exp u_p x_p)}$$

$$\cdot R^n_{\exp ux}S^1_{(\exp -u_p x_p)\cdots(\exp -u_1 x_1)}\Big)_{u_i=0;\,u=0}$$

that there exists $C_3(n, p) \geqslant 0$ such that

$$\left\|(\mathrm{ad}\,dS^1_D)^p(\mathrm{ad}\,dS^1_{X_1})^{l_1}(\mathrm{ad}\,dS^1_{X_2})^{l_2}\,dR^n_z\right\| \leqslant (C_3(n, p))^{l_1+l_2}|z|, \qquad z \in \mathfrak{j}.$$

From the relation

$$(\mathrm{ad}\,dS^1_x)^k(\mathrm{ad}\,dS^1_D) = (\mathrm{ad}\,dS^1_D)(\mathrm{ad}\,dS^1_X)^k - k(\mathrm{ad}\,dS^1_x)^k, \qquad x \in \mathfrak{r}.$$

We then deduce (4). Q.E.D.

2.2. Given a locally convex topological vector space $\mathcal{E}$, we denote by $\hat{\otimes}^n_\pi \mathcal{E}$ the completed projective tensor product of $\mathcal{E}$ by itself n times.

Given a Hilbert space E and a compact subset K in $\mathbf{R}$, we denote by $L^2(K, E)$ the space of square integrable functions from $\mathbf{R}$ to E (for the Lebesgue measure) the support of which is in K, by $L^2_c(\mathbf{R}, E)$ the strict inductive limit of the spaces $L^2(K, E)$ when K is any compact subset in $\mathbf{R}$, by $C^\infty_c(\mathbf{R}^n, E)$ the LF space of C^∞ mappings, with compact support, from $\mathbf{R}^n$ to E, by $\mathcal{D}_{L^2}(E)$ the Fréchet space of C^∞ functions from $\mathbf{R}$ to E which are in $L^2(\mathbf{R}, E)$ together with all their derivatives, and we set

$$\mathfrak{L}(\mathbf{R}^n, E) = \mathfrak{L}(\hat{\otimes}^n_\pi L^2_c(\mathbf{R}, E), L^2(\mathbf{R}, E)) \cap \mathfrak{L}(\hat{\otimes}^n_\pi C^\infty_c(\mathbf{R}, E), \mathcal{D}_{L^2}(E)).$$

We have the topological isomorphism $C^\infty_c(\mathbf{R}, E) = C^\infty_c(\mathbf{R})\,\hat{\otimes}_\pi E$ (see [2, II, p. 82]). We therefore have the following algebraic isomorphism:

$$\hat{\otimes}^n_\pi C^\infty_c(\mathbf{R}, E) = C^\infty_c(\mathbf{R}^n)\,\hat{\otimes}_\pi(\hat{\otimes}^n_\pi E) = C^\infty_c(\mathbf{R}^n, \hat{\otimes}^n_\pi E),$$

where the second equality is topological [2, II, p. 83].

If a function f is defined (almost everywhere) on $\mathbf{R}$ and $v \in \mathbf{R}$, we define $(\tau_v f) = f(x - v)$. The mapping $(v, f) \to \tau_v f$ is a continuous linear representation of $\mathbf{R}$ in $L^2(\mathbf{R}, E)$ (resp. $L_c^2(\mathbf{R}, E)$, resp. $C_c^\infty(\mathbf{R}, E)$, resp. $\mathcal{D}_{L^2}(E)$), and the tensor product $\otimes^n \tau_v = \tau_v \otimes \cdots \otimes \tau_v$ (n times) extends to a continuous linear representation of $\mathbf{R}$ in $\hat{\otimes}_\pi^n L_c^2(\mathbf{R}, E)$ (resp. $\hat{\otimes}_\pi^n C_c^\infty(\mathbf{R}, E)$). The space $\hat{\otimes}_\pi^n L_c^2(\mathbf{R}, E)$ is a strict inductive limit of the spaces $\hat{\otimes}_\pi^n L^2([-a, a], E)$ ($a > 0$) [2, I, Proposition 4], and a linear mapping A from $\hat{\otimes}_\pi^n L_c^2(\mathbf{R}, E)$ to $L^2(\mathbf{R}, E)$ is continuous if and only if for every $a > 0$, its restriction A_a to $\hat{\otimes}^n L_\pi^2([-a, a], E)$ is continuous. We then define the seminorm $p_a(A) = \|A_a\|$.

2.3. LEMMA. *Given* $B \in \mathcal{L}(\mathbf{R}^n, E)$ *such that, given* $a \in \mathbf{R}$, *there exists* $c_a \in \mathbf{R}$ *for which* $B(f)(x) = 0$ *for any* $f \in \hat{\otimes}_\pi^n L^2([-a, a], E)$ *and any* $x \geq c_a$, *there exists* $F \in \mathcal{L}(\mathbf{R}^n, E)$ *such that*

$$B(f) = \frac{d}{dx} F(f) - \sum_{i=1}^n F\left(\frac{\partial f}{\partial x_i}\right)$$

for any $f \in C_c^\infty(\mathbf{R}^n, \hat{\otimes}_\pi^n E)$, $F(f)(x) = 0$ *if* $x \geq c_{2a+1} + a + 1$, *and* $p_a(F) \leq 2(a + 1)p_{2a+1}(B)$.

If, in addition, there exists $c'_a \in \mathbf{R}$ *such that* $B(f)(x) = 0$ *for any* $f \in \hat{\otimes}_\pi^n L^2([-a, a], E)$ *and any* $x \leq c'_a$, *we have also* $F(f)(x) = 0$ *if* $x \leq c'_{2a+1} - a - 1$.

We choose two positive C^∞ functions χ^+ and χ^- on $\mathbf{R}^n$ such that $\chi^\pm(x_1, \ldots, x_n)$ is independent of $x_2, \ldots, x_n$; $\chi^+(x_1, \ldots, x_n) = 0$ if $x_1 \leq -1$; $\chi^-(x_1, \ldots, x_n) = 0$ if $x_1 \geq 1$; and $\chi^+(x_1, \ldots, x_n) + \chi^-(x_1, \ldots, x_n) = 1$ for every $(x_1, \ldots, x_n) \in \mathbf{R}^n$. We define $J_v^\pm = \tau_v B \chi^\pm(\otimes^n \tau_{-v})$. If $f \in \hat{\otimes}_\pi^n C_c^\infty(\mathbf{R}, E)$ (resp. $f \in \mathcal{D}_{L^2}(E)$) the mapping $v \to \otimes^n \tau_v f$ (resp. $v \to \tau_v f$) is C^∞ from $\mathbf{R}$ to $\hat{\otimes}_\pi^n C_c^\infty(\mathbf{R}, E)$ (resp. $\mathcal{D}_{L^2}(E)$).

As a consequence of the Banach Steinhauss theorem $v \to J_v^\pm$ is continuous (resp. C^∞) from $\mathbf{R}$ to $\mathcal{L}(\hat{\otimes}_\pi^n L_c^2(\mathbf{R}, E), L^2(\mathbf{R}, E))$ (resp. $\mathcal{L}(\hat{\otimes}_\pi^n C_c^\infty(\mathbf{R}, E), \mathcal{D}_{L^2}(E))$) for the topology of uniform convergence on compact sets. Given $a > 0$ take $f \in C_c^\infty([-a, a]^n, \hat{\otimes}_\pi^n E)$ (the space of functions in $C_c^\infty(\mathbf{R}^n, \hat{\otimes}_\pi^n E)$ with support in $[-a, a]^n$), then $J_v^+(f) = 0$ if $v \geq a + 1$ and $J_v^-(f) = 0$ if $v \leq -a - 1$. Moreover, $\chi^\pm(\otimes^n \tau_{-v})f \in C_c^\infty([-2a - 1, 2a + 1]^n, \hat{\otimes}_\pi^n E)$. We can, therefore, define $F^+ = -\int_0^\infty J_v^+ \, dv$ and $F^- = \int_{-\infty}^0 J_v^- \, dv$ which makes sense in $\mathcal{L}(\mathbf{R}^n, E)$.

We have

$$\frac{d}{dx} F^+(f) - \sum_{i=1}^n F^+\left(\frac{\partial f}{\partial x_i}\right) = -\int_0^\infty \frac{d}{dv} J_v^+ \, dv = B(\chi^+ f)$$

and

$$\frac{d}{dx} F^-(f) - \sum_{i=1}^n F^-\left(\frac{\partial f}{\partial x_i}\right) = \int_{-\infty}^0 \frac{d}{dv} J_v^-(f) \, dv = B(\chi^- f).$$

Then, $F = F^+ + F^-$ satisfies

$$\frac{d}{dx}F(f) - \sum_{i=1}^{n} F\left(\frac{\partial f}{\partial x_i}\right) = B(f).$$

If $|v| \leq a + 1$, $J_v^{\pm}(f)(x) = 0$ whenever $x - v \geq c_{2a+1}$ and consequently $F(f) = 0$ if $x \geq c_{2a+1} + a + 1$ and

$$p_a(F) \leq \int_0^{a+1} p_a(J_v^{\pm})\, dv + \int_{-a-1}^0 p_a(J_v^-)\, dv \leq 2(a+1)p_{2a+1}(B).$$

If, in addition, $B(f)(x) = 0$ if $x \leq c'(a)$, we then see, as above, that $F(f) = 0$ if $x \leq c'_{2a+1} - a - 1$.

3. Triviality of some 1-cohomology spaces of W. In the following, $(U, \mathcal{K})$ is a unitary representation of W in a (countable) Hilbert space which contains no one-dimensional representation of W. After a unitary equivalence we can suppose that $\mathcal{K} = L^2(]0, +\infty[, L^2(\mathbf{T}_1, \mathcal{T}))$, the space of square integrable functions (for the Lebesgue measure) from $]0, +\infty[$ to the space of square integrable mappings from the circle $\mathbf{T}_1$ (for the measure $d\theta/2\pi$) to a Hilbert space $\mathcal{T}$, and that U writes (with the identification $L^2(]0, +\infty[, L^2(\mathbf{T}_1, \mathcal{T})) \cong L^2(]0, +\infty[\times \mathbf{T}_1, \mathcal{T}))$:

$$U(\exp aM)\varphi(\xi, \theta) = \varphi(\xi, \theta + a),$$

$$U(\exp aD)\varphi(\xi, \theta) = e^{a/2}\varphi(e^a\xi, \theta),$$

$$U(\exp aX_1)\varphi(\xi, \theta) = \exp(ia\xi\cos\theta)\varphi(\xi, \theta),$$

$$U(\exp aX_2)\varphi(\xi, \theta) = \exp(ia\xi\sin\theta)\varphi(\xi, \theta), \qquad \varphi \in \mathcal{K}.$$

We define, in the complexified of $\mathfrak{w}$, $X^+ = iX_1 - X_2$ and $X^- = iX_1 + X_2$. Then, $dU_{X^+} = \xi\tau^+$ and $dU_{X^-} = \xi\tau^-$, $\tau^+\tau^- = \tau^-\tau^+ = I$. We denote by $\mathfrak{H}_K$ (resp. $\mathcal{D}(\mathfrak{H}_K)$) the algebraic direct sum $\oplus_{\lambda \in \mathbf{Z}} P_\lambda(\mathcal{K}^\infty)$ (resp. $\oplus_{\lambda \in \mathbf{Z}} P_\lambda(\mathcal{D}(\mathcal{K}) \cap \mathcal{K}^\infty))$ where $\mathcal{K}^\infty$ is the space of C^∞ vectors of $(U, \mathcal{K})$ and by $L_n(\mathcal{D}(\mathfrak{H}_K), \mathfrak{H}_K)$ the set of (algebraic) n-linear symmetric mappings from $\mathcal{D}(\mathfrak{H}_K)$ to $\mathfrak{H}_K$. Given an n-linear mapping f from $\mathcal{D}(\mathfrak{H}_K)$ to $\mathfrak{H}_K$ we denote by $f_{\lambda_1,\ldots,\lambda_n}$ the restriction of $f((\tau^+)^{\lambda_1} \otimes \cdots \otimes (\tau^+)^{\lambda_n})$ to $P_0(\mathcal{D}(\mathcal{K}) \cap \mathcal{K}^\infty)$ ($\lambda_i \in \mathbf{Z}$), and by $f^{a,b}_{\lambda_1,\ldots,\lambda_n}$, the restriction of $f_{\lambda_1,\ldots,\lambda_n}$ to $P_0(\mathcal{D}_{a,b}(\mathcal{K}) \cap \mathcal{K}^\infty)$. We then introduce the vector space $\mathfrak{L}_n^0(\mathcal{K})$ of elements $f \in L_n(\mathcal{D}(\mathfrak{H}_K), \mathfrak{H}_K)$ such that
   (a) there exists a finite set $I \subset \mathbf{Z}$ for which $P_\lambda f_{\lambda_1,\ldots,\lambda_n} = 0$ if $\lambda - \lambda_1 - \cdots - \lambda_n \notin I$,
   (b) for any $0 < a < b$, the mapping $f^{a,b}_{\lambda_1,\ldots,\lambda_n}$ extends to a continuous n-linear mapping from the Hilbert space $P_0(\mathcal{D}_{a,b}(\mathcal{K}))$ to $\mathcal{K}$,
   (c) there exist $h(a, b) \geq 0$, $c(a, b) \geq 0$ and $\alpha(a, b) \geq 0$ such that

$$\left\|\xi^p f^{a,b}_{\lambda_1,\ldots,\lambda_n}\right\| \leq h(a, v)(c(a, b))^p \exp(\alpha(a, b)(|\lambda_1| + \cdots + |\lambda_n|))$$

for any $p \geq 0$, $\lambda_1, \ldots, \lambda_n \in \mathbf{Z}$.

Given $f \in L_n(\mathfrak{D}(\mathfrak{H}_K), \mathfrak{H}_K)$, $x \in \mathfrak{w}$, we define $V(x)f = dU_x f - fd(\otimes^n U)_x$. We denote by $\mathfrak{L}_n^0(\mathfrak{H})_\infty$ the vector space of $f \in \mathfrak{L}_n^0(\mathfrak{H})$ such that $V^n(D)f \in \mathfrak{L}_n^0(\mathfrak{H})$ for every $n \geqslant 0$. The mapping $x \to V(x)$ is a representation of $\mathfrak{w}$ in $\mathfrak{L}_n^0(\mathfrak{H})_\infty$ to which we associate the space $Z_{\mathfrak{k}}^1(\mathfrak{w}, \mathfrak{L}_n^0(\mathfrak{H})_\infty)$ of 1-cocycles of $\mathfrak{w}$ with coefficients in $\mathfrak{L}_n^0(\mathfrak{H})_\infty$ vanishing on the Lie algebra $\mathfrak{k}$ of K; this is the vector space of linear mappings $x \to B(x)$ from $\mathfrak{w}$ to $\mathfrak{L}_n^0(\mathfrak{H})_\infty$ such that $B(k) = 0$ if $k \in \mathfrak{k}$ and $B[x, y] = V(x)B(y) - V(y)B(x)$, $x, y \in \mathfrak{w}$.

The space $B_{\mathfrak{k}}^1(\mathfrak{w}, \mathfrak{L}_n^0(\mathfrak{H})_\infty)$ of 1-coboundaries is the subvector space of elements $B \in Z_{\mathfrak{k}}^1(\mathfrak{w}, \mathfrak{L}_n^0(\mathfrak{H})_\infty)$ of the form $B(x) = V(x)Y$ for some $Y \in \mathfrak{L}_n^0(\mathfrak{H})_\infty$.

3.1. LEMMA. $Z_{\mathfrak{k}}^1(\mathfrak{w}, \mathfrak{L}_n^0(\mathfrak{H})_\infty) = B_{\mathfrak{k}}^1(\mathfrak{w}, \mathfrak{L}_n^0(\mathfrak{H})_\infty)$ when $n \geqslant 3$.

(a) We have first to find an n-linear mapping A from $\mathfrak{D}(\mathfrak{H}_K)$ to $\mathfrak{H}_k$ such that $B(x) = V(x)A$. Extending the linear mapping $x \to B(x)$ to the complexified of $\mathfrak{w}$, A must be a solution of the equations $V(M)A = 0$, $B(X^\pm) = V(X^\pm)A$ and $B(D) = V(D)A$. The first equality is equivalent to $P_\lambda A_{\lambda_1, \ldots, \lambda_n} = 0$, if $\lambda \neq \lambda_1 + \cdots + \lambda_n$ and for the remaining equations write

$$B(X^\pm)_{\lambda_1, \ldots, \lambda_n} = \xi \tau^\pm A_{\lambda_1, \ldots, \lambda_n} - \sum_{k=1}^n A_{\lambda_1, \ldots, \lambda_k \pm 1, \ldots, \lambda_n} \xi_k, \quad (3.1.1)$$

$$B(D)_{\lambda_1, \ldots, \lambda_n} = \xi \frac{\partial}{\partial \xi} A_{\lambda_1, \ldots, \lambda_n} - \sum_{k=1}^n A_{\lambda_1, \ldots, \lambda_n} \xi_k \frac{\partial}{\partial \xi_k} - \frac{n-1}{2} A_{\lambda_1, \ldots, \lambda_n}, \quad (3.1.2)$$

where ξ_k (resp. $\xi_k \partial/\partial \xi k$) is the mapping which to

$$\left(\varphi_1(\xi_1, \theta_1), \ldots, \varphi_k(\xi_k, \theta_k), \ldots, \varphi_n(\xi_n, \theta_n) \right)$$

associates

$$\left(\varphi_1(\xi_1, \theta_1), \ldots, \xi_k \varphi(\xi_k, \theta_k), \ldots, \varphi_n(\xi_n, \theta_n) \right)$$

(resp. $(\varphi_1(\xi_1, \theta_1), \ldots, \xi_k \partial \varphi_k(\xi_k, \theta_k)/\partial \xi_k, \ldots, \varphi_n(\xi_n, \theta_n))$).

The relation $B([M, X^\pm]) = \pm i B(X^\pm)$ rewrites $P_\lambda B(X^\pm)_{\lambda_1, \ldots, \lambda_n} = 0$ if $\lambda \neq \lambda_1 + \cdots + \lambda_n \pm 1$, the relation $B([X^+, X^-]) = 0$ rewrites

$$\xi \tau^+ B(X^-)_{\lambda_1, \ldots, \lambda_n} - \sum_{k=1}^n B(X^-)_{\lambda_1, \ldots, \lambda_k+1, \ldots, \lambda_n} \xi_k$$

$$= \xi \tau^- B(X^+)_{\lambda_1, \ldots, \lambda_n} - \sum_{k=1}^n B(X^+)_{\lambda_1, \ldots, \lambda_k-1, \ldots, \lambda_n} \xi_k. \quad (3.1.3)$$

The relation $B([M, D]) = 0$ rewrites $P_\lambda B(D)_{\lambda_1,\ldots,\lambda_n} = 0$ if $\lambda \neq \lambda_1 + \cdots + \lambda_n$, and the relation $B([D, X^\pm]) = B(X^\pm)$ rewrites

$$B(X^\pm)_{\lambda_1,\ldots,\lambda_n} = \xi \frac{\partial}{\partial \xi} B(X^\pm)_{\lambda_1,\ldots,\lambda_n} - \sum_{k=1}^{n} B(X^\pm)_{\lambda_1,\ldots,\lambda_n} \xi_k \frac{\partial}{\partial \xi_k}$$

$$- \frac{n-1}{2} B(X^\pm)_{\lambda_1,\ldots,\lambda_n} - \xi \tau^\pm B(D)_{\lambda_1,\ldots,\lambda_n}$$

$$+ \sum_{k=1}^{n} B(D)_{\lambda_1,\ldots,\lambda_k \pm 1,\ldots,\lambda_n} \xi_k. \tag{3.1.4}$$

If $\Omega_{\lambda_1,\ldots,\lambda_n}$, $\Gamma^+_{\lambda_1,\ldots,\lambda_n}$ and $\Gamma^-_{\lambda_1,\ldots,\lambda_n}$ ($\lambda_i \in \mathbf{Z}$) are n-linear mappings from $P_0(\mathcal{D}(\mathcal{H}) \cap \mathcal{H}^\infty)$ to $\mathfrak{H}_K$, we define the mappings

$$\Lambda(\Omega)_{\lambda_1,\ldots,\lambda_n} = \xi^2 \Omega_{\lambda_1,\ldots,\lambda_n} - \Omega_{\lambda_1,\ldots,\lambda_n}\left(\xi_1^2 - \xi_2^2 - \cdots - \xi_n^2\right)$$

$$- \sum_{k=2}^{n} \xi \tau^- \Omega_{\lambda_1,\ldots,\lambda_k+1,\ldots,\lambda_n} \xi_k - \sum_{k=2}^{n} \xi \tau^+ \Omega_{\lambda_1,\ldots,\lambda_k-1,\ldots,\lambda_n} \xi_k$$

$$+ \sum_{\substack{k,l=2 \\ k \neq l}}^{n} \Omega_{\lambda_1,\ldots,\lambda_k+1,\ldots,\lambda_l-1,\ldots,\lambda_n} \xi_k \xi_l$$

$$P(\Gamma^+, \Gamma^-)_{\lambda_1,\ldots,\lambda_n} = \sum_{k=2}^{n} \Gamma^+_{\lambda_1,\ldots,\lambda_k-1,\ldots,\lambda_n} \xi_k - \xi \tau^- \Gamma^+_{\lambda_1,\ldots,\lambda_n}$$

$$- \Gamma^-_{\lambda_1+1,\lambda_2,\ldots,\lambda_n} \xi_1.$$

From relations (3.1.1) we deduce

$$\Lambda(A)_{\lambda_1,\ldots,\lambda_n} + P(B(X^+), B(X^-))_{\lambda_1,\ldots,\lambda_n} = 0. \tag{3.1.5}$$

Conversely, suppose that we are given a set of (algebraic) n-linear mappings $A_{0,\lambda_2,\ldots,\lambda_n}(\lambda_2,\ldots,\lambda_n \in \mathbf{Z})$ from $P_0(\mathcal{D}(\mathcal{H}) \cap \mathcal{H}^\infty)$ to $P_{\lambda_2+\cdots+\lambda_n}(\mathfrak{H}_K)$ such that

$$\Lambda(A)_{0,\lambda_2,\ldots,\lambda_n} + P(B(X^+), B(X^-))_{0,\lambda_2,\ldots,\lambda_n} = 0 \tag{3.1.6}$$

and

$$B(D)_{0,\lambda_2,\ldots,\lambda_n} = \xi \frac{\partial}{\partial \xi} A_{0,\lambda_2,\ldots,\lambda_n}$$

$$- \sum_{k=1}^{n} A_{0,\lambda_2,\ldots,\lambda_n} \xi_k \frac{\partial}{\partial \xi_k} - \frac{n-1}{2} A_{0,\lambda_2,\ldots,\lambda_n}.$$

$$\tag{3.1.7}$$

If $\lambda_1 \in \mathbf{N}$, we define inductively

$$A_{\lambda_1+1,\lambda_2,\ldots,\lambda_n} = \xi\tau^+ A_{\lambda_1,\ldots,\lambda_n}\xi_1^{-1}$$

$$- \sum_{k=2}^{n} A_{\lambda_1,\ldots,\lambda_k+1,\ldots,\lambda_n}\xi_1^{-1}\xi_k - B(X^+)_{\lambda_1,\ldots,\lambda_n}\xi_1^{-1}$$

$$\tag{3.1.8}$$

and, if $-\lambda_1 \in \mathbf{N}$, we define inductively

$$A_{\lambda_1-1,\lambda_2,\ldots,\lambda_n} = \xi\tau^- A_{\lambda_1,\ldots,\lambda_n}\xi_1^{-1}$$

$$- \sum_{k=2}^{n} A_{\lambda_1,\ldots,\lambda_k-1,\ldots,\lambda_n}\xi_1^{-1}\xi_k - B(X^-)_{\lambda_1,\ldots,\lambda_n}\xi_1^{-1}.$$

$$\tag{3.1.9}$$

Using relation (3.1.3), we see by induction on λ_1 that (3.1.5) is satisfied and, as a consequence, that relations (3.1.1) are satisfied.

Given $l \in \mathbf{N}$, suppose that (3.1.2) is satisfied for $|\lambda_1| \leqslant l$. Then formula (3.1.4) writes, for $|\lambda_1| \leqslant l$,

$$B(D)_{\lambda_1\pm1,\lambda_2,\ldots,\lambda_n}\xi_1$$

$$= \xi\tau^{\pm}\left(\xi\frac{\xi}{\partial\xi}A_{\lambda_1,\ldots,\lambda_n} - \sum_{p=1}^{n} A_{\lambda_1,\ldots,\lambda_n}\xi_p\frac{\partial}{\partial\xi_p} - \frac{n-1}{2}A_{\lambda_1,\ldots,\lambda_n}\right)$$

$$- \sum_{k=2}^{n}\left(\xi\frac{\partial}{\partial\xi}A_{\lambda_1,\ldots,\lambda_k\pm1,\ldots,\lambda_n}\right.$$

$$\left. - \sum_{p=1}^{n} A_{\lambda_1,\ldots,\lambda_k\pm1,\ldots,\lambda_n}\xi_p\frac{\partial}{\partial\xi_p} - \frac{n-1}{2}A_{\lambda_1,\ldots,\lambda_k\pm1,\ldots,\lambda_n}\right)$$

$$- \xi\frac{\partial}{\partial\xi}B(X^{\pm})_{\lambda_1,\ldots,\lambda_n} + \sum_{p=1}^{n} B(X^{\pm})_{\lambda_1,\ldots,\lambda_n}\xi_p\frac{\partial}{\partial\xi_p}$$

$$+ \frac{n+1}{2}B(X^{\pm})_{\lambda_1,\ldots,\lambda_n}.$$

This relation combined with (3.1.8) and (3.1.9) gives

$$B(D)_{\lambda_1\pm1,\lambda_2,\ldots,\lambda_n} = \xi\frac{\xi}{\partial\xi}A_{\lambda_1\pm1,\lambda_2,\ldots,\lambda_n}$$

$$- \sum_{p=1}^{n} A_{\lambda_1\pm1,\lambda_2,\ldots,\lambda_n}\xi_p\frac{\partial}{\partial\xi_p} - \frac{n-1}{2}A_{\lambda_1\pm1,\lambda_2,\ldots,\lambda_n}.$$

As a consequence, relation (3.1.2) is satisfied for any $\lambda_1,\ldots,\lambda_n \in \mathbf{Z}$. We are, therefore, led to build a set of n-linear mappings $A_{0,\lambda_2,\ldots,\lambda_n}$ $(\lambda_2,\ldots,\lambda_n \in \mathbf{Z})$ from $P_0(\mathfrak{D}(\mathcal{H}) \cap \mathcal{H}^{\infty})$ to $P_{\lambda_2+\cdots+\lambda_n}(\mathfrak{S}_K)$ satisfying relations (3.1.6) and (3.1.7).

(b) Given a family $\{Q(m_2,\ldots,m_n); m_2,\ldots,m_n \in \mathbf{Z}\}$ of n-linear mappings from $P_0(\mathcal{D}(\mathcal{H}) \cap \mathcal{H}^\infty)$ to $\mathfrak{H}_K$, we define

$$
\begin{aligned}
F(Q)(m_2,\ldots,m_n) = &-\xi\tau^- Q(m_2 + 1, m_3,\ldots,m_n)\xi_2 \\
&-\xi\tau^+ Q(m_2 + 1, m_3 - 1, m_4,\ldots,m_n)\xi_3 \\
&+ \sum_{l=4}^{n} Q(m_2 + 1, m_3,\ldots,m_l - 1,\ldots,m_n)\xi_2\xi_l \\
&+ \sum_{l=4}^{n} Q(m_2 + 1, m_3 - 1, m_4,\ldots,m_l + 1,\ldots,m_n)\xi_3\xi_l,
\end{aligned}
$$

$$
\begin{aligned}
G(Q)&(m_2,\ldots,m_n) \\
&= \xi^2 Q(m_2,\ldots,m_n) - Q(m_2,\ldots,m_n)\left(\xi_1^2 - \xi_2^2 - \cdots - \xi_n^2\right) \\
&\quad - \sum_{l=4}^{n} \xi\tau^- Q(m_2, m_3,\ldots,m_l + 1,\ldots,m_n)\xi_l \\
&\quad - \sum_{l=4}^{n} \xi\tau^+ Q(m_2, m_3,\ldots,m_l - 1,\ldots,m_n)\xi_l \\
&\quad + \sum_{\substack{k,l=4 \\ k \neq l}}^{n} Q(m_2, m_3,\ldots,m_k - 1,\ldots,m_l + 1,\ldots,m_n)\xi_k\xi_l,
\end{aligned}
$$

$$
\begin{aligned}
H(Q)&(m_2,\ldots,m_n) \\
&= -\xi\tau^+ Q(m_2 - 1, m_3,\ldots,m_n)\xi_2 \\
&\quad -\xi\tau^-(m_2 - 1, m_3 + 1, m_4,\ldots,m_n)\xi_3 \\
&\quad + \sum_{l=4}^{n} Q(m_2 - 1, m_3,\ldots,m_l + 1,\ldots,m_n)\xi_2\xi_l \\
&\quad + \sum_{l=4}^{n} Q(m_2 - 1, m_3 + 1, m_4,\ldots,m_l - 1,\ldots,m_n)\xi_3\xi_l,
\end{aligned}
$$

$$
\begin{aligned}
\Lambda'(Q)(m_2, m_3,\ldots,m_n) = &\, Q(m_2 + 2, m_3 - 1, m_4,\ldots,m_n)\xi_2\xi_3 \\
&+ F(Q)(m_2, m_3,\ldots,m_n) + G(Q)(m_2, m_3,\ldots,m_n) \\
&+ H(Q)(m_2, m_3,\ldots,m_n) \\
&+ Q(m_2 - 2, m_3 + 1, m_4,\ldots,m_n)\xi_2\xi_3,
\end{aligned}
$$

and

$$
\begin{aligned}
\Delta(Q)(m_2,\ldots,m_n) = &\, \xi\frac{\partial}{\partial\xi}Q(m_2,\ldots,m_n) - \sum_{p=1}^{n} Q(m_2,\ldots,m_n)\xi_p\frac{\partial}{\partial\xi_p} \\
&- \frac{n-1}{2}Q(m_2,\ldots,m_n).
\end{aligned}
$$

We introduce a new parametrization of the set of n-linear mappings by $A'(m_2,\ldots,m_n) = A(0, m_2 + m_3, m_3,\ldots,m_n)$. Relation (3.1.6) rewrites

$$\Lambda'(A')(m_2,\ldots,m_n) + P(B(X^+), B(X^-))_{0,m_2+m_3,m_3,\ldots,m_n} = 0.$$

$$(3.1.10)$$

We notice that $F(A')(m_2,\ldots)$ (resp. $G(A')(m_2,\ldots)$, $H(A')(m_2,\ldots)$) contains only terms of the type

$$A'(m_2 + 1,\ldots) \quad (\text{resp. } A'(m_2,\ldots), A'(m_2 - 1,\ldots)).$$

Therefore the set $\Theta(m_2) = \{A'(m_2, m_3,\ldots,m_n); m_3,\ldots,m_n \in \mathbf{Z}\}$ is determined step by step from relation (3.1.10) once we are given the sets $\Theta(-1)$, $\Theta(0)$, $\Theta(1)$ and $\Theta(2)$.

By algebraic calculations we see that

$$\Delta\big(F(Q)\xi_2^{-1}\xi_3^{-1}\big)(m_2,\ldots,m_n) = F(\Delta(Q))(m_2,\ldots,m_n)\xi_2^{-1}\xi_3^{-1},$$

$$\Delta\big(G(Q)\xi_2^{-1}\xi_3^{-1}\big)(m_2,\ldots,m_n) = G(\Delta(Q))(m_2,\ldots,m_n)\xi_2^{-1}\xi_3^{-1},$$

$$\Delta\big(H(Q)\xi_2^{-1}\xi_3^{-1}\big)(m_2,\ldots,m_n) = H(\Delta(Q))(m_2,\ldots,m_n)\xi_2^{-1}\xi_3^{-1}.$$

Relation (3.1.10) implies then

$$\Lambda'(\Delta(A'))(m_2,\ldots,m_n) + \Delta(P')(m_2,\ldots,m_n) - 2P'(m_2,\ldots,m_n) = 0$$

$$(3.1.11)$$

where $P'(m_2, m_3,\ldots,m_n) = P(B(X^+), B(X^-))_{0,m_2+m_3,m_3,\ldots,m_n}$. Now, it results from (3.1.4) (by combining the equation with X^+ and the equation with X^-) that

$$\Lambda(B(D))_{\lambda_1,\ldots,\lambda_n} + \xi\frac{\partial}{\partial\xi}P(B(X^+), B(X^-))_{\lambda_1,\ldots,\lambda_n}$$

$$- \sum_{l=1}^{n} P(B(X^+), B(X^-))_{\lambda_1,\ldots,\lambda_n}\xi_l\frac{\partial}{\partial\xi_l}$$

$$- \frac{n+3}{2}P(B(X^+), B(X^-))_{\lambda_1,\ldots,\lambda_n} = 0. \quad (3.1.12)$$

With the change of variables

$$B'(D)(m_2, m_3,\ldots,m_n) = B(D)_{0,m_2+m_3,m_3,\ldots,m_n},$$

relation (3.1.12) implies

$$\Lambda'(B'(D))(m_2,\ldots,m_n) + \Delta(P')(m_2,\ldots,m_n) - 2P'(m_2,\ldots,m_n) = 0.$$

$$(3.1.13)$$

Suppose that we have found a set $A'(m_2, m_3,\ldots,m_n)$ of n-linear mappings from $P_0(\mathcal{D}(\mathcal{K}) \cap \mathcal{K}^\infty)$ to $\mathfrak{H}_K$ such that

$$B'(D)(m_2,\ldots,m_n) = \Delta(A')(m_2,\ldots,m_n) \qquad (3.1.14)$$

for $m_2 = -1, 0, 1, 2$ and $m_3, \ldots, m_n \in \mathbf{Z}$. It then results from relations (3.1.11) and (3.1.13) that relation (3.1.14) is true for any $m_2, m_3, \ldots, m_n \in \mathbf{Z}$.

Given any $\lambda \in \mathbf{Z}$, there exists a sub-Hilbert space $\mathfrak{T}_\lambda$ of $L^2(\mathbf{T}_1, \mathfrak{T})$ isomorphic to $\mathfrak{T}$ such that $P_\lambda(\mathcal{K}) = L^2(]0, +\infty[, \mathfrak{T}_\lambda)$. We define a unitary mapping π from $L^2(\mathbf{R}, \mathfrak{T}_\lambda)$ to $L^2(]0, +\infty[, \mathfrak{T}_\lambda)$ by $(\pi_\lambda f)(x) = x^{-1/2} f(\mathrm{Log}\, x)$. We have $d/dx = \pi_\lambda^{-1}\, dU_D \pi_\lambda$.

There exist constants $h(a, b) \geq 0$, $c(a, b) \geq 0$ and $\alpha(a, b) \geq 0$ such that

$$\left\| \xi^p B(x)^{a,b}_{\lambda_1, \ldots, \lambda_n} \right\| \leq h(a, b)(c(a, b))^p \exp(\alpha(a, b)(|\lambda_1| + \cdots + |\lambda_n|))$$

for $x \in \{X^+, X^-, D\}$. Then, on $P_0(\mathcal{D}_{a,b}(\mathcal{K}))$, we have

$$\left\| \xi^p B'(D)(m_2, m_3, \ldots, m_n) \right\| \leq h(a, b)(c(a, b))^p$$

$$\cdot \exp(\alpha(a, b)(|m_2| + 2|m_3| + |m_4| + \cdots + |m_n|)).$$

Using the mappings π_λ and Lemma 2.3, there exists a family $\{A'(m_2, \ldots, m_n);\ m_2 \in \{-1, 0, 1, 2\},\ m_3, \ldots, m_n \in \mathbf{Z}\}$ of n-linear mappings from $P_0(\mathcal{D}(\mathcal{K}) \cap \mathcal{K}^\infty)$ to $P_\lambda(\mathcal{K}) \cap \mathcal{K}^\infty$ where $\lambda = m_3 + \Sigma_{i=2}^n m_n$, which, for any $0 < a < b$, extend to continuous n-linear mappings from $P_0(\mathcal{D}_{a,b}(\mathcal{K}))$ to $P_\lambda(\mathcal{K})$, where $\lambda = m_3 + \Sigma_{i=2}^n m_i$, satisfying relations (3.1.14) and the inequalities

$$\left\| \xi^p A'(m_2, \ldots, m_n) \right\|$$

$$\leq h_1(a, b)(c_1(a, b))^p \exp(\alpha_1(a, b)(|m_2| + \cdots + |m_n|))$$

$$(3.1.15)$$

$(h_1(a, b) \geq 0,\ c_1(a, b) \geq 0,\ \alpha_1(a, b) \geq 0,\ m_2 \in \{-1, 0, 1, 2\},\ m_3, \ldots, m_n \in \mathbf{Z}, p \geq 0)$.

Solving then equation (3.1.10) step by step in m_2, we get inequalities (3.1.15) for new values of the constants $h_1(a, b)$, $c_1(a, b)$ and $\alpha_1(a, b)$; for any $m_2, m_3, \ldots, m_n \in \mathbf{Z}$, on $P_0(\mathcal{D}_{a,b}(\mathcal{K}))$. It then results from relations (3.1.8) and (3.1.9) that

$$\left\| \xi^p A_{\lambda_1, \ldots, \lambda_n} \right\| \leq h_2(a, b)(c_2(a, b))^p \exp(\alpha_2(a, b)(|\lambda_1| + \cdots + |\lambda_n|))$$

$(h_2(a, b) \geq 0, c_2(a, b) \geq 0, \alpha_2(a, b) \geq 0, \lambda_i \in \mathbf{Z}, p \geq 0)$, on $P_0(\mathcal{D}_{a,b}(\mathcal{K}))$. We then define $Y = A\sigma_n$. It is an element of $\mathfrak{L}_n^0(\mathcal{K})$ and $B(x) = V(x)Y$. Since $B(D) \in \mathfrak{L}_n^0(\mathcal{K})_\infty$, we have $Y \in \mathfrak{L}_n^0(\mathcal{K})_\infty$.

3.2. **LEMMA.** *Given* $B \in Z_t^1(\mathfrak{w}, \mathfrak{L}_2^0(\mathcal{K})_\infty)$, *suppose that there exists* $a' > 0$ *such that* $B(\mathfrak{w})(\mathcal{D}(\mathfrak{H}_K)^2) \subset \bigcup_{b' \in]a', +\infty[} \mathcal{D}_{a',b'}(\mathcal{K})$. *Then, there exists* $Y \in \mathfrak{L}_2^0(\mathcal{K})_\infty$ *such that* $B(x) = V(x)Y$ *and*

$$Y(\mathcal{D}(\mathfrak{H}_K)^2) \subset \mathcal{D}(\mathcal{K}).$$

Part (a) of the proof of Lemma 3.1 remains valid for $n = 2$. We have then to find a family $\{A_{0,\lambda_2}; \; \lambda_2 \in \mathbf{Z}\}$ of n-linear mappings from $P_0(\mathcal{D}(\mathcal{K}) \cap \mathcal{K}^\infty)$ to $P_{\lambda_2}(\mathcal{S}_K)$ satisfying relations (3.1.6) and (3.1.7) which we write here explicitly

$$\xi\tau^- A_{0,\lambda_2+1}\xi_2 - \left(\xi^2 A_{0,\lambda_2} - A_{0,\lambda_2}(\xi_1^2 - \xi_2^2)\right) + \xi\tau^+ A_{0,\lambda_2-1}\xi_2$$
$$= P(B(X^+), B(X^-))_{0,\lambda_2} \qquad (3.2.1)$$

and

$$\xi\frac{\partial}{\partial\xi}A_{0,\lambda_2} - A_{0,\lambda_2}\left(\xi_1\frac{\partial}{\partial\xi_1} + \xi_2\frac{\partial}{\partial\xi_2}\right) - \frac{1}{2}A_{0,\lambda_2} = B(D)_{0,\lambda_2}. \quad (3.2.2)$$

By Lemma 2.2, we see as in part (b) of the proof of Lemma 3.1 that there exist bilinear mappings from $P_0(\mathcal{D}(\mathcal{K}) \cap \mathcal{K}^\infty)$ to $P_{\lambda_2}(\mathcal{S}_K)$ for $\lambda_2 \in \{0, 1\}$, which for any $0 < a < b$ extend to continuous bilinear mappings from $P_0(\mathcal{D}_{a,b}(\mathcal{K}))$ to $\mathcal{K}$ satisfying relations (3.2.2) and the inequalities $\|\xi^p A_{0,\lambda_2}^{a,b}\| \leqslant h_1(a, b)(c_1(a, b))^p$, where $h_1(a, b) \geqslant 0$, $c_1(a, b) \geqslant 0$ and $\lambda_2 \in \{0, 1\}$. Using, in addition, the last part of Lemma 2.3, there exists $\delta(a, b)$ such that $0 < \delta(a, b) \leqslant c_1(a, b)$ and

$$A_{0,\lambda_2}\left(\mathcal{D}_{a,b}(\mathcal{K})^2\right) \subset \mathcal{D}_{\delta(a,b),c_1(a,b)}(\mathcal{K}).$$

It then results from relation (3.2.1) that A_{0,λ_2} is determined once we know $A_{0,0}$ and $A_{0,1}$ and it is seen inductively with new values of the constants $h_1(a, b)$, $c_1(a, b)$ and $\alpha_1(a, b)$ that

$$\left\|\xi^p A_{0,\lambda_2}^{a,b}\right\| \leqslant h_1(a, b)(c_1(a, b))^p \exp(\alpha_1(a, b)|\lambda_2|) \qquad (3.2.3)$$

and $A_{0,\lambda_2}(\mathcal{D}_{a,b}(\mathcal{K})^2) \subset \mathcal{D}_{\delta(a,b),c_1(a,b)}(\mathcal{K})$, for any $\lambda_2 \in \mathbf{Z}$. If we define

$$\Delta(A)_{0,\lambda_2} = \xi\frac{\partial}{\partial\xi}A_{0,\lambda_2} - A_{0,\lambda_2}\left(\xi_1\frac{\partial}{\partial\xi_1} + \xi_2\frac{\partial}{\partial\xi_2}\right) - \frac{1}{2}A_{0,\lambda_2}$$

we see that

$$\Lambda(\Delta(A))_{0,\lambda_2} + \xi\frac{\partial}{\partial\xi}P(B(X^+), B(X^-))_{0,\lambda_2}$$

$$-P(B(X^+), B(X^-))_{0,\lambda_2}\left(\xi_1\frac{\partial}{\partial\xi_1} + \xi_2\frac{\partial}{\partial\xi_2}\right)$$

$$-\frac{5}{2}P(B(X^+), B(X^-))_{0,\lambda_2} = 0. \qquad (3.2.4)$$

Relation (3.1.12) remains valid for $n = 2$; together with (3.2.4) it implies $\Lambda(B(D) - \Delta(A))_{0,\lambda_2} = 0$ $(\lambda_2 \in \mathbf{Z})$. This is rewritten, with

$$G_{0,\lambda_2} = B(D)_{0,\lambda_2} - \Delta(A)_{0,\lambda_2},$$

$$\xi\tau^- G_{0,\lambda_2+1}\xi_2 - \left(\xi^2 G_{0,\lambda_2} - G_{0,\lambda_2}(\xi_1^2 - \xi_2^2)\right) + \xi\tau^+ G_{0,\lambda_2-1}\xi_2 = 0.$$

Since $G_{0,0} = G_{0,1} = 0$, we have $G_{0,\lambda_2} = 0$ for any $\lambda_2 \in \mathbf{Z}$. This means that (3.2.2) is verified for any $\lambda_2 \in \mathbf{Z}$. It then results from relations (3.1.8) and (3.1.9) and inequalities (3.2.3) that there exist $h_2(a, b) \geq 0$, $c_2(a, b) \geq 0$ and $\alpha_2(a, b) \geq 0$ such that

$$\left\| \xi^p A_{\lambda_1, \lambda_2}^{a, b} \right\| \leq h_2(a, b)(c_2(a, b))^p \exp(\alpha(a, b)(|\lambda_1| + |\lambda_2|))$$

for any $\lambda_1, \lambda_2 \in \mathbf{Z}$, $p \geq 0$.

We then define $Y = A \circ \sigma_2$. Q.E.D.

3.3. LEMMA. *Given* $B \in Z_t^1(\mathfrak{w}, \mathfrak{L}_2^0(\mathcal{K})_\infty)$, *and* $Z \in \mathfrak{L}_2(\mathcal{K}_0, \mathcal{K}) \cap L_2(\mathfrak{D}(\mathfrak{H}_K), \mathfrak{H}_K)$ *such that* $B(x) = V(x)Z$ *on* $\mathfrak{D}(\mathfrak{H}_K)$, $x \in \mathfrak{w}$. *Then there exists* $Y \in \mathfrak{L}_2^0(\mathcal{K})_\infty$ *such that* $B(x) = V(x)Y$.

We denote by χ a C^∞ function from $]0, +\infty[$ to $[0, 1]$ such that $\chi(\xi) = 1$ if $0 < \xi < 1$, $\chi(\xi) = 0$ if $\xi \geq 2$. We define Z_0 by

$$Z_0(\varphi_1, \varphi_2)(\xi) = \chi(\xi)Z(\varphi_1, \varphi_2)\xi, \qquad \varphi_1, \varphi_2 \in \mathcal{K}_0.$$

We then define $Z_n = V_D^n Z_0$, $n \geq 0$. We have $Z_n \in L_2(\mathfrak{D}(\mathfrak{H}_K), \mathfrak{H}_K)$, and we see inductively that

$$Z_n = \left(\left(\xi \frac{\partial}{\partial \xi} \right)^n \chi \right) Z + \sum_{k=1}^{n-1} \binom{n}{k} \left(\left(\xi \frac{\partial}{\partial \xi} \right)^{n-k} \chi \right) V(D)^k (B(D)).$$

Since $Z \in \mathfrak{L}_2(\mathcal{K}_0, \mathcal{K})$ and $V^k(D)(B(D)) \in \mathfrak{L}_2^0(\mathcal{K}) \subset \mathfrak{L}_2(\mathcal{K}_0, \mathcal{K})$, we have $Z_n \in \mathfrak{L}_2(\mathcal{K}_0, \mathcal{K})$. Therefore, given any $0 < a < b$, there exists $\alpha_n = \alpha_n(a, b) \geq 0$ such that Z_n is continuous from $\mathcal{K}(K, \alpha_n) \cap \mathfrak{D}_{a,b}(\mathcal{K})$ to $\mathcal{K}$. This means that there exists $h_n = h_n(a, b) \geq 0$ such that

$$\| Z_n(\varphi_1, \varphi_2) \| \leq h_n \|\varphi_1\|_{\alpha,k} \|\varphi_2\|_{\alpha,k} \text{ for all } \varphi_1, \varphi_2 \text{ in } \mathcal{K}(K, \alpha_n) \cap \mathfrak{D}_{a,b}(\mathcal{K}).$$

If, in addition, $\varphi_i \in P_{\lambda_i}(\mathcal{K}(K, \alpha_n) \cap \mathfrak{D}_{a,b}(\mathcal{K}))$, we have $\|\varphi_i\|_{\alpha,k} = e^{\alpha\lambda_i}\|\varphi_i\|$, so that

$$\left\| \xi^p (Z_n)_{\lambda_1, \lambda_2}^{a, b} \right\| \leq h_n 2^p \exp(\alpha_n(|\lambda_1| + |\lambda_2|)),$$

i.e. $Z_n \in \mathfrak{L}_2^0(\mathcal{K})$. Consequently, $Z_0 \in \mathfrak{L}_2^0(\mathcal{K})_\infty$.

We now define $B'(x) = B(x) - dU_x Z_0 - Z_0 d(U \otimes U)_x$. We have

$$B' \in Z_t^1(\mathfrak{w}, \mathfrak{L}_2^0(\mathcal{K})_\infty) \quad \text{and} \quad B'(\mathfrak{w})\mathfrak{D}(\mathfrak{H}_K)^2 \subset \bigcup_{1 \leq b} \mathfrak{D}_{1,b}(\mathcal{K}).$$

By Lemma 3.2 there exists $X \in \mathfrak{L}_2^0(\mathcal{K})_\infty$ such that $B'(x) = V(x)X$. Consequently, $B(x) = V(x)Y$ where $Y = Z_0 + X$. Q.E.D.

3.4. A linear representation $(L, \mathfrak{L}_n(\mathcal{K}_0, \mathcal{K}))$ of W on $\mathfrak{L}_n(\mathcal{K}_0, \mathcal{K})$ is defined by $L_g f = U_g f(\otimes^n U_{g^{-1}})$. In the following we endow the space $\mathfrak{L}_n(\mathcal{K}_0, \mathcal{K})$ with the topology of uniform convergence on bounded sets. For this topology $(L, \mathfrak{L}_n(\mathcal{K}_0, \mathcal{K}))$ is not a continuous representation. The space $Z_K^1(W, \mathcal{L}_n(\mathcal{H}_0, \mathcal{H}))_\infty$ of differentiable 1-cocycles on W with coefficients in $\mathcal{L}_n(\mathcal{H}_0, \mathcal{H})$ is the set of C^∞ mappings F from W to $\mathcal{L}_n(\mathcal{H}_0, \mathcal{H})$ such that $F_k = 0$ if $k \in K$ and

$$F_{gg'} = F_g + L_g F_{g'}, \qquad g, g' \in W. \tag{3.4.1}$$

We then define $dF(x) = (d/dt)(F_{\exp tx})_{t=0}$ for $x \in \mathfrak{w}$. We have $dF(k) = 0$ if $k \in \mathfrak{k}$. We have, obviously, $dF(\lambda x) = \lambda dF(x)$, $\lambda \in \mathbf{R}$. Since

$$L_g dF(x) = \frac{d}{dt}(F_{g \exp tx})_{t=0}$$

and since the map $g \to U_g \varphi$, $\varphi \in \mathcal{D}(\mathfrak{H}_K)$, is C^∞ from W to $\mathcal{H}_0$, the map $g \to U_g dF(x)(\varphi_1, \ldots, \varphi_n)$, $\varphi_i \in \mathcal{D}(\mathfrak{H}_K)$, is C^∞ from W to $\mathcal{H}$. This means that $dF(x)(\mathcal{D}(\mathfrak{H}_K)^n) \subset \mathcal{H}^\infty$. We can differentiate the relation

$$F_{gg'}(\varphi_1, \ldots, \varphi_n) = (F_g + V_g F_{g'})(\varphi_1, \ldots, \varphi_n),$$

$$\varphi_i \in \mathcal{D}(\mathfrak{H}_K), \quad g = \exp tx, \quad g' = \exp ty,$$

at $t = 0$, and we get $dF(x + y) = dF(x) + dF(y)$ on $\mathcal{D}(\mathfrak{H}_K)$ and therefore on $\mathcal{H}_0$ since $dF(x) \in \mathcal{L}_n(\mathcal{H}_0, \mathcal{H})$. Using (3.4.1) to differentiate $F_{(\exp tx)(\exp ty)(\exp -tx)}$ at $t = 0$ we get

$$dF([x, y]) = (dU_x dF(y) - dF(y) d(\otimes^n U)_x)$$
$$- (dU_y dF(x) - dF(x) d(\otimes^n U)_y)$$

on $\mathcal{D}(\mathfrak{H}_K)$. Since $dF([k, x]) = dU_k dF(x) - dF(x) d(\otimes^n U)_k$ when $k \in \mathfrak{k}$, we have $dF(x) \in L_n(\mathcal{D}(\mathfrak{H}_K), \mathfrak{H}_K)$.

3.5. Suppose, in addition, that there exists $Y \in \mathcal{L}_n(\mathcal{H}_0, \mathcal{H})$ such that $F_g = U_g Y(\otimes^n U_{g^{-1}}) - Y$. The map $g \to F_g$ being C^∞ from W to $\mathcal{L}_n(\mathcal{H}_0, \mathcal{H})$, the map $g \to U_g Y(\otimes^n U)_{g^{-1}}$ is C^∞ from W to $\mathcal{L}_n(\mathcal{H}_0, \mathcal{H})$. Since, in addition, the map $g \to U_g \varphi$, $\varphi \in \mathcal{D}(\mathfrak{H}_K)$, is C^∞ from W to $\mathcal{H}$, this means that $Y(\mathcal{D}(\mathfrak{H}_K)^n) \subset \mathcal{H}^\infty$. By differentiation of F, we get $dF(x) = dU_x Y - Y d(\otimes^n U)_x$ on $\mathcal{D}(\mathfrak{H}_K)$ for $x \in \mathfrak{w}$. Since $dF(k) = 0$ if $k \in \mathfrak{k}$, we have $Y \in L_n(\mathcal{D}(\mathfrak{H}_K), \mathfrak{H}_K)$.

3.6. Suppose that $F \in Z_K^1(W, \mathcal{L}_n(\mathcal{H}_0, \mathcal{H}))_\infty$ and that there exists $Z \in L_n(\mathcal{D}(\mathfrak{H}_K), \mathfrak{H}_K) \cap \mathcal{L}_n(\mathcal{H}_0, \mathcal{H})$ such that $dF(x) = V(x)Z$ on $\mathcal{D}(\mathfrak{H}_K)$. Take $\varphi_1, \ldots, \varphi_n \in \mathcal{D}(\mathfrak{H}_K)$, define the functions

$$f_1(t) = \langle \psi, F_{\exp tx}(\varphi_1, \ldots, \varphi_n) \rangle$$

and

$$f_2(t) = \langle \psi, (U_{\exp tx} Z(\otimes U_{\exp -tx}) - Z)(\varphi_1, \ldots, \varphi_n) \rangle,$$

$x \in \mathfrak{w}$, $\psi \in \mathcal{H}^\infty$, which are C^∞ from $\mathbf{R}$ to $\mathbf{C}$. We have

$$df_1(t)/dt = df_2(t)/dt = \langle \psi, L_{\exp tx}(dF_x)(\varphi_1, \ldots, \varphi_n) \rangle$$

and $f_1(0) = f_2(0) = 0$ and therefore $f_1(t) = f_2(t)$. Then, $F_g = L_g(Z) - Z$ in $\mathcal{L}_n(\mathcal{H}_0, \mathcal{H})$.

4. Proof of Proposition 1.3. It is clear that (1) implies (2) by taking $X = f^2$. We now suppose that (2) is verified. We choose $A \in \mathcal{F}(\mathcal{K})$, $A^1 = I$, such that the representation $(T, \mathcal{K})$ defined by $T_g = AS_g A^{-1}$ satisfies properties (1)–(4) of Lemma 2.1. Define $X' \in \mathcal{L}_2(\mathcal{K}_0, \mathcal{K})$ by $X' = X - A^2$; we have

$$T_g^2 = S_g^1 X' - X'\left(S_g^1 \otimes S_g^1\right) \quad \text{on } \mathcal{K}_0.$$

The map $g \to F_g^2 = S_g^1 T_{g^{-1}}^2$ being C^∞ from W to $\mathcal{L}_2(\mathcal{K}_0, \mathcal{K})$, we have $F^2 \in Z_K^1(W, \mathcal{L}_2(\mathcal{K}_0, \mathcal{K}))_\infty$. It results from subsection 3.5 that $X' \in L_2(\mathcal{D}(\mathfrak{H}_K), \mathfrak{H}_K)$ and $dF_x^2 = dS_x^1 X' - X'd(\otimes^2 S^1)_x$ on $\mathcal{D}(\mathfrak{H}_K)$, $x \in \mathfrak{w}$. Property (4) of Lemma 2.1 together with the formula

$$\left(dS_{X^+}^1\right)^p\left(\left(\text{ad } dS_D^1\right)^l dF_y^2\right)$$

$$= \sum_{k=0}^{p} \binom{p}{k}\left(\text{ad } dS_{X^+}^1\right)^k\left(\left(\text{ad } dS_D^1\right)^l dF_y^2\right) d\left(\otimes^2 S^1\right)_{X^+}^{m-k}, \qquad y \in \mathfrak{w},$$

imply that $(\text{ad } dS_D^1)^l dF_y^2 \in \mathfrak{L}_2^0(\mathcal{K})$, i.e., $dF_y^2 \in \mathfrak{L}_2^0(\mathcal{K})_\infty$. Therefore $dF^2 \in Z_f^1(\mathfrak{w}, \mathfrak{L}_2^0(\mathcal{K})_\infty)$. It then results from Lemma 3.3 that there exists $B^2 \in \mathfrak{L}_2^0(\mathcal{K})_\infty$ such that $dF_x^2 = V(x)B^2$. Utilizing subsection 3.6, we get that

$$F_g^2 = S_g^1 B^2\left(\otimes^2 S^1\right)_{g^{-1}} - B^2 \quad \text{on } \mathcal{K}_0.$$

We define $\mathfrak{B}_2 = B^1 + B^2 \in \mathcal{F}(\mathcal{K}_0, \mathcal{K})$, B^1 being the canonical injection from $\mathcal{K}_0$ to $\mathcal{K}$. We have $(T_g \mathfrak{B}_2)^2 = (\mathfrak{B}_2 T_g^1)^2$ on $\mathcal{K}_0$. Given $n > 2$, suppose we have constructed a family $\mathfrak{B}_p \in \mathcal{F}(\mathcal{K}_0, \mathcal{K})$, $2 \leq p \leq n$, such that $B^p = \mathfrak{B}_p - \mathfrak{B}_{p-1} \in \mathfrak{L}_p^0(\mathcal{K})_\infty$ and $(T_g \mathfrak{B}_p)^p = (\mathfrak{B}_p T_g^1)^p$ on $\mathcal{K}_0$. It results from relations

$$\left(T_{gg'}\mathfrak{B}_n\right)^{n+1} = T_g^1\left(T_{g'}\mathfrak{B}_n\right)^{n+1} + \left(\left(\sum_{m \geq 2} T_g^m\right)T_{g'}\mathfrak{B}_n\right)^{n+1}$$

and

$$\left(\left(\sum_{m \geq 2} T_g^m\right)T_{g'}\mathfrak{B}_n\right)^{n+1} = \left(\left(\sum_{m \geq 2} T_g^m\right)\mathfrak{B}_n T_{g'}^1\right)^{n+1}$$

that F^{n+1}, defined by $F_g^{n+1} = T_g^1(\sum_{m \geq 2} T_{g^{-1}}^m \mathfrak{B}_n)^{n+1}$, belongs to $Z_K^1(W, \mathfrak{L}_{n+1}(\mathcal{K}_0, \mathcal{K}))_\infty$. By the same argument as for $n + 1 = 2$, we have $dF^{n+1} \in Z_f^1(\mathfrak{w}, \mathfrak{L}_{n+1}^0(\mathcal{K})_\infty)$. By Lemma 3.1. there exists $B^{n+1} \in \mathfrak{L}_{n+1}^0(\mathcal{K})_\infty$ such that $dF_x^{n+1} = V(x)B^{n+1}$, $x \in \mathfrak{w}$. It results from 3.6 that

$$F_g^{n+1} = S_g^1 B^{n+1}\left(\otimes^n S_{g^{-1}}^1\right) - B^{n+1}.$$

This can be rewritten with $\mathfrak{B}_{n+1} = \mathfrak{B}_n + B^{n+1}$, $(T_g \mathfrak{B}_{n+1})^{n+1} = (\mathfrak{B}_{n+1} T_g^1)^{n+1}$. Defining $B = \sum_{n \geq 1} B^n$, we have $T_g B = B T_g^1$ on $\mathcal{K}_0$. The formal power series $f = A^{-1}B$ satisfies $S_g f = f S_g^1$ on $\mathcal{K}_0$. Q.E.D.

REFERENCES

1. M. Flato, G. Pinczon and J. Simon, *Non linear representations of Lie groups*, Ann. Sci. École Norm. Sup. (4) **10** (1977), 405–418.

2. A. Grothendieck, *Produits tensoriels topologiques et espaces nucléaires*, Mem. Amer. Math. Soc., No. 16, Amer. Math. Soc., Providence, R.I., 1955.

3. M. A. Naimark, *Les representations linéaires du groupe de Lorentz*, Dunod, Paris, 1962.

4. J. Simon, *Non-linear representations of Poincaré group and global solutions of relativistic wave equations*, Pacific J. Math. **105** (1983), 449–472.

5. _____, *Non linear representations of the Euclidean group of the plane*, Amer. J. Math. (to appear).

LABORATOIRE DE PHYSIQUE MATHÉMATIQUE, FACULTÉ DES SCIENCES MIRANDE, UNIVERSITÉ DE DIJON, BP 138, 21004 DIJON CEDEX, FRANCE

Appendix

ABCDEFGHIJ –AMS/SP –898765